Universities in Imperial Austria, 1848–1918

A Social History of a Multilingual Space

Central European Studies

Universities in Imperial Austria, 1848–1918

A Social History of a Multilingual Space

Jan Surman

Purdue University Press
West Lafayette, Indiana

Printed in the United States of America.

Cataloging-in-Publication data is on file with the Library of Congress.

Paper ISBN: 978-1-55753-837-6
ePDF ISBN: 978-1-61249-561-3
ePUB ISBN: 978-1-61249-562-0

An electronic version of this book is freely available, thanks to the support of libraries working with Knowledge Unlatched. KU is a collaborative initiative designed to make high-quality books Open Access for the public good. The Open Access ISBN for this book is 978-1-55753-861-1.

Cover image: Graduates of the Institute of Austrian History Research, course year 1891 and 1893 (© Vienna University Archive, sign. 106.I.3018; Photo: K. K. Hoffotografin Rosa Jenik, between 1891 and 1893)

Contents

List of Illustrations

List of Tables

Acknowledgments

Dealing with multivocal narratives means engaging intensively with a broad range of literature and sources. Hence, I would like to thank those people and institutions that supported me over the years while this work was in progress. Among these are the Vienna Initiativkolleg "The Sciences in Historical Context," the Center for Austrian Studies in Minneapolis, the Institute for the History of Science of the Polish Academy of Sciences in Warsaw (Instytut Historii Nauki Polskiej Akademii Nauk), the Institute for the Contemporary History of the Czech Academy of Science in Prague (Ústav pro soudobé dějiny Akademie věd České republiky), the Center for the Urban History of East Central Europe in L'viv (Центр міської історії Центрально-Східної Європи), the Max Planck Institute for the History of Science in Berlin, and, finally, the Austrian Research Association (Österreichische Forschungsgemeinschaft), which generously awarded me a MOEL scholarship for research in Cracow, Warsaw, and L'viv. For support with archival work in Ukraine, I want to record my special thanks to Tarik Cyril Amar from the Center for the Urban History of East Central Europe in L'viv, whose generous help allowed me to overcome the obstacles involved in acquiring permission to use the State Archive of L'viv Oblast, as well as to Serhiy Osachuk (Сергій Осачук) from the Bukovina Center, for supporting my search of the holdings of the State Archive of Chernivtsi Oblast. I am also grateful to the Cultural Bureau of Carinthia's provincial government for help with obtaining copies from Bukovina. The Herder-Stipend, a Leibniz-DAAD Fellowship, and an appointment at the Leibniz Graduate School "History, Knowledge, Media in East Central Europe" at the Herder Institute for Historical Research on East Central Europe–Institute of the Leibniz Association provided me with the time and intellectual atmosphere to finish the manuscript.

My particular thanks also go to Peter Goller from Innsbruck for providing me with access to the Tyrolean scholarly past, and to Kurt Mühlberger from Vienna for generously providing me with materials about scholars at the philosophical faculty of the University of Vienna. I thank Gerald Angermann-Mozetič from Graz for continuous support over the years. Parts of this work have been presented in Boston, Budapest, Cracow, Darmstadt, Graz, L'viv, Prague, Sofia, Vienna, and Warsaw, and I want to thank all participants for their valuable comments, in particular my colleagues and faculty members from the Initiativkolleg "The Sciences in Historical Context" and the Doktoratskolleg "Austrian Galicia and Its Multicultural Heritage" (as well as the associated fellows Philipp Hofeneder and Börries Kuzmany) in Vienna, the Institute for History of Science in Warsaw, and the Leibniz Graduate School in Marburg. I am thankful to my colleagues in Marburg and at the working group for the history of science in Frankfurt am Main for stimulating talks and debates that helped clarify my ideas while I was writing the final version of the manuscript. I am also greatly indebted to the many archivists and librarians I continuously harassed, who allowed me to overcome my time constraints.

My particular thanks go to Mitchell G. Ash, Deborah Coen, Gary Cohen, Matthew Konieczny, and Soňa Štrbáňová, whose generous comments on the manuscript helped me conceptualize and organize the present study. Moreover, Mitchell Ash's support in recent years made this study possible in the first place. Johannes Feichtinger and Klemens Kaps supported me with their ideas and expertise, and their comments heavily influenced this text. Finally, I cannot thank my family enough for inspiring me and for always being there. I dedicate this book to my mother, who encouraged and supported me over the years but will not be here to cherish its publication.

Note on Language Use, Terminology, and Geography

Geographic or personal names were markers of identity and belonging in the nineteenth century (and remain so to some extent today) and thus were contested as elements of nationalist discourse. In many cases, individuals, especially those indifferent to nationalism, changed their names based on the context; for scholars who published in both the Cyrillic and Latin alphabets, changing transcription and translation rules mean that the names under which these scholars are currently known differ from those used during their lifetimes. To avoid unwieldy formulations, this work uses the English names currently in use when appropriate. For the sake of precision, in the case of cities that belonged to different states at different times, the name is given in the language of the given state at that time. Alternative names for people and places in other languages are noted at the first appearance of the name. This also applies to designations that are mentioned in the text and is used consistently for all the languages involved. Cyrillic names occasionally appear in the main text, which seems justified because many of the persons, places, and organizations dealt with here are in fact hard to identify if only a Latin transcription is provided.

For the sake of historical accuracy, this text includes a few terms that might be new to scholars not familiar with the Habsburg Empire of the nineteenth century or with the scholarly system of the time. Special terms referring to Habsburg universities (*Privatdozenten*, *Utraquisierung*, etc.) have been explained in the text or notes at their first appearance and, if possible, are replaced with English terms in the main text. The local geographic terms are best explained by means of a short overview of nineteenth-century central Europe.

The Habsburg Empire consisted of two halves, Cisleithania (the northern and western part, also called *Austria*) and Transleithania (the Hungarian Lands of the Crown of Saint Stephen). Cisleithania comprised fifteen provinces (crown lands); most important for this book are, from west to east, Tyrol, Styria (capital: Graz), Lower Austria (capital: Vienna, which was also the imperial capital), Bohemia (capital: Prague), Galicia (capital: L'viv), and Bukovina. In many of these provinces, more than one language was used: Tyrol included what is now South Tyrol, populated by German speakers and Italian speakers. In Styria German and Slovenian dominated, in Bohemia Czech and German, and in Galicia Polish and Ukrainian (nowadays western Galicia is part of Poland, and eastern Galicia is part of Ukraine). Finally Bukovina, now divided between Romania and Ukraine, was a multilingual province with German, Yiddish, Ukrainian, and Romanian as the most popular languages; it was home to Chernivtsi University.

One other differentiation deserves mention here—throughout the book I use the designation *Ruthenian* for the language that in the twentieth century became Ukrainian, and *Ruthenians* for the people who used it, for several reasons. First, it was the official designation for Ukrainian in the Habsburg Empire (Рутенський, Руський in Ruthenian, *Ruski* in Polish, and *Ruthenisch* in German). Second, Ruthenian identification differed from Ukrainian identification (which focused on unity with Ukrainians/Little Russians in the Russian Empire) and Russophile identification (which focused on unity with the Russian people and their religion, that is, Orthodox Christianity). Also, Polish speakers lived across all three central European empires: Habsburg, Prussian, and Russian. In the Russian Empire, they were the major population in the semi-autonomous Kingdom of Poland, which was formally stripped of its autonomy in 1867 and renamed Vistula Land. In Prussia most Polish speakers lived in the Province of Posen and in Prussian Silesia.

German, *Germany*, and *Austria* are very flexible terms and are used in the text in a few context-dependent meanings. *Austria* is the most widespread synonym for *Cisleithania*, although it sometimes also meant provinces with a German-speaking majority (i.e., the western part of Cisleithania); in Czech and Polish, *Austrians* were mostly Habsburg Germans. Especially in Bohemia and Galicia, German-speaking Habsburg subjects were also simply called *Germans* (sometimes with regional designations, like *Deutschböhmen* [Bohemian Germans]). These ethnonyms not only differed from language to language (and also depending on the speakers' political outlook) but also varied over time. To do justice to this complexity, but at the same time remain understandable, was one of the major obstacles this work had to face.

Abbreviations

AGAD, MWiO Archiwum Główne Akt Dawnych, C.K. Ministerstwo Wyznań i Oświaty = Ministerium für Cultus und Unterricht (Central Archives of Historical Records in Warsaw, collection Ministry of Religion and Education)

AT-OeStA/AVA Unterricht UM allg. Akten Österreichisches Staatsarchiv, Allgemeines Verwaltungsarchiv, Unterricht und Kultus, Unterrichtsministerium, Allgemeine Reihe, Akten (Austrian State Archives, General Archive of Administration, collection Education and Religion, General Section, Acts)

AT-UAW Archiv der Universität Wien (Archive of the University of Vienna)

AUC-HUCP *Acta Universitatis Carolinae—Historia Universitatis Carolinae Pragensis*

AUJ Archiwum Uniwersytetu Jagiellońskiego (Archive of the Jagiellonian University)

CDIAL Central'nyj deržhavnyj istoryčnyj archiv Ukraïny, L'viv (Central State Historical Archives of Ukraine in L'viv; Центральний державний історичний архів України, Львів)

DALO Deržavnyj arhiv L'vivśkoï oblasti (State Archive of L'viv Oblast; Державний архів Львівської області)

F. Fond (collection; фонд)

Fasc. Fascicle

FF NU Filozofická Fakulta Německé Univerzity v Praze (Collection of the Philosophical Faculty of the German University in Prague)

IAHR Institute of Austrian Historical Research (Institut für Österreichische Geschichtsforschung).

Inv.č. Inventární číslo (inventory number)

LF NU Lekarská Fakulta Německé Univerzity v Praze (Collection of the Medical Faculty of the German University in Prague)
Kart. Karton (box)
MED Medizinische Fakultät (Medical Faculty)
MF Medizinische Fakultät (Medical Faculty)
MZA Brno Moravský zemský archiv w Brně (Moravian Land Archive in Brno)
NA Národní archiv (National Archives, Prague)
NA, MKV/R Národní archiv, Ministerstvo kultu a vyučování Vídeň 1882–1918(1923) = Ministerium für Cultus und Unterricht (National Archives, Prague, collection Ministry of Religion and Education, 1882–1918[1923])
Op. Opys (inventory; Опис)
PA Personalakte, akt osobowy (personnel record)
PF Philosophische Fakultät (Philosophical Faculty)
PH Philosophische Fakultät (Philosophical Faculty)
SOA Litoměřice/Děčín Státní oblastní archiv v Litoměřicích, pobočka Děčín (State Regional Archives Litoměřice, Děčín Branch)
Spr. Sprava (file; справа)
Sign. Signatura (signature)
Sygn. Sygnatura (signature)
UAG Universitätsarchiv Karl-Franzens-Universität Graz (Archive of the University Graz)
UAI Universitätsarchiv Innsbruck (Archive of the University of Innsbruck)
ÚDAUK Ústav dějin Univerzity Karlovy a archiv Univerzity Karlovy (Institute of the History of Charles University and Archive of Charles University)
WF Wydział filozoficzny (Philosophical Faculty)
WL Wydział lekarski (Medical Faculty)
Z. Zahl (number of the file, in archival materials)

Introduction

A Biography of the Academic Space

Shortly before World War I, the professor of Romance languages at Innsbruck, Theodor Gartner, was completing a collection of Ladin folk songs, the outcome of an eight-year project intended to show that Ladinians are distinct from Italians.[1] During his career Gartner had studied in Vienna, then worked as a professor in Chernivtsi (Bukovina) and later in Innsbruck (Tyrol), a route well trodden by Cisleithanian academics. Always interested in Ladinian, he, after arriving in Bukovina, developed an interest in both the languages spoken there, Romanian and Ruthenian, subsequently publishing works on their vocabulary and grammar. Through his efforts, Gartner, a German Austrian with pan-German nationalist tendencies in his later years, thus influenced three national projects.[2] For Ruthenian in particular, Gartner's cooperation with Stepan Smal'-Stoc'kyj, a fellow Vienna graduate working as a professor of Ruthenian language and literature in Chernivtsi, was of utmost importance, marking a symbolic defeat of pro-Russian language reformists.[3] The ideas that they used to underscore the distinctiveness of Ruthenian from Russian were also applied to highlight the uniqueness of Ladinian: the official language was distinguished from any "contaminated dialects," an approach that closely followed the nationalist image of what the perfect language should be.[4]

Gartner's career, which led him from Vienna to Bukovina and Tyrol, was typical for the period analyzed in this book: imperial careering[5] was common among Cisleithanian academics of the time. But there were also other patterns: there were hundreds of unsalaried university lecturers (*Privatdozenten*) who worked at only one university, and a number of early twentieth-century scholars who migrated from Kiev or Warsaw to L'viv. This book tries to make sense of these patterns and proposes a concise view of the discourses and practices that shaped the Habsburg Empire, in particular its Austrian half, between 1848 and 1918. An analysis of imperial geography, in the modern sense of the social production of space, facilitates combining the centrifugal and centripetal moments that defined the empire: they become complementary rather than contrary processes.

Between 1848 and 1918, the universities of the Habsburg Empire underwent significant changes that corresponded closely with political and social developments in the state and its culture(s). Beginning with the 1848 revolution, a language-bound concept of identity gradually gained importance, slowly replacing loyalty to the state as the guiding political principle. These changes affected the Habsburg Empire (from 1867 the Austro-Hungarian Monarchy) in many ways. The autonomy of the Hungarian Kingdom and the Kingdom of Croatia-Slavonia (1867), the detachment of the Kingdom of Lombardy-Venetia (1859/1866), the collapse of the German Confederation (1866), the growing self-governance of Galicia, and multiple nationalistic conflicts shaped the region, its history, and its historiography. At the same time, the Habsburg Empire stood at the intersection of cultural projects that extended beyond its boundaries, most importantly, but not exclusively, the pan-German, pan-Slavic, Polish, and Ukrainian projects. The state borders marking political territory thus crossed other communicative and ideological entities.

The idiosyncrasies of the empire, often adduced when talking about its memory, are analyzed here from a unique angle, that of the institutional academic culture, at universities in particular. As institutions of higher education and scholarship that were closely connected but, I claim, far from identical, universities played a special role in central Europe.[6] Whether universities should produce civil servants or should rather promote scholarship was a key tension in these institutions' identity, which was shaped by complex and often conflicting social and political rules and expectations.

In an increasingly decentralized empire, two needs emerged—the need to educate loyal citizens and the need to foster a cultural identity—and although these were not necessarily contradictory, they increasingly grew apart. This tension was most visible in Galicia, as both Poles and Ruthenians/Ukrainians gravitated toward cultural identities extending beyond the empire; the fostering of these identities would inevitably end in conflict with the Crown. In contrast, the Czech, Hungarian, Slovenian, and other projects were geographically confined within the Habsburg borders and thus manifested themselves politically in different ways. Pan-German thinking, in versions up to 1918, also confronted the mainline policy of monarchic loyalty inscribed into the power relations of the monarchy, whose pluricultural[7] character contrasted with its politically induced monolingualism.

Shifting loyalties, malleable or multiple identities, nation building, tension, and conflict are the historical contexts on which this work is based. It is concerned, however, with a particular aspect of imperial reality, namely, academic institutions. More precisely, it follows the changes in the structure of academia in Cisleithania based on this region's imperial features. The original goal of this work was to analyze a network of university instructors over a period of sixty years (1848–1918); during this time, nationalists confronted empires, altering the imperial cultural pattern. But while political developments forged division, scholarly developments promoted contact and communication, moving toward internationality. However, to highlight the embedded nature of these processes and their long-lasting effects, I frame them with the dawn and afterlife of what I call here the *imperial academic space*; thus, the narrative of this book spans from the late eighteenth or early nineteenth century to the 1930s.

The focus here is thus the schizophrenic tension between supposedly supranational science and national scholarship.[8] This tension, one can argue, is the product of the inscription of science and scholarship into the cultural project of the nation. To a large extent, the present historiography follows the patterns developed during this time when the empire in its geographic totality was gradually becoming divided across linguistic, cultural, and historical entities, each following its own scientific exemplars. Viewed from the perspective of the now-dominant national historiographies, the empire became disentangled, which created loosely adhesive scientific narratives, with the prominent exception of analytic philosophy, whose analysis underscores its multinational existence.[9] At the same time, the "special conditions" characterizing the Habsburg multicultural space have gained more and more scholarly attention in recent decades, with academics tracing the patterns of the influx of cultural conflict.[10] The special conditions of these conflicts, paradigmatic of the Habsburg Empire, can be found across the globe at this time, and their importance for this particular empire is a product of cultural memory.

Thus, what seems to be a study of empire through the prism of scholarship is also a study of scholarship through the prism of empire, or rather through several prisms in the kaleidoscope of imperial memory. This proposed perspective therefore places a particular network in the foreground, concentrating on the several thousand careers spanning the historical moments of the empire, beginning with the institutionalization of philosophical faculties at universities following the 1848 revolution. In 1848 not only were national wishes expressed, but scientific integration and regulation also began. Until this time, research-based scholarship, except in medicine, had largely been excluded from the universities, finding its place in the seclusion of private or imperial institutions. The number of academies and universities did not change significantly over the subsequent years; from 1849 the so-called Thun-Hohenstein reform (discussed later) provided a solid basis for higher education even beyond the empire. By regarding the universities in Cracow, Chernivtsi (established in 1875), L'viv, Graz, Innsbruck, Prague (divided into two universities in 1882), Vienna, and Olomouc (closed in 1856) not as stable sites but as intersections of networks, I want to decenter the history of scholarship in imperial Austria. While most of the examples I discuss are from the universities in Vienna, Prague, Cracow, and L'viv, I argue that much can be discovered by regarding them as nodes within more broadly defined networks, both Habsburg and central European. Academic developments in Vienna or Cracow cannot be understood without taking those in Innsbruck or Chernivtsi into account, and vice versa. With the help of networks, I present a dynamic and changing space that encompasses all of Habsburg central Europe and, especially after 1918, reaches beyond it. The intellectual distance between Munich and Vienna, or between Warsaw and Cracow, was constantly being redefined, just like the distance between Vienna and Budapest, which grew rapidly in the 1860s.

The network analyzed here thus takes on a new aspect as part of a constantly changing academic structure across (at least) central Europe, closely interwoven with other empires and states that either shared cultural or linguistic traits or invited scholars from the Habsburg Empire to work at their institutions (e.g., the Principality of Bulgaria).[11] This analysis is therefore not only of an imperial space but also of a scholarly one; hence, I prefer to speak of *academic space* as the object of inquiry, with *space* defined as a social entity stretching across political boundaries and accommodating networks that supersede them. Moreover, this space was a dynamic entity; the changing relations among the state, culture, and science/education all affected the social components of the institutions examined here, which in turn influenced the exchange of knowledge. After the demise of the empire, Habsburg scholars migrated further, to universities in Ljubljana/ Leibach, Brno/Brünn, Warsaw, and Cluj/Klausenburg/Kolozsvár, as well as via Bratislava/Pozsony/Pressburg to Padua. This initial wave of academic mass mobility enlarged the network substantially and weakened its ties (a second wave followed the beginning of National Socialism and finally World War II only a few years later). The "Cisleithanization" of scholarship in central Europe, and the Habsburg legacy, with all its shortcomings and advantages, forms the final point of this narrative.

Intellectual Geographies

Recent decades have witnessed a growth in the importance of the geography of knowledge and spaces of knowledge in the history of science. With the established eminence of science as a social endeavor, lacking the universal claims of the mid-twentieth century, a growing literature on both the local appropriation of knowledge and the local conditions of its production has led to a reconsideration of scientific space and the processes under way within it.[12] Space as a new paradigm also aroused the interest of geographers. Most important, the spatial turn brought about a reevaluation of the influence of power relations in the scientific process. Concentrating on different sites where knowledge is produced, and the influence of spatial positioning on the shape of knowledge, the geography of knowledge extends the scope of the classic historiography of science and education.[13] Moreover, scholars emphasize that circulation is a site of knowledge formation, not simply a space between centers and peripheries, or between senders and receivers, that has no epistemic qualities of its own.[14] Yuri (Juri) M. Lotman, for whom the periphery is a space of increased intellectual productivity because it lacks the homogenizing power of the center, thus enabling cross-boundary relations impossible in the center, provided a metatheory for such conceptions of circulation.[15] Below I privilege Lotman's view over that put forward by Michel Foucault, for whom space was controlled by the center, while peripheries had only limited possibilities for innovation.[16]

One of the most important changes resulting from this approach is the notion that space is not something "out there" but an entity produced by repetitive actions that are influenced, but not determined, by social, cultural, and political contexts.[17] For instance, the production of space through the construction of railroads united vast regions of the United States and the Russian Empire, creating a sense of togetherness and state unity more decisively than any legal measures could have.[18] Recent work on higher education in the United States and Britain has highlighted universities as similarly unity-promoting institutions. In the nineteenth and early twentieth centuries, universities, although rooted in local circumstances, remained crucial parts of the unifying networks of education; norms and values were transferred at the same rate as scholars.[19] The tensions among the state/empire, ephemeral transnational science, and local cultural, social, and religious contexts were obvious, but skillful mediation created a network of institutions guided by the same norms, thus supporting the state that imposed them. As different as universities became, they were part of the project of intellectual unification—*e pluribus unum*, to use the slogan of the time.

While hierarchies and hegemonies influence the production of space, the spatial turn pays more attention to how people live in the space, exploring the possibilities offered by its contingency. This also means that the center-periphery structure is socially constructed, even if it is perpetuated by politics and accumulated prestige.[20] Works on the Spanish and German university systems clearly show how certain universities became centers, thereby influencing outcomes for the system as a whole.[21] However, while politics played an immense role, the structuring of academic space in Continental Europe into universities of entrance, universities of promotion, and final-station universities (*Einstiegsuniversität*, *Promotionsuniversität*, and *Endstationsuniversität*), as German historian Marita Baumgarten has named the different types of institutions, was a long-lasting process resulting more from the accumulation of cultural capital than from academic policy or financial issues.

The present work draws attention to another academic space: the university system of the late Habsburg Empire, and more precisely its Cisleithanian ("Austrian") part.[22] Not acknowledged as an empire *sensu stricto*, the area enclosed by Habsburg imperial boundaries witnessed in the sixty years between the "Spring of Nations" in 1848 and the "War of Nations" in 1914–18 a nexus of concurrent imperialism and nationalism, or of centripetal and centrifugal tendencies.[23] At the same time, it had to accommodate differing geographic projects, as stable "cultural nations" exceeded the monarchy's boundaries and became more and more bound to spaces defined by linguistic affinities. The identity issue of being a loyal national and imperial subject (either both or one or the other; the two were by no means mutually exclusive) was experienced both collectively and individually through inscriptions in everyday procedures, communication, and ideological networks as well as outbreaks of ceremonial patriotism.[24] While these identity projects differed depending on the historical situation and the cultural implementation (for example, the resuscitation of the idea of the Polish-Lithuanian Commonwealth or pan-German ideology), their interdependent development shared a common pattern subsumed under the banner of change from civic-cum-territorial to ethnocultural nationalism.[25]

Given its idiosyncrasies, the Habsburg Empire has recently been the subject of extensive research that has analyzed the contemporary nature of the putatively exclusive processes of state loyalty and ethnocultural nationalism. The history of science has, however, only recently taken note of this peculiar imperio-national space, previously confined to national narratives, and it has often merely produced recollections of particular institutional pasts in its function as an archivist of local memories. While the attention has recently shifted from nation to empire,[26] I argue that concentrating on the parallelism and interaction of national and imperial projects sheds more light on the sociogeographic character of knowledge in the central European "laboratory of world history" than does an either-or choice.[27] This work thus focuses on the development of science and scholarship in the space between the projects of empires and the projects of nations. The mediations and tensions that occurred between the needs and demands of scholarship and those of education serve as an example of scientific interacademic mobility, through which such spatial ambiguities can best be visualized.

Academic mobility did not stop with the end of the empires. Even if the sociocultural contexts are different, an analysis of the Habsburg scholarly peregrinations can say much about when policies of exchange bear the most fruit and how long-term the effects of these policies are. The Erasmus mobility program and the Bologna Process have, in different ways, been acknowledged as tools for bringing Europeans together and fostering a common, if not unitary, identity.[28] To a large extent, these programs intend to reconcile schisms that the nineteenth century produced.[29] Indeed, many parts of this book are concerned with how and why universities became national outposts, but also when they started to be international again.

Contrary to historians of nationalism, I argue that the nationalization of the peripheries was itself a reaction to processes that began in Vienna, the intellectual center.[30] Just as in the nineteenth century Slavic activists opposed the politically induced prevalence of German as the medium of education (not the traditional role of German as the language of publication), in the twentieth and twenty-first centuries scholars from universities that utilized, for example, German or French as their academic language are reacting to the imposition of English as the lingua franca of scholarship.[31] They do not oppose publishing in English so much as *having* to publish in English, including in disciplines that are intrinsically local, like regional historiography.

Habsburg Space(s)

The Habsburg space was occupied by the irony of contesting spatiality. After this area was divided in 1867 into territories centered on the "Garden" (Vienna) and the "Workshop" (Budapest),[32] the increasing number of nationalities brought about new forms of spatial conflict, between staging the empire and staging the nation.[33] This duality had developed slowly over time. When in 1851 the professors at the Jagiellonian University greeted Franz Joseph in their traditional togas instead of the prescribed clerk uniforms, stressing their independent traditions, this was met with serious political consequences. Less than thirty years later, however, Galicians took part in the commemoration of the Siege of Vienna of 1683, with separate festivities in Cracow and Vienna that underscored the different perceptions of the historical importance of this event.[34] Throughout the nineteenth century, the university buildings across Cisleithania represented intellectual unity visually and publicly, but in the second half of the century, they increasingly did so only in German-language universities, including Chernivtsi. The Collegium Novum in Cracow (completed in 1887) and a new building at the University of L'viv (conceived in 1912 but never realized) were purposefully designed to include "Polish" elements.[35] The space changed with shifting political affiliations as well; in 1907 universities throughout the empire protested the violation of university autonomy in the case of Ludwig Wahrmund, which also provoked the first demonstration by Czech and German students since 1859. Here, the existence of a common enemy—conservative clerics—largely overcame national differences, uniting the empire.

During the nineteenth century, the Habsburg space also gradually moved from the unity of an empire held together by the monarchy and the German language toward the political dualism of one monarch and two distinctive parliaments for its respective halves, characterized by different state languages, German and Hungarian. The fabric of languages and politics, including the language of education, grew apart not only along the divisions between Cis- and Transleithania but also within these semi-autonomous entities. National languages increased in importance, and German, the de jure nonnational language of the empire that was endowed with imperial and national allure, witnessed a decrease in practicality in the face of opposition by nationalists.[36] Academia was directly included in this process, influencing it and being influenced by it. Moreover, the spatial projects of different nationalist activists overlapped to create hierarchies, particularly in Galicia, where Poles controlled the provincial Diet, creating micro-imperialisms.[37]

The growing influence of nationalist discourses meant that projects to consolidate imperial space could no longer be induced by the center.[38] The empire's policy-driven structure led to conflicts, for example, the Badeni Crisis of 1897. The introduction of compulsory bilingualism in Bohemian government offices led to serious opposition from German-speaking politicians and nationalist activists, who saw this measure as undermining their privileged position, not as promoting equality or improving communication for Czechs.[39]

At the same time, the national space was increasingly represented as different from the imperial space, having its own boundaries as well as a distinct history and culture. The eminent Prague historian František Palacký created, for example, an ethnicity-based history of Bohemia, in which Czechs and Germans constituted historically disparate factors, divided by language, religion, and folklore.[40] Polish-language scholarship energetically pursued research based on the space of the Commonwealth despite political restrictions.[41] The legal distinctiveness of some Habsburg provinces and historical non-Habsburg state traditions had already been the subject of treatises in the first half of the nineteenth century. A similar strategy was seen in the late nineteenth century for Ruthenians/Ukrainians, whose historical ethno-spaces were divided between the Russian Empire and the Habsburg Empire.[42] In comparison to Czech nationalists, who imagined autonomy within the Habsburg Empire, both Polish and Ruthenian nationalists' imagination went beyond Galicia's boundaries; in particular, the Polish nationalists early on envisaged the reunification of the Polish-Lithuanian Commonwealth. Recall, however, that the Commonwealth generally did not mean an independent national state but rather an autonomous entity within the Habsburg Empire, as Austro-Slavism and loyalty to the emperor were popular in Galicia, in large part because of the threat of Russian imperialism, which was often referred to and was commonly codified in writing and popular culture.

The strengthening of national projects, which influenced all areas of cultural life, took place within the framework of Habsburg culture and the empire's intellectual atmosphere. What was, however, the Habsburg imperial scientific space as imagined and practiced by scholars? A brief glance at its strategies and institutions should clarify this. The role of scholarship-related policy in structuring the Habsburg academic space can be illustrated by the opening of the Imperial Academy of Sciences and Arts (Kaiserliche Akademie der Wissenschaften und Künste) in Vienna in 1847. Klemens Wenzel Metternich, the minister of state (1821–48), saw it as both a state-controlled "valve" for scholars—fulfilling their wish to have an institution to further their work and thus easing political tensions previously fueled by the lack of such a place—and a means to improve Habsburg's standing internationally.[43] During the discussions on the creation of the academy, its supraregional character was somewhat disputed both by proponents of a strong Viennese center for science and by those who wanted the Viennese academy to reach the same level as the provincial learned societies of the time. Among the nominees in 1847 and early 1848 were not only Viennese scholars (who constituted about half the nominees) but also Czech-Bohemian, Hungarian, and Italian scholars, signifying the unity of the Habsburg scientific community at that time.[44] Galicia, symbolically incorporated through Josef Russegger, a geologist and the administrator of the salt mines in Wieliczka/Großsalze (a corresponding member[45] of the academy in 1848), was officially excluded owing to the political turmoil in Galicia. Michał Wiszniewski, a professor of Polish literature in Cracow, was proposed as a corresponding member in 1848, but his nomination was rejected by the emperor.[46] The first Polish and Ruthenian scholars were chosen only in the late nineteenth century.

The academy was to be imperial, as its name indicates; in reality, it never was. Non-German-speaking authors rarely published in its periodicals or participated in its book series. Creating the image of a united monarchy, the series Fontes Rerum Austriacarum (Austrian historical sources) included sources on imperial spaces that, although centered on Vienna, also included Bohemia in the fifteenth century (see volume 20 of the version edited by František Palacký in 1860).[47] Apart from a number of works on various Habsburg monasteries, the most attention was paid to Veneto, a part of the monarchy that the Habsburgs were gradually losing at the time. One can also find documents on and from Carniola, Istria, and Transylvania but not Galicia. Indeed, the series Fontes Rerum Austriacarum, Bohemicarum,

Polonicarum (!), Hungaricarum, and Italicarum[48] were planned, but the suggestion of a state history encompassing local histories was soon replaced by an Austriacarum rather than a Habsburgicarum. The introduction and description of the objectives of the series, despite occupying several pages in the first ten volumes, were soon removed. Nationally oriented editions of sources appeared outside of the series, such as Augustyn Bielowski's six-volume *Monumenta Poloniae Historica = Pomniki dziejowe Polski* (Polish historical monuments, 1863–92), which opened with documents on Slavs in the Vistula region, and Antoni Zygmunt Helcel's *Starodawne prawa polskiego pomniki* (Monuments of old Polish laws), published from 1856 on, envisaging an empire-transgressing space. *Monumenta historiae Bohemica* (Bohemian historical monuments) (with a secondary title in Czech, *Staré paměti českých dějin* [Bohemian/Czech historical monuments]) was later published under the supervision of Anton (Antonín) Gindely in Prague from 1865 on.

While the imperial academy was intended to synthesize the forces concentrated in local academies, its mutation into an "Austrian" academy proved to be an obstacle to communication. To begin with, it had different competences than the local proto-academies (i.e., the scientific societies), not to mention the national academies (e.g., the French and British ones). As James E. McClellan has discussed, academies across Europe shared similar structures, competences, and scopes.[49] However, while the imperial academy was in many ways similar to other academies across Europe, the most important proto-academies in the Habsburg monarchy were in fact structured differently, and they had different aims. Regional proto-academies of science such as the Cracow Scientific Society (Towarzystwo Naukowe Krakowskie) and the Patriotic Museum in Bohemia (Vaterländisches Museum in Böhmen / Vlastenecké muzeum v Čechách, known after 1848 as the České museum [Bohemian/Czech Museum] and from 1854 as the Museum Království českého [Museum of the Czech Kingdom])[50] concentrated on the development of science and scholarship in their national tongues after 1848. The Society of the Patriotic Museum in Bohemia (Gesellschaft des vaterländischen Museums in Böhmen, established in 1818) began life as a multicultural Bohemian institution, but under the reign of Palacký, it soon turned to publishing predominantly on the past and present of Czechs in Bohemia. From its inception, the Cracow Scientific Society (established in 1815, incorporated in 1846 in Galicia) aimed to expand Polish-language scholarship through literary research and the development of a scientific language. While membership in the Society of the Patriotic Museum in Bohemia was limited to Bohemians, especially members of the aristocracy, the Cracow society consisted mostly of professors from the Jagiellonian University. Nevertheless, these organizations did not actually function as societies of a multicultural space because their concentration on the national language restricted publishing and lecturing opportunities for other scholars. The reorganization of these societies into fully developed academies (both named after Franz Joseph, of course) supported the empire's division into national spaces. Members of the Franz Joseph Czech Academy for Science, Literature and the Arts (Česká akademie císaře Františka Josefa pro vědy, slovesnost a umění, established in 1890) were forbidden from publishing in languages other than Czech in the academy's journals. The Academy of Arts and Sciences (Akademia Umiejętności, from 1919 the Polish Academy of Arts and Sciences [Polska Akademia Umiejętności]), which was born out of the Cracow Scientific Society, was in an even more awkward position, as the region within which it could recruit faculty members exceeded the empire's borders, while the legal system differentiated between state-defined "provincial" (*krajowy*) and "foreign" (*zagraniczny*) members, with both sections limited in numbers. Here, the imperial boundary intersected with the national geography; one of the main criticisms of the academy was that it did not include the most renowned Polish scholars and thus did not represent the entire Polish cultural space. Similarly, the Ševčenko Scientific Society in L'viv (Naukove tovarystvo imeni Ševčenka, established in 1873) was formally restricted to Galicia, although it in fact included Ukrainians from both the Russian and Habsburg Empires. In 1907 an identical scientific society opened in Kiev; its first head was Mychajlo Hruševs'kyj from L'viv, who not only transferred the structure of the society but also created a parallel set of journals.

The transimperial character of the Ševčenko Scientific Society after 1907 may be considered an exception, but nationalist efforts to exceed the imperial space had symbolic importance. One of the most important ideas was the symbolic assertion of their nonimperial space, for example, through cooperation in matters related to printing. The dissemination of books from other empires was often restricted; thus, many works were printed in two or three publishing houses in different empires. Helcel's *Starodawne prawa polskiego pomniki*, for instance, was published in Warsaw but using type from Cracow.[51]

This symbolic creation of a space for scholarship cannot be restricted to national spaces, however. In the first half of the nineteenth century in particular, the idea of a Slavic brotherhood united the Slavs of the Habsburg Empire. Perceiving a lack of an educated public within national spaces, several journals addressed "Slavs" as an existing public capable of reading each other's languages. The *Kwartalnik naukowy, wydawany w połączeniu prac miłośników umiejętności* (Scholarly quarterly, edited in cooperation with lovers of knowledge), edited by Helcel from 1835 to 1837, included Slavic and German scholars in its board of editors. With an openly antinationalist viewpoint, it strove to review as many works from Slavic literature as works written in other languages.[52] The Czech-language journal *Krok: Weřegný spis wšenaučný pro wzdělance národu Česko-Slowanského* (Krok: Public general scientific journal for the educated people of the Czech-Slav nation, 1821–40) similarly addressed a non-German space, oscillating between a Czech (ethnic) space, a Czech-Slovak (language) space, and a Slavic space. It was also ironic that the Slavic space lacked a precise definition. In the introduction to the journal, Jan Svatopluk Presl defined Slavs in opposition to Germans but acknowledged that this was a foreign definition, because Slavs also differed internally.[53] The term *pan-Slavic*, initially as a counterpart to *pan-German*, introduced another space of interaction, which was subsequently tightened to create a space reminiscent of the German Confederation. The pan-Slavic movement did not go beyond this definition; it lacked not only a mythology but also a communicative basis and, most important, regular interaction. At the first Slavic Congress of 1848, it was already visible that the nationalists' focus on national languages threw the claim of the unity of the Slavic language into oblivion. Subsequently, pan-Slavism not only failed in practice but was criticized as a cheap substitute for internationalism;[54] pan-Slavic academic interaction perhaps did not cease to exist,[55] but it became of only tertiary importance, after its heyday in the *Vormärz* (Pre-March) period and during neoabsolutism.

Despite their concentration on nationality as their primary point of reference, most Habsburg institutions retained international and thus intercultural components. On the one hand, this was driven by the membership of foreign (i.e., nonnational) scholars in local academies, awarded mostly to prominent scholars but also to scholars who had a particular political alignment within the empire. For example, the Academy of Arts and Sciences in Cracow nominated Heinrich Zeissberg, a former professor of history in L'viv and a specialist on the "Polish" Middle Ages, as well as Eduard Suess, a geologist and politician who before becoming president of the imperial academy in Vienna opposed the existence of the University of L'viv.[56] On the other hand, the imperial academy in Vienna organized pan-Habsburg projects and commissions, aiming to include scholars representing all of the Cisleithanian provinces. In contrast, provincial organizations that had previously been transcultural mostly became battlefields of conflicting interests and slowly turned into monolingual organizations; for them, an exchange with scholars with different cultural allegiances was itself a form of internationalism.

Overview of the Chapters

To do justice to the differing spatial projects in the empire, this book takes the perspective of academic institutions and their governing body, namely, the Ministry of Religion and Education (Ministerium für Cultus und Unterricht). I follow a biographical perspective, looking at the gestation, birth, maturation, and demise of the academic system in the monarchy. The story does not end with the dissolution of the monarchy, though, since the successor states drew not only their academic cadres but also their models for a university system from their shared past.

I begin my narrative with a description of the Habsburg scientific landscape of the late eighteenth and early nineteenth centuries, showing how certain seeds of cultural differentiation were planted (but did not bloom) under Metternich's regime. After the revolution in 1848, the immediate changes in university policy implemented many liberal measures within Habsburg scholarship. These were systematized and put into practice under the minister of education Leo Thun-Hohenstein,[57] with whom chapter 2 is concerned. Both in theory and in practice, this period was instrumental in not only producing a common Habsburg academic space but also filling it with a particular ideologically laden approach to knowledge; the scholarly appointments made during this time meant that this approach remained influential throughout the century. This policy also introduced institutions that became instrumental in promoting the disintegration of the common space; in particular, the philosophical faculties changed universities from producers of civil servants to producers of culture, which made that faculty an easy object of nationalist agitation. The linguistic disintegration that began in

1848, however, encountered a serious backlash because of the neoabsolutist political atmosphere.

I argue in chapter 3 that the most important changes took place in the 1860s, when, after Thun-Hohenstein's resignation, subsequent ministers practiced a much more liberal policy than had been possible during neo-absolutism. They allowed university autonomy to be implemented, which affected both scholarship and the language of instruction. The discussions over language also show how the initially imperial idea of Kultur-Bildung (culture-education) became inscribed into the national rhetoric of the German-language elites of western Cisleithania and how it was translated into national claims by other Habsburg cultures.

It is precisely this process, along with the onset of liberalism in the linguistic subsystems of Cisleithania, that I deal with in chapters 4 and 5. All three spaces—Czech, German-Austrian, and Polish—developed in different directions over time. The German-language universities, initially included in all pan-German networks, became more isolated after the Austro-Prussian War. The empire thus grew more reliant on its own graduates, who were mostly educated in Vienna and eventually sent out to work at provincial universities. A hierarchy of universities stabilized toward the end of the nineteenth century: at the top was Vienna, overrun with Privatdozenten but appointing only well-known scholars as professors, whereas Innsbruck and Chernivtsi were at the bottom: they had almost no Privatdozenten, and professors frequently spent only a few years there before being appointed to a larger university. Galicia, however, was open to scholars from abroad from the 1870s on. Through the appointment of scholars from the Russian and German Empires as well as frequent habilitations by graduates from these two states, its universities became monolingual but multicultural. By contrast, the Czech University of Prague drew from Bohemian and Moravian institutions and, except during the period immediately after the university split into two, experienced almost no exchanges with the rest of the empire or abroad. It did, however, seek to retain international cooperation through different means. At the same time, the universities in Prague and Galicia were undergoing a process of intrafaculty differentiation across ideological lines, which grew stronger toward 1900.

Importantly, the spatial processes described here were vital for shaping scientific advancement in the Austrian half of the Habsburg Empire. They led to diminishing movement of scholars across the Czech, German, and Polish subsystems and the intensification of other forms of exchange. However, spatial issues also determined the development of a disciplinary nexus in the empire, as the durable (i.e., codified) diversification of disciplines was also hierarchical, and thus connected to the spatially determined hierarchy of universities, as were the migratory networks.

With the ongoing division of academic spaces, issues of religious denomination, which I discuss in chapter 6, remained problematic for universities. First, Jewish scholars, although admitted as Privatdozenten, were underrepresented in higher positions. Increasing anti-Semitism, which occasionally turned violent in Innsbruck, Graz, and Prague, inhibited the appointment of Jewish scholars from Vienna, where numerous Privatdozenten were Jewish, creating glass ceilings and "invisible ghetto walls" that hindered their careers. At the same time, Jewishness was redefined from a religious to an ethnic and cultural category. While conversion represented a possible loophole in the anti-Semitic legal policy of the 1850s, the boundaries of Jewishness were defined more in terms of ethnicity in the late nineteenth century. While being Jewish and German was hardly a contradiction for most people, the populist discourse across the empire tended toward exclusive definitions.

World War I led to institutional disintegration and division across the intellectual landscape of central Europe. As I show in chapter 7, not only did the legacy of the empire dominate the many possible models of university education, but scholars from Cisleithanian universities shaped the institutions of the interwar period, with regard to both science and organization. However, this postwar Cisleithanization of central Europe, which brought forward fascinating innovative trends (e.g., analytic philosophy throughout the space in question), cannot be understood without the changes already set in motion in the Thun-Hohenstein era.

Finally, I want to mention two groups who are not heroes of my story but are indeed largely touched by it. First, women's academic careers were obstructed and made impossible for many years. It was only in 1905 that the first woman habilitated at a Habsburg university—Elise Richter. Indeed, it was precisely the atmosphere I described in chapter 6 that reinforced this exclusion.[58] The second group is the geographically immobile scholars, who make up the majority of the scholars I examine when looking at career patterns.[59] In the later nineteenth century, this group also faced the negative effects of the mobility requirement. While I describe how this group came into being and offer a more optimistic view of their careers than their exclusion would imply, I do not engage with their lives and careers in detail. I see their story, however, in terms of different career choices, not academic failure, and I offer examples illustrating that a university professorship was not always the preferred career choice. Especially given the recent situation in the global academic job market, the story of the academic precariat is probably more necessary than ever, and this book should serve as an invitation for future scholars to tell it.

Chapter 1

Centralizing Science for the Empire

There is no freedom of discussion and of thought; for each science there is one compulsory . . . textbook, from which nowhere and never, not even in oral commentaries, one is allowed to drift. A student's memory is strengthened at the cost of his intelligence; his head is filled with an abundance of unbeneficial, unpractical things, so that there is no room left for thinking, —his character, his moral education are totally neglected. . . . That is why one finds few or no students at the Austrian schools who were called there by the love of science, or an interest in the things one can learn. Almost all attendees see their studies as a necessary evil, as an unavoidable means to arrive some day at an official function, or rather at the remuneration that all of them envision in the distance as the only aim of their golden dreams.

—Viktor Andrian Werburg, Österreich und dessen Zukunft[1]

Austrian Universities were created by the sovereign as autonomous corporations, endowed with constitutional privileges and laws of property. With time, they largely lost their autonomous positions and are organized now as state institutions, although their position as juridical persons has not been rescinded by legal means.

—Ministry of Religion and Education, 1897[2]

The assessment of Cisleithanian universities published anonymously by the liberal politician Viktor Andrian Werburg (see epigraph) introduces the topic of the structure of the scholarly landscape before 1848. During the nineteenth century, questions of what "science and scholarship" meant, what place they would have in universities, and what the function of universities would be were raised several times, leading to a variety of solutions. Some of the most influential changes were the reforms of 1849, when the new Ministry of Religion and Education not only reformed the universities but also rewrote their histories.[3] The connection between politics and history writing was particularly evident in 1853, as the conservative faction of the Habsburg Parliament pilloried the liberal reforms, while historians and publicists allied with the ministry crafted a gloomy picture of pre-1848 academic misery. Many later historians, up to the present day, have accepted this picture rather uncritically, repeating the story of how Count Leo Thun-Hohenstein triggered the takeoff of higher education immediately after the revolution of 1848.[4]

In this chapter I challenge this view. I claim that the criticisms of pre-1848 Habsburg scholarship are often linked with a conceptual imposition of the post-1848 idea of academia and that, instead, one has to accept the functional dualism of scholarship during the first half of the nineteenth century. Early nineteenth-century scholarly endeavors can tell us much about how different political activists perceived the role of scholarship in the Habsburg Empire. At the same time, this period shows two different models of spatial structure in Habsburg scholarship: one accentuating a decentralized and multilingual monarchy and one promoting the primacy of Vienna and the German language.

Before 1848 Habsburg universities were institutions for the production of loyal subjects, while the primary places for the production of scientific knowledge in the empire included museums, state collections, libraries, botanical and zoological gardens, pharmacies, and a number of more or less formal societies and clubs. The latter, especially, played a prominent role by hosting and financing renowned scholars. The imperial cabinets in Vienna, as well as the imperial library, held resources that attracted researchers from all over the empire, and the state supported such endeavors by awarding positions to the most scholarly and politically suitable individuals. While these positions were mostly administrative, for example, as a head librarian or curator, they allowed enough time for research, making them crucial for the production of new knowledge. Universities were at the time far from the importance they achieved in the second half of the century. They were rather like high schools, concerned more with the education of civil servants than with the development of scholarship. Although fostering scholarly interest among students was not their primary aim, university professors were still often internationally renowned scholars, especially in the sciences and medicine.

Even the University of Vienna, located amid formidable imperial collections, "did not enjoy a good reputation in the learned world."[5] The exception was the medical sciences, for which Habsburg universities were renowned well beyond central Europe.[6] Lorenz Oken, the famous natural scientist and foremost organizer of pan-German scholarly communication through his journal *Isis* (established in 1816) and his role in the creation of the Congresses of German Natural Scientists and Physicians (Versammlungen Deutscher Naturforscher und Ärzte), wrote in 1818 a fitting description of the problems Habsburg scholarship encountered, commenting on the inauguration of the Patriotic Museum in Bohemia. Praising the collections in Graz, Prague, and Vienna as some of the most interesting in Europe, he stated that they would not lead to scientific development if they were not included in the communication network of science: "What do you do with it? Nothing. Nothing. And once more nothing."[7] In particular, he blamed repressive censorship for the passivity of Habsburg scientists: "But why do the scholars do nothing? There is the rub. Here we come to our old song. Restraint of the press, restraint of mind. . . . Do you not realize that everything in the world is so reciprocal, that scholar stimulates scholar. If you had a lively general literary life and work . . . they [the scholars] would be allowed to write everything that the wind whispers in their ears."[8]

Censorship, which inhibited intellectual exchange within the monarchy as well as with scholars in other countries, figured in critical writings almost universally as the main hurdle to scientific flourishing. However, a second factor, the lack of scholars in the centralized scientific institutions, was also seen as a serious obstacle, not only by Habsburg scholars but also by foreigners, such as the British surgeon William Wilde.

Reporting on his journey to the empire in 1843, Wilde portrayed Vienna as a city with a lively scholarly production, especially in medicine (pathological anatomy and ophthalmology), and a profound scholarly history. He wrote, "It is more than Egiptian blindness in them [the Austrian monarchy and the ruling house] to remain passive spectators of the overpowering efforts of the Sclaves [Slavs] and Magyars, and not to strengthen and bind together . . . the German elements of the constitution." He continued, "Is it not an unaccountable and unwarrantable neglect of the German race, whose scientific worth and capability is so much underrated in comparison to the Hungarians, Bohemians, and Italians, to whom academies are permitted."[9]

Wilde denounced what the German-Austrian scientific landscape lacked in comparison to international (here, British) standards. First, despite the existence of scientific productivity, this was not channeled through journals under the auspices of a centralized academy that could place its stamp of approval on them. Nor was it possible to coordinate the work of different institutions. For example, there were no meetings for "mutual instruction" by scholars, where they could exchange ideas and steer joint projects.[10] Second, Wilde saw Habsburg scholarship as an outcome of networks of scholars from the varying cultures, which he called races. Vienna, a symbol of German culture in the empire and thus of the German Confederation, lagged, in this Briton's eyes, behind Pest, Prague, Milan, and Venice in intellectual productivity. For observers trained in the British Empire, by 1843 the Habsburg Empire was already characterized by ongoing conflict among clearly defined cultures rather than being a multicultural ensemble embodying peaceful cooperation.

Wilde clearly grasped some of the main characteristics of the empire, in which multiple languages coexisted but scientific communication was limited by scholars' lack of linguistic skills. The ongoing development of national bibliographies and dictionaries, and the growing scholarly and literary production in national languages, prevented an overview of the empire's cultural production as a whole; this production was attributed to the different linguistic groups, not to the empire.

But the problem was not the growing number of publications in Slavic languages but the hegemonic structure of language competence. While Slavic scholars read and used German (among other languages), German scholars could read French, Italian, or English but rarely the other languages of the empire. In 1830 the influential journalist Franz Sartori criticized this German-centrism of the empire, reminding his colleagues that "the *German* language is *not the sole* language in the Austrian Empire"[11] and arguing for cultural cooperation and the overcoming of linguistic boundaries. Although the idea of the *Gesammt-Monarchie* (lit., Whole-Monarchy, i.e., a unified monarchy) was supported in various ways, this rarely went so far as to include educational multilingualism; there was no acknowledgment of the multitude of literary languages suitable for higher education. Sartori was also unique in showing an interest in the cultural life of the periphery while himself being part of the political center; he stressed the Habsburg ideals of cultural autonomy and productivity to his German-speaking readers. Most scholars preferred to look toward other centers, France or the other lands of the German Confederation, disregarding what was happening in different languages within their own state. Habsburg scholars participated in the Congresses of German Natural Scientists and Physicians, with the twenty-first congress even taking place in Graz in 1843.[12] However, there was no congress of Habsburg science to foster a common identity, as the congresses in other states or empires did, or even the congresses that spanned state boundaries, as in Scandinavia.[13] In addition, it seems that only a few people such as Sartori even desired such a gathering.

Composite Scholarship in a Composite Monarchy?

With the support of Maria Theresa and Joseph II, in the course of the late eighteenth century German became the primary language of the empire. This met with opposition from Magyar and Slavic language activists, who were increasingly expressing their desire for their languages to be treated on a par with German. The last quarter of the eighteenth century saw an increasing number of apologies for the Slavic languages, which aimed to reevaluate the linguistic hierarchies within the public and political spheres.[14] A centralization process during the reign of Maria Theresa, intended to unite the empire, did just the opposite, instead forging patriotic identities that increasingly aligned themselves with the different languages of the provinces. In turn, interest in the humanities in general began to grow among the provincial elites, resulting in the creation of scholarly societies.

Intending to forge interest in regional histories and languages, from the early nineteenth century the aristocracy began bringing forward and supporting various scholars, who, paid and partly sheltered from governmental policy by the aristocracy, could publish and travel with fewer constraints than scholars employed at the imperial institutions. This new aristocratic interest in scholarship also led to the establishment of the first scholarly societies in the Habsburg Empire. While a large number of such societies survived for less than a year, and several lingered longer, a few began to evolve into small academies of science.[15] Similarly, the aristocracy founded provincial museums, such as the Patriotic Museum in Bohemia (Prague), the Hungarian National Museum (Magyar Nemzeti Múzeum) in Pest, the Joanneum in Graz (Styria), the Moravian-Silesian Museum (Mährisch-Schlesisches Museum) in Brno, and the Lubomirski Museum (Muzeum Książąt Lubomirskich, a branch of the Ossoliński Scientific Institute [Zakład

Narodowy Imienia Ossolińskich]) in L'viv, with the principal aim of forging both scholarship and local patriotism.[16] In the late eighteenth and early nineteenth centuries, these provincial institutions were still linked to a strong sense of patriotic regionalism, rather than to the resuscitation or invention of nations. In most cases, this local patriotism was also not linguistically exclusive but rather inclusive, seeking to unite regional peoples from all social and linguistic groups. The aristocratic patronage enabled the museums to be active internationally and encouraged scientific development irrespective of political limitations.[17] In fact, the scholars and institutions supported by aristocrats enjoyed to a certain extent a better situation than those financed directly by the empire, which were under closer scrutiny from Vienna. The learned societies in Bohemia and Galicia were able to realize various versions of provincial scholarship in the first half of the nineteenth century.

In Prague the Private Society in Bohemia for the Development of Mathematics, the Fatherland's History, and Natural History (Private Gesellschaft in Böhmen, zur Aufnahme der Mathematik, der vaterländischen Geschichte und der Naturgeschichte), an aristocratic organization founded around 1771, included representatives of several noble Bohemian families. It was strictly a regionally bound institution that aimed to foster research on provincial and regional topics and to catch up with "German" cities, where academies had already reinforced universities, as Ignaz Born wrote in the introduction to the first volume of the society's proceedings.[18] In 1784 Joseph II and the Studienhofkommission (the Aulic Educational Commission, serving as the de facto Ministry of Education) denied the society status as a learned academy. The society was, however, allowed to use university facilities; it received one room in the Prague Carolinum (from 1828, two rooms), and its bylaws were approved. In 1791 Leopold II awarded the society royal status, and from then on it was known under the bilingual name Königliche böhmische Gesellschaft der Wissenschaften / Královská česká společnost nauk (the Royal Bohemian Society of Sciences), uniting Bohemian scholars regardless of their language or religious affiliation.[19] The society's links with the aristocracy ensured a stable financial situation, allowing it to grant awards, subsidies, and scholarships and to publish *Gelehrte Nachrichten* (Learned news, 1771–72) and, later, *Abhandlungen* (Treatises).[20]

In Galicia, in contrast, the first provincial learned society was established only in 1827, when Count Joseph Maximilian (Józef Maksymilian) Ossoliński, the imperial librarian in Vienna, opened the Ossoliński Scientific Institute (Ossolineum) in L'viv after ten years of preparation. Ossoliński was an amateur historian, primarily interested in source research;[21] however, he was internationally known and was one of only three Habsburg scholars invited to become members of the Society for Older German History (Gesellschaft für ältere deutsche Geschichtskunde), which edited the prominent series Monumenta Germaniae Historica.[22] The Ossolineum, devised as a provincial institution, increasingly became a Polish one, however. In the 1830s the institute printed conspiratorial writings and edited sources on the November Uprising (1830–31); as a result, it was placed under police control, and its activities were severely limited. It was revived only after 1848. Despite its struggles, it continued to forge an understanding between the speakers of the two Galician languages, bringing together the allegiances of Polish and Ruthenian scholars.[23] The Ossolineum was also linked to other Polish institutions in Cracow, Warsaw, and Poznań/Posen, and its publications clearly envisioned a space different from the Galician one.[24] The Cracow Academic Society Linked with the University of Cracow (Societatis Litterariae cum Universitate Studiorum Cracoviense Conjunctae / Towarzystwo Naukowe Krakowskie z Uniwersytetem Krakowskim połączone) became a cradle of Polish-language scholarship after 1815, even if it was of only local importance because it was part of the Free City of Cracow (1815–46).

In the period before 1863, however, it was in the Grand Duchy of Posen and the Russian Empire's Kingdom of Poland (from 1867 Vistula Land) that Polish-language scholarship thrived, escaping Metternich's censorship.[25] In particular, the Russian Empire provided, until 1831, very favorable conditions for universities under the protection of the tsar and the local aristocracy, allowing them to teach in Polish.[26] In Prussia chairs of Slavic languages were created at the universities in Berlin and Wrocław/Breslau, and societies concentrating on Slavic languages and history emerged; several of the émigrés from the Habsburg Empire who were teaching in Prussia moved back to the Habsburg Empire after 1848 and were instrumental in Habsburg government measures to strengthen loyalty after that time.[27]

While the Ossolineum was an independent, private institution, Ruthenian scholarship flourished around state-sponsored institutions, namely, the Studium Ruthenum (Студіум рутенум), established in 1787, and the Stauropegion Institute (Stavropihiys'ky Instytut, or Ставропігійський інститут), established in 1788 as the Greek Catholic successor to the Orthodox Dormition Brotherhood (Uspens'ke Bratstvo).[28] Both were closely associated with the Greek Catholic Church, and both educated and organized

Ruthenian elites around it, including hosting a printing house for Ruthenian literature. The Studium was an autonomous part of the university that offered lectures in Church Slavonic.[29] The institute, headed by the historian and archivist Denys Zubryc'kyj (Денис Зубрицький), had a high scholarly profile and served as a meeting place for L'viv's Ruthenian intellectuals.

Zubryc'kyj's works illustrate, however, the political essence of the debates about Ruthenian culture. While striving to underscore Ruthenians' distinctiveness from Poles, Zubryc'kyj saw Ruthenians as a branch of Rus' culture, united by the use of Church Slavonic. A new generation of Ruthenian nationalists, however, pleaded for cultural development based on the vernacular spoken in Galicia.[30] However, the church's influence also hindered such vernacular-language ideologies: *Rusalka Dnĕstrovaja* (The nymph of the Dniester), published anonymously in Buda in 1837 by three Studium students, set the standards for late nineteenth-century vernacular Ruthenian.[31] Nevertheless, strong opposition from church authorities prevented it from finding as many supporters as intended. *Rusalka Dnĕstrovaja* was published in Buda to escape Galician censorship (it had been rejected by a Galician censor for Ruthenian literature, the professor of moral theology Venedykt Levyc'kyj [Венедикт Левицький]). Yet its circulation was hampered by the L'viv metropolitan Mychajlo Levyc'kyj (Михайло Левицький), who bought almost the entire run of the first edition.[32] Moreover, church authorities exiled all three authors to small villages as priests, which impeded their future activities. While the language issue for Galician Greek Catholics was not set before 1848, it was clear that the gap between different groups was increasing and was being translated into ethnic terms. Indeed, the idea of introducing a Polish-based alphabet to write Galician Ruthenian attracted only a few—predominantly, but not exclusively, Polish nationalists claiming Ruthenian as a Polish dialect.[33]

The development of provincial societies concentrating on language and history shaped both the Austrian and Hungarian parts of the monarchy. In the latter, Ferenc Széchényi founded a museum and library as early as 1802 but succeeded in creating the Hungarian Learned Society (Magyar Tudós Társáság) only in 1825. In line with other learned institutions, this society concentrated in its early years on developing a Hungarian scientific language and literature as well as modernizing scholarship in the Hungarian part of the monarchy.[34] The society clearly supported the idea of cultural distinctiveness for the Hungarian Crown, although this was not its primary aim; this was also not the same as supporting the goal of political autonomy.[35]

With time, regional societies, initially pluricultural and not tied to a particular national group, were increasingly inscribed into nationalistic policies, and their resources were used to propagate different national positions. Paradigmatic here is the Patriotic Museum in Bohemia. In the article advertising the opening of the museum in 1818, Franz Graf von Kolowrat clearly depicted science and scholarship as a means to forge a transcultural understanding: "The history of all people [*Völker*] identifies epochs in which the energy of nations, directed outward, excited by long tempests, when calmness returns, reclaims itself, reconciles bedraggled muses, and elevates the arts and sciences to flourish."[36] However, in due course, the museum contributed substantially to the establishment of Czech nationalism by opening its publications to Czech-speaking authors. From 1827 the Patriotic Museum in Bohemia published the *Monthly of the Society of the Patriotic Museum in Bohemia*, in Czech and German versions (*Monatsschrift der Gesellschaft des Vaterländischen Museums in Böhmen* and *Časopis Společnosti wlastenského museum w Čechách*), both edited by František Palacký. Although both journals were established to "foster enlightened knowledge among the people [*líd*],"[37] their content differed: *Časopis* dealt mostly with Czech literature and history (publishing analyses as well as, for example, poems). Indeed, the editorial for the first edition stated, "Often proclaimed and felt in our nation was the need for such a journal, which, adapted to the knowledge of the more enlightened [people] among the folk, fills the gaps and deficiencies existing in our language and literature. . . . [T]he content of the journal will be: firstly the broad scope of useful sciences and arts, then the knowledge of the homeland, and finally and especially the answer to the needs of our language and literature."[38]

The German-speaking publication also included a wide range of historical and philological studies concerned with the Czech nation and with Slavic culture but met with only marginal interest, with fewer than two hundred readers per issue. In 1830 it began to appear quarterly, and by 1832 it had been canceled; readers were informed that the journal would appear irregularly, which heralded the end of its existence.[39] The Czech journal was renamed *Časopis Českeho Museum* (Journal of the Bohemian Museum), and financial problems forced it under the patronage of the Czech Foundation (Matice česká), an autonomous branch of the museum concerned with literature that also owned a printing house specializing in Czech-language publications. Scholars gathered around these early museum-built networks of Czech patriotic scholars and educated a public desperate to hear spoken

Czech as a scientific language and to become involved in fostering patriotic scholarship. By 1847, 685 people had the highest and most expensive membership status, *zakládatel* (founder), with a growing percentage of them coming from the bourgeoisie.[40]

The establishment of Czech and Ruthenian as literary and scholarly languages, and their use in scholarly publications, remained largely unfinished business in 1848. Their use, together with an ever-growing number of publications in Polish, did begin to create an intellectual disruption in Habsburg cultural life, however. "Culture," previously limited to elites and transregional social groups, extended to a broader population within geographically delimited nations. The nineteenth century followed the model of eighteenth-century cameralism, which had abandoned Latin-based scholarship and introduced new ways to popularize knowledge for the public, thus inducing a growing rejection of the republic of letters and moving more toward a science for the people as part of provincial well-being.

The change from transnational Latin to state languages had been perceived differently among different groups, since from the late eighteenth century languages were variously seen as either a neutral tool of communication or a symbolically laden medium. German and Polish were representational languages of loyalty in the Habsburg Empire and the now nonexistent Commonwealth, respectively, as well as for ideologies of (ethnic) nationalism, which manifested itself only much later. Publishing in a language other than that of the state slowly built up a sense of belonging to something other than Habsburg society. In most cases, however, in 1848 it remained unclear what the new community would be. Czech activists had the option to be Bohemians (different from Moravians), Czecho-Slavs, or Czechoslovaks, among others. Ruthenians could opt for Russian, Little Russian, Rus', Ukrainian, or local Galician/Ruthenian projects, with each movement using different, yet mutually understandable, vocabularies and having its own corresponding alphabet. Whether Austrians were just another Germanic people who needed a distinct language and whether Poles should modify their language to include groups regarded as minorities were fiercely debated in the early nineteenth century, although political identities still varied considerably.

Scholarship conducted in vernacular languages was mostly locally oriented, encompassing descriptive and ethnohistorical disciplines and aiming for a broader fostering of culture. However, it lacked a public, an issue that came to light only later in the century. Still, in the early nineteenth century, nationalized scholarship did not offer fierce opposition to state institutions, which were tuned toward other educational scientific models, to the dismay of many who envisioned freedom and liberalism, irrespective of their cultural or ideological background. It was rather a complementary system separate from state-supported institutions and turning toward a new public. Clearly, many scholars saw the problem of lack of communication across the empire and proposed statist solutions, such as the creation of an academy of sciences, a place uniting scholars from throughout the monarchy and offering them opportunities for communication.

Centralizing Science: The Imperial Academy

Because the regional aristocrats were investing in local societies, and the central government remained disinterested in forging new knowledge, interest in a centralized scholarly institution was limited. The aristocracy even openly complained in the 1840s that the creation of a central learned society would diminish the importance of the well-functioning regional societies and lead to unwanted centralization.[41] Provincial elites were clearly opting for a monarchy where cultural distinctiveness was cherished, and scholarship was one means to support this. The creation of a Viennese academy, which had already been proposed by Gottfried Wilhelm Leibniz around 1700,[42] was opposed not only by many aristocrats but by Metternich as well, who initially did not support the idea of autonomous science and scholarship. He would allow the academy only if it were in the political interests of the empire, and this was not the case until after 1845, when pressure against censorship and an oppressive regime grew stronger. The Imperial Academy of Sciences and Arts, inaugurated in 1847, served, however, not only as a meeting point for scholars but also as a project structuring the scholarly geography of the empire, centered on the capital city. The absence of the word *royal* (*königlich*) from the academy's name symbolized that the Cisleithanian part stood at the center, thus securing Hungarian distinctiveness at the scholarly level.

Speakers at the inauguration of the academy underscored its political role beyond any doubt. Its aim, apart from forging scholarship, was "to secure the . . . beneficial knowledge and experience . . . as well as to support the government's functions through answering questions and problems that belong to the scope of scholarship."[43] Metternich saw the institution as both a state-controlled outlet for scholars and a means to better the empire's standing in international competition, as notable academies were already highly valued.[44] To guarantee state control over the academy, Archduke John of Austria served as its curator, and the academy was subjected to censorship of both its publications and correspondence. However, on 13 March 1848 the government freed the academy from censorship owing to its inefficiency.

The first president of the academy was the famous diplomat and pioneer of oriental studies Joseph von Hammer-Purgstall. Before the creation of the academy, he clashed with politicians over his involvement with a famous 1845 memorandum, *Die gegenwärtigen Zustände der Zensur in Österreich* (The present conditions of censorship in Austria).[45] During his tenure as president (1848–49), his political views became milder, and he argued that the academy should be neither a political nor an educational body but rather ought to deal with science itself. Under his presidency, the withdrawal from political involvement was immediate: for instance, the academy refused to lend its support to political gatherings such as the Frankfurt Parliament.[46]

Although its pan-imperial character remained contested, the academy aimed to serve as a supraregional meeting place for scholars across the empire. The reality, as described in the introduction to this book, lagged behind these ambitious plans. While regional societies contested the primacy of Vienna, the academy itself turned to fostering Austrian, that is, German/Habsburg, science.

The empire's two scholarly spaces, the provincial and the imperial, clearly began to grow apart in the early nineteenth century, and the imperial academy was, in a way, a last resort to unify them again. Now I turn to the universities to show, first, how these institutions dealt with the problem of spatial disparities before 1848. Then I discuss how the 1848 revolution changed the universities' outlooks and brought forward new agendas, which led to the Thun-Hohenstein–Exner reforms of 1848–49.

The Vormärz University

During the Enlightenment, universities were restructured from autonomous corporations into state agencies, in which "scholarly education [*gelehrte Ausbildung*] turned into a form of 'state production.'"[47] Throughout Europe, including in other states in the German Confederation, Vormärz was an epoch in which universities came under increasing supervision from governments, which feared, in particular, student unrest.[48] Also in Russia, where universities traditionally had a strong corporate character, the government was trying to limit them, although, ironically, with much less success than in the Habsburg Empire or Bavaria.[49] Similarly, in the Habsburg Empire the imperial administration closely scrutinized the universities. Universities were defined primarily as places of education and discipline, not as places where the *artes liberales* should thrive. Joseph II wrote in his resolution of 25 November 1782 that

> the youth must not be taught things they would use in a strange way or in a way that does not serve the well-being of the state, since the essential studies at the university serve the education of state functionaries, and are not dedicated to breeding scholars. They [scholars] should acquire scholarly qualifications by themselves, once they acquire the first principles. One should not believe that one can find a single example of someone becoming [a scholar] merely through a lectern.[50]

Four decades later, Francis II formulated similar ideas, reasserting universities' role as educational institutions: "I will have my subjects learn all those things that are useful in common life, and likely to keep them attached to our persons and their religion. I do not want teachers who fill the heads of my students with that nonsense which turns out the brains of so many youths in our days."[51]

The above-mentioned dualism between education and scholarship was pivotal for the imperial/statist understanding. Through their corporate character, Habsburg universities also had a firm link with the city where they were located and the regional public. *Doktoren-Collegien*, the colleges of doctors[52] and professors (both active and retired), were part of the university and had the crucial right to award doctorates (*Promotionsrecht*); they also had members in the academic bodies (faculties, academic senates, etc.). At the same time, they were compulsory representatives of all graduates, similar to the Chamber of Labor, controlling accreditation for practice, especially for jurists and medical students.[53] From 1818 the office of the dean was also under the control of the Doktoren-Collegien, and professors were not permitted to hold this position as it would keep them from teaching.[54]

The corporate character of the universities did not mean that there was no place for science within the university walls. A glance at the names of, for example, the physicists or chemists, especially in Vienna, reveals modern and well-acknowledged scholars, who were also well linked internationally.

However, the lack of funding for new institutes and research opportunities hampered innovation. At the same time, universities in other German states—not only in Prussia (Berlin or Halle) but also in Bavaria (Munich), the Kingdom of Hannover (Göttingen), and the Grand Duchy of Hesse (Gießen)—gained more of a reputation, turning toward new educational methods and experimental science. Even the Russian Empire was more liberal toward universities at this time, allowing them considerable autonomy in order to facilitate the modernization of the state; it both invested in foreign professors and sent leading Russian academics abroad.[55] Habsburg scholars knew this and demanded changes to bring their universities up to par with the provincial academies. As in other states, supervision by the *Studiendirektoren*,[56] the censorship of schoolbooks, and strong political control over the subject matter (both the curriculum and the content of each lecture) were among the factors blamed for academic misery. As a result, university reform was one of the most prominent demands during the 1848 revolution.

The number of Habsburg universities and faculties varied over time, but they remained closely linked to the existing educational premises of the central government. Most universities (apart from those in Vienna, Prague, and Pest) were demoted to *Lyzeen* (lyceums) in the late eighteenth century, but in the early nineteenth century Francis I reinstated universities in L'viv (1817), Innsbruck (1826), Graz (1827), and Olomouc (1827), but without medical faculties. In the provinces, medical studies were taught in university-connected medical-surgical academies (*mediko-chirurgische Lehranstalten*); these had a limited number of teachers, and the courses were oriented toward the practical education of midwives and surgeons (*Wundärzte*). The Imperial and Royal Medical-Surgical Joseph's Academy (k.k. medizinisch-chirurgische Josephs-Academie) in Vienna, established in 1785, had the same practical orientation; in the 1820s it became de facto the second medical faculty of the university, serving as an important place for teaching and practicing medicine, even if it was not formally incorporated into the university. The medical faculties themselves were divided into a two-year surgical course of study for civil physicians and surgeons (*Chirurgisches Studium für Civil und Wundärzte*, including courses for midwives), structured similarly to the courses at the medical academies, and a five-year study of pharmacology and the higher surgical arts (*Studium der Arzneykunde und höheren Wundarzneykunst*); this reflected the duality between practical education and "higher" education.

The philosophical faculty (*Philosophicum*), reformed throughout the empire in 1805, had the same semi-university status as the medical academies, forming a preparatory level between the gymnasium and the university.[57] The philosophical faculty taught a wide range of disciplines, including humanities and the sciences (except medicine), but with special consideration to philosophy, which was defined as a "medium of high intellectual culture" and a "groundwork science [*Wissenschaft*] for all other vocational sciences"[58] and was clearly denoted as preparation for the subjects taught at the university.

University lectures were held based on the so-called *Vorlesungsbücher*, textbooks that had to be approved by the Ministry of Education and which were literary read aloud. Disobedience was severely punished; some notable scholars were removed from their universities for violating this rule.[59] Although professors were allowed to submit their own books as the basis for their lectures, only a few decided to do so, as this path was highly complicated and uncertain. It wasn't until the late 1820s that free lectures based on the lecturer's own manuscripts were allowed for noncompulsory subjects.[60]

The restrictions within the Habsburg monarchy also influenced the ways in which universities could interact with scholars and institutions in other countries. The possibility of studying abroad (including in the non-Habsburg parts of the German Confederation)—which was especially tempting for non-Catholic students since Habsburg universities were Catholic institutions—was restricted greatly in 1829; foreign courses and diplomas were not accepted, and students attempting to cross the border required police authorization.[61] The government was seemingly alarmed that the freedom of learning and teaching introduced at some foreign universities could open a channel through which liberal or anti-absolutist ideas could travel.[62] Students who wanted to study outside the empire but were not members of the privileged aristocracy[63] could bribe functionaries, but this could bring its own problems with the police.[64]

Restrictions on the exchange of ideas were reinforced in other areas as well. From 1815 on, libraries produced lists of banned books; these could not be read in the library and included Johann Gottlieb Fichte's *Staatslehre* (Doctrine of the state, 1813) and Joseph von Hormayr's *Taschenbuch für vaterländische Geschichte* (Pocket book of the history of the fatherland, 1811–48). Further, authors such as Goethe, Schlosser, and Kant could be read only *erga schedam*, that is, with permission from the local police department.[65] Moreover, because of his paranoia toward liberalism, Metternich banned universities from corresponding with foreign schools.[66]

The development of Galician universities was more complicated. The Cracow Academy (Akademia Krakowska, later renamed the Jagiellonian University) was the provincial university (*Landesuniversität*) for Galicia in 1805–17, while during the same period the University of L'viv was closed, and only a lyceum operated in that city. After 1817, when Cracow became a free city, L'viv's lyceum was given the status of a university under the name Francis I University; it was structured along the lines of other Habsburg universities, with German as the language of instruction. A chair of Polish language was created in 1817 but filled only in 1827 by Mikołaj Michalewicz, neither a good scholar nor a gifted teacher.[67] The Cracow Academy was at that time a semi-autonomous body controlled by protector states (Habsburg, Prussia, and Russia), with extended rights that included the possibility of accepting students from other regions of the pre-partition Commonwealth. This privilege was revoked in the aftermath of the November Uprising, because the university was regarded as an important place for forging revolutionary nationalist ideas and contacts.[68] At this time, the academy was still a small provincial institution, with some two hundred students, compared with the fourteen hundred at L'viv. The curriculum was based on that of Habsburg universities, with a preparatory philosophical faculty. Only the law faculty worked according to a slightly altered curriculum from the University of Berlin. After the Cracow Uprising in 1846, the Habsburg Empire incorporated Cracow, and the Cracow Academy began to be restructured on the Austrian model. While initially there were plans to close it, the government decided to retain it, thanks to the goodwill of the government's minister plenipotentiary Stephan Ladislaus Endlicher, a Viennese botanist. Its restructuring was completed during the reforms of 1849, which unified education across the monarchy.[69]

The language of instruction was the most important binding element in the pre-1848 empire: Latin in all subjects in the secular faculties and German in the philosophical faculties. Even lectures on vernacular literatures were held in Latin in L'viv and Prague. The only exception was the practical teaching of foreign languages (readerships) and the first year of education for midwives and surgeons, which took place in the local language. Since civil servants and physicians dealt with the local population, which in many cases knew neither German nor Latin, inclusion of the vernacular in the university system was necessary to enable interprovincial transfer of staff. Some knowledge of the local language was also required to obtain teaching positions at certain universities.[70] In Cracow from 1833 onward, the language of instruction was Latin, with the exception of practical subjects and lectures at the philosophical faculty, in which instructors had a free choice of language, except in the subjects of religion, philosophy, and the classical languages (taught in Latin) and Polish literature and popular mechanics (taught in Polish).

Linguistic uniformity at the faculties enabled lecturers to be mobile and reinforced the standardization of the Habsburg intellectual space. To level the chances of scholars from all provinces, standardized open contests (*Conkursverfahren*) were introduced, consisting of an exam with three questions and an open lecture. Teachers who already held an appointment at another university were exempt from the exam. The *Studiendirektoren* compiled the results into a standardized list (the *Kompetenztabelle*), less often naming only the three best candidates in hierarchical order (the *terna*), and forwarded it to the Studienhofkommission together with the opinions of the provincial government. The final appointment by the emperor was provisory for three years (the *Probetriennium* or *Provisorium*) and at the end of that term had to be verified to become a permanent position.[71] In this way, the open contests allowed scrutiny of the political and ideological appropriateness of the candidates. The process of appointing professors was indeed somewhat similar to that for officers in the army: applicants had to not only comply with the political ideology of the monarchy but also be able to resist, or even appease, any nationalistic feelings at the universities. As for military personnel, this meant moving teaching staff across provinces.[72] In the case of universities, however, the circulation was hegemonic: only a few scholars who had not been educated at the main universities could get a position there, while staff from the universities in Vienna or Prague were widely represented at universities in other provinces.

The rules for appointments and the actual practices both supported the centrality of Vienna. Early nineteenth-century lawmakers foresaw that senior professors should be appointed to the University of Vienna as a reward for their long service and as a guarantee of high scholarly standards at the central university of the empire.[73] In fact, most scholars teaching in the capital were nominated in this way.[74] This led to criticism of the low research standards in Vienna, because older professors usually concentrated more on teaching than on scientific production. Critical intellectuals spoke of Vienna as an "honorable house of invalids,"[75] and Ernst von Feuchtersleben, responsible for the universities for a short time during the chaos of 1848, made the rejuvenation of the Viennese medical faculty one of his priorities.[76]

While scholarly quality was not the main priority at the universities, the government still pursued academic professionalization. From 1811, universities included *Pflanzschulen zur Bildung künftiger Professoren* ("nurseries for the education of future professors"), which consisted of assistants, adjuncts, prosectors, and so on. In the medical faculty, the *Pflanzschule* consisted of, more or less, all scientific personnel assigned to professors, both at the university and at the hospital, including assistants and secondary physicians. The other faculties had a limited number of young academics: the theological and philosophical faculties each had two, and the law faculty had one.[77] The main aim of the *Pflanzschule* was to prepare scholars for a professorship, and professors were officially forbidden to treat their younger colleagues as servants (*Handlanger*), which could impede their academic progress.[78]

While they did not serve as a meeting place for international scholars, Habsburg Vormärz universities were an interesting mixture of social and cultural backgrounds. At the Viennese medical faculty, for example, immediately before the revolution, most professors were the offspring of lower state officials and members of the bourgeoisie. Aristocrats were rare; similarly underrepresented were peasants, although one can find sons of millers and village judges.[79] However, even more impressive examples of social mobility were possible: Antoni Bryk was officially a serf until 1848; he illegally obtained a university education in Vienna and ignored repeated requests by his lord to return to Galicia as a military physician. After the revolution, already a free man, he was appointed a professor of forensic medicine at Cracow.[80]

Given their educational and practical orientation, pre-1848 universities and intellectuals played an important role in discussions on the ideology of the state and/or nation, as their position was certainly privileged in comparison with that of private scholars. Simply through elaborations on linguistics, several university scholars gained respect within national groups, although they were rarely in the first ranks of patriots or nationalists. The brothers Jan Svatopluk Presl and Karl Bořiwog Presl, professors of zoology and mineralogy and of natural history and technology in Prague, respectively, who were also active Czech nationalists, can be regarded here as rare exceptions to the rule. To a large extent, however, universities effectively remained tertiary institutions intended to forge patriotism among state officials, producing subjects loyal to the empire and the throne. It must also be noted that many professors indeed participated in the 1848 revolution and that their ideas on the role of the university were not in direct conflict with those of the students, as was later claimed. Thus, even if the Studienhofkommission had succeeded in keeping nationalists of all sorts outside the university walls, 1848 proved that it had not eliminated liberalism.

More important, universities, like the other scholarly institutions discussed in this chapter, were not universally accepted by political groups within the monarchy. The use of German as the language of instruction was not a problem only for the increasingly nationalized provinces. By predominantly nominating German-speaking scholars, universities failed to include provincial residents as teachers, estranging the universities from the city elites, especially in Galicia.[81] One exception to this rule was the historian Joseph Mauss (born in Tengen, now in Baden-Württemberg but until 1806 part of the Habsburg Empire), who enjoyed celebrity status in L'viv and is said to have encouraged his L'viv students to participate in the November Uprising in 1830–31.[82] Scholars' adaptation to the urban culture they encountered played an even more important role after 1848, often deciding entire careers.

Scientific excellence clearly did not necessarily correlate with openness to nationalism, even if later generations did remember many scholars who united these characteristics. Yet, even in the Vormärz, the public was increasingly involved in regional scholarly endeavors linked to linguistic projects, such as the Patriotic Museum in Bohemia or the Ossolineum. In the prerevolutionary discourse, these two assets apparently began to merge, especially among non-German elites. Universities, highly esteemed as vital institutions of cultural and intellectual life, especially in smaller cities, were seen as places whose potential had yet to be fulfilled. By 1848 students and significant parts of the city public in L'viv, Pest, and Prague were also certain that the solution to academic misery was not only greater freedom but also the inclusion of local languages as the medium of instruction. As a result, the 1848 revolutionaries requested linguistic equity, which should not be hastily interpreted as only a nationalistic claim.

On the Barricades: Universities in 1848

The revolution of 1848, often seen as a turning point in the history of the Habsburg Empire, brought far-reaching changes for universities and intellectual life in central Europe. First, the short-term liberal government remodeled the universities based on the Prussian system, although with variations reflecting the cultural particularities of the empire. Universities began to teach humanist subjects at the academic level, in accordance with liberal and nationalist demands, but with the same aim as in the Vormärz, that is, promoting a loyalist narrative, a plan that ultimately backfired. Second, the revolution spawned various regional demands: Bohemia sought a reassessment of the boundaries of the German Confederation, the Hungarians wanted changes in the structure of political relations, and the Kingdom of Lombardy-Venetia demanded federalization and secession. All this illustrates the instability of the imperial space and political structures, across the empire as well as within the provinces themselves, requiring new modes of spatial governance. Third, the constitutional reforms, as well as the liberalization of cultural life, although brief and followed by a neoabsolutist regime, reconfigured the political structure of the monarchy as well as the discourse of loyalty and culture's place in it. The Frankfurt Parliament, the Krems Parliament, the Prague Slavic Congress, the April Laws in Hungary, the Petition of Liptovský Mikuláš (Liptau-Sankt-Nikolaus, Liptószentmiklós), and other events did not result in changes to the laws, but they publicly presented the points of agreement among the different parties. This, along with the abolition of censorship, enabled the creation of an active public sphere and an open discussion of how the monarchy should be structured. For universities, and scholarship in general, changes in the political sphere did not mean a complete revolution but rather a set of gradual transformations facilitated by the atmosphere of 1848, including the free flow of literature, the accentuation and acceptance of cultural diversity, and a relaxation of border policing, which elevated the importance of cultural-cum-linguistic spaces while lessening the influence of state borders.

As the wave of revolutionary movements and outbreaks in 1848 shook the Habsburg monarchy, students were among the first on the barricades in Cracow, Prague, and Vienna (see figure 1).[83] Their teachers often joined in or even led the political reaction against absolutism, proving that political supervision during the Vormärz was either unsuccessful or not as grim as often claimed. This was, of course, not the first openly political movement against the government in which scholars participated. In Cracow, for example, scholarly political activism had a long-standing tradition. During the uprising in the Free City of Cracow in 1846, the professors of the medical faculty had cared for the wounded insurgents on the battlefield. The professor of Polish language and literature Michał Wiszniewski was even, for a day, the self-proclaimed leader of the rebellion in Cracow, although

FIGURE 1 University Square in Vienna during the night of 13–14 March 1848 and the establishment of the Academic Legion. (Archive of the University of Vienna, 106.I.584. Artist: R. Swoboda.)

he strove to conclude the rebellion through political mediation, against the will of the nationalistic organizations.[84] In other regions, groups of scholars and intellectuals fueled political liberalism, demanding the liberalization of public and cultural life, but without engaging in open antigovernment action.

In university cities, students formed so-called *Studentenlegionen* (Student Legions), whose aim was to aid the revolutionaries through active participation. At the beginning of the movement, national issues were decidedly in second place behind political calls for coups d'état against Metternich's oppressive regime, in favor of liberalism. In Prague, Bohemian students who identified as Czech or German fought together, forgetting their cultural conflicts and differences and turning against the government. Paradoxically, this meant turning their rage against Leo Thun-Hohenstein, who shortly before had been named governor of Bohemia. The young count was held captive in the Carolinum and was released only through the mediation of the language scholar and historian Pavel Josef Šafárik (also written Šafařik), who later had a massive influence on Thun-Hohenstein's appointment policy in Bohemia.[85] Alliances across linguistic and cultural-political borders were forged. Viennese students signed a petition calling for lectures in Czech at the Charles-Ferdinand University in Prague and lectures in Polish at the Jagiellonian University in Cracow. In Galicia, however, the supranational idea of political revolution lost out to national divisions, as Ruthenian nationalists fiercely rejected cooperation with the Polish national party and vice versa.[86]

Professors also manned the barricades, demonstrating the ineffectiveness of Metternich's attempts to forge uncritical loyalty to the universities. Even before the revolution, the Viennese Juridical-Political Reading Association (Juridisch-Politische Leseverein) had united intellectuals of all estates, including students and professors. They played an eminent political role in promoting anti-absolutist policy, lobbying the court for, among other things, the abolition of censorship.[87] In Innsbruck the professors Albert Jäger and Alois Flir, among others, stood at the center of the struggle over the question of Tyrolean autonomy.[88] In Cracow academic legions were organized by the professor of library sciences Józef Muczkowski and the physiologist Józef Majer; in L'viv the librarian Franciszek Stroński and the chemist at the technical academy, Friedrich Rochleder, led the academic legion.[89] And in Pest professors were involved in the revolution on the side of the Hungarian party and supported independent reforms of the universities.[90] However, political participation also brought negative outcomes for the universities: for example, the university buildings in Vienna and L'viv were closed, the first owing to a political decision seeking to counter the possibility of student gatherings in the city center, the latter owing to serious damage during the bombardment of the city.[91] Prominent supporters of the Hungarian Revolution, including some university lecturers, had to leave the country after the revolution failed. Most professors were, however, swiftly reinstated, as were other officials who initially experienced repercussions after 1848–49.[92]

Petitions remained the most useful and effective tool in the revolution, following the growing success of political negotiation, which gradually took the place of the mutiny-oriented revolutionary outbursts that had been issuing unconditional but barely acceptable demands. Even though the appeals raised in the petitions were not entirely successful, the mediation of multiple interests showed more promise than did military actions, although both the success of dialogue and the subsequent changes remained closely connected to the government's assessment of the revolutionary demands.

Determining what to include in the petitions led to dissension both between professors and students and between faculties; the discussions brought to light the variety of approaches to the function of universities and scholarship. At the same time, an analysis of the petitions shows that while some demands were common across the whole empire, the views from the capital and the provinces differed in many respects. The regional disparities heightened once liberal possibilities were in sight, and the ministry had to negotiate among differing interests and unify the structure of the academic space.

The proceedings at the Jagiellonian University, where several drafts were discussed, help to illustrate the problem of restructuring universities in a monarchy with different academic traditions. The first petition to the emperor, composed by the rector Józef Brodowicz and accepted by the students and professors in March 1848, aimed to reintroduce university autonomy according to the 1818 bylaws, encouraging freedom of teaching and learning and granting the university exclusive legal control over students—*intra* and *extra moenia* (within and outside of university walls). Furthermore, the project pleaded for the restitution of funds and lands (including those from the parts of the Commonwealth now under Prussian and Russian rule) and for the subsumption of all educational facilities in the city under the university's governance with a guarantee that "apart from the university and establishments linked to it, no other educational institutions would be established without its knowledge and explicit consent."[93] This was a particular concern for religious corporations that were responsible for their own schools. The petition demanded, furthermore, "that no Jesuit or ex-Jesuit ever finds himself in any teachers' corporation, and moreover, that this order, most fatal for human kind, never sets foot on this soil."[94] This project thus aimed to reclaim the privileges the university had enjoyed in the eighteenth century, when it controlled virtually the entire Polish part of the Commonwealth and successfully hindered the establishment of other academic institutions. This resolution, however, never left the building owing to a subsequent conflict between Brodowicz and the students.

The next petition, proposed in the autumn of 1848 by Józef Majer, included the abolishment of courses on religion, the use of Polish as the medium of instruction in all subjects, and the introduction of the history of Poland among the courses taught, as well as, similar to Brodowicz's proposal, financial demands. This project also met with opposition, especially because of the questions of religion and language it raised. The canonical jurist Feliks Leliwa Słotwiński, for example, opposed it, stating that religion should guard students from the "errors of philosophy" and that the exclusive use of Polish not only would negatively affect disciplines such as Austrian, Roman, and civil and church law but would also "attest national hate . . . and affront the first rule of Christian religion."[95] Majer's petition was finally presented to the new governor of Galicia, Wacław Zalewski, and incorporated suggestions for new chairs, including for the history of Poland, Polish law, and the languages of eastern Europe. Some of these demands were fulfilled, especially the use of Polish, acknowledged on 11 October 1848 by the governor: professors who did not know Polish could remain at the university, but Polish-speaking assistants would be appointed to support them.[96]

Several months earlier, Franz Stadion, the governor of Galicia and later minister of the interior, had already allowed the partial use of Polish in L'viv by Privatdozenten, but the main language of instruction was to remain German, or possibly Ruthenian, which was apparently envisioned to slowly replace German as the language of instruction in Eastern Galicia.[97] The partial privileges for Polish in this part of the province were abandoned shortly after a change in prime ministers at the end of 1848, with the argument that the majority of the inhabitants of Eastern Galicia were more averse to Polish than to German.[98] The issue of language use at secondary schools became one of the critical questions for the Prague Slavic Congress, where Polish and Ruthenian nationalist organizations each envisaged their respective language as a leader in cultural matters in L'viv and achieved no binding agreement.

The issue of cultural equity was also at stake in Prague. The students who prepared the petition, which the faculty accepted and supported, placed freedom of religion and teaching at the forefront of their demands but included university autonomy in legal questions, inclusion of the technical schools as part of the university (as the fifth faculty), and freedom of assembly according to the laws of the University of Munich.[99] The petition, forwarded to the government in late March 1848, was answered on 2 April: as in L'viv, Privatdozenten[100] were allowed to teach in Czech, German, "or any other language";[101] freedom of teaching and religion was approved; and students were allowed to study at foreign universities.

While the equality of languages was widely discussed at the provincial universities, and was seen as part of the liberalization of academia, in Vienna the political reorganization and structural liberalization of the educational system were central. This restructuring also, however, included multilingual instruction as a means of stabilizing loyalty. Between the beginning of the revolution in 1848 and June 1849, the minister of education changed several times, depending on political alliances: first, Franz Freiherr von Sommaruga, then Ernst von Feuchtersleben (de jure *Unterstaatssekretär*, that is, undersecretary of state), followed by several interregna during which the ministry was subordinated to or joined with other departments, and, finally, Leo Thun-Hohenstein, who arrived in office in July 1849, directly after his rather unfortunate time in Bohemia. Before appointing Thun-Hohenstein, the government considered the ministry as a possible concession to the Slavic subjects of the empire. Among possible candidates for the office, František Palacký attracted the most interest. Palacký, a renowned historian and an acknowledged Bohemian patriot, was (in)famous for his refusal of an invitation to the Frankfurt Parliament and was a critic of Habsburg alignments with the German Confederation; he was also a signee of the Slavic Congress in Prague and a Lutheran.[102] Franz Pillersdorf, the minister of state from May to June 1848, was willing, however, to include Palacký in his government, probably as a symbolic recognition of the political influence of the loyal Slavic spokesman. The German conservatives as well as the Catholic press regarded this as "insane" and a "mockery of sanity and reason"; in their view, Pillersdorf's government had offered the position to "the most impossible of impossibles, the man . . . who is responsible for the lion's share of the current Bohemian tumults."[103] It was, for them, a symbol of the "assassination of our great German fatherland,"[104] which was threatened by such appointments, which were turning Austria into "a Slav state."[105] Palacký, however, rejected the nomination, stating that he could serve the fatherland better on other fronts. Even though the project of including Palacký in the government failed, Habsburg politicians awarded several educational concessions to the Slavs to promote loyalty in the direct aftermath of the upheavals. These included appointments of Slavic scholars and permission to use Slavic languages in teaching.

Among state officials, the idea of university reform went through several stages during the revolution and its aftermath. The initial step was political advancement in the freedom of teaching and learning in late March 1848,[106] followed in June by the announcement of plans to reform the education system, formulated by Feuchtersleben and Franz Exner, a Prague professor of philosophy and pedagogy who had been responsible since April 1848 for the preparation of educational reforms in the Ministry of Education. They envisioned universities as part of the cultural but not the political arena, thus breaking with the pre-1848 withdrawal of academia from public life. Feuchtersleben also supported corporate ideals of the university as a unity of professors and academics. In his eyes, the "caste-like enclosure" of professorships should especially be avoided: "the necessity of a connection with the scientific folk life [*wissenschaftliches Volksleben*] . . . is to be adamantly defended and fought for."[107] In November 1848 Feuchtersleben resigned, leaving countless projects unfinished; only two were partially completed, namely, the renewal of the Viennese medical faculty through the pensioning off of five, in his eyes, overage professors and the reorganization of philosophical study into a faculty.[108]

Shortly after Feuchtersleben's resignation, the government published two laws on 11 December 1848 changing the appointment rules for professors and on 19 December a law concerning those for Privatdozenten. The academic senate remained officially responsible for preparing proposals for new professorships and sending them to the Ministry of Education. Instead of the *Kompetenztabelle*, faculties were now obliged to prepare terna proposals, which were much less formal in style.[109] Once a chair was unoccupied, the university had to ask the provincial government to issue a public tender with deadlines; it was, however, by no means obliged to include in the terna those scholars who applied. Rather, the proposal should discuss scholars appropriate for the post, both domestic and foreign. Only in exceptional cases were *Conkursverfahren* allowed, held not by the faculty but by the ministry. The ministry could also hold its own *Conkurs*, if unsatisfied with the proposal. Also, the three-year probationary period (*Probetriennium*) was retained, leading later to protests by the universities, which regarded it as demeaning academic dignity.[110] Importantly, the ministry also established the minimum remuneration for full professors. Associate professors—scholars permanently appointed for disciplines that were not part of the curriculum, who thus could be specific to a single university—negotiated their salaries on a case-by-case basis until 1918. In this, Vienna remained the best-paying university, with Prague in second place, followed by Cracow and L'viv and, finally, Graz, Innsbruck, and Olomouc, where the regular salary was only two-thirds of the salary in the capital (see table 1). This salary structure had an immense influence on the career paths of professors until the end of the empire in 1918.

The law concerning Privatdozenten superseded the local regulations, which had often been provisional and chaotically enacted. While these had stressed university autonomy and had given academic bodies control over the habilitation procedures, the new law privileged the ministry. In addition to being accepted by the faculty, a candidate for Privatdozent had to go through a public examination, a test lecture, and confirmation by the ministry before being officially permitted to teach.[111] The *Privatdozentur* was limited to the faculty and the university that approved it; any change in either

TABLE 1 Salaries of full professors at Cisleithanian universities (in guldens)

	1849	1870	1898
Vienna	1,600	2,200	3,200
Prague	1,300	2,000	3,200
Cracow	1,200	1,800	3,200
L'viv	1,200	1,800	3,200
Graz	1,000	1,800	3,200
Innsbruck	1,000	1,800	3,200
Olomouc	1,000	n/a	n/a
Chernivtsi		1,800[1]	3,200

Sources: "Erlaß des Ministers des Cultus und Unterrichts, womit die mit Allerhöchster Entschließung vom 26. October 1849 genehmigte provisorische Vorschrift über die künftige Regulirung der Gehalte und des Vorrückungsrechtes der Facultäts-Professoren an den Universitäten zu Wien, Prag, Lemberg, Krakau, Olmütz, Gratz und Innsbruck mitgetheilt wird," *Allgemeines Reichs-Gesetz- und Regierungsblatt für das Kaiserthum Österreich 1849* (Vienna: Hof- und Staatsdruckerei, 1850), 811–13; "Gesetz von 9. April 1870 betreffend die Gehalte der Professoren an den Weltlichen Fakultäten der Universitäten und das Quartiergeld der Facultäts-Professoren in Wien," Reichsgesetzblatt, 12 April 1870, 75–76; "Gesetz von 19. September 1898 betreffend die Regelung der Bezüge der Professoren an Universitäten und denselben gleichgehaltenen Hochschulen und Lehranstalten," Reichsgesetzblatt, 20 September 1898, 295–96.
Note: n/a, not applicable.
[1]Data is from 1875.

of these meant that the process had to be repeated (there were exceptions to this rule, however).[112] Moreover, Privatdozenten had to receive permission to use teaching aids, demonstration materials, and seminar libraries, which made their position dependent on the full professors who controlled these resources. The subject (*Fach*)[113] covered by a Privatdozent depended on a syllabus submitted during the habilitation process, and it could be expanded only with the ministry's approval. Thus, this law favored professionalization and political supervision instead of the previous principles of autonomy. In the direct aftermath of the granting of autonomy in 1848, several universities appointed scholars as *Dozenten* without the ministry's authorization; after the new regulations were enacted, these scholars had to habilitate to achieve the status of Privatdozent.[114] Formal habilitation procedures and ministerial control led to a considerable reduction in the number of instructors, especially in Prague, but the ministry harshly reminded the faculties that they were responsible for controlling the teaching and political behavior of their instructors in accordance with the new rules.[115]

Another important change also occurred in 1848: the appointment of several Slavic scholars, especially for the chairs of Slavic languages. These included, most prominently, Franc Miklošič (Franz von Miklosich) and Jan Kollár for chairs in Vienna (the latter for Slavic archaeology), František Ladislav Čelakovský and Jan Pravoslav Koubek in Prague, and Jakiv Holovac'kyj in L'viv (for the Ukrainian language), most of whom were very likely supported by Šafárik.[116] In the appointment papers for Čelakovský that were handed to the emperor, the ministry openly stated that such appointments were political, without clarifying, however, what political direction was intended.[117] In this way the ministry not only supported the Austro-Slavic movement but also appointed intellectuals who were openly anti-Hungarian (Kollár and the Lutheran theologian Karol Kuzmány) or anti-Polish (Holovac'kyj). It was an important change from the policies of Vormärz, which had kept nationalists out of the universities. The inclusion of a number of Slavic scholars aimed to appease nationalist activists, but at the same time it lessened the universities' uniting role by allowing political dissent to enter the professorship.

The most important manifestation of the 1848 commitment to liberalism was, however, the proposal prepared by Exner during Feuchtersleben's ministerial term. The proposal was overtly liberal and oriented to university models in other German states, but it remained true to the function and position of the university in the tradition of the Vormärz. It was, in fact, built largely on the 1830s discussions about university reforms, in which Exner had had a leading role.[118] According to the draft published in the government's own *Wiener Zeitung* (Viennese newspaper) late in July 1848, the education system was to remain a representation of the *Volk*. Its main function was to prepare functionaries and teachers for future careers. Universities thus represented not scholarship but the political and national needs of the provinces. Moreover, universities, Exner wrote, "are in the first place educational establishments. It is of utmost importance not to impose on them any services, which would endanger their primary purpose."[119] He proposed an educational structure based on the pedagogy of Johann Friedrich Herbart, centered on gymnasia, with universities clearly subordinated to the needs of secondary education. Together with the nominee from Szczecin/Stettin, the Protestant classical philologist and educational reformer Hermann Bonitz, he also remained responsible for gymnasium curricula, which shaped secondary education until the late nineteenth century.[120]

Exner's role in the implementation of these reforms diminished over time, and he died prematurely in 1853. He remained popular among university professors, however, and his projects have been acknowledged as more liberal than those that were ultimately introduced. Franz Krones formulated a metaphor for the change in the political atmosphere between 1848 and 1849, stating that the final reform related to Exner's project as "the imposed constitution [of 1849] [did] to the April Constitution."[121] This reform implementation was already marked less by Exner than by Thun-Hohenstein, the "conservative savior"[122] of Habsburg education, who saved education both *for* and *from* the conservatives. As a moderate politician, he fiercely rejected the neoabsolutist turn toward complete subjection of universities to the government but at the same time pursued a statist and Catholic appointment policy, discussed in the next chapter. As I argue, while conducted with conservative ideologies in mind, Thun-Hohenstein's modifications and appointments in fact paved the way for the developments in the late nineteenth century, including spatial disintegration along linguistic lines.

Chapter 2

The Neoabsolutist Search for a Unified Space

An einen Unterrichtsminister.
Einen Selbstmord hab' ich euch anzusagen.
Der Cultusminister hat den Unterrichtsminister todtgeschlagen.

—Franz Grillparzer, around 1855[1]

[What is't, Mephisto?] Why such hurry?
Why at the cross cast down thine eyes?

—Georg-Emmanuel Haas, criticizing the relaxation of conservative Catholic education under Thun-Hohenstein in 1853[2]

After the turmoil of the revolutionary year of 1848, in mid-1849 the conservative Catholic reactionary Leo Thun-Hohenstein, with his like-minded entourage in the *nomen est omen* Ministry of Religion and Education, initiated the final steps in the major educational reforms and ensured their implementation.[3] In this chapter I offer an interpretation of his policy and the ideas behind it. I argue that universities did not change considerably under Thun-Hohenstein (1849–60); they remained an instrument of state policy and were only secondarily scientific institutions. Nevertheless, the changes implemented between 1848 and 1860 were pivotal for the Habsburg universities, bearing fruit, however, only after the liberalization of higher education in the 1860s. Moreover, as I demonstrate in later chapters, organizational regulations established in 1849, including scholarly appointments in the 1850s, largely defined Habsburg science and scholarship well into the late nineteenth century and even the early twentieth century.

The ongoing reform of the educational system was a pivotal step in the gradual stabilization and control of the various university regulations enacted in 1848, which often applied to a single province. Corresponding to Franz Exner's Herbartian *Bildungs* (education) ideology, the reforms envisioned a system of educational continuity, encompassing establishments from *Volksschulen* (primary schools) to universities; the latter would serve primarily to educate teachers and prepare textbooks. Education, in the sense of the development of individual talents, especially through humanistic disciplines, was supposed to guarantee both loyalty to the throne and scholarly quality.[4] Ministerial policy throughout this period walked a tightrope between Thun-Hohenstein's desire to establish Catholic-based scholarship and the lack of appropriate scholars, which forced him to acknowledge the need to appoint academics from the non-Habsburg parts of the German Confederation.

On the spatial level, three major changes characterize this period. First, the Habsburg universities drew closer to the universities of the German Confederation on both the symbolic and personal levels. Second, the unification of university space through the reintroduction of German as the language of instruction in 1853 was a largely mythologized and politicized process. The assessment of this change varied widely and also depended on one's national orientation.[5] Finally, Thun-Hohenstein clearly followed a path of modernization, which included opening the universities to scholars from different national backgrounds. This opened a path to the developments in the 1860s and 1870s, when universities began to drift apart, forming subsystems defined by the language they used in teaching.

Toward the *Ordinarienuniversität*

Thun-Hohenstein took up office in July 1849, in the midst of the final period of the educational reorganization; only a few days after his inauguration, the law concerning the organization of the universities was enacted. The new law reorganized the academic body into an autonomous faculty controlled by the full professors (*Ordinarienuniversität*), which weakened the corporate character of the university. It also permitted freedom of teaching and learning, at least to an extent, and standardized the curricula.

The central issue remained the question of autonomy, which liberal scholars and universities saw as a prerequisite to modernization.[6] The 1848 laws on Privatdozenten and professorial appointments had strengthened the universities but retained the paramount role of the Ministry of Religion and Education. Conservative politicians preferred, however, tighter control than the 1849 law provided. The Kremsier Constitution, prepared in 1849 but never implemented, intended to place the universities under the strict control of regional governors.[7] Andreas Baumgartner, an influential conservative politician and respected physicist, proposed that the church should have direct control over the universities.[8] Yet another project was discussed in 1853, when the minister of the interior, Alexander Bach, and the minister of finance, Phillip Krauß, pleaded for the reinstallment of *Studiendirektoren*.[9]

Faced with the tensions between liberal university proposals and conservative desires to tighten the political supervision of universities, Thun-Hohenstein chose a middle way, awarding autonomy to the universities, with the provincial governments and church authorities retaining the right to comment on nominations and with the ministry having the final say.[10] He strengthened the faculties by giving them the exclusive right to propose deans and rectors, and he emphasized that he wanted distinguished active professors to be chosen as rectors.[11] Thun-Hohenstein, however, opposed the corporate idea of a university as a community of professors, colleges of doctors (Doktoren-Collegien), and students, which the faculties preferred; this was the main discrepancy between the reforms and the faculties' wishes. Students' status as a corporation was swiftly removed, and they were subordinated primarily to the civil code, with only a few matters remaining under academic jurisdiction.[12] University teachers were threatened with sanctions if they did not inform the police of abnormal student absences or of their meetings and associations. From 1849 on, such meetings and associations were usually illegal.[13]

Similarly, the ministry limited the influence of the Doktoren-Collegien, fiercely criticized as radical organizations trying to "dominate [the universities] anew."[14] After Exner, in his outline of the new legislation, proposed their complete abolition, first Ernst von Feuchtersleben and then Thun-Hohenstein pleaded for some of their functions to be retained. Ultimately, Doktoren-Collegien remained only at the universities in Prague and Vienna, playing a central role in graduating students and proposing rectors but losing the privilege of accrediting graduates for practice.[15] In these two cities, the deans of the Doktoren-Collegien remained members of the academic senate, although full professors outnumbered them two to one.[16]

This strengthened autonomy made Habsburg universities into *Ordinarienuniversitäten*, in other words, universities controlled by full professors. The new organizational reforms gave full professors the majority in all administrative bodies in the universities, in addition to control over teaching aids and seminars; they were also required to supervise students and Privatdozenten. The ministry retained, however, several means of repression, such as the *Probetriennium* and the right to relocate professors (*Versetzung*), force a retirement, or terminate a contract, measures that had been used for political reasons at various times. Decision-making about Privatdozenten was even more centralized. The ministry could reject a habilitation without cause; propose changes, for example, in the scholarly discipline for which the habilitation was approved; or award remuneration based on a petition by the university. The ministry was also in no way obliged to grant faculties' requests, or even to react to them; this privilege was used later to prevent undesired habilitations, professorial promotions, and chair appointments. Thun-Hohenstein also requested the protocols from the proceedings of the university senates and faculties, at first under the pretense of supervising the reform's progress, as the reform was to be revised after three years; however, ministerial review of the protocols continued until 1918, as indicated by notes in the archives.[17]

How much autonomy the universities would be permitted in practice thus remained the sole responsibility of the ministry, which could either decide to interfere in university matters or just confirm the academic senates' decisions. The high officials in the ministry, in charge of making recommendations to the emperor, were not necessarily professional politicians, however, and were often scholars themselves; in many cases the ministry consulted other academics about the quality and moral behavior of the persons in question.

Like autonomy, the meaning of freedom of teaching and learning (*Lehrfreiheit* and *Lernfreiheit*), a cornerstone of the reforms, also remained a matter of contention among political pressure groups.[18] "Freedom" certainly did not mean unconditional self-government in what would be taught and learned; it was constructed and presented more as the antithesis of the politicization of universities before 1848. *Lernfreiheit* implied partial freedom in the choice of lectures in the curriculum, the free choice of lecturers, and a relaxation of the exam system, with general instead of semester and annual exams. "The freedom [to choose] the lectures, the time, and the instructor whom they want to hear"[19] was, although eloquently formulated, hardly realizable in practice. In subsequent discussions, the universities, especially the medical faculties, criticized it as impracticable because medical subjects had to be learned in the proper order; this was swiftly regulated by the ministry through the prescribed curricula. Compared to *Lehrfreiheit*, *Lernfreiheit* was certainly in second place, as in the case of the replacement of *Unterrichtsgeld* (tuition fees paid per semester) by *Collegiengelder* (tuition fees paid per lecture). This change was a means of not only supporting the Privatdozenten but also assuring that students did not take unnecessary lectures: "The fees will be, as the freedom of learning continues, a barrier to youthful improvidence, which one cannot do without anywhere where it [the freedom of learning] exists."[20] In addition, professors and deans were obliged to take attendance at lectures, a requirement that the ministry repeated on several occasions, signaling its importance for the successful disciplining of students.

Lehrfreiheit was also limited by concessions to state authorities: the ministry oversaw the lecture catalogs and could cancel lectures, remove teachers, or transfer them to universities in which their ideological or political opinions would find little or no resonance. As I argue below, Thun-Hohenstein frequently used these measures to discipline professors. Further, the ministry, based on faculty proposals, regulated who should teach which lectures at specific universities. For instance, professors who in 1849 were allowed to teach "every topic of their scientific field" could from 1851 onward teach only "related subjects"[21] in their faculty; any change was subject to ministry approval. Similarly, Privatdozenten remained under ministerial control. Furthermore, the ministry later controlled the lectures, rejecting those whose syllabus or designation was politically suspect.[22] That said, the extent to which the authorities actually did (or could) supervise the content of seminars and lectures in practice remains open to conjecture.

A Catholic Counterbalance to Prussia

Ministerial decrees and speeches depicted the universities in other parts of the German Confederation, especially those enjoying academic freedom, as the ideal of scientific and social development. This idealization remained, however, more in the sphere of rhetoric and was by no means unconditional. Rather, bits and pieces of regulations from various states of the German Confederation were adjusted to fit Austrian regional peculiarities, in particular, religion, which was certainly the largest issue in the process of reform.

The idea of a local model based on "German" universities had begun already before Thun-Hohenstein. The minister of education between March and May 1848, Franz Freiherr von Sommaruga, announcing the abandonment of censorship and the introduction of freedom of learning and teaching, saw "German universities" as models but clearly stated that their structure should be adopted "only as much as the conditions in the fatherland allow."[23] At the same time, in Exner's view, the success of "non-Austrian German universities" supported the introduction of their system, which was even seen as necessary because "future cross-boundary communication between them and the Austrian universities requires it."[24] This pointed in the direction of exchange but also redefined the desired boundaries of the scientific space. Thun-Hohenstein's confidant Carl Ernst Jarcke, an influential Prussian-born jurist, in a memorandum in 1849 also pleaded for free exchange, arguing that Prussia "owes its influence in Germany, which reaches far beyond its material power, mostly to the fact that it was able to obtain, if it wished, any higher talent from every corner of Germany."[25] However, academic reciprocity was not without its limits: "I would recommend that inviting Protestant teachers to Austrian universities should at least not be the rule," wrote Jarcke in the same text.[26] Making Austria a "Catholic counterbalance to Prussia" was hailed in 1853 as one of the major tasks of the university system.[27]

The development of the philosophical faculties hints at the role of scholarship as a means of both external propaganda and internal popularization of the state ideology. Their foremost duty was the education of teachers and the production of textbooks. If one considers the number of foreign scholars appointed, new seminars created, and books bought for the libraries, philosophical faculties were ridiculously expensive, especially since student numbers were low. Directly after the completion of the reforms in 1853, the philosophical faculties in Cracow, Graz, and Innsbruck each had fewer than 20 students, L'viv had 75, and Vienna and Prague each had slightly fewer than 100. The medical and law faculties, in contrast, witnessed growing demand.[28] In 1855 the philosophical faculty in Vienna had 24 professors and 275 students, while its medical faculty had 19 professors and 579 students through most of the 1850s; even the theological faculty was more popular than the philosophical faculty.[29]

Defending the reforms, Thun-Hohenstein often expressed his conception of science as a panacea for the national and social problems of the Habsburg "composite state."[30] One could say that science and scholarship, and thus universities, became one of the favored channels of propaganda and a source of arguments to legitimize certain claims, be they loyalist, patriotic, nationalistic, or whatever. In Thun-Hohenstein's eyes, the monarchy could thrive only with the acceptance of a particular shared narrative, which would counter nationalistic claims. This narrative included not only loyalty, cultural reciprocity, and Catholicism as cornerstones but also the claim that the empire was the only guarantor of cultural progress: an idea in which universities had a pivotal role and which later (for example, under the minister of education Karl Stremayr, 1870–80) seamlessly mixed with German cultural imperialism. Thun-Hohenstein and his supporters powerfully mobilized a picture of free, unbound scholarship leading the state to a cultural paradise. This image also served to demonstrate the improvements that political changes had brought about compared to the situation in the Vormärz.

In particular, historical disciplines such as the history of law, national histories, the history of languages, and archaeology were to be mobilized and supported, which brought about considerable changes: not only new chairs but also the introduction of seminars. (Seminars were research-oriented courses based on intensive cooperation between a professor and his students, the predecessors of modern seminars. As they were given room within the university buildings, and increasingly included more professors, they also became the precursors of today's institutes.) Through concentration on minute source work, Thun-Hohenstein intended to promote "unbiased science" (*voraussetzungslose Wissenschaft*). This went hand in hand with the renunciation of nationalist historical narratives, on the one hand, and of the philosophy of history, legal philosophy, and natural law, on the other.[31] The ministry denounced all kinds of philosophy, from Georg Wilhelm Friedrich Hegel to Immanuel Kant and Johann Friedrich Herbart,[32] and in their place proposed a yet nonexistent "philosophy which enjoys public acceptance by both science and the church."[33] "In the meantime," wrote the ministry in 1853, "it remains the duty of the ministry to direct policy toward this aim as far as possible, and to prevent every manifest and veiled impulse against the [divine] revelation."[34]

Catholicism and its relationship with the freedom of teaching and learning was one of the most delicate issues in the reform movement. While this was not an issue for Thun-Hohenstein, whose philosophy of ideal scholarship involved the Catholicization of the most important matters at the university, especially in the humanities and law, it was a central question for the general character of universities. Although the equality of religious denominations was part of the constitution and not directly addressed in the academic laws, the subsequent decrease in the equality of Jews and the Concordat of 1855 made non-Christians unwelcome. Even the universities themselves were not sure how to deal with this issue. For example, they were uncertain whether to consider Privatdozenten as state functionaries, who had to be Christian.[35] Although Thun-Hohenstein allowed chairs of Hebrew in L'viv, Prague, and Vienna, none of those appointed became full professors.[36]

Non-Catholics were also legally prohibited from becoming deans and rectors in Vienna,[37] although the interpretation of this rule was far from straightforward. In 1852 the university consistory[38] challenged the nomination of the Lutheran Hermann Bonitz as dean of the philosophical faculty in Vienna, forcing Thun-Hohenstein to reject his application and underscore the Catholic character of the university.[39] The philosophical faculty's choice was also fiercely discussed in public, with mostly negative opinions underscoring the historically Catholic character of the university. Notably, Sebastian Brunner, the dean of the theological Doktoren-Collegium, launched a fierce campaign against the nomination; shortly afterward, Brunner was appointed the university's main priest, demonstrating once more the entanglement of church and state, which made the issue of religion complex.[40] Non-Catholic university officials were first elected, and their elections confirmed by the ministry, only after Thun-Hohenstein's resignation on 20 October 1860.

But assessments of Thun-Hohenstein's denominational policy varied, showing the difficulty of his position. Franz Grillparzer, one of the leaders of the liberal movement before 1848,[41] criticized the minister for becoming increasingly subservient to the Catholic Church. For others, like Georg Emmanuel Haas, quoted at the head of this chapter, he was not a Catholic savior but rather a Mephisto who nominated Protestant foreigners instead of Catholic Habsburg citizens.

Language(s) for the Empire

Like his teacher Bernhard Bolzano, Thun-Hohenstein, influenced by the ideas of the Enlightenment, was skeptical about the political and cultural hegemony of the German language in the empire.[42] Shortly after his nomination as the minister of religion and education, in a Czech-language pamphlet published 1849, he underlined the necessity of the "real equal status" of Slavs and their languages.[43] Since he saw the interests of the state as paramount, superseding nationalistic interests, he rejected the federalization proposed by adherents of Austro-Slavism and depicted an idealized multicultural empire.[44] In particular, he criticized the nationalism of the Poles, who in his eyes were striving to regain the independence of the Polish-Lithuanian Commonwealth.

Thun-Hohenstein's insights on education, however, were based on a hierarchical cultural dualism that appeared in a stronger form in the German nationalist discourse in Bohemia; there, "German" was equated with "culture and civilization." These ideas conflicted sharply with the demands of the Slavic nationalist movements, which challenged the universality of such opinions. Correspondingly, Czech nationalist liberals regarded the pamphlet as a direct assault on their policy.[45] The following quotation shows the main traits of both Thun-Hohenstein's *Staatsnationalismus* (state nationalism) and his policy as minister of education:

> The conditions obtaining until now have had the effect—and the Slavs are not at fault for this—that the number of Slavic men who unite solid scholarliness with the ability to disseminate it in their mother tongue is still low, whereas nobody—especially in Bohemia—reaches scientific maturity without completely understanding at least German. It is thus of great importance for the intellectual upswing of the Slavs in Bohemia that all men who are able to teach competently in the Czech language in any subjects be given the chance to do so. It is, by the way, no less in their interest to seek scientific education in German lectures. If people are satisfied with this, the number of Czech chairs will still be quite low initially, but it will be higher every year, cultivating and expanding the national forces. If, on the other hand, a completely misguided conception of equality is imposed, and a Czech chair is created beside each German one, or if complete gymnasia and university faculties in the Czech language are founded with consideration only of the sizes of the populations, the national cause may be illuminated with what seems to the ignorant eye to be a dazzling glamour, but with each passing year it will fade away. And even more important, true *Bildung* will be strongly impeded, even repressed. . . . Moreover, such a foolish and jealous conception of the principle of equality, which snatches only at equality of appearance, would have the consequence that, whenever means were lacking, German *Bildung*-institutions would be destroyed as Czech ones were constructed alongside them. . . . We must oppose such pernicious aberrations and perversions, which are useful to nobody. . . . Wherever real rights are in question, equal laws should be applied, and

> the state should watch over and ensure that such rights are never injured or limited for the sake of national sympathy or antipathy.[46]

Despite the affirmative tone on the nationality question, these words constituted a denial of the national cultural autonomy in educational matters that nationalists had demanded during the revolution. The centralist apologist Thun-Hohenstein not only regarded state regulations as more beneficial than subordination to nationalist sentiments but also claimed, in a slightly paternalistic tone, that German cultural superiority should continue, rebuking and contesting both the ideas of Austro-Slavism and the nationalists' formulation of this issue. At the time of the publication of the pamphlet and his nomination to high office in the summer of 1849, Thun-Hohenstein's ideal policy of national equalization was far from being generally successful, and he acknowledged his painful experiences during the 1848 June Uprising in Prague, when he became the object of attacks by liberals of both nationalities, including his previous allies.[47]

The language of education was one of the most important topics at the Slavic Congress in 1848; the representatives demanded language equality not only in secondary education but also in tertiary education in Cisleithania and Transleithania. The final petition to the emperor, written by František Palacký, called for a number of universities to be made bilingual and also proposed the addition of new universities for some minority groups, such as the Slovaks and Serbians.[48] A special appendix concerning Galicia, whose Polish and Ruthenian representatives hardly agreed, recommended the freedom of teaching in both languages in Galicia. In another petition, aimed at the general public, science and scholarship as a whole were elevated as panaceas for cultural development, conjoined with the concept of Slavic reciprocity: "The convergence and fraternization of Slav peoples could bring only benefits to humanity and glory to us, when it occurs in a peaceful way and with defense of freedom. Therefore, to begin with, the revival of literary reciprocity and cultivation of collaboration in science and the arts are in our interest. We only follow this path, when we ask for the teaching of all Slav dialects at each Slav academic institution. The annual scientific congresses should inspire us Slavs, like the other peoples, to a higher intellectual life and should facilitate the exchange of ideas."[49]

These words of František Alexandr Zach show clearly that the value that Thun-Hohenstein attributed to science in his pamphlet[50] was not far from Czech views. In fact, Thun-Hohenstein evaluated the congress as essentially positive: "In my opinion such congresses are not threatening to the state, as long as they remain limited to influencing public opinion and [influencing] through public opinion and [through this] bring the Slav peoples in the monarchy to a clear awareness of what their national interests demand."[51]

Apart from a complete change in the language of instruction, from German to provincial tongues, more moderate demands were put forth by Czech and Polish nationalists once their more excessive demands had failed. However, such a tempering of demands was also a product of the political atmosphere; the revolutionary zeal had faded somewhat. Whereas in Bohemia support for Czech-German bilingualism was widespread,[52] nationalists in Galicia sought monolingual universities. The removal of the German language was not the only objective: the Polish nationalists also fiercely rejected the introduction of Ruthenian as a medium of instruction at the gymnasia and universities, repeatedly claiming that that nation and language were only the ideas of Franz Stadion, the governor who had enacted privileges for Ruthenian to weaken Polish in Galicia.[53] Polish nationalists attempted to legitimize their rejection of Ruthenian culture and language by claiming that Ruthenians were not a separate cultural entity. Critics of Ruthenian also emphasized the low cultural development of the language and its similarity to Russian, arguing that political support for Ruthenians' national claims would lead to alignment with the Russian Empire.[54]

However, the proposed language changes were not simply part of the political process; they also caused the Viennese government to invest in the publication of specific vocabularies for gymnasia,[55] textbooks, and even scholarly publications, such as those in Ruthenian by Vasyl' Voljan (Василь Волян).[56] After the constitutional guarantees for the ten provincial languages (*Landessprachen*) had been granted, these languages' inability to seamlessly cover the issues of administration, which hindered the implementation of bilingualism in institutions, became obvious, leading to the creation of a commission whose aim was to prepare the Legal and Political Terminology for the Slavic Languages of Austria (Juridisch-politische Terminologie für die slavischen Sprachen Oesterreichs).[57] The requirement of a "developed language" for educational purposes was thus not abstract; both regional and imperial politicians as well as many intellectuals were, with varying intensity, working on this idea.

The claim that a language of instruction had to be developed to enable a university to fulfill its functions took various forms at different times. In the early nineteenth century, the communication value of language was seen as paramount, rather than its symbolic value.[58] After 1848 this issue polarized scholars, but nationalist activists, fueled by the 1848 congress, regarded questioning the level of a language's linguistic development as an antinational act. In the first three years after the 1848 revolution, the idea of language's significance for the cohesion of the educational system was pushed to the background. Of the universities touched by the language question, Cracow employed the most scholars lecturing in a local language, in this case Polish. Prague had several lecturers capable of teaching in Czech, while L'viv and Olomouc had almost no instructors teaching in local languages (Polish or Ruthenian in L'viv and Czech in Olomouc).[59]

In most appointments, Thun-Hohenstein looked for scholars with knowledge of provincial languages. This was, however, not always possible because of the changed curricula, which required the minister to search for candidates in non-Habsburg parts of the German Confederation. For Cracow and L'viv, most of the qualified scholars who knew Polish or Ruthenian were living abroad and/or could not be hired for political reasons.

In 1852–53 the political atmosphere in the monarchy concerning multilingualism changed. German was reinstated as the language of secondary schooling and bureaucracy. This measure also influenced the universities. Their links with gymnasia and the civil service meant that non-German-language universities would produce highly educated officials who were not conversant in German, the language they now had to use in their professional careers. Given that historians regard the language change of 1853 as a symbolic act, the uncertainty about the reasons behind it may be surprising. The widespread story of a forced Germanization is full of flaws. For Prague, there is no single document confirming that the government or the ministry forbade Czech as a medium of instruction. More plausible is the thesis that individual scholars' decision to cease teaching in Czech was purely pragmatic: there were simply not enough students who spoke Czech and no established terminology, especially in the sciences and medicine.[60] In a petition in 1864 arguing for Czech lectures at the university, Czech students pointed out that these had been abolished in 1852 owing to an "unfavorable time" (*Ungunst der Zeit*).[61] For Cracow, the acts concerning the language change are missing from the university archive, and the documents related to the process allow divergent interpretations. The following discussion of the proceedings at this university will illustrate not only this "unfavorable time" in the early 1850s but also the changes in the understanding of science and scholarship during this period.

Uniting through German: Cracow

In the early 1850s, the Jagiellonian University found itself at the center of attention because it openly supported Polishness among its faculty. The situation was aggravated in 1851 when the professors greeted Emperor Franz Joseph during his visit there in their traditional togas instead of the official Habsburg uniforms worn by all civil servants. Wearing of the official uniform had not been legally required but was made law shortly after the emperor's visit.[62] Following local government reports on the revolutionary sympathies of some professors, the provincial government of Galicia ordered that Polish professors at the university be supervised, suspecting them of propagating political separatism.[63] These suspicions led to the disciplinary discharge of Antoni Helcel, Józefat Zielonacki, Wincenty Pol, and Antoni Małecki in January 1853; in addition, Franz Joseph revoked the university's autonomy and also ordered the appointment of a curator.[64] In the *Ministerkonferenz* (Ministerial Conference), Thun-Hohenstein, confronted with the suspension of autonomy, which had taken place without his knowledge, unsuccessfully defended the equality of languages, which in his eyes encouraged Polish loyalty. He succeeded, however, despite opposition from centralists such as Alexander Bach and the minister of justice Karl Krauß, in securing his preferred candidate for the office of curator, Piotr Bartynowski, the president of the k.k. Oberlandesgericht (Higher Provincial Court) in Cracow and a professor of Roman law, whom conservatives in the government regarded with skepticism as a "national Pole" (*Nationalpole*).[65]

At the same time, the situation also changed in Cracow. The newly appointed professor of German literature, František Tomáš Bratranek, himself a bilingual Moravian, penned in early 1853 a *pium desiderium* (pious wish) for the introduction of German as a language of instruction. Bratranek wrote that the university, the smallest in the empire, could not, for political reasons, host the best Polish-speaking professors and that all students already spoke fluent German after attending the gymnasium. He therefore considered it to be "in the students' interest" that "already from the next semester all matters which are in any way connected to their competence for the civil service should be instructed at our university in the German language."[66] Bartynowski, together with the deans who had likewise been installed without taking the faculties' wishes into account, seems to have supported Bratranek's petition, as did some of the faculty.[67]

The Moravian scholar also drafted the final text of the petition, which gives insight into how Habsburg scholars perceived scholarship and its social role at the time. Apart from the arguments raised initially by Bratranek, the petition emphasized the locality of education and the universality of science: "The university is primarily to be considered a nursery and a base for the development of science; science is, though, of a universal nature; thus, its development will be held back by such establishments which are turned toward special and, besides that, very local [circumstances]."[68] As most scientific texts were written in German, French, or English, reliance on translations for teaching slowed the free flow of knowledge. Not only did translations lag behind the originals, but not everything could be translated. Moreover, Polish did not possess a developed scientific terminology at the time, according to the petition, and even leading Polish scientists published in German owing to the lack of a Polish-reading public.[69] The petitioners thus claimed that for the sake of science, it should be instructed in a world language, in this case German.

The universality of science, as put forward in the petition, was not a mere argumentative device to legitimize the language change. The argument here was that the scientific process necessitated the communication of results in the international arena, independent of language: "The scientific literature differs most sharply in its universality from the belles-lettres. While one has to appreciate that it perfectly demonstrates the nationality, and also the individuality of its bearer, the desirable thriving of science requires a strongly objective attitude, which rejects all national and individual sympathies."[70]

This put the educational function of the university behind the imagined universality of a *République des Lettres* and of the dominant "world languages." At the same time, science here was deprived of its locality; it became a cosmopolitan, transnational occupation, reserved only for elites. Local publication and circulation were not only secondary but also unimportant for the production of scientific knowledge per se because they did not take place in the "learned languages." Polish was nevertheless prominently mentioned in the petition as a language of science and scholarship, suggested as having a "lively future that was not to be doubted."[71] However, the petition continued, "it is of importance for students that their swift advancement in their scholarly development is not impeded through philological work on the perfection of [Polish scholarly] terminology." Further, while the university should address universal issues, the question of Polish science should be dealt with within the "peculiar" (*eigenthümlich*) institution of Cracow's Scientific Society, "whose members are for the most part professors of the Royal and Imperial University and which made the further enhancement of national interests as its primary goal. If Polish scientific literature has a germ of a viable future, it will be most suitable to commit it into the care of [Cracow's Scientific] Society, whose enthusiasm seemed so far most laudable, and will certainly suffice to foster the beginnings of terminological accounts to prosperous development, which by no means should be the duty of the university."

The petition heralded the official introduction in December 1853 of the use of German in Cracow and L'viv, "for the duration of martial law [in Galicia]," which, however, ended already in 1854.[72] The removal of Polish lectures was not complete, as the ministry allowed two professors of the medical faculty, Józef Majer and Antoni Kozubowski, to teach their classes in Polish; this privilege was awarded at first for one year and then renewed on an annual basis until 1861, when regular lectures in Polish resumed.[73] However, at the same time, German-speaking professors held parallel lectures, and the Polish ones became optional. Thun-Hohenstein's memorandum also proposed that "to give attention to the development of the Polish language, a distinct chair of Polish language and literature be appointed and that it be left to the discretion of Privatdozenten to read allowed disciplines in the Polish language, and, inasmuch as a vital necessity exists, to cover this or that subject in the Polish language."[74] Both Thun-Hohenstein and the academic senate of the Jagiellonian University clearly strove to fill the position of the chair of Polish language and literature; the latter also urged the University of L'viv to appoint a corresponding chair.[75]

Further contradicting the story of a forceful and unwelcome Germanization, the conservative Cracow journal *Czas* (Time), in several articles, accepted the language change as serving practical purposes well.[76] Furthermore, Bratranek, whose petition had begun the process of introducing German, also remained at the university after the language of instruction changed back to Polish. This probably resulted from a university petition showing the professors' support for Bratranek but also from his popularity in Galicia.[77] Because he published widely on Polish-German relations in the newspaper *Dziennik Polski* (Polish daily) in 1869, he was viewed as someone who almost became a Pole.[78]

Walking a tightrope, the ministry was seeking a means to accommodate national claims while guaranteeing that universities would serve as imperial institutions. This balancing act was not unlike those related to questions of religion and relations with non-Habsburg parts of the German Confederation, which I discuss later in this chapter. More important, the ministry advanced a language model differing from, but not contradictory to, that promoted by nationalist activists. It pleaded for multilingualism in social life, one part of which, education, should take place in German. Since Latin had been the lingua franca only a few years previously, this was not a novelty for scholars. Even the nationally oriented academics and politicians used the same argument to reject lectures in Ruthenian a few years later.[79]

The language issue was perhaps the most important element in the implementation of the reforms. It was closely connected, however, to changes in the disciplinary nexus and in appointment patterns, which I turn to now. This also brings us back from the provinces to Vienna and to its practical relation with the idea(s) of German science.

Tradition, Locality, and the Natural and Medical Sciences

After the reforms, the empire, unsurprisingly, lacked scholars capable of carrying out the new academic and political projects. Additionally, Thun-Hohenstein's ministry publicly presented the restructuring of the academic landscape as a thrust toward a new knowledge, and thus a break with not only the legal cornerstones of Vormärz universities but also the scholars who had shaped the prerevolutionary academia. This meant, initially, the inclusion of previously marginalized scholars in faculties.

The first appointments at Habsburg universities after 1848 were indeed directed toward the transformation of the professorate, especially in the humanities and historical legal subjects.[80] Since the development of historical or philological scholarship had been limited in the Habsburg Empire before 1848, it was hard to find appropriate specialists within the country, especially ones acquainted with new methodological approaches. In addition, the reforms introduced the new concept of seminars, which were created for modern languages as well as for history, the latter in the form of a philological-historical seminar that focused on the classical languages required for teachers at the gymnasia.[81] Simultaneously, the natural and medical sciences followed a rather different path of change, marked by gradual rejuvenation and inter-Habsburg migration, although state interests still predominated.

While reformist tendencies in the humanities and philosophy were developing at the universities, the natural sciences and medicine were far from being inundated by non-Habsburg scholars at this time, with some prominent exceptions, such as Ernst Brücke. Inter-Habsburg migration and appointments from other scientific institutions, like technical academies (*polytechnische Institute*), were common. This confirms that in Thun-Hohenstein's education policy, the humanities and legal subjects played an enormously important role; in these areas, the ministry was prepared to appoint scholars from abroad despite protests from faculties and conservative critics. However, this trend also demonstrates that the natural sciences of the Vormärz, even though absent from the universities before 1848, were much more highly developed and that new scholarly ideas did not mean a rupture in their development, as happened in the humanities.

There were, however, three additional reasons for the ministry's support for the appointment of Habsburg scholars in the sciences. First, geography, mineralogy, zoology, and biology were local sciences at this time. They linked a theoretical background with a descriptive analysis of the local environment. Thus, even in the latter part of the nineteenth century, both the faculties and the ministry regarded specialization or interest in the particularities of the natural world in the local province as an asset. Second, appointments of scholars from abroad frequently meant higher salaries, and except in the period directly after the revolution, the Ministry of Finance demanded that Thun-Hohenstein cut expenses. Newly appointed professors would also have to accept research equipment that was either insufficient or outsourced to independent institutions. In particular, celebrated scholars, pleading for extensive research opportunities and needing to relocate equipment and assistants, were less likely to be appointed because of the cost to the universities. Less prominent, younger, and local scholars were simply cheaper in many cases.

Third, the atmosphere, fueled by Catholic conservatives, was unfavorable to both foreigners and the natural sciences themselves. The university found itself embroiled in the conflict between the Catholic Church and the sciences, the *Materialismusstreit* (conflict over materialism), revolving around the question of whether, and to what degree, the new developments in the sciences, especially the biological sciences, conflicted with Catholic doctrines.[82] Shortly after the controversy over the nomination of Bonitz as the dean of the philosophical faculty, Brunner published a series of articles depicting the university as infiltrated by followers of materialist doctrines and people "in part religiously indifferent, in part Josephine-superstitious, in part humanistic-anti-Christian liberal."[83] The official position of the ministry was not far from Brunner's antimaterialist views,[84] and Thun-Hohenstein, even if hesitant, removed scholars who favored materialism.[85]

With such critics in the government, the universities' consistories, and the public sphere, further appointments from abroad for chairs in which local research traditions existed were certainly a risk for the ministry, especially given the difficulty of presenting such appointments as aiming to prevent further revolution, as Thun-Hohenstein argued in his nominations of professors in the humanities.

That the natural sciences did not command as much political interest in the post-1848 era as the humanities does not mean that they stagnated. The innovation taking place in the humanities, prompted by imported scholars, certainly did not occur here; however, supporting education in the gymnasia, where the natural sciences were better represented after 1848, also required the speedy filling of chairs. Thun-Hohenstein made it clear that the gymnasia stood at the forefront of these changes; in April 1851 he asked those teaching the natural sciences to pay special attention to the preparation of teachers when choosing the topics covered in their lectures.[86] Finally, professorial duties at the university were often linked to responsibilities in other institutions, especially the directorships of botanical gardens, observatories, and Viennese institutions such as the Central Bureau for Meteorology and Terrestrial Magnetism (Zentralanstalt für Meteorologie und Erdmagnetismus) and the Geological Survey (Geologische Reichsanstalt). These positions had to be filled swiftly, which created disparities in the treatment of universities that had such institutions (Vienna, Prague, Cracow) and those that did not (Graz, Innsbruck).

Much thus needed to be done if the universities were to equal those in Prussia. With the reorganization of the philosophical faculties, the natural sciences were in many cases institutionalized academically and professionalized in form and content for the first time. For instance, the new chairs of botany, mineralogy, and zoology replaced the chair of natural history; a new chair in geography was created, although initially only at the universities in Vienna and Cracow. Chemistry and botany had been taught as one subject at the medical faculty before 1848, without seriously taking into consideration the scholarly interests of the teachers, who were required to cover a broad range of increasingly disparate matters in their lectures. In this case, it is rather unsurprising that, contrary to the ministerial goal of specialization, the scholars nominated for the new chairs had a much broader education and a variety of degrees (although this was not always required). One of the newly appointed professors from Prussia summed up the chaos: "Doctor of medical science, magister of obstetrics, Moravian corporate full public professor of general natural history and agricultural economics, plus deputizing professor of Bohemian language and literature. In this written title you have the typical representative Austrian scholarly figure."[87]

In several other disciplines, such as meteorology and astronomy, scientific traditions existed, particularly at the technical schools. Transferring teachers in these disciplines to the universities, as well as modifying the research infrastructure, was a step toward turning universities into research institutions. Here, however, another problem arose: the technical academies and universities covered a similar range of subjects, raising the question of how to reform both without creating conflict. In several cases, the ministry accentuated the importance of the natural sciences as the transmitter between theory and practice at the university, spanning the symbolic boundary between technical education and the humanities-led universities. This boundary was especially visible in the division of the practical secondary education provided in the *Realschule* from the humanistic education of the gymnasia. In this way, the natural sciences were included in the idea of the cultural development of the monarchy, in which the universities were supposed to have a pivotal role in all areas of scholarship. To achieve "the aim of contemporary development of industrial activity," professors should not only be theoreticians but also be familiar with "practical requirements."[88]

Although the technical academies, in contrast to universities, concentrated on a practical approach, the strengthening of the philosophical faculties at the universities triggered questions about merging the technical academies and universities or else differentiating them more clearly.[89] Doubling the institutes of science would require costly infrastructure, the critics pointed out. However, the technical academies had a political value beyond the simple education of engineers: for example, in Brno the technical academy was the only tertiary school in Moravia after the dissolution of the University of Olomouc.[90] After teachers of foreign languages were added, the technical academies not only aimed to produce engineers who would work locally but also imagined exporting them abroad, like physicians, whose influence had even reached the Ottoman Empire.[91] This was of utmost political significance to counterbalance the international influence of Prussian and French engineers. Although Thun-Hohenstein's ministry did not complete the reform of the technical academies, they were awarded a professional status similar to the universities'. The *Realschulen*, which had been incorporated into the technical academies until 1852, became a type of secondary school The ministry began allowing Privatdozenten to teach at the technical academies and enlarged the number of instructors. While professorships at a university were more prestigious than those at a technical academy, their salaries were equal; thus, scholars in the academies were not necessarily interested in moving to a university post.[92]

Nonetheless, the technical academies experienced a sort of brain drain in the 1850s, because the ministry frequently nominated their experienced scholars for posts at universities. Other institutions also provided the philosophical faculties with professors for the natural sciences, however. In Vienna the imperial cabinets (*Hofkabinetten*) were the main source of professors for the natural sciences.[93] The Joanneum in Graz and the Bohemian Museum in Prague were other prominent organizations from which scholars came.[94] Since the pre-1848 medical faculties also included professors of chemistry and biological sciences, several scholars were moved from these faculties to the philosophical faculties, with a changed chair designation. Only a handful of scholars from abroad were nominated, and if local scholars were available, the ministry turned to them even if the faculties wished otherwise.[95]

For this reason, Habsburg scholars were employed, and the ministry clearly favored the students of only a few prominent natural scientists. In chemistry, for instance, Thun-Hohenstein appointed the students of the Viennese professor Joseph Franz von Jacquin throughout the empire, although most of them had also worked with the pioneer of organic chemistry, Justus Liebig, in Gießen.[96] Stephan Endlicher in biology, Franz Zippe in mineralogy, and Karl Kreil in physics had a similar influence. Since these four scholars taught in Prague or Vienna, their influence reproduced the centralization of Habsburg education, which, contrary to traditional narratives, was not confined to the Habsburg capital.

Remaining within one's own tradition had, however, some negative consequences. First, older professors mostly concentrated on teaching and writing textbooks rather than conducting research.[97] Frequently, they also remained within the scholarly traditions of the Vormärz, such as in their insistence on descriptive approaches. For example, Zippe, an adherent of the conservative geognosy of Friedrich Mohs, used a descriptive approach borrowed from zoology and biology, consisting of a systemization based on exterior characteristics.[98] Second, because nominations and the establishment of new chairs were not happening simultaneously at all universities but depended on local conditions (e.g., natural history was divided into chairs of biology, mineralogy, and zoology only after the last professor in that field had died), some professors moved from university to university numerous times within a short period. Moreover, the regular deaths and retirements of older scholars at both universities and technical academies increased the turnover further still.

The story of the chair of physics and mathematics in L'viv illustrates the chaos in the natural sciences at the time, with regard to both geographic and disciplinary mobility. The mathematician Victor Pierre moved to the University of L'viv from the L'viv Polytechnic in 1853 and took over the chair of Alexander Zawadzki, a biologist who had taught at the philosophical faculty in Przemyśl and who, after 1848, was a professor of physics and mathematics at the University of L'viv. Zawadzki was removed from the university[99] and transferred to the *Realschule* in Brno, where he served as the vice president of the Naturalists' Society in Brno (Naturforschender Verein in Brünn)[100] and actively supported Gregor Mendel.[101] By 1857 Pierre had been appointed to Prague to replace the deceased Franz Adam Petrina (František Adam Petřina). Wojciech Urbański, who had been a Privatdozent for mathematical physics in L'viv from 1850 on, served as a replacement lecturer but two years later became the main librarian of the university library and ceased teaching. Finally, in 1860 a recent graduate from Vienna, the twenty-three-year-old Alois Handl, was appointed to the chair of physics and mathematics, only to leave the university because of the language change in 1872. After a short period at the Military Academy (Militär-Akademie) in Wiener Neustadt, Handl became a professor in Chernivtsi.[102] Such movements frequently involved new linguistic environments; because L'viv's and Prague's polytechnics were strong in scholarship but financially weak, scholars at these academies were more likely to move to a university in another city than were the lecturers at the Viennese technical academy.

While there were conflicts of interest concerning personnel in the philosophical faculties, the medical faculties in Vienna, Prague, and Cracow, as well as the medical-surgical academies, experienced more continuity than breaks with tradition. In particular, the possibility of habilitation was taken more seriously than at the philosophical faculty. Because the clinical and hospital facilities were concentrated in the capital and the number of students soared, the University of Vienna profited most from the possibility of including young scholars in teaching and research. By 1852 more than twenty scholars had attained positions as Privatdozenten there, with the same number of scholars habilitating in 1853–60, whereas Prague had fewer than ten throughout this period. In 1860 only eight scholars were teaching as Privatdozenten in Prague, while in Vienna there were twenty-one.[103] For reasons unknown, until 1862 no physicians habilitated in Cracow (or the ministry did not confirm any); similarly, none were confirmed at the philosophical faculty, where political factors hindered some scholars' careers.[104]

Owing to the lack of young academics, a result of the underdevelopment of assistantships before 1848, the first appointees for professorships after 1848 included mostly practicing physicians, eventually complemented by promoted Privatdozenten. Because transfers to and from other Habsburg academic institutions were limited by the practical orientation of such institutions, and the university preferred theoretically versed physicians over practicing ones, almost no scholars changed their affiliation during Thun-Hohenstein's ministry. The few who did were rarely influential and stable assets for their faculties, changing positions frequently.

As in the philosophical faculties, the ministry was cautious about hiring foreign physicians in the medical faculties. Only one non-Habsburg scholar, Ernst Brücke, was appointed in Vienna in 1849; he was nominated during the time when Franz Stadion was responsible for nominations. The ministry favored, however, scholars returning to the Habsburg Empire from other parts of the German Confederation. But their numbers were not overwhelming, with just one such scholar appointed in Vienna and three in Prague, among them Jan Evangelista Purkyně, the eminent Czech-Bohemian physiologist from Wrocław/Breslau.[105]

At the universities in Prague and Cracow, medicine remained closely tied to the language question, once again complicating the appointment procedures. The Prague faculty requested that the ministry appoint only scholars who knew both provincial languages, pleading also for the creation of parallel chairs in practical disciplines.[106] In Cracow knowledge of Polish was essential for newly appointed staff since professorships were linked to clinics; thus, the ministry resorted to Galician-born scholars.[107] The ministry appointed no Polish speakers from abroad and rejected all proposals for Privatdozenten, probably because the scholars in question had a troublesome political past.[108] But the ministry appointed only three professors who did not speak Polish: for the chair of pathological anatomy and, as German-speaking counterparts of Majer and Kozubowski, in physiology and anatomy.[109] All other scholars nominated at the time had been born in Galicia, and they remained at the university after the language changes in 1861. At least two of these, Józef Dietl and Antoni Bryk, admitted that German was their primary language when giving their acceptance speeches. However, Dietl swiftly became a Polish nationalist activist, while Bryk taught in Polish and participated in Polish-language scholarly endeavors.

The medical and natural sciences were the exception rather than the rule, however. In the humanities, the period between 1848 and 1860 witnessed a real revolution, setting the scene not only for major developments within the universities but also for an enormous change in the intellectual atmosphere throughout the empire. In the following, I illustrate these developments in three disciplines that were reformed with a Habsburg distinctiveness from "German" ideas in mind; in the late nineteenth century, these disciplines' trajectories united the Habsburg space. First, historiography was attuned to show Habsburg commonalities, as well as linkages among the provinces; it simultaneously fostered provincial histories and the narrative of state unity. Its central institution, the Institute of Austrian Historical Research (Institut für Österreichische Geschichtsforschung [IAHR]), produced the most central European historians well into the twentieth century. Moreover, while scholars at the provincial scholarly societies turned to national histories,[110] the narratives emanating from the universities, even from the Slavic ones in Prague and Cracow, were far more conciliatory.[111] Second, the concentration on comparative theories in all branches of linguistic research challenged ideas of national distinctiveness, bringing forward the linguistic entanglements of the past and the present and hailing them as beneficial. The scholars nominated in this period showed a marked disinterest in both the linguistic purism so treasured by nationalist activists and the histories of literatures, the main component of the imagining of nations.[112] Finally, the vision of philosophy that Thun-Hohenstein followed in his nominations opened Habsburg academia to a range of Catholic approaches, like Karl Christian Krause's panentheism and Anton Günther's speculative theology. At the same time, Thun-Hohenstein fought against Hegelian or Kantian ideas, blaming them for stimulating revolutionary events such as 1848.[113] This, on one hand, left a void within secular approaches, which was filled in the 1870s by positivist and neopositivist philosophy and, on the other hand, ensured the prominence of Catholic philosophies at the universities well into the fin de siècle.

The Habsburg Empire as a Conservative Space: Historiography

The importance of historiography for the new narrative of the monarchy was signaled already in 1847 when Joseph Chmel gained support for his pan-Habsburg projects and began to lead the historical commission at the Imperial Academy of Sciences and Arts in Vienna. The universities, whose main function, according to Joseph Alexander Helfert, was "the fostering of the humanities and familiarity with the institutions and history of the fatherland,"[114] followed closely. Helfert, who before 1848 was a jurist and historian of church law and in 1848–60 served as *Unterstaatssekretär* (undersecretary of state) in the Ministry of Education,[115] pulled the strings in the ministry throughout Thun-Hohenstein's tenure, especially in the historical disciplines. In his eyes, a patriotic, statist direction in education was the only way to create a feeling of nonethnic national unity, an outlook Thun-Hohenstein clearly agreed with.

Searching for uniting origin myths, Helfert directed historians' attention especially toward the Middle Ages and early modern history to find common enemies of the central European populace, such as the Mongolians.[116] He also embraced the marriage policy of the Habsburgs, which in his eyes created larger states that could better protect the population, as in the case of Albert II's unification of Austria, Hungary, and Bohemia in 1438.[117]

Such a construction of the Habsburg monarchy as a state brought into being by a historical imperative also required writing the histories of particular provinces to substantiate their development as naturally leading to the creation of "Greater Austria."[118] However, such an analysis first required the historical sources for all provinces to be collected and edited, which Helfert, in agreement with Chmel, saw as necessary before any attempt at analysis. For this purpose, the ministry created an up-to-date institute for source research, the IAHR; the preparations for this included an examination of the leading European historical institutes.[119] The past was reduced to the "glorious" Middle Ages, while the more recent history of the eighteenth and nineteenth centuries remained clearly in the background.[120] Despite Helfert's declarations about the linearity of historical development, cultural memory was selective, excluding, for instance, Josephinism and emphasizing the uniting force of Catholicism, promoted by the conservatives. In accordance with Helfert's view of historiography as a patriotic, and thus

Catholic, activity, Thun-Hohenstein created two new chairs of history at the universities, one for general history and one for Austrian history.

Interestingly, this division of chairs denoted the preferred orientation of the nominees. While the chairs of Austrian history at the various universities were filled with Catholic historians born in the Habsburg Empire, Thun-Hohenstein did not hesitate to promote foreign and Protestant scholars to cover lectures in general history. The most prominent chairs, as well as the directorship of the IAHR, were clearly reserved for Catholics, including exponents of Catholic conservativism who, owing to their activism in favor of Catholic conservatism, had had to leave other universities in the German Confederation. The best known were a few members of the Görres Circle, an antiliberal movement propagating political Catholicism in Munich, who lost their positions because of their involvement in conservative Catholic protests against Ludwig I's affair with the Irish dancer Lola Montez.[121]

Especially with regard to specialists in the auxiliary sciences of history, almost absent from Habsburg scholarship in the Vormärz, Thun-Hohenstein was flexible in applying the denominational rules. He was even willing to nominate Protestant scholars to the University of Vienna, which was primarily a Catholic institution. In 1849 he stressed that the chair of history in Vienna should be awarded to Catholics,[122] but in 1857 he appointed a specialist in auxiliary sciences of history, Theodor Sickel, as an associate professor and a leading member of the IAHR, even though the young scholar had been born in Prussia, was Protestant, and was even suspected of spying for France.[123] One must add, however, that Sickel was not Thun-Hohenstein's first choice and that he had been living in Vienna for a few years before his appointment, teaching paleography at the IAHR.[124] Through Sickel, the IAHR became the central European institution for the critical discipline of document research, its proclaimed aim being to rebuff the teleological-philosophical approaches that had predominated in historical research before 1848.

Such nominations were possible only through personal contacts and protection, something, as I show later, that was vital for developing careers in the 1850s. To ensure that candidates had the proper ideological consciousness, Thun-Hohenstein relied on a network of trusted sympathizers, who in turn corresponded with scholars abroad. One such person was Johann Böhmer, a famous historian working in Frankfurt am Main, known for his strong aversion to Prussian Protestantism.[125]

The universities in Prague and Galicia also experienced new trends in historiographical research. Thun-Hohenstein searched for bilingual scholars who would support his idea of Catholic state patriotism, and thus finding candidates was not always easy. It proved especially complicated in Galicia, as most known Polish-speaking historians had either been involved in Polish uprisings or actively supported Polish nationalism and thus were unsuitable. Thun-Hohenstein also shunned nominating the towering figure of historiography in Prague, "the historian of the Czech nation," the Hussite promoter Palacký.[126] Nevertheless, the minister also clearly hesitated to appoint scholars to these universities who could be regarded as cultural or national foreigners or could spark nationalistic conflicts. Even in L'viv, fluency in all three languages of the city—German, Polish, and Ruthenian—was seen as a prerequisite.[127]

While agreeing with the need for knowledge of local languages, Thun-Hohenstein appointed only those men who supported his ideological principles. The universities protested vehemently, claiming that more prominent historians were available. In Cracow, ignoring faculty protests, Thun-Hohenstein decided to nominate Antoni Walewski, a conservative loyalist with no major historical publications and no formal education in history.[128] Walewski was rumored to be a formal secret agent of the Habsburg government whose aim was to undermine the Polish character of the Jagiellonian University, for example, in the actions against the allegedly nationalist agitation of several professors in 1853.[129]

With cultural conflict looming in Bohemia, the ministry decided on a two-professor solution in Prague: one chair of history would be associated with German culture, the other with Czech culture. As the Czech chair, Thun-Hohenstein appointed his close friend, Václav Vladivoj Tomek, who not only was conservative and Catholic but also promoted a positive picture of German-Czech relations, making him an apt candidate for a university position in the bilingual province.[130] As I demonstrate later, Tomek was a loyal supporter of Thun-Hohenstein's administration and philosophy regarding the development of the university. His "German" counterpart was Constantin Höfler, the *Großdeutsch*[131] ultramontane historian of Catholicism recently dismissed by Munich, who seemingly also cherished Tomek's support.[132] In his first years in Prague, Höfler published a broad range of studies on Bohemian history, and he befriended and cooperated with Palacký and Šafárik, managing the cultural tensions well.[133] While German culture and a German civilizing mission had always been at the forefront of his writing, from the late 1860s he began to pursue a clear anti-Czech narrative and became one of most energetic pro-German nationalist activists, founding several German-Bohemian cultural institutions.[134]

The Habsburg Empire as an Entangled Space: Philologies

With the strengthening of philological and historical education—united in one seminar—classical philology grew in importance. Based on the model of the non-Habsburg German Confederation universities, the classics were elevated to become a main humanist subject in the Habsburg Empire, serving as a point of departure for humanist education.[135] Here, private recommendations by Thun-Hohenstein's network of trusted men were crucial, although the ministry, cautious of Franz Joseph's reactions, always highlighted its choices with reference to the nominee's religious denomination. Emblematic here is the reasoning presented in the nomination of Ludwig Lange in 1855. Lange was placed second on Prague's philosophical faculty's short list behind the Catholic Karl Halm, but he cherished the support of his predecessor, Georg Curtius: "Despite his outer religious commitment [to Protestantism], he [Lange] lacks nothing of genuine Catholic conviction," reads the ministerial document. Meanwhile, Halm was described as Catholic only in denomination, and the document stated that his influence on the students would be "more alarming than that of a Protestant."[136]

As in the historical fields, scholars from abroad were valued highly, but, in contrast to historiography, in philology several chairs had been occupied by Habsburg scholars already since the Vormärz. Nevertheless, young scholars from abroad were nominated from the outset, while older scholars were either transferred to smaller universities or, if nearing retirement, pensioned off. Bonitz, who served as Thun-Hohenstein's confidant for classical philology and who consulted with the philologist Friedrich Haase in Wrocław/Breslau,[137] played a critical role here. Even in Cracow, where matters were again complicated owing to language issues, the ministry appointed the historian of classical literature Antoni Małecki, who had graduated from Berlin and taught at a gymnasium in Poznań/Posen. While the deaths or retirements of older professors meant that philology showed more mobility among scholars than did the other humanistic disciplines, the younger generation of both professors and students had already been educated by scholars who came from abroad.

The new philology, accentuating exegesis and grammar in place of the previous mechanical learning of vocabulary and translation,[138] played a prominent role in the cultivation of classical values but was also an eminently political issue. This was not only because of the stress on researching minutia and the rejection of grand narratives. Paradigmatic here is the chair in Prague. Georg Curtius was appointed professor and director of the Philological Seminar (Philologisches Seminar), and August Schleicher was appointed shortly thereafter as an associate professor of comparative linguistics. These nominations were important for two reasons: they counterbalanced the long-serving but unproductive full professor Michael Canaval, and both newly nominated scholars worked on comparative linguistics, which owing to its emphasis on similarity and contact among languages was of political importance in the multinational monarchy. One can clearly perceive the political dimension of this innovation in both Schleicher's linguistic *Stammbaumtheorie* (family-tree theory) and Curtius's research on classical philology. While Schleicher promoted the close kinship of Lettish-Slavic and Germanic as Indo-Germanic "sister languages,"[139] Curtius wrote that "comparative linguistics has proven that countless centuries before the beginning of Greek and Italian history, the common ancestors of the Indians, Persians, Greeks, Romans, Germanic people, Slavs and Celts constituted one folk."[140]

This vision was strongly reminiscent of the narratives of the past that historical research was to provide, according to the political imagination of the conservatives. Emphasis on the political value of philology was quite common. Curtius's successor, Ludwig Lange, even included a version of a political program for nationalities, which sought to unite them in spite of their cultural differences in the pursuit of the higher aim of humanity. In his inaugural lecture, he described Greek and Roman ideals as a "spiritually refining force . . . in a present dampened by materialism, especially for youth, [who are] receptive to all things good and beautiful." Moreover, he continued, "we can learn from the Romans how one can remain fully national and nonetheless achieve humanity. As Romans did not become Greeks, the new nations [*Völker*], be they Slavs or Germans, should not dismiss their national peculiarities, if they are valuable; nationality should only be cleansed of the muck in the acid test of attempts at humanity."[141]

The prominent role of comparative linguistics in the appointment policy of the 1850s was not visible only in the cases of Schleicher and Curtius; comparative studies was a popular political device for accentuating national interconnections. In particular, it highlighted the role of research on the original language of the Slavs, seen, depending on the author, as Old Church

Slavonic or Old Church Slavic. Compared with research on particular language formations and vernaculars, writing on Old Church Slavonic as the basis from which the Slavic languages evolved brought the common elements shared by these languages into the foreground. Unsurprisingly, Thun, unofficially, even requested that scholars nominated for chairs in national philologies know this language and entrusted the new professor of Slavic philology in Vienna, Franc Miklošič, with making the final decisions on this issue.[142]

While Old Church Slavic was regarded as the antithesis of national particularism,[143] several projects launched in the 1850s in Vienna pointed toward a rejection of the vernacular nature of Slavic languages. Both the series Legal and Political Terminology for the Slavic Languages of Austria (Juridisch-politische Terminologie für die slawischen Sprachen Österreichs) and two Slavic journals, edited in Vienna and supported by Thun and Helfert, *Slovenské noviny* (Slovak news) and *Vídeňský denník* (Viennese journal), proposed approaches that softened the differences among languages instead of encouraging their divergence.[144] Similarly, the Ruthenian conservative journal *Věstnyk'* (Вѣстникъ, or Herald) was published in Vienna, although it remains unclear to what extent it received the support of the Ministry of Education.[145] Thun-Hohenstein also backed two Prague scholars, Jan Pravoslav Koubek and Jan Erazim Vocel, antagonists of Palacký in the reform of the Czech alphabet in 1848; they preferred an alphabet that would minimize the disparities between the Slavic languages.[146]

Clearly, the scholars working on these Vienna-based projects were mainly university professors appointed in and after 1848. They included Šafárik, who was pulling the strings in language-based subjects and who took scholars' academic qualities as much as their linguistic-political alignment into account.

While most chairs for Slavic languages and literatures had been nominated before Thun-Hohenstein was appointed, his own decisions in these matters were quite controversial. In Prague, when František Ladislav Čelakovský died in 1852, the ministry found nobody suitable to take his place. Martin Hattala, an associate professor and the author of textbooks and grammars of Slovak, won a full professorship only after Thun-Hohenstein resigned.[147] In L'viv, despite countless pleas from the university, Thun-Hohenstein refrained from raising Jakiv Holovac'kyj's salary, leaving it at the 1848 level, which was below the normal salary for full professors.[148] The chairs of Polish language and literature in Galicia, a traditionally difficult issue, remained unoccupied until 1856, in Cracow from 1848 and in L'viv from 1851, when the position holders were removed for political reasons.[149] The universities were trying to push through Polish luminaries, most of whom were politically unacceptable to the ministry.[150] In turn, the scholars whom Thun-Hohenstein wanted to appoint to these positions declined.[151] In the case of renowned writer and journalist Józef Ignacy Kraszewski, although the ministry, the university, and the nominee agreed on terms, the Russian government refused to issue him a visa, thwarting the appointment. Finally, with some hesitation on all sides, Thun-Hohenstein agreed to promote the long-term auxiliary professor[152] of the Jagiellonian University, Karol Mecherzyński, to a full professorship.[153] Negotiating directly with Thun-Hohenstein, Antoni Małecki was relocated from Innsbruck to L'viv the following year, changing his primary designation from classical to Slavic philology.[154] Both scholars were in line with the ministerial ideas on language research, clearly not fulfilling the nationalist activists' hopes that these chairs would be conduits of national propaganda.[155] Małecki, educated as a grammarian, learned Old Church Slavic and then used Miklošič's formal approach to languages as the basis for his own grammatical texts on Polish.[156] Mecherzyński's research concentrated more on language than on literature, and his publications on the influences of Latin and German on the development of Polish confirmed his interest in comparative studies.[157]

The question of chairs of German literature and language arose only after professors of Slavic languages had been appointed. Since German was the language of the monarchy, this might come as a surprise; however, language teaching was the domain of readers (*Lektoren*), and lectures on German were also held by professors of comparative philology. Moreover, German had been taught under the guise of aesthetics or rhetoric (*Beredsamkeit*) at Habsburg universities. German studies, in its newer philological form, was also very political, concentrating on luminaries and historical continuities and facilitating the spread of a pan-German consciousness to which the Habsburg monarchy was averse.

The strategy of depoliticizing linguistic disciplines in German studies was similar to that used in Slavic studies. First, only grammarians and philologists from abroad were nominated, with the focus clearly more on language than on literature. In most cases, the ministry preferred scholars with clear research interests in Catholic topics.[158] Here, some Catholic writers were appointed even if they lacked a formal education.[159] Second, Thun-Hohenstein's first choices were local scholars or scholars with long-standing contacts in Vienna, seemingly certifying the prime importance of both quality and the correct ideology.

The differences in the handling of German language and literature are best highlighted by comparing Cracow and Innsbruck. In Galicia, where German was hailed by government media as a guarantor of cultural progress and an asset uniting this recent addition to the empire, the chairs in Cracow and L'viv were swiftly filled and refilled. In Innsbruck, in contrast, Joseph Novotny taught both Italian and German as a titular professor (i.e., neither tenured nor receiving a regular salary), although from 1854 only German was mentioned on the lecturers' list beside his name. Only in 1858 did the ministry propose as a professor of German language and literature Ignaz Zingerle, a Merano-born gymnasium teacher and librarian, known for his collection of Tyrolean tales and his interest in the culture and ethnology of the province. Yet, even here, the ministry stated that it was hardly possible to find "an individual with the necessary scientific education for this discipline in Innsbruck among Habsburg scholars."[160] (Thun-Hohenstein appointed only two non-Habsburg professors in Innsbruck.) Initially, Franz Joseph rejected the proposal; a second proposal, accompanied by an illustration of the situation of literature studies in the monarchy, was accepted without delay.[161] That the University of Innsbruck gained the chair so late is even more surprising if one considers that from 1854 the university had a chair of Italian language and literature.

While appointments in Galicia occurred swiftly, they were not without their problems. The original nominee for Cracow, Karl Weinhold, asked after only a few months to be relocated from a city he considered to be culturally deficient (he had also lost a number of manuscripts in the city fire of 1850).[162] Thun-Hohenstein swiftly appointed in his place a student of Exner, Bratranek. Although established and valued as an innovative scholar and as an Augustinian friar with the correct mind-set, Bratranek was nevertheless atypical of the appointments for this chair. Not only was he openly Hegelian, but his work also concentrated not on the Middle Ages but on nineteenth-century literature, especially Goethe, and aesthetics. He and the philosopher Józef Kremer, who was working in Cracow, were the most famous Hegelians at any of the Habsburg universities at the time, although both were conscious of the boundaries set by the ministry and accordingly rarely published their work.[163] As the above-quoted *pium desiderium* from 1853 certifies, Bratranek was also a convenient appointment for the government in political matters.

The Habsburg Empire as a Catholic Space: Philosophy

While the historical and philological disciplines served as mediators of state unity and cultural diversity by supporting certain narrative strains, the choice of scholars for the chairs of philosophy shows the importance of this discipline in the conservative Catholic project of the alteration of intellectual culture. In contrast to the situation in the other humanistic disciplines, however, the mistrust toward recent philosophical systems and the accentuation of historical matters led to the continuation of the Habsburg philosophical tradition instead of the importation of professors from abroad.[164] Even local scholars were scrutinized, however, and professors who favored speculative philosophy were supervised and/or removed from influential positions. The chair of philosophy, usually linked to pedagogy at the time, was directed toward the history of philosophy or moral philosophy. Although scholars active in these fields also worked on logic or aesthetics, the professionalization of philosophy as a separate academic discipline was hardly discernible, especially at smaller universities. While much happened in Vienna and Prague, other universities, with just one chair of philosophy, were mostly out of the minister's view. At the peripheral university in Cracow, even Hegelianism was accepted, although only in its Catholic version.[165]

The ministry's comment, quoted above, that philosophy should become a Catholic domain should be taken literally. The ministry actively supported this, using spurious arguments. When Hermann Rosenberg, a Jewish scholar from L'viv, applied for habilitation there in 1854, the legal obstacles were largely overcome by stating that the habilitation process should not consider the person's religious denomination. Nevertheless, the ministry's final answer was short and precise: Rosenberg's appointment could not be considered given that "the teaching position in philosophy can only be granted to a man of Christian belief."[166]

Also, the appointments between 1849 and 1860 show the clear dominance of Catholic philosophy, although without a clearly discernible prevalence of one of its different and conflicting versions. Since the chair of philosophy was to be a showpiece of ideology, scholarly production was less important than teaching, especially because the position covered both philosophy and pedagogy. This prominence of pedagogical functions also explains the large number of continuities with the pre-1848 situation. Although modern philosophy entered Habsburg academia in the 1870s, several scholars appointed by Thun, in particular Robert Zimmermann, shaped the development of this discipline in the Habsburg Empire well into the fin de siècle.

The first appointments, before Thun-Hohenstein's inauguration as minister, supported the philosophy of Johann Friedrich Herbart.[167] Thun-Hohenstein initially maintained this direction, especially since most Herbartians were also students of his own teacher, Bolzano. Most notably, Zimmermann, Bolzano's *Herzensjunge* (favorite pupil), had an astonishing career, starting in Olomouc but moving swiftly to Prague and then, in one of Thun-Hohenstein's final decisions, to Vienna.[168] A productive author of compulsory textbooks, Zimmermann was the most important advocate of Herbart and Bolzano in those initial years but in particular supported Thun-Hohenstein's campaign against Hegelianism and Kantianism.[169] In 1850, however, and in particular after Exner's death, this direction grew less popular, and the proponents of a pronouncedly Catholic philosophy, supported by Thun-Hohenstein's confidants, replaced it.[170] As a counterbalance to Herbartianism, Thun-Hohenstein appointed followers of Krause's panentheism, who propagated an idea of God as an all-encompassing essence, visible in the material and nonmaterial worlds.[171] The most influential Catholic philosophy of the time was, however, the philosophical theology of a supporter of Bolzano, Anton Günther, at the time a private scholar in Vienna.

Günther's philosophy strove to overcome the division between knowledge and faith, creating an anthropocentric and philosophical theology, balancing theological dogmas and scholarship. In 1857, however, this balancing act failed, and the Catholic Church declared Güntherism to be heresy and put his work on the List of Forbidden Books (*Index Librorum Prohibitorum*). Before this papal intervention, Günther's ideas had flourished, even if Günther himself had not been nominated for a professorship. The archbishop Friedrich Schwarzenberg, a student and friend of Günther, helped introduce this philosophical direction at both the theological and philosophical faculties in Prague.[172] In Vienna the most noteworthy nominee was Georg Schenach, who worked on a system of Catholic-based metaphysics that incorporated materialistic systems. His "philosophical walk on eggshells"[173] merged Günther's speculative theology with another Habsburg tradition, Friedrich Jacobi's sensualism. Significantly, Schenach, who died several months after his nomination, was also Thun-Hohenstein's personal philosopher (*Leibphilosopher*) in Vienna. Since followers of Günther fought bitterly against Exner's interpretation of Herbart and fiercely criticized Thun-Hohenstein for promoting other philosophical discourses,[174] these nominations were a clear sign of changing influences with regard to education.

The difficulty of finding professors acceptable to all parties appears in the case of the first scholars of philosophy teaching in Prague after 1848, Augustin Smetana and Ignaz Jan Hanusch (Ignac Jan Hanuš). When Exner left a professorship in Prague to join the ministry in Vienna, his provisory replacement was Smetana, his student and assistant in Prague, expected to be his successor despite his open Hegelianism. As a provisional teacher at the university, Smetana read Kant directly after the revolution; however, the ministry canceled his lecture on Hegel, which had been planned for the second semester and which he published the same year as a brochure.[175] Notwithstanding his extensive networks, including Exner (whom Smetana asked directly for support) and scholars abroad, the young philosopher could not find a position, neither at a university nor at a gymnasium. His position certainly worsened after his conflict with church authorities when he left the Catholic Church, which also distanced him from Exner.[176]

Smetana's failure, however, opened the door for his close friend Hanuš, formerly a professor in L'viv, who was appointed to the chair of philosophy in May 1849.[177] The careers of both friends remained intrinsically linked, however. Stricken with financial difficulties, and disappointed about the withdrawal of political and religious liberalism, Smetana, who had long-standing health problems, died in 1851 at the age of thirty-seven. According to rumors, in his final hours, Hanuš protected his bed from church officials; the dying philosopher had asked his friends to ensure that the church authorities would not be able to spread rumors that he had returned to the church in the last moments of his life.[178] Despite political and church antagonism, Smetana's funeral turned into a sympathetic display of liberalism, causing problems and political consequences for the participants as well as for the university. Smetana was a member of the Doktoren-Collegium, and the faculty thus had the right, or even the moral obligation, to send representatives to the funeral, which must have caused some friction, to say the least.[179] Despite the risk, Hanuš was present at the funeral, and he also asked his students to participate; the funeral turned into a demonstration against the church and its influence on the state, making Smetana a memorable figure until today (see figure 2). This caused problems in Hanuš's relations with both the police and the Catholic professors in Prague.[180] Accused by the local priests of antireligious sentiments and Hegelianism, accusations that

FIGURE 2 Augustin Smetana's struggle for liberalism and against the domination of the Catholic Church not only became a point of reference for his contemporaries but remains in the collective memory even today. Here, the scene of his burial. (From Galerie NE: Galerie osobnosti, které se nebály jít proti proudu [Galery-No: Gallery of people who were not afraid to swim against the current]. Drawing by Jaroslav Ježek.)

were confirmed by none other than Tomek, Hanuš was suspended shortly after the funeral. Thun-Hohenstein dismissed Hanuš at the beginning of the winter term of 1852.[181]

Intentions and Results

Given Thun-Hohenstein's strategy of balancing local and foreign appointments, one can discern differences in the nominees depending on the faculty and locality. The medical faculties remained predominantly Habsburg. Correspondingly, they were also bound to the local language situation, both in Prague and in Cracow; at these universities in 1860, only a few scholars who had been born outside of the respective province likely spoke only German.[182] In Vienna in 1860, all but one scholar at the medical faculty had been born in the Habsburg Empire. At the philosophical faculty in Vienna, in contrast, a third of the lecturers had been born outside the monarchy. Three non-Habsburg-born scholars (i.e., 30 percent of all professors) taught in Graz, and one in Innsbruck, where around half of the scholars were Tyroleans. Faculties within multilingual provinces also show less concentration on local scholars, as in the case of medicine. In Prague in 1860, almost half of the lecturers had been born outside Bohemia or Moravia, and the number of scholars capable of lecturing in Czech was considered to be seriously low and fell within a decade.[183] In L'viv 60 percent of lecturers came from outside Galicia (at least five scholars out of twelve spoke Polish, and one spoke Ruthenian). In Cracow and Pest, however, most professors were bilingual, once more pointing to the special position these two universities held.[184]

The continuity of scholarship varied as well. At the medical faculties, around half of the scholars teaching in 1860 had been there since 1848, with the exception of Cracow, where all but one scholar had been appointed after the revolution. Philosophical faculties, in contrast, had been thoroughly reformed. The philosophical faculties were not uniform, however. The chairs of languages of course differed across the monarchy. The chair of bibliography, linked to the directorship of the university library, existed only in Cracow. Having the largest philosophical faculty and the most Privatdozenten, the University of Vienna also offered the greatest range of subjects, clearly privileged compared with other universities, a situation that would be discernible later as well.

That Vienna and Prague had slightly different roles in the nexus of the monarchy was also indicated by the position of readers of modern languages. While these two universities hosted representatives of most languages spoken in the monarchy (including Hungarian, although not Ruthenian, Russian, or Slovenian), with the aim of encouraging language competences among future bureaucrats and officials, smaller universities taught only the local languages. For most of the period, Innsbruck and L'viv entirely lacked modern languages apart from German and their respective local languages; in Cracow, Ruthenian and French were taught; and in Graz, French, Italian, and Slovenian. This division, certainly disadvantageous to students in Galicia, Styria, and Tyrol, was influenced by infrastructural differences in the cities themselves, as teachers were mostly not full-time employees of the universities but, for instance, worked primarily in official posts in the court or administration.

When rumors spread in early 1860 that Thun-Hohenstein would be resigning from his position, the atmosphere at the universities was uncertain. Many considered him the reformer of the university system and their savior from the conservatives. This was true of Galician scholars, who openly lamented the news in the Cracow daily *Czas*, and of the Catholic Croatian bishop

Josip Juraj Strossmayer, later instrumental in founding the Royal University of Franz Joseph I in Zagreb (Sveučilišta Franje Josipa I. u Zagrebu) and the Croatian Academy of Sciences and Arts (Hrvatska akademija znanosti i umjetnosti).[185] Others remembered the Habsburg universities of Thun-Hohenstein's time critically, writing about the tense atmosphere at the universities and the police supervision of professors.[186] Because Thun-Hohenstein's time as minister was concomitant with the neoabsolutist regime, it is hard to say whether the critical voices requesting another intellectual and political restart were directed against him or against the political atmosphere in general.[187] It is clear, however, that the assessment of Thun-Hohenstein's tenure depended on one's ideological position: positive voices came from the conservative and liberal Catholic milieus, and critical voices from the non-Catholic and also ultra-Catholic sides.

It has often been claimed that Thun-Hohenstein's plans were far from fully realized. For example, Alphons Lhotsky claimed that Thun-Hohenstein deliberately strove for a conservative and Catholic university through his appointments.[188] Thun-Hohenstein's admirers, in contrast, both at the time and later in the nineteenth century, claimed that his openness and liberal planning were hindered by neoabsolutism, stating that his reforms were a milestone in the academic policy of the empire and its successors.[189]

The impression that the reforms of 1848–49 were Thun-Hohenstein's work was not only an outcome of his impressive propaganda campaign. Thun-Hohenstein became a symbol of Habsburg policies, one that was applied at different times and in the service of different needs.[190] Some later reformers highlighted him as a protector of academic autonomy; that only those chosen by him experienced such autonomy was not important. Thun-Hohenstein also towered above Habsburg universities in a literal sense as well. In 1893 a monument for Thun-Hohenstein was unveiled in the Arkadenhof (arcade court) of the University of Vienna (figure 3), where famous university scholars are commemorated. Notably, it is the only full-figure monument in the university courtyard.

By considering the university before and after Thun, one can certainly note a range of differences. The financial support universities received from the state allowed facilities such as libraries, institutes, observatories, and clinics to be enhanced considerably. Professors from universities abroad brought with them not only scientific knowledge but also a practical orientation as to what resources the libraries should include and how seminars should be organized. In the 1850s, though, the function of the universities did not change considerably; they remained teaching facilities and were only

FIGURE 3 Busts of Hermann Bonitz (left) and Franz Exner (right) with the monument of Leo Thun-Hohenstein (middle). The inscription on Thun-Hohenstein's monument reads: "COMES LEO DE THUN-HOHENSTEIN MINISTER PUBLICUS 1849–1860 QUI AUSPICIIS IMPERATORIS AUGUSTISSIMI FRANCISCI JOSEPHI I UNIVERSITATES ET GYMNASIA NOVIS LEGIBUS INSTITUTISQUE FELICITER REFORMAVIT IN EA RE CONSILIO EXIMIORUM VIRORUM / EXNER ET BONITZ STRENUE ADIUTUS." (Leo Thun-Hohenstein, public minister 1848–1860, who under the auspices of the venerable emperor Franz Joseph I reformed the gymnasia and universities through new legislation and institutions in a very fruitful manner. In this he enjoyed the active help of the extraordinary men Exner and Bonitz.) (Archive of the University of Vienna, 106.I.3002. Sculptor: Karl Kundmann.)

secondarily concerned with research. As in the Vormärz, they produced loyal state officials, not independent scholars, even if the official propaganda said otherwise.

The success of creating Habsburg universities sympathetic to the monarchy and to German as the language of culture was short lived, precisely because of the changes of 1848–60. From the moment the humanities entered the university, this institution remained at the forefront of nationalists' interest. Benefiting from the liberal appointment policy that began in the late 1860s, universities later became the foremost producer of difference, whether linguistic, historical, or even artistic. In many fields, the scholars nominated by Thun-Hohenstein could, however, pursue their projects further, be they ideological or not. Since many of his nominees were appointed while they were still in their twenties or early thirties, they had several decades to make their mark. Zimmermann was not pensioned until the summer of 1896, probably the longest living of Thun-Hohenstein's appointees.[191] After the 1860s, when ministers put academic autonomy into practice, the scholars appointed by Thun-Hohenstein nominated their own successors, perpetuating certain traditions well into the twentieth century.

Success in imitating Prussian universities was limited. Thun-Hohenstein held up this aim to his adversaries, but the commitment to achieve it was limited by finances, by the retention of Catholic values, and by support for the local traditions of scholarship. The positive figure of the Habsburg scholar who became "German" (including Czech nationalists like Purkyně and Čelakovský, who had both lived in Wrocław/Breslau until 1848)[192] was not mere rhetoric but also a proclaimed aim of ministerial policy, in which "Germany" served as an idealized paradise, especially for the humanities. Since Thun-Hohenstein strove to nominate Catholics, despite looking for models in Prussia, he was importing scholars directly from Bavaria.

It also became clear that the smaller universities in the monarchy, including the Charles-Ferdinand University in Prague, were not competitive enough with other German-language universities, in terms of both finances and research facilities. These smaller institutions offered career advances for foreign scholars, but such appointments were very often quite short-term. Newly appointed foreign professors could receive a call back to their home university, and this was clearly perceived as a threat to the universities, which was intensively discussed in subsequent decades. Indeed, most of the scholars Thun-Hohenstein recruited from abroad eventually left the monarchy, often achieving considerable influence at universities in the German Empire.

The humanities were the field in which the ministry was most willing to invest; the sciences and medicine clearly remained secondary, with a number of rather accidental appointments because there was no clear ministerial strategy as there was in the humanities. This is something of a paradox, since those were the fields that flourished in subsequent decades. Similarly ironic is that the conscious choice of lecturers often introduced developments that contradicted the ministry's intentions. The withdrawal from the abstract, and the corresponding turn to the concrete, as Thun-Hohenstein wrote in one of the appointment records,[193] opened the door to all sorts of positivist approaches in the humanities and philosophy, as the Viennese historian Johannes Feichtinger has pointed out.[194] The philosophical approaches of Zimmermann and Schenach did not remain widely influential, and this led to conflicts in the faculties. In historiography, the nonteleological auxiliary sciences of history strengthened, especially with the establishment of the IAHR in Vienna. However, the creation of a grand narrative largely failed, at least at the university level. The Slavic appointees who were supposed to create such grand narratives, Tomek and Walewski, were ultimately unsuccessful; the former was unable to enforce his idea of writing a "shared" Slavic-German history, and the latter was severely criticized for his glorification of the "Austrian" and "Catholic" elements in history and finally ostracized by both the university and the public.[195]

With regard to spatial policy, Thun-Hohenstein opened the empire to outside scholars but only those from the German Confederation. This policy, however, stimulated the Poles and Ruthenians, whose scholars often lived in the Russian Empire, to argue for the privilege of appointing them. The encouragement of pan-Habsburg mobility for scholars meant that for the first time there was also a united space, including Pest and Galicia. While this space still centered on Vienna, and the most important location for entering a career remained the University of Vienna, exchanges among provincial universities were possible, including of scholars who identified as Poles or Czechs. This strain of mobility dried out in the liberal period, challenged by nationalist conflicts.

In the 1850s the hierarchical structure of the university system did not change. While the period is too short for me to sketch more than a few career paths, the structure laid down in 1848, along with its salary regulations, was clearly decisive. There were exceptions, though, since salaries could also be individually negotiated and could be higher than the official ones. Although no full professors moved from Vienna to Innsbruck, for example, a few transferred from Cracow to Graz, even though, according to the law, Cracow scholars earned two hundred guldens per year more. It is also important to note that the Vienna-centric legacy of the Vormärz was reinforced. The IAHR became the central Habsburg institution in historiography, and most future professors had studied there. In turn, the Medical-Surgical Joseph's Academy, which had reopened in 1854, was able to appoint full professors from any Habsburg university.[196] This moved Carl Rokitansky, the foremost Viennese physician, who from 1863 served as a counselor in the Ministry of Education, to request, shortly after Thun-Hohenstein's resignation, the lessening of salary differences between universities, a measure that he saw as absolutely necessary for an efficacious university system.[197]

Chapter 3

Living Out Academic Autonomy

We do not want to always be your pupils! We feel in us . . . enough power and energy to finally stand on our own, to cultivate knowledge and science on our own.

—Jan Evangelista Purkyně, 1862[1]

Dear friend and Ritter! We are searching for a minister of education. Perhaps you would like to take this office? This position is not so evil. You can accomplish much good; one should pursue great deeds of culture. It is curious that so many people point out trivia: one only wants everything liberal, primary schools, tolerance for all religious denominations, improvement of spiritual development—but the Concordat has to be preserved. Universities are to be flirted with; sciences are to be boosted—only the Catholic character cannot be touched; the archbishop has to remain the university chancellor as afore. You can fire all the people who bewail the archaic laws of the faculties; a lot of new things could be formed here—but the old doctors' council guilds have to be preserved. Much is to be organized, not only in Vienna, to build a university, establish various scientific institutes, double the number of teachers, as the whole of Hungary and the Danube principalities want to obtain their culture from us—but it should cost no money!

—Theodor Billroth to Wilhelm Lübke, 30 January 1870[2]

Leo Thun-Hohenstein's resignation in 1860 and the end of neoabsolutism meant yet another significant change for the Habsburg university system, all within just over a decade. It would be wrong to call it a revolution, but it brought a realization of the promises of the 1848 revolution. The Habsburg universities had to wait another few years for the liberalization of education; however, the early 1860s paved the way forward more than Thun-Hohenstein could have envisaged. Two major alterations in Habsburg politics deserve mention here. First, the government's handling of university matters between 1861 and 1867, when there was no Ministry of Religion and Education, symbolically strengthened the scholarly community in relation to the politicians. While during Thun-Hohenstein's tenure the ministry made most decisions without consulting the faculties, from 1863 on university scholars had an important voice, if not necessarily the final say. An intermediate body composed of selected Cisleithanian academics, the *Unterrichtsrath* (education council), initially became the pivotal body for educational matters in Vienna. After its abolition in 1867, subsequent ministers rarely disagreed with the faculties' appointment proposals; in this way academic autonomy, prescribed in the postrevolutionary legislation, became more of a reality. Second, language changes in Pest (from German to Hungarian), Cracow (to bilingual Polish-German instruction), and, finally, L'viv (to Polish-Ruthenian instruction) changed the intellectual geography of the empire. From this point on, linguistically codified academic subsystems began to develop, and these in turn created their own spaces of mobility.

Universities developed their own dynamics, even if framed by the political, legal, and social contexts. In this chapter I discuss the most important legal changes and show how they influenced the cohesion of the imperial university space.

From the point of view of monarchical academic space, the change in the language of instruction mattered most. Therefore, I look at this change, considering the role ascribed to universities and scholarship in general. Skeptical German-speaking politicians bemoaned the lack of control over non-German institutions and claimed that they had become cut off from the Habsburg system.[3] Slavic scholars countered that the language change did not necessarily mean the dissolution of the empire and that contacts should be kept. As I argue below, all these voices have to be read in context. For instance, the criticisms from the German-language press and politicians bore traces of the imperialistic equation of German and culture. And, as I have previously argued, in discussions in L'viv, the Poles used a similar argument to withhold Ruthenian as the language of instruction. But even if the nature of the contacts between Cisleithanian universities with different languages of instruction changed, such contacts were maintained. For example, Cisleithanian universities prepared joint legislative initiatives and joint petitions. Here, shared interests played a larger role than cultural differences, and the faculties spoke mostly with one voice.

Further, I look at the implementation of university autonomy and the effect it had on the structure of academic space. The Ministry of Education, as I will show, still meddled with nominations but mostly served as a regulative body that had to take the whole empire into consideration. This pertained in the first place to disciplines that were awarded their own chairs and to habilitations, where ministerial decrees influenced disciplinary specialization. The ministry often criticized specialization and requested that Privatdozenten cover a broader area of teaching. Similarly, appointments, especially for full professorships, had to take into account their effects on other universities: organizational, financial, and symbolic. Since the universities, which were well informed about ministerial decisions, used developments in other parts of the empire to support their own demands, the ministry had to be cautious about its every step. Not unlike in cultural politics, this strengthened conservative policies.

The Ministerial Interregnum: The Unterrichtsrath and the Realization of Autonomy

With Thun-Hohenstein's resignation from the position of minister of religion and education, universities were for a short time administered by Joseph Alexander Helfert. In 1863 Helfert was dismissed, and the government founded the Unterrichtsrath, based on French models and composed of selected Cisleithanian academics. This now became the key body in university affairs, tasked with preparing expert reports on academic matters, and was an important intermediary for the minister of state, who signed all papers before they reached Franz Joseph.[4] The idea that professors would oversee appointment procedures not only led to the replacement of the Ministry of Education by the Unterrichtsrath in the short term but also resulted in a considerable symbolic enhancement of universities' position in the decision-making process in the long run. The Unterrichtsrath was not an authoritative institution, as Thun-Hohenstein envisioned the ministry to be, but rather a consultative body offering expertise on university proposals.

The creation of the Unterrichtsrath exhibited a strong continuity with Thun-Hohenstein's ideological ideals, and the few key decisions it made were in line with the ministerial policies of the 1850s. Liberal scholars criticized its members for coming from the conservative Catholic end of the academic spectrum and for preferring even more conservative policies than Thun-Hohenstein had.[5] Some decisions clearly support that view, and sometimes the Unterrichtsrath commented on issues beyond the scholarly achievements of the candidates. Like Thun-Hohenstein, it also discussed the methodology that the scholars in question applied, favoring conservative epistemologies. For instance, according to one of the records from 1865, Josef Bayer's habilitation for "Aesthetics and the Newer History of German Literature" not only combined a philosophical and a philological discipline in a problematic way but also applied the "wrong" methodology: to habilitate, he should have applied an "analytical and historical" approach.[6] While the humanities remained under the eye of other ministers of education, none scrutinized them so deeply as Thun-Hohenstein and Unterrichtsrath did in regard to methodological matters.

During the four years of the Unterrichtsrath's existence, there were few appointments and habilitations, apart from those made necessary by language changes at the universities; the initial phase of the faculty change had been completed under Thun, and no alterations in the curriculum required additional personnel. The most serious problem of the 1860s, the relocation of scholars after the language changes in Cracow and Pest, had mostly been solved before the Unterrichtsrath was established. Just over thirty professors, predominantly from Pest, changed their place of teaching within a few years in the single most intensive migration wave in Habsburg university history.

Graz profited the most from the relocated scholars, although it was rarely their first choice.[7] Because releasing permanent professors from the civil service was difficult, the government intended to relocate them immediately to other universities to support teaching.[8] Although the universities were consulted about the candidates to be relocated, some appointments took place despite the faculties' opposition.[9] Even negative opinions from the Unterrichtsrath did not count for much.[10] Unsurprisingly, these relocations followed markedly nationalistic patterns, with universities even letting go of local scholars they considered unacceptable for linguistic reasons.[11] Those marked as foreigners could not stay even if they promised to learn the appropriate language.[12]

Science for the People: Polonizing Galicia

The discussion on the language of instruction in Cracow intensified after the fall of neoabsolutism and occupied many column inches for several months in the leading journal *Czas*.[13] On 20 October 1860—the same day on which the October Diploma was issued, a decentralizing document that gave certain powers back to the provincial aristocratic elites—Franz Joseph directed a letter to the minister of the interior, Agenor Gołuchowski, underscoring the importance of higher education for Galicia. In the letter Franz Joseph, clearly working on the image of a benevolent kaiser of the Habsburg peoples, requested an expert opinion on the change in the language of instruction, which led to the sending of delegations to Vienna.[14] It is ironic that Piotr Bartynowski, who had been employed to prevent the rise of nationalistic tendencies at the university in the first place, headed the delegations. This process was concomitant with similar changes in the Russian Empire, where in 1857 a Polish-language medical-surgical academy was established. That academy was restructured in 1862 into the Warsaw Main School (Szkoła Główna Warszawska), which was, in effect, an imperial university.[15] This development in the Russian Empire played no role in the discussions, but it is imaginable that Franz Joseph wanted to be as forthcoming with his Polish subjects as Alexander II of Russia was with his.

While the importance of German as the state language dominated the 1853 deliberations in Galicia, several years later an argument arose on the value of a person's native language for science and education. According to *Czas*, the use of Polish at the university was "natural, just, useful, for the youth, as well as for science," and public education in Polish was "natural and inborn."[16] However, this "science" was not actually science and scholarship in a narrow sense but rather education, as the Polish word *nauka* includes both meanings. While in the texts mentioned earlier the terms *Wissenschaft* and *nauka* can be read as synonyms, scientific content was not within the bounds of the later debate.

The Jagiellonian University in Cracow became a matter of national pride, and nationalist activists perceived it as the most important means to achieving national autonomy. Students' need and right to be taught in Polish were equated with popular education, which would be fueled by the atmosphere of the university. While the needs of science and opportunities for employment were mentioned as decisive in 1853, in 1860 the needs of vertical communication between professors, students, and the population of the province as a whole were stressed. In this case, the University of L'viv was included on equal terms in petitions as the "younger brother," with clear statements that the language change at the university in the capital of the Polish-dominated region of Galicia would be as vital as that at the Jagiellonian University.

The most interesting apologies for Polish were written by Józef Dietl, the former rector of the Jagiellonian University and a foremost nationalist activist, and Antoni Helcel, a legal historian, who (re)defined the nationalist narrative through the question of the educational purpose of the language of instruction.[17] In both cases, the German language was clearly described as foreign, hindering schoolchildren's and university students' ability to master the materials taught and representing a clear turn toward the folk-based linguistic theories of Johann Gottfried Herder and others. With the axiom that Polish was sufficiently developed to be a learned tongue (even surpassing German in its syntactic flexibility or diversity of vocabulary), the communication value of world languages was acknowledged but given secondary importance. At the same time, both scholars argued that the Ruthenians (derogatorily described) needed to use Polish as a language of culture; they thus turned the previously adopted position upside down and here disregarded the symbolic and educational component. Ruthenian might be accepted by rights only when it had developed sufficiently through contact and exchange with Polish, which in turn reminds one of German-speaking scholars' argument against the equity of languages in the empire;[18] Dietl did, however, argue that gymnasium pupils should be educated in both provincial languages.[19]

Although Dietl enlarged the scope of university education in Ruthenian to four practical subjects and included Privatdozenten, who could freely choose the language of their lectures, the contradiction between the arguments relating to Polish and Ruthenian is obvious. In fact, Dietl's proposal for practical implementation was in its rhetoric not far from that written by Thun-Hohenstein in 1849 for the introduction of German in Cracow in 1853, with similar arguments about achieving peaceful coexistence and linguistic duality through the preponderance of one language. But now it was the Ruthenians who should have contact with scholarship through the vehicle of the Polish language, and only a few exceptionally gifted scholars could be accepted as Privatdozenten teaching in Ruthenian. In contrast to Thun-Hohenstein's view, though, in Dietl's narrative the aim of developing both cultures was to strengthen the Slavs in opposition to the German element in Galicia. Nevertheless, this argument was very mild compared with the strong assimilationist movements openly endorsed by many intellectuals and politicians, who pleaded for the assimilation of Ruthenians to the Poles.[20]

On 4 February 1861 the Jagiellonian University was given bilingual status. The lectures in the medical faculty were to be held in Polish (apart from the history of medicine and the so-called medical encyclopedia, that is, a cursory overview of medicine early in the course of study), although with special attention to German terminology and literature. Further, the philosophical faculty was to have German lectures in German language and literature, history, and classical philology (for the sake of future teachers). Nevertheless, these subjects would have parallel Polish chairs, with lectures and seminars in both languages. The law faculty remained de facto separated into "general legal subjects," such as statistics, economics, and Roman law, with lectures in Polish, and "positive Austrian and German subjects," encompassing civil and criminal law, administration, the history of German and Austrian law, and so on, with lectures in German. Moreover, professors teaching in German were expected to know Polish terminology, and those lacking it were to be replaced within a year.[21]

These language changes did not entirely fulfill the hopes of the nationalists, however. Nationalists from *Czas* and the university's deputations pleaded for complete Polonization and did not stop trying to achieve this aim. At the same time, the issue of Ruthenian as a medium of education was still on hold, confirming the strengthening Polish dominance in the province.[22] Despite the efforts of Hryhorij Šaškevyč (Григорій Шашкевич)—the ministerial official in charge of Ruthenian schools, the author of the Ruthenian grammar book for gymnasia, and a member of the Supreme Ruthenian Council (Holovna rus'ka rada, or Головна Руська Рада)—Ruthenian first became a teaching language in gymnasia in 1867. Further, it was used only in the first four classes (at the Imperial and Royal Academic Gymnasium[23] in L'viv), as the ministry considered that Ruthenian "did not reach the level of development" necessary for dealing with scientific issues, according to the official statement on this issue in 1849.[24] Similarly, Ruthenian university chairs were to be created only in accordance with Ruthenians' linguistic and cultural development, which had all the consequences that such an imprecise idea embodies—an issue that I will show remained controversial until 1918 and beyond.

Galician Politics and Ruthenian Cultures: The University of L'viv

The tensions surrounding the issue of Ruthenian as a language of university education in the 1860s were still far from the violence that would ensue from the 1890s onward. Moreover, the issue of language was not solved at this time, nor was Ruthenians' own belief in the ability of their native language to function as a scientific language clear. Even fierce patriots in the early 1860s doubted whether the time was ripe to regard Ukrainian as an independent scholarly language in the Russian Empire.[25] During the parliamentary discussion on the school reforms of 1869, the Ruthenian advocate Stepan Kačala (Степан Качала, also Stefan Kaczała) partly agreed with the Polish criticisms but stated that the lack of literature and the imperfection of the language should not be a reason for excluding Ruthenian from higher education. On the contrary, only through the equity of languages in education could this deficiency be removed.[26] In addition, the petition on the regulation of the school question put forward by the Ruthenian politician Julian Lavrivs'kyj (Юліян Лаврівський) did not foresee a swift restructuring of the University of L'viv into a bilingual one, mentioning only a few subjects to be taught in German "for now," in particular those essential for teacher education and careers in the bureaucracy.[27] While Ruthenian politicians criticized the Poles, mentioning among other documents the memorandum of the Prague Slavic Congress, where equality of rights had been accepted, the ministry's decision in October 1869 to preserve the current language situation at the University of L'viv was seen as satisfactory. Although German was retained as the language of general instruction, with lectures in Polish and Ruthenian in the law faculty and the chairs of languages and literatures, this represented a failure of Ruthenian claims. The ministry's decision also rejected the official petition of the Galician Diet (drafted by a Polish majority) of September 1868 to replace German with Polish while continuing to allow Ruthenian for a few subjects.[28]

The *Staatsgrundgesetz* (Basic Law) of 1867 included the equalization of language rights "in schools, offices, and public life,"[29] fueling nationalists' hopes that universities would automatically undergo a language change. It took some years and a change of government to fulfill these hopes, however. Within a month of the nomination of Alfred Józef Potocki, a Galician nobleman, as minister of state in the spring of 1870, the government realized that the Poles could boycott the Parliament, as the Czechs had been doing since

1867.[30] Since the geopolitical situation had also changed, and the Warsaw Main School had closed in 1869, there was once again no "Polish" university in central Europe, which was an important issue for the nationalists. To prevent a boycott, the government declared Polish the sole teaching language at the Jagiellonian University on 30 April 1870, fulfilling one of the main wishes of the Polish parties. On 4 October the same was announced for L'viv's technical academy. Some politicians as well as professors felt that the academy made the existence of the university in L'viv unnecessary, proposing to move the university to Opava (Troppau, Opawa), to bind Silesia more closely to the monarchy.[31]

The next minister of state, Karl Sigmund Hohenwart (February–October 1871), had to secure support for his cabinet from the Polish parliamentarians who had united into the so-called *Polenklub* (Polish Club) and was willing to make further concessions, supported by the minister of education, Josef Jireček (February–October 1871). Instead of moving the university to Silesia to keep German as the language of instruction, on 4 July 1871 Polish and Ruthenian were made de jure equal languages of instruction in L'viv, making Polish the de facto language of instruction. Strengthened by this measure, the Polish majority at the University of L'viv repeatedly requested that the ministry regulate the language question, that is, acknowledge Polish supremacy by not increasing, or even by decreasing, the number of professorships with Ruthenian as the prescribed medium of instruction. Finally, shortly after the division of the Charles-Ferdinand University in Prague in 1882, the Polish professoriat succeeded in its demands. In a ministerial decree on 5 April 1882, Polish was declared the language in which lectures should be taught "as a rule," with Ruthenian lectures held only with the approval of the ministry.[32] It is clear that the Cisleithanian minister-president Eduard Taaffe (1879–93) fulfilled the nationalists' demands regarding higher education as a means to appease the Czech and Polish parties and gain their support for his government.

The political assertion of the Poles' cultural and educational supremacy had, however, other effects than those intended by Galician nationalists; it resulted in the intensification of Ruthenian intellectual life and support for demands for independent academic institutions. This was even more important since at the time Ukrainian was banned in the Russian Empire; therefore, numerous supporters of Ukrainian language and scholarship were moved to give their patronage to Galician institutions.[33]

From the 1860s, different Ruthenian political groups established their own educational organizations that represented their political alignment and language projects. In the end, populist Ruthenian organizations, including the educational literary society Enlightenment (Prosvita, established in 1868) and its scientific branch, the Ševčenko Scientific Society (Naukove tovarystvo imeni Ševčenka, established in 1873), became extremely influential.[34] By the late 1880s, however, they were being seriously challenged by the Russophile Kačkovs'kyj Society (Tovarystvo imeni Kačkovs'koho).[35] It is nonetheless reasonable to assume that the political conflicts among Ruthenian cultural projects did not slow the development of scholarly institutions. The Enlightenment society and Kačkovs'kyj Society frequently cooperated since their anti-Polish sentiment and the issue of cultural demarcation of Ruthenians from Poles clearly outweighed any internal divisions. Both were also instrumental in building a larger Ruthenian-speaking public, which would later benefit as the main recipient of Ukrainian scholarship.

I turn later to the question of scholars' patriotic engagement in the process of cultural boundary work, but certain characteristics of Ruthenian arguments from around the turn of the century require more careful analysis here. As noted earlier, two main arguments were commonly mobilized for and against language change in the empire: from the viewpoint of cultural dominance, instruction in a national language could be allowed only if that language was sufficiently developed, whereas from the viewpoint of a national culture, only instruction in the national language would allow a national culture to develop. The Ruthenian (and also Czech) arguments followed the latter, arguing that a national university would not be the result of cultural development but rather a means to achieve it. Ruthenian professors stated, for example, in an open memorandum in 1907 that a Ruthenian university could "bring the conditions that favor the peaceful development of science and further cultural development of our nation."[36] Moreover, the press saw it almost as a panacea to cure all the problems Ruthenians were facing in Galicia. In 1907 the daily *Svoboda* (Freedom) argued that with a Ruthenian university "economic development will be easier, and Moscowphilism will melt like wax in fire. The university will be the final aim and center of the political struggle for the independence of the nation. From the university the great voice of the nation will resound."[37]

The most prominent proponent of Ukrainian nationalism, Mychajlo Hruševs'kyj, from 1894 a professor of general history with special consideration to eastern Europe at the University of L'viv, discussed establishing

Ukraine as an autonomous cultural nation both in his historiography and in his popular writings. In his vision of cultural separatism, science/scholarship/education—наука (*nauka*)—had the aim of not only demonstrating cultural strength but also increasing the self-awareness of the Ukrainian population in Galicia and Ukraine: to use his own metaphor, it would help in the process of renouncing "the culture of the knife."[38] The triple meaning of наука, encompassing science, scholarship, and education, is clearly evident, but наука here also evokes culture and civilization and is a synonym of progress, both as an aim and as a means:

> One of the main questions regarding cultural language and the fruition of national life is the question of academic education in this language. Until a language finds entrance to higher education institutions, until it is a language of university or other academic lectures, until it is a tool of scientific work in lectures and books, a nation [народність] that speaks this language will feel as if it were a "low-grade," culturally handicapped nation. It will receive from all a suspicious look, supposing that they consider it neither a cultural nation, nor its language as a cultural language. Academic, university science in one's [own] language attests culturality; it gives a stamp of cultural entitlement to a given nation, in the eyes of contemporary man. Independent of the size of the nation, or the dimensions of its political, economic, and cultural, practical and intellectual talents, the nation considers itself then a cultural nation, and senses the moral right to request such attention from other [nations]—that she will be respected as a cultural nation, culturally equal with the other nations. Hence, we see that all nations that appeared so far, or are just coming to their national rights or to a reputation as a cultural nation, struggle for an independent academia [вищу школу], with lectures in their language, and when that is not possible, then at least lectures on several subjects in their language at a university.[39]

In the conflict over the University of L'viv, two claims turn up repeatedly. The Polish side claimed that the freedom of learning and the possibility of habilitation had given the Ruthenians concessions that they had not taken advantage of because of a lack of qualified scholars.[40] After 1900 this argument, expressed vehemently in the brochures of Dietl and Helcel I discussed above, took a more nationalistic turn, in which Ruthenian development in general was negated. The lack of acknowledged scholars was seen as evidence of the unviability of Ruthenian culture; however, at the same time, the university was a political arena in which professors obstructed Ruthenian claims, arguing that the laws were on their side. The ban on Ruthenian as a language of university affairs (*Geschäftssprache*), the rejection of enrollment certificates (*Inskriptionsscheine*) in Ruthenian, the opposition to new Ruthenian chairs and habilitations, and, finally, the problematic participation of several professors in the right-wing nationalistic National Democratic Party were widely commented on in the Ruthenian press, and this led to a series of violent conflicts. Thus, the Polish argument of freedom was confronted by a Ruthenian claim that the Polish (nationalistic and chauvinistic) majority restricted access to legally prescribed privileges, thus hindering Ruthenian cultural development. In many instances, Poles were presented as imperialists speaking with a forked tongue: on the one hand, criticizing Prussia for blocking Polish in the Province of Posen (Provinz Posen, Prowincja Poznańska) and, on the other, hindering Ruthenians' demands for equal opportunities.[41]

Emancipation and Dependency: Doubling Bohemia

The structure of the arguments in the Czech-German discussion on university education has common traits with the Polish-Ruthenian case. Throughout the nineteenth century, Czech nationalists (*patriots* in the parlance of the day) strove to emancipate themselves from German language and culture. Jan Evangelista Purkyně put the feelings of many Czech activists toward German succinctly, addressing in 1862 in Karlovy Vary/Karlsbad the participants of the thirty-seventh Congress of German Natural Scientists and Physicians with the words quoted at the beginning of this chapter. Notably coinciding with the end of neoabsolutism, the claim that Czech and German should be treated equally returned after a hiatus in the 1850s.

For Czech scholars, the 1860s did not lead to many changes, however. Underrepresented at the university in Prague, Czech scholars were grouped at other scholarly institutions, most notably the Museum of the Czech Kingdom (already then known in Czech as Národní muzeum, that is, the National Museum), the Prague Archives, and, especially, the Czech technical academy. The latter was created through the division of the Prague Polytechnic in 1869 (which thus preceded the split of the university by more than a decade).[42] A number of Czech scholars also moved to universities abroad, thus becoming vehicles of scholarly transfer in the humanities and mathematics (Croatia, Bulgaria) and medicine (the Russian Empire).[43] In fact, more scholars identifying with the Czech project had chairs abroad than in the empire. Most of them returned in 1882, forming the basis for the faculties of the Czech Charles-Ferdinand University, although some decided to continue their careers abroad.

The dissolution of Bohemia into Czech and German public spheres was a gradual process, and most Czech scholars saw steady but inevitable emancipation as the guarantee of progress. Even nationally oriented scholars like Purkyně did not plead for an abrupt division but opted for the cohabitation of languages within scholarly institutions as the ideal state.[44] The issue at stake was how to achieve this cohabitation and how to strengthen a language that by the 1860s was scarcely being used in academic matters.

To guarantee the development of scholarship in Czech, several organizations were established; among the first were the Union of Czech Mathematicians (Jednota českých matematiků), established in 1862 as the Society for Lectures on Mathematics and Physics (Spolek pro volné přednášky z mathematiky a fysiky); the Society of Czech Chemists (Spolek chemiků českých), established in 1872; and the Society of Czech Physicians (Spolek lékařů českých), established in 1862. As nationalist institutions, these societies published Czech-language journals, adding to existing German-language revues.[45] These developments highlighted, or made visible, the division between the two linguistically codified scientific landscapes as well as adding to the linguistic division through the conscious choice to nationalize their proceedings and publications. Bohumil Eiselt, for example, wrote only for the main Czech-language medical journal, Časopis lékařův českých (Journal of Czech physicians), after its establishment, although he had previously published eagerly in the German-language journals of the Prague faculty. He was also responsible for making this journal purely Czech, translating a great number of articles that had been sent to him in German for the journal.[46]

Strikingly, most of these organizations began as Bohemian societies and underwent a process of nationalization in a few years. The Society of Czech Physicians included in its early years a broad range of Bohemian scholars; however, it conducted its activities in Czech, and its later development led toward cultural exclusivity. Also, the Union of Czech Mathematicians developed from a multicultural to a linguistically monolithic organization.

Initially, most of its lectures were conducted in German. However, the number of Czech lectures gradually rose, and within five years, with the exception of several prominent scholars who lectured in German, lectures were held almost exclusively in Czech. Curiously, the trajectories of the founders' careers reflect an early Bohemian dualism. Those who entered academic careers taught across the empire, in institutions with various languages of teaching. Some later published only in German, remaining, however, members of the Union of Czech Mathematicians and thus symbolically aligned with Czech scholarship.[47]

An analogous step to the creation of bilingual Bohemian scholarly organizations to strengthen the Czech language were proposals to provide legal guarantees for Czech lectures at the university. Some politicians already wanted a separate institution in the early 1860s, but they constituted a minority.[48] Student petitions from this time argued for a few chosen lectures in Czech and did not favor a complete division.[49] The Bohemian Diet preferred this proposal, although German representatives of the university diligently reminded them that habilitation was open to scholars of all nationalities; the government, however, rejected the proposal.[50] In subsequent years, Czech politicians and scholars several times proposed making the university bilingual (*utraqustisch*),[51] but German Bohemians, who saw the university as a historical monument of German culture, fiercely opposed this.[52]

As in other discussions about language change, arguments about natural rights to education in one's own language and the role of the university for Habsburg subjects stood at the forefront. The petition of Czech medical doctors in 1872 summarized their political claims: "The Bohemian Nation has an entitlement to a Bohemian university not only through natural law, but also because of its intellectual development and education."[53] Nationalists argued that a Czech university would be epoch making for Czech culture[54] and would bring peaceful coexistence to Bohemia.[55] Demands for a second Czech university in Brno or Olomouc repeated similar arguments: "The second university would bring more freedom for the students and, to a certain extent, also for professors, [and] would accelerate and strengthen scientific [*vědecký*] development";[56] it was generally hailed as a "cultural necessity" (*Kulturnotwendigkeit*) for Czechs.[57]

Even in 1880, however, Czech scholars wondered whether the early opening of a separate Czech university would be premature and do Czech culture more harm than good. And if the political situation made it necessary, they argued, it should happen no sooner than in ten to twenty years, owing to a lack of qualified scholars.[58]

In contrast to Czech claims that they were ready for emancipation, German discussions of the time clearly evidenced the rhetoric of dependency and cultural underdevelopment. While German Bohemians favored the idea of Czech-language chairs, they stressed incessantly that this should come from Czechs' cultural work and not through political machinations. Official writings from the faculties before the division of the university into Czech and German institutions also expressed this.[59] A plethora of writings from the German professoriat and German cultural activists argued that there were always legal ways to achieve habilitation. Their aim was clearly to underscore the picture of the University of Prague as liberal and open to scholars of all languages. At the same time, such writings confirmed German dominance over the university, reaffirming that Czechs were thus far not represented in this institution.

Other memoranda showed hegemonic stereotypes even more clearly. A memorandum by the German professors of the medical and philosophical faculties in 1879 argued that accepting Czech as a language of instruction would show favoritism toward nationalist thinking rather than science.[60] The professors believed that Czechs would lose more than they gained through such a change, not only because most scholarly works were published in German, but also because most Czech students could understand German, while only a few German students could understand Czech.

Unsurprisingly, according to a memorandum published by Czech scholars in response to the German one, it was precisely the German-speaking professors who were obstructing the development of Czech academic activities. Moreover, their ideal of science was described as a "dead printed letter" ("todtes bedrucktes Papier") that ignored the fact that the scientific betterment of the university could be achieved only through the multiplication and (cultural) diversification of the teaching staff. Finally, the memorandum stressed that science as described in the German memorandum included the nationalistic claim that, as scholars, Germans do not need to read Czech literature, whereas Czechs should read German literature.[61]

The claims for German hegemony also took a more critical tone, especially among non-Bohemians. Leopold Wittelshöfer in his *Wiener Medizinische Wochenschrift* (Viennese medical weekly) was particularly critical of Czech culture and published a series of anti-Czech articles throughout the 1880s and 1890s. Shortly before the inauguration of the Czech university, Wittelshöfer stated that "there could be no doubt on which side the 'stronger lineage' [*das 'stärkere Geschlecht'*] is, and some names, which one hears as the future professors of the Czech medical faculty, appear to us very incredible. There are times in which also the professors are scarce as hen's teeth."[62]

The argument of scientific underdevelopment can be found throughout the century and beyond, but it was not the main thorn in German Bohemian sides. To quote Wittelshöfer once more: "To try to take possession of the oldest *German* university through ruses and through completely unnatural coalitions in the Diet is an assassination attempt on nineteenth-century science and culture, a pillage and destruction of a 500-year-long strenuously acquired intellectual property."[63] With such an accentuation of science and culture, it is quite clear that Wittelshöfer was defending "German" science and its main institution in Bohemia, the Charles-Ferdinand University. This was also a clear claim that politics was endangering *Wissenschaft*, which would otherwise sustain its leading role in Bohemia: "Not that we fear that German science could be dimmed by the Czech one, but she will be put to death through these influences, which originate in lack of knowledge, greed, and quarrelsomeness."[64] In 1882, when the university in Prague was divided into two, the *Wiener Medizinische Wochenschrift* sadly complained that this meant "the end of the German university."[65] It is ironic that this argument, the downfall of the German Charles-Ferdinand University, is present even in the works of scholars critical of German nationalism. I will discuss later how there was not much support for this argument. In fact, it is likely that this was an unconscious continuation of German nationalist discourse—present, for example, in Wittelshöfer's words—rather than a reflection of reality.

The Czechs of the East and the Ruthenians of the West

The similarities and entanglements in the Ruthenian and Czech struggles for cultural independence are well known.[66] Both opposed leading cultures—Polish in Ruthenian cultural areas and German in Czech areas—that controlled the university system, which saw itself as a source of intellectual and cultural well-being. At the same time, adherence to these leading cultures and, to a large extent, common intellectual and cultural roots made emancipation attempts akin to tilting at windmills. Despite rhetorical claims and placatory actions, such as the Galician Diet's subsidy for the society Enlightenment, neither Czech nor Ruthenian nationalists could count on the fulfillment of their demands. The geographic overlap in these national projects, with L'viv being the desired cultural center for both Poles and Ruthenians, and Prague for both Czechs and Germans, as well as the statistically proven national heterogeneity and the national pasts of both cities and institutions, confirmed by historical studies, made political influence crucial in decisions on the legitimacy of the wishes expressed.

Language change at both universities was inextricably linked to securing political stability. Since the autonomous universities were now under the control of a majority of scholars identifying with one national project, only political initiatives made a national balance possible, through acts establishing professorships and chairs. As noted before, the Ministry of Education was responsible for maintaining the Ruthenian chairs in L'viv when German was the language of teaching. Two other significant positions—a chair for eastern European history and a second chair of Ruthenian language and literature—came into being out of political expediency, the first as the outcome of the Polish-Ruthenian Agreement of 1890, the second at the initiative of the governor of Galicia Kazimierz Badeni (1888–95), against fierce opposition from the university.[67] Similarly, in Prague the most significant gain for Czechs between 1850 and 1882 (when the university was divided) came in 1871. At this time, Minister of Education and Religion Jireček succeeded in appointing professors of physics, zoology, botany, and mineralogy who would teach in Czech. These nominations proceeded without consultation with the philosophical faculty, since, as the minister stated, the professors in the subjects in question at the Charles-Ferdinand University did not speak Czech and were thus unable to evaluate the writings of those he proposed.[68]

Also, the arguments for and against creating new universities showed a number of similarities. The objective trinity—law, history, and statistics—was mobilized by all the parties. Each of them utilized the "facts" of the existing legal order and "just" historical claims to the building and name, as well as statistical data, to support their own claims. Thus, according to the Ruthenians, the University of L'viv had been established in 1784 as a provincial (i.e., nationally neutral) institution; the Poles, in contrast, claimed that it was founded in 1661 by the "Polish" king Jan II Casimir.[69] While neither Czechs nor Germans questioned that the University of Prague had been established in 1348, they fiercely debated the identity and aims of its founder, Charles IV. It remained highly controversial whether Karl IV founded the

Latin university to create "a center for German scholarliness in Prague" or whether Karel IV was motivated by a love of Czech literature, "which was nearest to his heart."[70] The impossibility of deciding whom the university belonged to finally led to the division of the Charles-Ferdinand University. Both universities created in this way were legal successors of the Charles-Ferdinand University and retained its name, with the addition of "Czech" or "German."[71] In the twentieth century, this decision led to further disputes. In 1920 the famous Mareš Law (Lex Mareš) stated that the Czech university was the only legal successor of the ancient Charles-Ferdinand University. In 1934, when the German University in Prague refused to hand over the insignia (the symbol of historical continuity) to the Czech University, street fights called the Insigniáda (the fight over insignia) broke out.[72]

Further, statistics proved prone to different readings. Discussing Ruthenian scholarship in the 1860s, Dietl criticized that official statistics equated religion and nationality, and commented sarcastically on the rapid growth in the number of Ruthenian students in 1856–57, stating that "what was in 1856 still a Pole remade itself in 1857—or rather was remade."[73] In the following years, Czechs and Ruthenians used census statistics to support their rights to have new institutions of higher education.[74] The counterargument, used by supporters of dominant groups, derived from the statistics on students attending gymnasia or on the nationality of university students, which in their view confirmed the cultural inequality.[75] This was a double-edged sword: for German nationalist statisticians in 1913, who compared the numbers of students with the provinces' contributions to the state budget, the same statistics showed that "the non-German intelligentsia was nursed at the cost of Germans."[76]

In the end, neither a Ruthenian university nor a second Czech one was created, the only concession in Cisleithania being the Alma Mater Francisco Josephina Czernovicensia,[77] established in 1875 in a city whose name, if one takes the statistics seriously, should be written טשערנאָוויץ.[78] To illustrate the mythical (and mythologized) multiculturalism of Chernivtsi: the university, with German as the medium of instruction, was hailed as an oasis of civilization and a German outpost in Slavic "Half-Asia,"[79] a Ruthenian refuge from the Polonization of the University of L'viv,[80] and the only university for the Romanian minority in Bukovina. The Greek Orthodox theological faculty was placed in the residence of the Greek Orthodox metropolitan of Bukovina, a masterpiece built by the Czech architect Josef Hlávka, a prominent philanthropist, whose support was decisive in the establishment

FIGURE 4 Residence of the metropolitan of the Greek-Orthodox Church of the Bukovina, now the main building of Chernivtsi University. (Private collection. Author unknown.)

of the Franz Joseph Czech Academy for Science, Literature and Arts (Česká akademie císaře Františka Josefa pro vědy, slovesnost a umění).[81] Habsburg cultural eclecticism had peaked (see figure 4).

A German Outpost in the East: Chernivtsi

Given the multiplicity of languages spoken and heard in Chernivtsi, the choice of the language of instruction was central in petitions. The person behind most of them, Constantin Tomaszczuk, was predestined to represent the cultural variety of the city. Born of parents with a similar Bukovinian cultural background but (according to the secondary literature) different national allegiances, Tomaszczuk stylized the planned university as an oasis of different nationalities held together by one shared language, German. His plans insisted that only "common education" (*gemeinsamer Bildungsgang*) could create the "political nationality of Austrianness" (*politische Nationalität des Österreicherthums*). This hinted at the direction that education should take: "German science has the claim of universality. And only because German education has universal standing, the non-German sons of Bukovina strive for a German university."[82] This argument for the universality of German fell on fertile ground, especially after the linguistic changes in Galicia. Tobias Wildauer, the speaker of the parliamentary budgetary commission on this issue, argued that after the Galician universities "lost their universal significance and took on the character of camp establishments . . . the whole widely stretched East of the Empire lacks a universally accessible site for fostering science."[83] The minister of education and religion at that time, Karl Stremayr, who not only supported the project but also considered himself one of its driving forces, similarly saw the Austrian mission as bringing culture to the East.[84] In a petition to Franz Joseph, he stressed once more the importance of German *Bildung* in the linguistically mixed regions, discussing, among other locations, Olomouc, Brno, Opava, and Bielsko/Bielitz/Bílsko).[85]

Stremayr stressed that while all these cities would profit culturally from a university, Chernivtsi had one particular asset: a university in this city would be an instrument of foreign policy. According to him, it would profit Romanians, both those living in Bukovina and those from abroad. Since the 1860s the University of Iaşi (Romania), the nearest university to Chernivtsi, had actively attracted the Romanian-speaking population of the region. Thus, with the establishment of the university in Bukovina, "especially the Romanians of neighboring countries will be pulled once more strongly toward German 'Bildung', and thus a step will be taken toward the retrieval of the historical Austrian influence on this nation."[86] One should bear in mind that at this time Romania was still a province of the Ottoman Empire despite striving for independence and had a pro-Prussian *Domnitor* (hereditary ruler), Carol I (Karl von Hohenzollern-Sigmaringen); the political implications of asserting this "cultural significance" should not be underestimated.

In the end, German was made the language of instruction and of administrative affairs (except in several subjects at the Greek Orthodox faculty). Indeed, notwithstanding the presence of peoples of many allegiances in the assembly hall, the speakers at the opening ceremony clearly accentuated the superiority of German culture and the German spirit.[87] While other languages were also represented within the university's walls, this was neither initially planned nor achieved in large numbers. It was only thanks to the petitions of Ruthenian and Romanian deputies that the ministry agreed to create special chairs for both languages and literatures. The Romanian chair took, however, the place of the chair of "oriental languages," which in many cases meant Hebrew.[88] The Jewish population, according to statistics the most numerous group in the region, thus was not represented among the university professorship. The existence of different chairs for language and literature facilitated the later creation of various national organizations, in which intellectuals played an important role. The growth of associations such as the Romanian Arboroasa (The Woodland), the Ruthenian Січ (Sich), the German Verein der christlichen Deutschen (Society of Christian Germans), and the Jewish German-speaking Hasmonäa meant, on the one hand, nationalist/religious mobilization across imagined boundaries but, on the other, the beginning of modern nationalist movements in Bukovina. Ruthenian and Romanian professors were active in the creation of these organizations and attracted nationalist students as well as German ones; the Jewish organizations were linked more to the former L'viv Privatdozent Lazar Elias Igel (at the time the chief rabbi of Bukovina).[89] Chernivtsi was indeed an appealing place for professors to train as public intellectuals.

Not every group welcomed German as the medium of instruction. Since the university tried never to favor any national group, nationalist activists increasingly regarded it as a foreign body and a source of German nationalism. It was, for example, the only university that rejected the Ruthenian students who left the University of L'viv in 1901 because of a yearlong Ruthenian boycott of the Eastern Galician university.[90] But opposition outside of the province was also active: the anti-Semitic press in Vienna bemoaned the creation of a university in a far corner of the empire where most of the adherents of German culture were Jews.[91]

O Trieste, o nulla! The Italian University Question

Although the Slav question remained the most important national issue in the late Habsburg Empire, western Cisleithania did not remain immune to cultural tensions. While in Galicia the "Tyroleans of the East" struggled for their university, in Tyrol German nationalists imagined the *Welschtiroler* (Italians, or "Welsch-Tyroleans") as Slavs who wanted to challenge German cultural boundaries in the province.[92]

After the cities of Pavia and Padua freed themselves from the Habsburg Empire, Italian-speaking Habsburg citizens could study only at the University of Innsbruck. In particular, the importance of Italian legal studies was discussed throughout the nineteenth century; serious proposals for the creation of a law academy or faculty in Trieste remained unresolved (its creation was decided shortly before World War I but not carried out).[93] Since the cities in question comprised a variety of cultural groups, the ministry had to ensure that no one cultural group opposed the creation of a university for any other cultural group. Slovenes were the main challenge to the establishment of an Italian institution in Trieste, whereas Czech groups contested the location of such an institution in Vienna, otherwise an elegant solution that would avoid clashes with Tyrolean and Istrian nationalists.[94]

In Innsbruck there was also debate about the languages in which students should be taught and examined. Since 1867 the civil service in Tyrol had been multilingual, so this issue concerned the law faculty the most. Before 1864 students could take Italian *rigorosa* (oral exams) at several law faculties across the country, and the University of Innsbruck offered several Italian lectures. Although both the Tyrolean Diet and the university claimed the equality of the Italian language at the university (but without a fifty-fifty division of chairs), the number of Italian lectures gradually diminished, and the political atmosphere around them grew tense.

In the 1860s the creation of parallel chairs in Italian at the law faculty resulted in projects proposing the university's reorganization, with additional rights for the Italian language—but only to such an extent as to "not imperil the unity of the German faculty [and to] exclude the lame incubus of bilingualism [*Utraquisierung*],"[95] as German-speaking professors argued. In contrast, in the 1870s the ministry ignored demands by the medical faculty and the Diet for a continuation of bilingual instruction in midwifery in Innsbruck.[96] The final straw was the habilitation of Francesco Menestrina in Austrian civil law in the Italian language in 1901. This realized what had legally been possible for decades, but as a direct consequence, both the German nationalist professors at the university and the influential Tyrolean *Burschenschaften* (student fraternities) raised the alarm. Not only was cultural conflict within the province a problem, but so was the possibility of strengthening pro-Risorgimento Italian activists. The intensity of Tyrolean nationalization could be perceived even at the level of nationally indifferent groups such as the Ladinians, whose language became a cause of disagreement in the later nineteenth century: whether it was distinct from Italian or a dialect of that language.[97]

In the end, serious clashes among students and the interpellations of strong German Catholic parliamentarians led in 1904 to the withdrawal of all privileges for the Italian language and the conversion of its chair into a readership position.[98] Since the government was legally obliged to guarantee the possibility of legal studies in Italian, especially doctoral exams in juridical disciplines, a law faculty was supposed to be opened in Vienna. However, this was also hindered by protests by German nationalists.[99] Finally, the "Trieste or nothing" position bore fruit. However, while both the "Tyroleans of the East" and the "Welsch-Tyroleans" had national universities promised (and budgeted) for the late 1910s, neither of these projects was realized owing to the outbreak of World War I.

On the Road to Autonomy: Liberalizing Academic Policy

Language policy was, of course, just one of many policies that changed in the 1860s, but its impact on both the spatial policy in the empire and the way science and scholarship were perceived is hard to underestimate. This change went hand in hand with other measures strengthening the autonomy of both universities and provinces. However, for the ministry, the empire was still one academic space, and the measures enacted for one university mattered for the others, notwithstanding the language of instruction. Now I want to turn to an analysis of how ministerial policy changed within the late nineteenth century and how the ministry dealt with the constantly changing university system—changing not only in matters of language but also in hierarchies and disciplines.

The increasing academic autonomy included in the first place matters related to habilitation and the conditions of entry into academic environments. Here, the ministry limited its involvement and mediated only in some contentious issues between faculties and scholars striving for habilitation. Similarly, in questions of promotion, the ministry delegated its responsibility to the respective faculties, retaining, however, a decisive voice. While in the 1850s several scholars had applied directly to the ministry (either to the minister or to the responsible *Sektionschef* [section chief]) for an appointment or an increase in rank, later the ministry sent such proposals back for consultation within the faculties; it followed the recommendations of the university and did not impose its own decisions. On a few occasions, politicians intervened directly without consultation, but universities protested fiercely (including involving the press) against any limit on their independence, which was protected by law.[100] Although the dominance of the ministry was still indubitable, the growing importance of nineteenth-century civil society, the press, and multiple political parties in Parliament restrained active interventionism in the field of university policy.

The autonomy of universities also changed their appointment procedures, from being governed by the ministry to being influenced and guided by it. Nominations were based on documents from the faculties or additional expertise (although private contact and audiences with the minister or Sektionschef were not unusual). With the exception of relocations after changes in the language of instruction at universities, the appointment of scholars who were not included in the faculty proposals was rare; if this happened, it was for ideological reasons. In addition, appointments of scholars who were not the first choice in the faculty proposal (terna) were mostly linked to financial issues or the appointment of scholars from abroad. With universities reclaiming their autonomy, appointments that the ministry made in spite of the terna mostly led to conflict and, as the minister of education Sigmund Eybesfeld put it regarding the University of Cracow in 1882, to the degradation of university prestige as well.[101]

Even when the negotiations with all of the proposed scholars failed, or the scholars proved unacceptable for some reason, the ministry asked for a new proposal rather than making a decision on its own. Between 1861 and 1918, in 83 percent (418) of the appointments at the medical and philosophical faculties of German-language universities, a scholar proposed in the first faculty proposal was appointed,[102] and in 58 percent (295) of cases, the scholar appointed was the faculty's first choice. Cases in which the ministry appointed scholars based on a second proposal made up only 6 percent (29). Scholars not listed in the proposal were appointed without consultation with a faculty or deliberately against a faculty's recommendations in just 11 percent (54) of cases. While the latter were more frequent in the first years of the new ministry—appointments of scholars not included in the faculty proposal amounted to 17 percent under Stremayr (1870–79) and 14 percent under his successor, Paul Gautsch (1879–93)—after 1893 the ministry's compliance with faculties' recommendations increased considerably, and most short-term ministers agreed with the universities' proposals. This was followed by a low rate of appointments from outside the faculty proposal under Wilhelm von Hartel (7 percent; 1900–5) and Max Hussarek (10 percent; 1911–17).

Clearly discernible here is also the different value the ministers placed on different subjects. While appointments in medicine, the natural sciences, and mathematics mostly conformed with faculty proposals, the humanities displayed a residue of the tradition of political involvement in disciplinary development. Most appointments from outside the faculty proposal took place in the subjects that had been seen since 1848 as crucial in the process of controlling education: in philosophy, 55 percent of the appointees in 1861–1918 had not been included in the faculty proposal; in history, it was 20 percent, and in classical philology, 15 percent.

The new approach to relations between the ministry and the universities meant a turn toward participative politics in appointments, which granted more influence to faculties and scholars. The realization of Exner's "Entwurf" went even further than initially proposed. Not only were the highest officials in the ministry, the Sektionschefs, appointed by the universities and professors themselves, but the ministry was also successively supported by deputies from Galician universities and Czech-speaking scholars. This institutionalized a consultative agency in university matters, making the Sektionschefs primarily responsible for conducting the appointment procedures in the ministry.

This change is even more striking when one considers the changes affecting the German and Russian Empires during the same period. In Prussia, Friedrich Althoff tightly controlled the nominations of university staff through political maneuvering, guided by a network of informants, similar to Thun-Hohenstein's methods a few decades earlier.[103] In Russia, ministers were constantly trying to meddle in university affairs, and this clearly intensified over the nineteenth century, although the precise effects of all this have yet to be examined in detail.[104] It seems that the Habsburg Empire was swimming against the current in university matters, clearly allowing universities more autonomy than its neighbors did.

That said, the division of labor and the influence of individuals on the final proposals are hard to determine, since ministers also had formal informants within the universities.[105] Correspondence could have been directed through one of the Sektionschefs or the minister himself, or they could have held private meetings, speeding up the appointment procedures and clarifying the content of the proposals; however, the records of such meetings were not preserved, unless described in letters, articles, or memoirs. The sources used here do not rule out that there was a Habsburg equivalent to Althoff who pulled the strings in the ministry but was not visible in the official documents, although this would seem rather improbable.

University Autonomy and Its Enemies: The Road of the Terna

With the abandonment of the Vormärz practice of standardized open contests (*Conkursverfahren*), appointment procedures usually took several months, with several steps between the formation of the proposal commission and Franz Joseph's signature. Still, for several years after 1848 faculties turned to open contests in order to prepare proposals, which suggests that lesser-known scholars would also have had the opportunity to be included in the proposal; reliance on the faculties' own information would have reduced such scholars' chances.[106] In later years, however, this practice was abandoned, and in 1875 Stremayr finally explicitly forbade holding contests for the chair of geography in L'viv and requested that the regulations on faculty proposals be followed, emphasizing faculty autonomy regarding the way in which they chose scholars for the terna.[107]

The procedures for nominations were strictly regulated, leading to dissent not only within commissions but also among intermediaries between the commission and the ministry. In the first place, the faculty (in which full professors were always in the absolute majority, while the Privatdozenten were represented by only two scholars) chose representatives to form a commission, which then prepared the proposal. A commission typically consisted of three full professors in the discipline in question and/or neighboring disciplines, and this directed the process in a particular disciplinary direction from the beginning. Although this method confined the choice to scholars known to the commission members, the faculties strove to ensure variety by advertising new positions in the press and through their own personal networks. In many cases, as soon as a position was made public, or even as soon as the death of the holder of a chair was announced, scholars directed letters to the faculty proposing themselves or their students for the position.[108] While many letters found their way to the commissions, it is imaginable that most of this correspondence did not, making it impossible to trace any patterns.[109]

Once the commission had prepared a proposal, the faculty voted on its content. In crucial cases, experts in the field were asked for their opinions;

these experts normally included professors from Vienna and Habsburg scholars teaching in the German Empire. In particular, smaller universities resorted to these means; not only did they generally lack specialists who could reliably judge the abilities of candidates, but outside opinions also offered symbolic support for the candidates they were proposing.[110]

Finally, the faculty could vote either on the proposal or, in just a few cases, on each of the proposed scholars, which gave all its members the ability to alter the shape of the proposal. A majority opinion or possibly a minority opinion (*Minoritätsvotum*) could include completely different scholars, or the same scholars in a different order; in one case, a scholar proposed *primo loco* (in the first place) was even proposed by the minority to be the only scholar included (a so-called *unico loco* proposal).[111] Each professor, whether on the commission or not, could also propose his own *votum separatum* (separate opinion), which the dean had to forward to the ministry with his comments on the division of the votes in the faculty. Deans also had the freedom to include their own opinions, presented in the form of recommendations, although they rarely used this option.[112]

Before the proposal reached the minister of education, the provincial government also weighed in, in most cases simply by forwarding the proposal with additional reports on the moral behavior of the candidates but, on occasion, providing decisive judgments. In some cases in the German-language universities—especially if the chair was linked to a function in which the provincial government was included, mostly in medicine (e.g., the directorship of the psychiatric hospital)—the governor's opinion on the proposed appointment was also included. The provincial government thus influenced various appointments, such as that for a professor of psychiatry in Graz,[113] or the establishment of a chair in the history and theory of music in Prague, where the provincial government argued that the records of the Prague Conservatory needed supervision.[114]

The dual function of professors as academic and provincial officials could also be detrimental. For example, Ludwig Kleinwächter's conflict with the provincial government following a scandal over the Tyrolean Provincial Birth and Foundlings Institute (Tiroler Landes Gebär- und Findelanstalt) in Innsbruck caused his dismissal from the university.[115] The scandal was probably promoted by the Catholic Church, since he was a pro-choice practitioner and a known theoretician of abortion. Disciplinary procedures also led to Kleinwächter not being taken into consideration for appointments at other universities later on.[116]

The provincial government, however, not only offered assessments of the faculty's candidates but could also directly influence appointment procedures within faculties. In Innsbruck, for example, conservative Catholic politicians, particularly the provincial governor Theodor Kathrein (governor 1904–16), strove to influence the faculty to promote the scholars they preferred. They achieved, among other things, the appointment of conservative Catholic historians.[117]

Unlike at German-language universities, governors took a lively interest in the appointments at both Bohemian and, in particular, Galician universities. Both Galician and Bohemian governors created special commissions for assessing nominations for L'viv in 1871 and the Prague medical faculty in 1882, arguing that the current faculties were incapable of preparing proposals owing to their linguistic incompetence.[118] While Bohemian governors carefully read and commented on the acts prepared by the faculties but avoided direct involvement, Galician governors were involved more often. This ranged from establishing an extra commission in case the university was unable to find suitable specialists[119] to giving decisive votes in cases where faculties were divided. Some of these decisions were indeed controversial. In 1906 the Galician governor, Andrzej Potocki, interfered in a nomination for the professorship of Polish history when the majority of the faculty proposed Szymon Askenazy, a Jewish Polish historian of the early nineteenth century. Potocki supported the candidate of the minority, who was more convenient because of not only his religious denomination but also Askenazy's ideological views. Askenazy argued for an active struggle for independence, in contrast to the mainstream view of loyalty to the Habsburgs, and criticized the dominant but pessimistic view of the Polish Lithuanian past held by the Galician Cracow school of history. But even here Potocki ensured that the ministry awarded the well-respected historian the chair of modern history.[120] The ministry and the governors most often became involved in decisions in the humanities, which remained an important element of symbolic policies in the provinces.

Most proposals were prepared with the knowledge that the scholars named were willing to join the faculty. Prospective candidates were also informed of the financial benefits and the facilities available. This was accompanied in some cases by a possible visit to the university so the prospective candidate could judge the conditions at the institute. Scholars' demands, including renovation plans and the costs (or proposed expenses) of acquiring the necessary books, were forwarded with the faculty proposal, while in the Ministry of Education direct negotiations were conducted with both the proposed scholars and the Ministry of Finance, requiring a careful financial balancing act.

The inclusion of the Ministry of Finance in the decision-making process was not merely symbolic but rather allowed the finance minister a direct route to reject candidates. The list of foreign scholars not appointed for financial reasons is quite long and includes well-qualified candidates and even celebrities.[121] In such cases, the ministry preferred younger, and cheaper, Habsburg scholars, even if the faculties opposed them as detrimental to the quality of the faculty.

The Ministry of Finance could also influence whether a scholar would be granted an associate or a full professorship. The complications are visible in the appointment of Rudolf Brotanek as an associate professor of English philology at the German University in Prague. While the faculty proposed two scholars from abroad as the top candidates, the ministry decided on the third-choice Brotanek because

> the . . . foreigners would with high likelihood expect instant appointment to full professor; however, as highlighted in the subservient submission with respect to [Alois] Pogatscher's appointment to Graz,[122] in the refilling of the vacant chair of English philology only an associate professor should be appointed, owing to the necessary savings from the appointment of [Karl] Luick to Vienna,[123] on which the minister of finance made dependent at that time the second full professorship at . . . the University of Vienna.[124]

Although the direct exchange of information between the two ministries is hardly visible to historians' eyes—most often this was hidden behind ominous formulations such as "mit Einvernehmen" (in agreement) and "im kurzen Wege" (meaning brief, internal communication)—financial reasons were the most often cited cause for not adopting a faculty's proposal.

The relationship between universities and the ministry was for a time so unbalanced that the faculties slowly ceased proposing a list of three scholars in every case and began issuing so-called unico loco (i.e., single-candidate) proposals, thus deciding for themselves who should be appointed. Indicative of these power relations is that between 1870 and 1909, out of forty unico loco propositions, all but four led to an appointment.[125] Finally, in 1909, the overuse of this practice led to a conflict between the medical faculty of the German University in Prague and the ministry. The Prague medical faculty proposed unico loco an anatomist from Freiburg, Ernst Gaupp, pointing out that the medical faculty of the University of Vienna had recently appointed another anatomist, Ferdinand Hochstetter, using an unico loco proposal; that is, they had only been able to find one suitable candidate. The ministry, however, rejected the proposal from Prague, stating that the University of Vienna had proposed only Hochstetter because he was "important" for them but that there were many candidates other than Gaupp. In the correspondence that followed, the Prague faculty accused the University of Vienna of using the unico loco too often, and the ministry fiercely defended its position that it was used rarely and only when there were no other qualified candidates in the empire. The ministry inadvertently confirmed the imbalance between the capital city and Bohemia, however, by finally appointing a young scholar from Vienna instead of Gaupp to the medical faculty in Prague.[126] Nevertheless, in most controversial situations, the ministry corresponded with the faculty and either asked for a new proposal, accompanied by a comment explaining why the appointment of a scholar from the previous proposal could not be realized, or asked the faculty to vote on the inclusion of other scholars in the proposal.[127]

A number of private individuals, networks, and institutions might also have influenced appointments in various ways. Chairs connected with other institutions were especially crucial. This was the case for meteorology in Vienna, since the chair was linked to the directorship of the Central Bureau of Meteorology and Terrestrial Magnetism, where the Imperial Academy of Sciences and Arts asserted its rights.[128] The academy's voice was seldom authoritative—it usually only confirmed the faculty's choices—but it still gave the candidates a better standing with respect to the ministry.

The final step in the appointment process remained the privilege of Emperor Franz Joseph, who took advantage of his legal right to refuse his signature on only a few occasions. In most cases when he refused his signature, the emperor asked for all the documents needed and granted his signature afterward. The emperor did not sign the minister's proposal twice in 1872, following the appointment of Stremayr, a member of the German Liberal Party (Deutschliberale Partei), as the minister of religion and education.[129] Both of the rejected nominees were Prussian, and in one of the rejection notices, Franz Joseph criticized the number of professors from Prussia who had recently been appointed, a clear signal for the minister to limit this practice.[130] However, since the universities were increasing in size in this period, appointments from abroad were the only means of ensuring the quality of universities, and Stremayr could not and did not abandon them.

From the faculty proposal to Franz Joseph's signature, many things could change. Below I go into more detail about how agreement on the scholar to be appointed was reached, how his position (full or associate professorship) was determined, and even how his disciplinary designation could affect things. The legal road was complex, and many factors could influence the final outcome. However, since politicians gradually accepted that universities were not a political body any more, they rarely interfered with the nominations. They did so mostly in a few special disciplines that were still considered vital for provincial and state policies.

This depoliticization went hand in hand with another process, namely, the professionalization of university teachers. As I argue below, from the 1870s Habsburg scholars began to have stable careers, beginning with the achievement of habilitation and ending, if they were successful, in a professorship. This, of course, did not mean that a Privatdozent would go all the way up the ladder, but if a scholar wanted to be successful, certain steps at the right time would facilitate this. Professionalization had two serious repercussions. First, even renowned scholars from outside of academia had limited access to professorships if they had not habilitated. Second, professionalization strengthened linguistic boundaries because the system of rewards was bound to the language of publications. Scholars habilitating at a Habsburg university had to apply with a special publication, the *Habilitationsschrift* (habilitation thesis). This was a book in the humanities and a serious research article in the natural sciences and medicine, written in the main teaching language of the institution the scholar intended to habilitate at. While exceptions can be found, this increased the pressure on scholars to choose early on which language they would publish in, which affected their choice of career.

Habilitation between Professionalization and Patronage

With the growing autonomy of the universities, the critical issue for a Habsburg scholar was the conditions of entry into universities, regulated by the laws on habilitation. It is striking that although competition for promotions within a given faculty was certainly fierce, career advancement (including a change of university) was rather a question of mediation, strategic presentation of one's knowledge and, of course, personal connections, although certain factors, such as a scholar's religious denomination, impeded it. In contrast, the rejection of a habilitation by the faculty or the ministry was seen as denying a scholar's academic competence and thus any possibility for a university career. Reactions to such rejections were often very emotional. They could result in a quiet ending to a scholar's career[131] or prolonged confrontations in the press and courts.[132] In problematic cases, a rejection could be contested with appeals to the faculty or directly to the ministry, or even by trying one's luck at a different university, although the latter was rare and undertaken only in cases of obvious personal or political conflict at the first university. In the appeal procedures, both sides often turned to external experts for an assessment.[133] Since the *Ordinarienuniversität* promoted strong teacher-student relations, some professors felt offended by the rejection of a habilitation and took the side of their students.[134]

A "strong tie" in the student-teacher relationship was in fact a prerequisite for habilitation, especially because the social capital within the faculty was mostly concentrated in a few hands, as Pavel Kolář has demonstrated for the historical disciplines.[135] Nonetheless, habilitation involved all of the professors in the faculty, first on the commission and then in the examination and public lectures. Thus, "weak ties" to all professors, or rather the absence of "negative ties"[136] with other scholars, to retain the terminology of network theory, were significant. The habilitation of Władysław Natanson in Graz was, for example, supported by Ludwig Boltzmann; given the latter's uncertain health as well as his possible move to Munich, the young scholar corresponded with the Graz sociologist Ludwik Gumplowicz about choosing the best moment for filing the papers. Natanson failed to answer questions on Kelvin's theory of vortex motion (*Wirbelbewegung*), and both he and Gumplowicz accused the questioner, Heinrich Streintz, of German nationalism and fear of competition.[137]

In this regard, professors were in a privileged position, especially the chairs of seminars and clinics, who controlled the resources a Privatdozent would need in teaching. Conflict with the head of the Institute for Physiology (Physiologisches Institut) in Vienna, Hermann Widerhofer, caused the Privatdozentur of Leopold Unger to be terminated: he had written an article on the misery of the Privatdozenten in Vienna, which provoked a harsh reaction from Widerhofer, who had been directly criticized in the piece. Even though the majority of the faculty stood behind Unger, Widerhofer succeeded in having him dismissed. The young physiologist continued his career, however, habilitating once more as soon as Widerhofer retired in 1901.[138]

Heads of institutes were also legally allowed to refuse a Privatdozent the use of teaching aids, which would have effectively ended a career before it even started. A letter concerning such permission was to be enclosed in every habilitation proposal. Although no instances of such a refusal were noted, this certainly supported the trend of habilitating under one's own teacher, as other professors might oppose younger competitors' access to materials, instruments, and research aids they had gathered, especially if they had assistants striving for a career as well. In one case, the withdrawal of the right to use an institute's facilities led to the exclusion of a scholar from the university, effectively ending his academic career: in 1905 the archaeologist Arthur Mahler was forbidden from using the facilities at the archaeological institute directed by Wilhelm Klein in Prague. The reasons had, as the rector wrote,

> to do neither with the person [of Mahler], nor with his scientific or teaching activity. The reasons [for forbidding Mahler to use facilities of the archaeological institute] are caused by special[139] conditions at the University of Prague, which have been hard or impossible to eliminate. Professor Klein saw it as his duty to ascertain that potential conflicts among students over the question whether a docent of a non-German nationality is acceptable or unacceptable at the German University in Prague are not carried out in the presence of his precious collections.[140]

It is clear in spite of the veiled terminology that Klein's denial of access resulted from the protests and even assaults by German-national students on Mahler, a politically active Zionist intellectual. I return to the influence of street conflicts on universities in more detail in chapter 6; for now it should be clear that professors could end the careers of Privatdozenten if they wished, as Klein obviously did in the case of the unfortunate Mahler.

Most habilitation records are very short and formal and refer to paragraphs of the law in cases of rejection; the reason for rejection was usually the poor quality of the candidate's scientific publications or his lack of suitability for teaching. Seldom are the reasons more thoroughly explained. For example, in the case of the Tyrolean inventor Anton Nagy, his paper on the therapeutic use of a combustion turbine and his wording in the documents moved the referents to conclude that the applicant was not a "mentally normal person."[141]

The dry style prevailing in documents sent to the ministry points to another feature of the habilitation system, which was its gradual professionalization and, hence, the importance of personal connections. Those seeking habilitation were seldom unknown at the university; in most cases they were already active within its walls as assistants or demonstrators.[142] It was also quite common for students (including, although to a lesser extent, Privatdozenten) to move with their teachers to another university. Such moves could transgress both state and cultural boundaries.[143] In Galicia and Bohemia, the search for assistants was in many cases a search for a successor; the older professors would support their assistants in gaining scholarships and developing the necessary contacts.[144] The primary selection was thus made when choosing and promoting graduate students.[145] At the same time, especially at medical faculties, many Privatdozenten moved after graduation, predominantly from the capital city to smaller universities, often in positions as assistants. But contacts with the faculty were clearly also important. Except in one case, all rejected habilitations in the empire were by scholars who were not working as assistants.[146] And even in that case a second try was successful.[147]

At the same time, professionalization lessened the chances of entering academia for scholars who were outsiders. After the 1850s, transitions from a long-term position in a profession not tied to a university (including as a gymnasium teacher) to a university position were rare; scholarly ability demonstrated in specialized publications gained weight over teaching abilities. By 1884 teachers who had not previously held university positions could not be directly appointed as full professors but only as associate professors.[148]

The professionalization of academia can also be seen in the fact that habilitation took place rather swiftly after graduation, that is, after 5.8 years (at an average age of thirty) in philosophical faculties and after 8.7 years (at an average age of thirty-three) in medical faculties. Scholars who habilitated by the average age were promoted two and a half years sooner in both faculties. Those older than the average took three years longer to be promoted; however, for those who habilitated around the ages of thirty-five and thirty-eight, respectively, the period between habilitation and promotion was considerably longer, distorting the statistics. Around 10 percent of habilitations were concluded after the age of forty, substantially longer after these scholars' graduation, raising the question of what motivated older scholars to habilitate. Although some older Privatdozenten succeeded in gaining professorships, most remained in the role of Privatdozent, and it is rather improbable that scholars habilitating at an older age aimed to have an academic career.[149]

The working conditions of young graduates seeking a university career favored speedy habilitation.[150] The number of scholarships was limited, and the payment low, and assistants were bound by some rigorous rules: a maximum of four years in the same position[151] and, at least at some universities, a ban on marrying.[152] Clearly, the time between graduation and habilitation was financially exhausting.[153] This process offered social advancement but favored those whose families had a good financial situation. The sons of civil servants, the urban bourgeoisie, and scholars constituted the majority, however.[154]

There was also a financial aspect to habilitation, making it attractive even for scholars with no intention of entering a university career. Especially for physicians and jurists who had their own practices and lawyer's offices, being a Privatdozent (or carrying any title of *Dozent*) increased their status and thus their income. That such titles were acquired (or even used without formal habilitation) for reasons of prestige was a continual source of criticism.[155] Because in the late nineteenth to early twentieth century roughly 40 percent of Privatdozenten—mostly in practical disciplines—did not have an occupation listed, one can imagine that they had private practices (the other 60 percent were most often chiefs of clinics, assistants, or chief physicians).[156] Further, these Privatdozenten seldom achieved a promotion. However, since the title was rescinded if a scholar was not actively teaching (even though it was not necessary to teach every semester), most Privatdozenten remained in their positions, especially in the capital, thus aggravating the image of an overcrowded first step on the career ladder.

To return now to the factors facilitating the academic careers of Privatdozenten, their function as a reservoir of scholars was reminiscent of the idea of a *Pflanzschule* before 1848. Promoting Privatdozenten was seen not only as supporting local scientific traditions but also as being vital to sustain the function and attractiveness of the Privatdozentur. It is clear that the faculties consciously used these arguments to promote local scholars, especially in more debatable cases.[157] But one also finds evidence in support of local tradition in ministerial notes, where "tradition" conceals the fact that the nomination of a young Privatdozent was simply the cheapest option to fill a professorship.[158]

Medical faculties were especially torn between supporting local scholars and hiring external candidates. Habsburg medical institutions had strong local traditions but also strove to obtain the best possible scholars. They also had to convince the ministry, which valued tradition and finances more than innovation, to appoint a given candidate.[159] At the Viennese medical faculty, for example, of the ninety full professors in 1848–1918, a third spent their academic careers exclusively in the Austrian capital. Another third were educated in Vienna, subsequently held a professorship at another Habsburg university, and were eventually transferred back to the capital. Holding to Habsburg tradition was not achieved without conflict, however. The dominance of a few disciplines went hand in hand with a lack of specialization in others. Although Habsburg medical faculties had, with a few exceptions, the most advanced specializations among the German-language institutions,[160] sometimes they struggled to find appropriate specialists among local scholars.

One sees this conflict most prominently in pediatric medicine. In the second half of the 1880s, the ministry had to resort to nominating Habsburg general physicians instead of specialists (which the empire lacked at the time) for the chairs in Prague and Graz.[161] The prominent pediatrician Hermann Widerhofer protested this measure, claiming that pediatrics was an established and specialized discipline and that the appointment of inexperienced general physicians caused bafflement and "harm[ed] the scientific dignity" of specialized doctors.[162]

One can only speculate about what effect the concentration on locality had on these appointments, since locality was hardly an objective measurement of the quality of the scholars under consideration. But there was a growth in the use of words such as *tradition*, *continuation*, and *student*. This allows one to speak, especially with regard to the ministry, of a strategy that promoted local scholars or, with the same idea of local improvement, of foreign scholars who could help establish a new subject in the empire.

In cases of regional rivalries (Germany vs. Austria, Polish Galicia vs. Austria, Czechs vs. Germans), universities and political institutions gradually rejected the importance of exchange across linguistic boundaries. Even if this was not explicitly expressed, academic autarchy within linguistic subsystems was the aim. The addition of the legal issues of citizenship and national identifications created a kind of hierarchy of foreignness. While for Austrian universities this was, in descending order, "Austrian"—Cisleithanian—Habsburg—German-speaking—others, in Galicia the top positions were reserved for Polish-speaking Galicians and (Habsburg) Silesians, followed by Polish-speaking scholars from Russia and Prussia, other Slavs, and, finally, German-speaking Austrians. These hierarchies were supported by the accentuation of nationality or mother tongue by scholars from multinational regions (especially Bohemia, Galicia, and Moravia, as well as, less often, Transylvania and Carniola), who often included it in their curriculum vitae, frequently adding information about their religious denomination as well.

The importance of a scholar's identification did not end with the habilitation proposal. In the appointment process, the mother tongue, as an indication of nationality, was considered a more important criterion than citizenship. This was true not only in Galicia and Bohemia but also at German-speaking universities with regard to scholars from Transleithania, as the Hungarian part of the empire had separate citizenship from 1867. Although no formal rules were adopted for scholars born in "Greater Hungary," the ministry clearly favored them over scholars from abroad and was also willing to offer them high salaries.[163] Nevertheless, most scholars born in Greater Hungary who worked at Cisleithanian universities in fact had Austrian citizenship; the children of civil servants serving across the empire were accredited (*zuständig*) to their fathers' municipality, and since many civil servants from Cisleithania served in Hungary, a number of their sons were subject to this rule.[164]

Through the focus on locality and its frequent equation with language, legal practices caused Habsburg scholarship to grow apart. But the structure of disciplines, codified and decided on by the ministry, held the different universities together. Once more, the Privatdozenten were the first people whose careers were influenced by the ministerial decisions concerning disciplinary specialization. A glance at ministerial practice shows that the hierarchy, with Vienna as the main university, also had a major impact on disciplinary differentiation across the empire.

Disciplinary Networks

While the ministry restricted itself to affirming habilitations and avoided direct involvement in faculty procedures, it retained the right to decide in cases where contention arose over which discipline/area the habilitation would be awarded for. From 1888 onward, in particular, the rules were imprecise, leaving open the question of the demarcation between a discipline and a subdiscipline. For example, between 1888 and 1892, the Cracow philosophical faculty and ministerial experts debated whether a scholar could be habilitated for the narrow field of the morphology and biology of thallophytes and the biology and morphology of cryptogams or whether it had to be broader, with the final decision in favor of plant anatomy being made only after four years of discussion.[165] More than a decade later, the Viennese medical faculty and the ministry clashed over a habilitation for public medical service with the inclusion of knowledge on inoculation (Öffentliches Sanitätswesen mit Einschluss der Impfkunde). While the faculty regarded it as too narrow a specialty, the ministry decided that this disciplinary designation was indeed correct and should be accepted.[166]

In such instances, the ministry limited itself to questions concerning the designation of the discipline. Even if an external expert disagreed with the faculty's opinion on the quality of the author's publications, the ministry did not follow up, leaving such decisions to the faculties.[167] The question of how to deal with differentiation of knowledge was mostly answered through the addition of a specialization to a more general area. This included disciplinary enlargement (e.g., "philosophy with special consideration of sociology" or "balneology and hygiene of health resorts"), period denotations (especially in literature studies and historiography), and specialization, such as "experimental psychology and methodology of natural sciences." However, more exotic designations were also allowed, such as "infinitesimal calculus and its use for geometry" and hydrobotany.

This acceptance of partial specialization in law and in practice was yet another outcome of the pervasive construction of the university as both a teaching and a research institution. According to the 1888 habilitation law, a Privatdozent could acquire the right to teach (*venia docendi*, henceforth *venia*) only "for the whole discipline, or a larger area of it, which can be regarded as an integrative whole."[168] Moreover, Privatdozenten were allowed to offer lectures and seminars only in the areas covered by their habilitation. As a result, the choice of the disciplinary designation reflected their teaching duties and their potential income from Collegiengelder, rather than their scholarly interests. At the same time, the widening of habilitation areas was also problematic. If a scholar decided on, or was forced to apply for, a broader discipline, he not only had to demonstrate wider knowledge in the habilitation process but subsequently had to cover it in lectures.

This regulation particularly disadvantaged smaller universities and the Privatdozenten teaching there. There, young scholars competed with professors for students to attend their lectures, leading to questions about the division of lectures in order not to lower professorial earnings. This led to the informal practice of awarding habilitations only for disciplines not covered in regular lectures. Privatdozenten could thus either choose a narrow specialization, leading to fewer students and thus less money, or move to a larger university.

The final obstacle to the professionalization of the Privatdozentur was thus financial. Privatdozenten were approved to teach, but their remuneration remained limited to Collegiengelder, in rare cases improved by regular salaries, if proposed by the faculties. This regulation limited young scholars to serving as assistants at the institutes of the university or to being employed and paid externally, unless, of course, their families were well off.[169]

Further, the regulation that Privatdozenten could not work or live far from the city in which they held a position caused particular problems for the philosophical faculties.[170] While in the medical faculties doctors generally had positions in hospitals, which were concentrated in the large cities, or turned to private practice, teachers (the main group from which Privatdozenten were recruited and/or who worked in the philosophical faculties) had much more scattered and unregulated positions. This issue, like many others, was handled differently by different universities and in different cases. While one can find an instance in which three hundred kilometers separated the university and the gymnasium where a Privatdozent taught,[171] usually faculties accepted only teachers from neighboring cities.[172] Obviously, this practice influenced young scholars' careers, leading them both to and from the capital; faculties were also quite accommodating in this regard, not causing problems if Privatdozenten moved owing to professional relocations and allowing simplified procedures for habilitation at the new university.[173]

Teaching was, however, not the only para-university occupation of Privatdozenten. There was great diversity in their positions, which, depending on the discipline, could be linked to different institutions, such as archives for historians, central bureaus, and so on. In fact, for a number of Privatdozenten, the university was not their primary place of work. They linked their teaching with directorships or curatorships at various institutions or taught at technical academies or semi-academic institutions (e.g., the School of Commerce, the School of Industry, and the School of Brewery in Vienna; the Industrial School and Academy of Fine Arts in Galicia; and the Academy of Agriculture in Dubliany/Dublany). The official staff catalogs at the end of the nineteenth century listed only around half of the Privatdozenten at philosophical faculties as lacking an additional occupation, although this source is not particularly reliable.

The occupational structure of universities displays an interesting spatial differentiation. In Vienna, Prague (especially at the Czech University), and

Cracow (but, surprisingly, not in the provincial capital L'viv), Privatdozenten who worked primarily in nonuniversity scholarly institutions outnumbered those who worked as teachers, but at smaller universities this ratio was reversed. Through the curriculum vitae submitted with a habilitation, one can see that a large number of scholars had worked as teachers before their habilitation, and a gradual distinction between pedagogical and scientific specialization is discernible. With the stronger professionalization of the teaching profession, and numerous scientific organizations that granted scholarships on which scholars could live during the prescribed two-year minimum gap between graduation and habilitation, the distinction between academic scholarship and school teaching became more pronounced. Nevertheless, although there were regulations lessening the workload of gymnasium teachers who were also lecturing as Privatdozenten,[174] their precarious situation was the subject of many debates.[175]

The regulation of the habilitation process and professorial appointments brought about a strong unification in the structure of the faculty across the empire, defined by the curriculum. Similarly, habilitations retained disciplinary consistency between 1848 and 1918, with the humanities and the sciences granting the majority of habilitations. After 1848, there were more habilitations in the humanities than in the sciences, except between 1860 and 1869. From 1880 onward, the number of habilitations in the sciences grew, and habilitations in the humanities stagnated. Only from 1900, and only if one includes the biosciences, did habilitations in the sciences outpace those in the humanities. In Galicia and at both universities in Prague, however, the dominance of the humanities over the sciences with respect to habilitations was greater than at other institutions. This had to do with a large number of habilitations in nation-building areas (history, language, literature) and the peculiarities of these universities' location in regions with overlapping nationalities.

Still, there were noticeable differences at the local level. Such local traditions included a preponderance of philology in Vienna, with eighty-two habilitations, constituting 75 percent of all habilitations in this field in the Habsburg Empire (and 21 percent of all habilitations in Vienna). Such concentrations were also possible at provincial institutions: Innsbruck developed a particularly strong school in historiography, led by Julius Ficker; fourteen scholars habilitated in this discipline, accounting for 16 percent of all habilitations in historical disciplines in Cisleithania and 29 percent of all habilitations in Innsbruck. Moreover, this particular Tyrolean cluster of excellence had an immense influence on Habsburg historiography: most of the Innsbruck Privatdozenten were appointed as professors throughout the monarchy.

The hierarchically oriented regulative system had, however, a serious consequence: the symbolic centralization of disciplinary boundaries, largely defined in relation to the central universities in Prague and Vienna. In 1904, as the philosophical faculty of the Czech University in Prague applied to appoint Jindřich Matiegka as the chair of anthropology, the ministry took into consideration that neither such a chair nor such an institute existed in Vienna. Hence, it opposed creating an official chair but granted Matiegka the title of associate professor (but no salary). Then in 1908, shortly after an associate professorship was created in Vienna, Matiegka was granted a paid associate professor position.[176] A similar case occurred in the field of hygiene, for which a chair was established first in Vienna (1875)[177] and then almost ten years later in Cracow, Graz, and Prague.[178]

In the most important disciplines at the medical faculty, the universities in Prague and Vienna were almost always among the three faculties in which new disciplines first appeared and were sanctioned by professorships. Dermatology (together with syphidology) and bacteriology were the only ones where they were not the first (see also appendices 1 and 2). The former, however, already existed under the name of syphidology, and the latter was used to denote habilitation disciplines only in Slavic universities. A similar picture emerges if one considers fields that did not become formally established disciplines but advanced as areas of habilitation: radiology, electrotherapy, and orthopedics.

In philosophical faculties, the situation was more complicated because of the much more flexible designations, but the central universities were again the disciplinary precursors. Only the historical disciplines, with early specialization in Innsbruck, and Slavic historiographies and languages showed a slightly different picture. Surprisingly, a large number of disciplinary pioneers were unsuccessful and ceased to teach after only few years. While some of them were in fields that never really achieved the formal status of an academic discipline, others failed in disciplines that became common academic subjects only a few years later. While it is impossible to say whether this was due to the personalities of these scholars or the conservatism of university structures, it is clear that acceptance of a new discipline was a delicate matter.

The pioneers of academic disciplinary differentiation who did succeed were those who enlarged or changed their designated specialization during their careers. The system of disciplines, which largely defined the conditions of academic advancement, was prescribed in the curricula and viva voce (*Rigorosum*) rules, which were not particularly flexible; the curricula were changed about every twenty years, apart from in medicine, where the curriculum from 1833 was in force until 1872. Although the universities themselves were more or less flexible in the designation of lectures, higher up the ladder the situation became more complicated. While Privatdozenten could teach quite freely within their respective areas, designations of professorships were linked to the possibility of including the subject in the *Rigorosum*, that is, completing the commission and making rules for the exam. Thus, while Privatdozenten were limited more by the possibility of finding students willing to pay them, their road to a professorship went through the ministry, which had to accept the existence of a discipline that other universities could then apply to have. Such enlargements were usually a long-term process stimulated by the appointment of scholars with a high reputation and accompanied by written opinions on the necessity of a new designation or the division of a chair, which resulted from the "development of science" and/or the establishment of such a chair at foreign faculties.[179] The most elaborate act of this kind was a collective petition by the philosophical faculties for a third systemized chair of mathematics in 1907, which not only referred to scientific progress, teaching load, and the growing importance of mathematics as an auxiliary science but also included comparative statistics and a list of professors of mathematics in several European countries.[180]

Structures and Diversities: Coping with the Branching of Knowledge

Growing pressure from universities to increase the number of professors and promote more and more specialized Privatdozenten made the ministry look for ways to amend academic positions without incurring a considerable financial burden. There were two principal modes of diversification: introducing titular full professorships (mostly for associate professors but including several cases of Privatdozenten with the title of full professor) and granting a so-called *ad personam* (by individual appointment) professorship. The latter meant that the scholar was acknowledged as a luminary in his specialty, but the ministry was not willing to grant him a normal tenured position, because that would mean that his position would be filled after him or that other universities, being on equal terms, would argue for such a chair as well. Medieval history, balneology, and comparative anatomy and plant physiology were fields where the ministry accepted habilitations but refused to establish normal professorships.[181] More "exotic" or specialized disciplines—such as entomology, organic chemistry, paleontology, petrography, plant physiology, neurology and neuropathology, and urology, to name only those that were sanctioned and not-infrequent areas of habilitation—were either changed in the appointment process to cover more general areas or added to general disciplines (e.g., "psychiatry and neurology"). Although there was obvious specialization among professors in the same discipline, which was also required during the appointment process and visible in the lectures they taught, this system inhibited rather than promoted specialization, not only restricting the career opportunities of scholars in nonofficial disciplines but also requiring increasingly broader knowledge.

Owing to its large number of parallel chairs, the University of Vienna provided the most possibilities for specialization within its existing structures. These included unofficial specializations, which were, however, clearly taken into account when preparing the proposals for professorships. The most famous is the division of the two Viennese chairs of surgery into one concerned with "small" surgery, the specialty of Johann Heinrich Dumreicher, and one concerned with "large" surgery, the specialization of Theodor Billroth.[182] For smaller universities, though, the possibility of specializing was limited by the teaching load, making faculties seek pedagogues rather than researchers; also, paradoxically, these universities would apply for new chairs not because of student overflow but because of the impossibility of lecturing at a suitable scientific level. This resulted in the growth of personal and institutional infrastructure at the University of Graz, the University of Innsbruck, and the German Charles-Ferdinand University, but at the expense of the University of Cracow, the University of L'viv, and the Czech Charles-Ferdinand University.[183] This was hardly justified by the number of students. For example, in Vienna there were twenty-six students for each professor at the philosophical faculty, while in Innsbruck there were six. Although the statistics seem similar across the universities if one includes the Privatdozenten, smaller universities still had lighter teaching loads (see table 2).

Table 2 Student-professor (S-P) and student-instructor (S-I) ratios at Cisleithanian universities, 1866–1910

		1866	1880	1890	1900	1910
				Medical faculty		
Vienna	S-P	28.6	18.0	32.0	18.8	26.4
	S-I	13.0	9.0	13.9	6.4	6.6
Graz	S-P	15.7	5.6	24.9	12.0	13.2
	S-I	13.3	3.8	16.1	6.8	7.1
Innsbruck	S-P	n/a	6.2	12.7	8.1	9.1
	S-I	n/a	4.7	10.6	6.9	7.7
Prague	S-P	14.3	11.8	n/a	n/a	n/a
	S-I	9.5	6.3	n/a	n/a	n/a
Prague: German	S-P	n/a	n/a	29.8	10.8	11.1
	S-I	n/a	n/a	17.7	6.3	5.5
Prague: Czech	S-P	n/a	n/a	NDA	17.6	18.8
	S-I	n/a	n/a	NDA	11.0	11.3
Cracow	S-P	9.7	12.4	20.8	5.0	17.3
	S-I	7.6	8.2	13.9	2.9	10.8
L'viv	S-P	n/a	n/a	n/a	7.7	15.1
	S-I	n/a	n/a	n/a	4.5	6.0
				Philosophical faculty		
Vienna	S-P	16.2	10.2	6.7	13.1	26.8
	S-I	9.4	6.3	3.2	6.0	10.9
Graz	S-P	4.0	4.5	5.3	3.8	7.5
	S-I	3.3	3.3	3.4	3.0	5.7
Innsbruck	S-P	5.7	4.6	3.2	5.0	5.7
	S-P	4.5	3.7	2.3	3.8	4.4
Prague	S-I	13.1	7.7	n/a	n/a	n/a
	S-P	11.6	5.5	n/a	n/a	n/a
Prague: German	S-I	n/a	n/a	3.2	4.5	10.9
	S-P	n/a	n/a	1.8	2.9	6.7
Prague: Czech	S-I	n/a	9.8[1]	NDA	13.8	20.3
	S-P	n/a	6.3[1]	NDA	8.3	11.1
Cracow	S-I	2.1	3.4	4.7	7.3	19.7
	S-P	NDA	2.5	3.6	5.6	13.5
L'viv	S-I	10.6	NDA	8.6	7.2	23.4
	S-P	10.6	NDA	4.6	4.3	14.0
Chernivtsi	S-I	n/a	3.7	2.4	NDA	8.2
	S-P	n/a	3.7	2.4	NDA	6.7

Sources: Printed lecturer catalogs for individual Cisleithanian universities, including student statistics.

Note: n/a, not applicable; NDA, no data available.

[1]Data is from 1882.

The situation in 1910 clearly shows that Vienna was the preferred place for specialization. Here, for example, geology was divided from paleontology, systematic botany from plant anatomy and physiology, and English philology from English language and literature, while full professorships were established for mineralogy, musicology, pedagogy, and German-language history, in addition to German literature. In 1910 the philosophical faculty at the University of Vienna presented in its lecture catalogs full professors in thirty-eight disciplines and associate professors in twenty-two. Twelve disciplines taught by associate professors were not covered by full professors. In contrast, Graz had only twenty-four full professorships and eleven associate professorships (six of whom taught disciplines not covered by full professors), Innsbruck had seventeen full professorships and ten associate professors (five of whom taught disciplines not covered by full professors), and Cracow had twenty-six full professorships and thirteen associate professors (seven of whom taught disciplines not covered by full professors). Cracow also included agricultural studies.[184]

While most of the disciplinary areas that were different at provincial universities than in Vienna were more general, a few can be considered to be specializations. For example, in Cracow there were associate professors for anthropology, economic history, the history of natural sciences, and experimental psychology and theory of science; in Innsbruck there was a professorship for the history and culture of the ancient Orient. The other divergences in disciplines resulted from local conditions: Italian language and literature in Innsbruck, Slovenian philology in Graz, Ruthenian language and literature in L'viv and Cracow, and böhmische/Česká (Bohemian/Czech) history and Czech language and literature in Prague.

At the formal level, it was almost impossible to rise from under the shadow of Vienna. Considering that most institutional innovation apart from that at the central university took place at universities deregulated through language (and power), the reforms had interesting theoretical implications. While networks of supervision and comparison tightly linked the German-speaking universities, with the University of Vienna seeking to sustain its superiority and centrality, this power structure was less coherent in Galicia or at the Czech University in Prague, where diversification followed different paths. Since institutional and disciplinary innovation was supervised by the ministry, in most cases originating from Vienna and later from other universities according to their respective status (Cracow, Graz, and Prague and, finally, Innsbruck, L'viv, and Chernivtsi), "peripheral" innovations rarely resulted in systemic change, for two reasons. In the first place, institutional innovation was inhibited at smaller German-language universities, which had to follow the capital city. Second, as the flow of information between universities with different languages weakened, the possibility of specialization and disciplinary innovation did not result in a financial burden because other universities did not demand the same concessions.

To put it more theoretically, while "Austrian" universities conformed to the center-periphery models of Michel Foucault, Galician universities and the Czech University in Prague followed the model of Yuri (Juri) M. Lotman.[185] Innovation at the "Lotmanian peripheries" was more common but had no repercussions in the center and hardly translated into systemic innovation. The predominance of a norm-making center, here Vienna, inhibited innovation in the Foucauldian peripheries, that is, those that continued to be closely supervised.

For political reasons, this differentiation took place after centralized power had deteriorated and universities became parts of new networks, intensifying especially after 1918. Foucauldian peripheries were deprived of influence. For instance, Chernivtsi University (Universitatea Regele Carol II din Cernăuți) in Romania was subordinated to the University of Bucharest (Universitatea din București); the German University in Prague, after defending its move to Liberec/Reichenberg, not only gradually lost importance but also switched its orientation from Vienna to Berlin.

In contrast, Lotmanian peripheries were able after 1918 to become central without undergoing serious internal change. The Czech University in Prague was the only university on which the "Czechoslovak" model could be based, and thus it had no competition. In the process of creating universities in Poland, the Habsburg model was chosen from among several models of academic education (e.g., with regard to disciplines, academic grades, organization of universities, and their relations with the state), although not without serious opposition.[186]

While the issue of disciplinary differentiation was a question of power, it was only one of the spatial issues of Cisleithanian universities. This should not, however, be taken as suggesting that the universities were pulling in all directions and the ministry was the only common denominator. To conclude this brief overview of the changes in the liberal period, I want to turn to those issues in which common space was most manifest: legal initiatives.

Retaining Common Space: Legislative Initiatives

The change from a ministry that imposed centralized university policies to a ministry that served as an administrative and supervising body involved many legislative questions. Stremayr had already requested opinions on habilitation procedures and on the admission of women to universities in 1873.[187] The same consultations also happened a few decades later, with a similar request for the opinion of the faculties.[188] Universities also tried to increase their influence, not only proposing improvements to single faculties but also strengthening the symbolic capital of academia as a whole by organizing and preparing joint expert reports, especially on salaries or new chairs. Between 1891 and 1896, an informal commission on the remuneration question, initiated by and based at the German University in Prague, developed a petition to improve remuneration, gathering, among other information, data on the salaries and Collegiengelder of all faculties and organizing meetings of university representatives.[189] In 1907 delegates from all of the universities, led by the philosophical faculty at the University of Cracow, prepared a memorandum on the improvement of mathematical education at universities.[190] In the same manner, Privatdozenten as well as assistants organized collective petitions to support their claims.[191] Interestingly, discussions about such cooperation were widely circulated in the academic and semi-academic press in different languages during the late nineteenth century, confirming that not only universities but also university matters as a whole were understood as matters of state in learned circles.[192]

However, when joint bodies were institutionalized to provide expertise for further policies, linguistic divisions again became visible. In 1898 the universities created a legislative support mechanism called the Academic Conferences (*Hochschulkonferenzen*) for the German Empire and German-Speaking Cisleithania, thirteen years later renamed the Austrian Conference of Rectors (Österreichische Rektorenkonferenz).[193] The organization of universities into networks transgressing the empire's boundaries underscored the dualism between state and culture and the drifting apart of scholarly cultures and networks. Although they maintained common interests and political structures, their separation implied changes in their perception of cultural needs, often exceeding Habsburg boundaries and thus coming into conflict, as the broadly perceived interests of the empire did not always match the needs of a language community.[194] Even discussions about developments within the whole state were more frequently held within the local organizations of teachers of higher education than within Habsburg organizations. The local organizations included university instructors and were clearly determined by linguistic boundaries, both in their legal status and in the language used for publishing.[195] This cemented discussion groups with shared interests even though cooperation in the legal initiatives mentioned above and meetings of representatives were still taking place.

The *Schillerfeier* (Schiller Celebration) of Friday, 11 November 1859, was, for students at the universities, a day of political demonstrations and the reiteration of demands for the abolition of neoabsolutism after the Habsburgs had been defeated in Sardinia; the freedom of student associations was on the agenda. While it was, as the Czech legend says, the last shared rally of Czech and German students in Prague, on the same day the Polish patriot and German-speaking Jew Moritz Rappaport lauded Schiller at the University of L'viv.[196] To this, another Jewish Polish nationalist, Ludwik Gumplowicz, bluntly commented, referring to Rappaport, "He's such a prick!"[197] At the same time, the German nationalist Tobias Wildauer in Innsbruck spoke vividly: "From his [Franz Joseph's] hand the German spirit gained complete freedom across all the parts of the vast Reich. It will march through them and accomplish the mission that the spirit of history so doubtlessly assigned it."[198] The polysemy of "the German poet" in Innsbruck and L'viv, separated by a thousand kilometers (almost the width of the monarchy), can be taken as a symbol of the variety of cultural loyalties and nationalization projects at the time.

The failure of the idea of empire-uniting German Habsburg loyalty is obvious, even if one can find remnants of it in Chernivtsi. German as a symbolically hegemonic language was hardly practicable in an empire in which nationalists had more and more say. Here, Habsburg governments practiced different policies than both the German and Russian Empires, which at the time were strengthening language-led state unification processes and removing the last bits of autonomy that linguistic minorities had cherished until the 1860s. While in the Habsburg Empire the languages of education were proliferating in order to secure subjects' loyalty, in other empires subjects were channeled toward monolingualism to create unity.

Contrary to what some politicians had feared, the language changes did not mean the dissolution of the empire. Slavic scholars—even those with clear-cut definitions of national identity—claimed in their writings that a complete linguistic separation in university education was neither possible nor wise and accentuated practical bilingualism.[199] This was due to the prevailing idea of science as a universal endeavor but was also championed for purely practical reasons. I discuss this further in the chapter on Slavic universities. In fact, knowledge of German was necessary for a university career, and this was an obstacle for scholars from the Russian Empire in Galicia, whose German was mostly deficient. In addition, scholarships were seen as obligatory, and thanks to the central institutions in Vienna, this was the first city of choice. Thus, the empire and its resources remained a vital reference point for scholars at the non-German-language institutions.

Street and university conflicts were major topics in the press at the time and have strongly influenced the historiography of Cisleithanian universities up to the present. These overshadowed the contacts and commonalities between universities. Such commonalities were influential not only at the time but also after 1918, creating, for instance, a common space of disciplinary assignments. The next two chapters examine in more detail the spatial structures that emerged from the liberal-national policies described here. I begin by discussing German-language universities and the career patterns there. As I argue, the careers of scholars there were increasingly directed toward the German Empire and less and less toward the Habsburg Empire, creating a system in which scholars from Slavic universities hardly had a place. But this system was also closing toward the German Empire, owing mostly to large numbers of graduates and staff at the University of Vienna. This in turn affected the role that different German-language Habsburg universities played in scholars' careers.

Chapter 4

German-Language Universities between Austrian and German Space

What should a minister of education [do], when every smaller or bigger province wants to teach in its own language, when he can neither freely command universities nor schools nor teachers' appointments? In all provinces one wants to have only natives at the universities; for [Privat]Dozenten here, there are no aims, no career, because the bridges to Germany were previously dismantled owing to arrogance, and not all can be professors in Vienna. . . . Withal, there is much talent here.

—Theodor Billroth to Wilhelm Lübke, 24 December 1867[1]

The restructuring of Habsburg universities as described in earlier chapters, the centralization of German-speaking academia, and the structural disentanglement of Slavic universities as a result of ministerial ordinances and academic practice went hand in hand with processes of internal specialization, which in turn influenced academic spatial practice. Scholarly mobility within the monarchy—conditioned by internal differentiation and linguistic affinities with neighboring regions or, in the Czech case, by their absence—exemplified the concurrence of the processes of internal specialization and academic spatial practice. Three examples provide interesting insights into how circulation among academic institutions was entangled with infrastructural, political, and cultural factors: (a) German-speaking

Habsburg universities, with numerous institutions and the possibility of exchange with the German Empire; (b) Galician universities, associated with large Polish-speaking communities in Prussia and Russia that lacked a Polish-language university in their home country; and (c) Czech-language universities, backed up by technical academies. The "superstructure" of the monarchy and ministry remained important, even though universities defined themselves as increasingly independent within linguistically defined networks. This coexistence of the space of the state and of the space defined by language affiliation—both spaces that had their own internal differentiations—will be the topic of this chapter and the next.

German-language Habsburg institutions present a particularly interesting case, oscillating between being imperial, Austrian, and German institutions. At different times, the ministers, influenced by the Habsburg and European sociopolitical contexts, favored one space or another. However, this was a more complicated situation than at first appears. During the nineteenth century, Habsburg-Prussian tensions were at an all-time high, but the growth of Habsburg universities and the simultaneous lack of young scholars made transfers from Prussia inevitable. Even an unhappy Franz Joseph could hardly stop them.

When one looks at the statistics, one can find patterns in scholarly careers. These patterns help to question findings based on scholars who had exemplary careers, which have so far dominated the research on Habsburg universities. Mobility is a personal experience but is structured by systemic pull-push factors, such as formal regulations and informal conventions on how to achieve a career. The latter grew in importance in the late Habsburg Empire, as the overabundance of young scholars was not accompanied by an enlargement of the professoriat, leading to more intense competition and pushing large numbers of scholars out of the university system.

German-language universities in the Habsburg Empire were becoming increasingly Austrian; they had their own hierarchy and a career path distinct from both the imperial and the pan-German models. This was specifically tuned to the needs of the University of Vienna and had a clear hierarchical structure. Aspects favoring Vienna included legal regulations, the practices of the Ministry of Education, and, not least, the choices of the scholars themselves. It is not surprising that most scholars regarded Vienna as the academic pinnacle of their careers.

Careers and the Formation of Scholars

A brief glance at the scientific personnel active at Cisleithanian universities reveals a network dominated by the University of Vienna, which had the largest number of professors and Privatdozenten, at times making up more than half of the scholars in each academic rank at German-language universities (see table 3). At the medical faculty, most scholars were Privatdozenten, with fewer associate professors and still fewer full professors; at the philosophical faculty, Privatdozenten outnumbered professors. The network of personnel thus formed a pyramid at the medical faculty, with a large number of instructors[2] at its base and a diminishing number of scholars toward the peak, and an hourglass at the philosophical faculty. These two structures, favoring competition at all levels and producing a broad stratum of underpaid or even unpaid teaching staff, still called in German the *Mittelbau* (midlevel faculty), was characteristic for the University of Vienna. The Prague universities, the Galician universities, and the medical faculty in Graz were also slowly changing to a pyramidal structure, which corresponded to the need for Privatdozenten to cover lectures. A pyramidal structure indicated a steep career path, and while Privatdozenten at the Czech Charles-Ferdinand University in Prague and the Jagiellonian University could hardly switch universities, many young scholars in Vienna decided to move to other universities in the Habsburg Empire or abroad (the latter was common in medicine), or to nonacademic institutions. In contrast, Chernivtsi, Innsbruck, and the philosophical faculty in Graz usually employed fewer Privatdozenten than professors, and the structure of the teaching faculty would have formed an inverted pyramid. These universities thus had a limited influence on the education of scholars at the beginning of their careers.

Even though many scholars chose to habilitate to further their careers outside of universities, the Privatdozentur was, in most cases, the first step on the academic ladder. And in Vienna, where nonuniversity academic jobs were abundant, turnover in the Privatdozenten was still high. Even though the number of older habilitated scholars in the capital city was substantial, the average age of Privatdozenten, measured every ten years, did not vary significantly across universities.[3] One cannot say with certainty what reasons led young scholars to leave the university. But the average age was distorted by the exponential growth of the Viennese Privatdozentur, and it underscores the quantity of well-educated habilitated scholars the capital

TABLE 3 Number of university instructors in Cisleithania, 1850–1910

		1850	1860	1870	1880	1890	1900	1910
					Medical faculty			
Vienna	Full professors	15	15	18	17	20	24	22
	Associate professors	6	7	15	38	32	31	40
	Privatdozenten	22	36	49	55	68	107	187
	Total	43	58	82	110	120	162	249
Graz	Full professors			9	10	11	14	12
	Associate professors			3	8	9	10	14
	Privatdozenten			3	8	11	18	22
	Total			15	26	31	42	48
Innsbruck	Full professors			9	8	11	11	14
	Associate professors			2	5	4	7	8
	Privatdozenten			2	4	3	3	4
	Total			13	17	18	21	26
Prague up to 1882	Full professors	10	12	13	13			
	Associate professors	12	8	8	11			
	Privatdozenten	12	9	17	21			
	Total	34	29	38	45			
Prague: German	Full professors				11	12	15	15
	Associate professors				10	10	12	18
	Privatdozenten				15	15	19	33
	Total				36	37	46	66

Total: German-language universities	Full professors	25	27	49	48	54	64	63
	Associate professors	18	15	28	62	55	60	80
	Privatdozenten	34	45	71	88	97	147	246
	Total	77	87	148	198	206	271	389
Prague: Czech	Full professors				6	11	14	15
	Associate professors				5	5	11	21
	Privatdozenten				4	9	15	24
	Total				15	25	40	60
Cracow	Full professors	8	10	13	9	11	12	14
	Associate professors	1	1	2	5	9	9	14
	Privatdozenten			4	7	10	15	17
	Total	9	11	19	21	30	36	45
L'viv	Full professors						11	14
	Associate professors						3	4
	Privatdozenten						10	27
	Total						24	45
Total: Galician universities	Full professors	8	10	13	9	11	23	28
	Associate professors	1	1	2	5	9	12	18
	Privatdozenten	0	0	4	7	10	25	44
	Total	9	11	19	21	30	60	90
Total: All Cisleithanian universities	Full professors	33	37	62	74	76	101	106
	Associate professors	19	16	30	82	69	83	119
	Privatdozenten	34	45	75	114	116	187	314
	Total	86	98	167	270	261	371	539

Continued

TABLE 3 Number of university instructors in Cisleithania, 1850–1910—cont'd.

		1850	1860	1870	1880	1890	1900	1910
					Philosophical faculty			
Vienna	Full professors	22	26	31	42	43	46	50
	Associate professors	4	4	9	11	11	15	18
	Privatdozenten	15	15	18	32	60	73	99
	Total	41	45	58	85	114	134	167
Graz	Full professors	7	6	14	17	19	24	27
	Associate professors	1	2	4	11	7	13	11
	Privatdozenten	1	1	5	10	15	9	12
	Total	9	9	23	38	41	46	50
Innsbruck	Full professors	7	10	13	17	19	25	23
	Associate professors	1		1	4	6	6	12
	Privatdozenten		1	1	5	11	10	10
	Total	8	11	15	26	36	41	45
Prague up to 1882	Full professors	15	16	19	39			
	Associate professors	6	7	3	7			
	Privatdozenten	4	3	2	19			
	Total	25	26	24	65			
Prague: German	Full professors				20	25	29	30
	Associate professors				6	7	7	8
	Privatdozenten				9	14	14	20
	Total				35	46	50	58
Chernivtsi	Full professors				10	15	17	18
	Associate professors				6	2	1	4

	Privatdozenten						2	5
	Total				16	17	20	27
Total: German-language universities	Full professors	51	58	77	145	121	141	148
	Associate professors	12	13	17	45	33	42	53
	Privatdozenten	20	20	26	75	100	108	146
	Total	83	91	120	265	254	291	347
Prague: Czech	Full professors				14	18	27	28
	Associate professors				6	8	8	16
	Privatdozenten				11	13	23	36
	Total				31	39	58	80
Cracow	Full professors	8	15	14	15	19	25	29
	Associate professors	4		3	7	2	10	15
	Privatdozenten			4	8	6	11	20
	Total	12	15	21	30	27	46	64
L'viv	Full professors	8	9	11	12	13	19	24
	Associate professors	3	3	2	1	4	8	7
	Privatdozenten	2	1		6	15	18	21
	Total	13	13	13	19	32	45	52
Total: Galician universities	Full professors	16	24	25	27	32	44	53
	Associate professors	7	3	5	8	6	18	22
	Privatdozenten	2	1	4	14	21	29	41
	Total	25	28	34	49	59	91	116
Total: All Cisleithanian universities	Full professors	67	82	102	186	171	212	229
	Associate professors	19	16	22	59	47	68	91
	Privatdozenten	22	21	30	100	134	160	223
	Total	108	119	154	345	352	440	543

Table 4 Places of graduation for scholars habilitating at different universities, 1848–1918

	Graduated from				
Habilitating at	**Vienna (%)**	**Graz (%)**	**Innsbruck (%)**	**Prague (%)**	**Prague: German (%)**
			Medical faculty		
Vienna	81	4	1	3	4
Graz	24	49	8	0	3
Innsbruck	54	8	35	0	4
Prague	8	0	0	86	n/a
Prague: German	10	5	0	6	77
Prague: Czech	1	0	0	16	0
Cracow	10	0	0	1	0
L'viv	5	0	0	3	0
			Philosophical faculty		
Vienna	69	4	3	1	2
Graz	24	53	3	1	0
Innsbruck	20	9	57	0	0
Prague	7	0	0	70	n/a
Prague: German	23	6	0	12	40
Prague: Czech	4	0	0	8	0
Cracow	7	0	0	0	0
L'viv	8	0	0	0	0

Note: n/a, not applicable.

university produced. This growth strengthened the dominance of Vienna as the training university for young scholars throughout the empire, even if only a portion of them left for academic positions at other universities.

In contrast to the widespread assumption that it was a cosmopolitan hub, the University of Vienna was throughout this period still a very local institution in relative terms. In absolute numbers it hosted a plethora of scholars from across the empire and abroad. The high number of nonlocal instructors also had to do with the pyramidal structure and the number of nonlocal graduates, the source of future Privatdozenten. Still, in relative terms, Vienna had the highest percentage of its own graduates among its habilitations (69 percent in the philosophical faculty and 81 percent in the medical faculty; see also table 4).[4] The same can be said for the percentage of Vienna's own Privatdozenten among its professoriat (76 percent in the philosophical faculty and 88 percent in the medical faculty), although the proportion is lower the higher one goes up the ladder, dropping to below 50 percent for full professors in the

TABLE 4 Places of graduation for scholars habilitating at different universities, 1848–1918—cont'd

Habilitating at	German Empire (%)	Russian Empire (%)	Prague: Czech (%)	Cracow (%)	L'viv (%)	Total (%)[1]
	Graduated from					
	Medical faculty					
Vienna	0	0	0	0	0	93
Graz	5	0	0	0	0	88
Innsbruck	0	0	0	0	0	100
Prague	3	0	n/a	0	0	97
Prague: German	0	0	0	0	0	99
Prague: Czech	0	0	76	1	0	95
Cracow	1	2	0	80	1	96
L'viv	3	14	0	54	16	95
	Philosophical faculty					
Vienna	12	0	0	0	0	91
Graz	11	0	0	0	0	92
Innsbruck	7	0	0	0	0	91
Prague	9	0	n/a	0	0	85
Prague: German	0	0	0	0	0	81
Prague: Czech	1	0	79	0	0	92
Cracow	14	9	0	53	7	89
L'viv	18	3	0	7	42	78

Note: Chernivtsi was excluded owing to its low number of habilitations. Only the first habilitation was considered. The percentage includes all Privatdozenten, i.e., including those with an unknown place of graduation or who graduated at other academies; to calculate the percentage of missing cases, subtract the numbers in the right column from 100. Russian Empire magister/candidate degrees are counted as graduations.

n/a, not applicable.

[1]The totals do not sum to 100 because places of graduation are unknown for some scholars, and a few graduated at other universities. To calculate the percentage of missing cases, subtract the numbers in the right column from 100.

philosophical faculty (see table 5). At the level of full professors, the faculties of the University of Vienna aimed to appoint more scholars with varying educational backgrounds, a phenomenon I scrutinize in more detail below.

Vienna was also the largest exporter of young academics (both graduates and Privatdozenten) to other universities. Similarly, Viennese graduates constituted a considerable number of the Privatdozenten at the other German-language universities in the empire, accounting for almost a quarter of the habilitations in Graz and more than half of the habilitations in medicine

TABLE 5 Percentage of own offspring among the professorship, 1848–1918

University	Degree or position gained at the university	Medical faculty: Associate professor (%)	Medical faculty: Full professor (%)	Philosophical faculty: Associate professor (%)	Philosophical faculty: Full professor (%)
Vienna	PhD	81	71	67	45
	Privatdozent	86	65	82	46
	Without habilitation	7	7	7	10
Graz	PhD	26	12	21	10
	Privatdozent	50	17	29	15
	Without habilitation	13	n/a	11	2
Innsbruck	PhD	19	6	27	21
	Privatdozent	24	6	39	23
	Without habilitation	6	4	5	27
Prague: German	PhD	73	35	27	9
	Privatdozent	75	30	50	17
	Without habilitation	n/a	n/a	0	9
Cracow	PhD	75	45	26	31
	Privatdozent	84	47	51	31
	Without habilitation	13	37	23	33
L'viv	PhD	43	10	22	15
	Privatdozent	43	10	40	21
	Without habilitation	n/a	15	25	45

Note: The categories are nonexclusive. That is, if a scholar graduated from Vienna and then worked as a Privatdozent there, he is included in both percentages. Also, several universities are omitted owing to special conditions that make their situation not comparable to the others.

n/a, not applicable.

in Innsbruck (see table 4). The exception is Prague, whose own graduates made up a high number of the Privatdozenten teaching there because of its traditionally strong medical faculty. Further, scholars infrequently returned to the province or city in which they had been born; indeed, any return would not have been seen as providing a career boost. Mobility between graduation and habilitation had no significant influence on whether scholars achieved professorial positions in either a faculty or a university.[5]

In absolute numbers, however, moving to a different university before habilitating was rare, except that the Cracow and L'viv universities attracted a large number of graduates from beyond the monarchy who subsequently habilitated there (see table 4). The trend of remaining at the university where one had graduated had, of course, financial and career-related consequences, and it seems to have been closely connected to the availability of extra-academic occupations in the university cities. Given the competition, career advancement was tightly connected with finding support and networks even before graduating. Within a faculty, students were often promoted by their teachers, but professors could also influence the nominations of their students at other universities, using informal networks that linked faculty members and even extended into the ministry.

Several university-led factors influenced the career and mobility choices of young graduates and Privatdozenten. On the one hand, Privatdozenten at smaller universities had a better chance of academic promotion compared with those in Vienna, since the number of academics competing for professorships was comparatively high in the capital. On the other hand, leaving the central university, that is, Vienna, meant less money, both from lectures and, especially in the case of practicing physicians, from nonacademic and semi-academic occupations. Moreover, for physicians, a smaller university meant fewer opportunities for practical work, which was highly valued in future appointments, as chairs were linked to hospital duties. Thun-Hohenstein had already stressed that medical scholars at smaller universities had to have experience in both practice and teaching, and he favored those working in the capital.[6] Subsequently, the ministry regarded practical ability as more important than scientific research for the small medical faculty in Innsbruck.[7] Since some chairs were heads of clinics, legal approbation for medical practice was a necessity, favoring Habsburg candidates.[8] These arguments should, however, be taken with caution. Almost throughout the whole period in question, the various ministers of education applied a particular combination of practical, institutional, and ideological arguments to support the export of personnel from the Vienna Medical School and reaffirm its dominant role in central Europe.

Salaries, Prestige, and the Habsburg Hierarchies

During the nineteenth century, it became increasingly rare for a scholar who had worked at only one university to be nominated for a full professorship; therefore, the question of geographic mobility remained crucial for scholars within the empire in regard to both their personal careers and any development policies at the faculties. The differences among faculties were evident early on. At the philosophical and medical faculties, those who changed university during their careers made up 45 percent and slightly fewer than 30 percent of scholars, respectively. These figures are biased by the number of immobile Privatdozenten, in particular in the Viennese medical faculty. In 1910 around 50 percent of the full professors at the University of Vienna were products of that university and had spent their whole career in Vienna, while at Graz and the German University in Prague, more than 80 percent of professors had come from a different institution. Crucially, it was the University of Vienna that imported and exported the majority of staff working at medical and philosophical faculties (see table 6). While the absolute number was high, however, the imported scholars constituted only around 10 percent of the teaching scholars and around 50 percent of the full professors at Vienna between 1848 and 1918, although the latter were largely Viennese offspring returning from other Austrian universities. With a few exceptions, scholars who left Vienna and pursued careers at other universities were Viennese products, having studied, graduated, and habilitated there. While the movement of scholars was determined by a variety of personal, cultural, and scientific factors, the system remained largely centered on Vienna: other universities profited from graduates from Vienna, and Vienna could choose the best scholars from across the empire in its appointments.

Transfers between Habsburg universities were, in most cases, career advancements. Most scholars were promoted (by one rank or two) during the change of university or were moved to universities higher in the hierarchy, with higher salaries. The increase in salary was either obligatory (by law) or individually negotiated. Even if salaries were subject to negotiation during the appointment procedure, the legally codified differences in regular salaries were partially responsible for the Vienna-centric nature of transfers throughout the nineteenth century. Throughout this period the ministry opposed appointments of scholars from universities with higher regular salaries for scholars of the same rank, as this would burden the budget and create legal precedents.[9] The salary discrepancies also made the Ministry of Finance one of the most important agencies controlling appointments. With the regulations of 1849, the salary structure was built around Vienna as the center: professors in Vienna not only earned more (see table 1 in chapter 1) but also received additional money for housing. (New regulations lessened this discrepancy in 1870, and egalitarian salaries were finally introduced in 1898; Viennese professors retained, however, most of their additional

Table 6 Transfers between Habsburg German-language faculties, 1848–1918

Transfer from	Transfer to					Total
	Vienna	**Graz**	**Prague and Prague: German**	**Innsbruck**	**Chernivtsi**	
			Medical faculty			
Vienna	n/a	22	19	31	n/a	72
Graz	15	n/a	8	7	n/a	30
Prague and Prague: German	30	6	n/a	10	n/a	46
Innsbruck	10	18	7	n/a	n/a	35
Chernivtsi	n/a	n/a	n/a	n/a	n/a	n/a
Total (% of the total faculty)	55 (10)	46 (31)	34 (14)	48 (50)	n/a	183
			Philosophical faculty			
Vienna	n/a	42	27	24	22	115
Graz	21	n/a	12	9	8	50
Prague and Prague: German	29	5	n/a	8	6	48
Innsbruck	22	11	5	n/a	3	41
Chernivtsi	3	7	10	8	n/a	28
Total (% of the total faculty)	75 (13)	65 (32)	54 (21)	49 (35)	39 (50)	282

Note: n/a, not applicable.

privileges owing to the higher cost of living in that city.) And in individual negotiations over salaries, they were clearly privileged, achieving salaries much higher than the standard ones.

The differences were enhanced by the student fees (Collegiengelder) for enrolling in a lecture or seminar series, because professors in Vienna could count on more participants (see table 7). Only after 1898 were professors prohibited from charging for their lectures, an issue that had been fiercely discussed from the moment when student fees were first enacted. Discussing the salary reforms, and in particular the proposal to abandon the fees, the faculties opposed any change to previous practice. They argued not only that Collegiengelder ensured student attendance at lectures but also that they enabled competition among professors, who, if student fees ceased, would lack the motivation to prepare interesting lectures and would return to being civil servants.[10] The issue of medical theoreticians was also raised, since they could not earn money via private practice. If they received no Collegiengelder, this would deter young scholars from specializing in this area.

Throughout the late Habsburg period, numerous brochures, petitions, and committees addressed the issue of unequal salaries within the empire, the privileging of scholars at the University of Vienna in individual salary negotiations, and, more rarely, the discrepancies between Habsburg salaries and those abroad.[11] In the 1860s Carl Rokitansky had already addressed this issue in his brochure *Die Conformität der Universitäten mit Rücksicht auf gegenwärtige österreichische Zustände* (On the conformity of

Table 7 Percentage of professors receiving a given amount of Collegiengelder at philosophical faculties in Cisleithanian universities, 1892–93

		Collegiengelder (guldens per year)				
	Number of professors	**>1,000 (%)**	**500–1,000 (%)**	**100–500 (%)**	**50–100 (%)**	**<50 (%)**
Vienna	56	16	9	30	21	23
Graz	33	3	18	27	15	36
Innsbruck	29	3	7	34	14	41
Prague: German	31	3	10	29	10	48
Chernivtsi	18	0	6	22	28	44
Prague: Czech	30	23	7	20	23	27
Cracow	27	7	19	37	22	15
L'viv	19	11	16	53	21	0

Source: Petition der philosophischen Fakultäten an den k.k. Universitäten um Regelung der Bezüge Ihrer Professoren, February 1894, p. 2, ÚDAUK, FF NU, Sign. K/a (Profesoři), Inv.č. 186–93, Kart. 9.

the universities, concerning the current Austrian situation, 1863), in which he pleaded to reduce the monopoly of the University of Vienna in regard to professors' salaries. He fiercely opposed the idea of a central university with satellites serving only as "nursery or transit schools for other universities, or even [as] institutions for accommodation and provisioning of deficient talents and workforces."[12] While Rokitansky wrote from a double position as a Viennese professor and an official in the ministry, most disputants took a more one-sided stance. While professors at provincial universities strove to level salaries and criticized Vienna's predominance,[13] Viennese professors opposed any equalization of salaries, stating that this would "severely damage larger universities."[14]

One of the points often raised was that equalization of salaries would disadvantage the University of Vienna because of the higher cost of living in the city; professors, especially those with larger families, would then prefer to remain at smaller universities, where the cost of living was less expensive. Smaller cities seeking to have a university established there, such as Salzburg, saw exactly this as being to their advantage.[15] As some writers claimed, scholars in university cities were even unable to find apartments befitting their social standing, especially near their institutes.[16] More drastic were descriptions of professors with families who were "hindered in [their] spiritual development owing to concerns about food."[17] Such descriptions were surely slightly dramatized, but living conditions were in fact a problem for all members of the Habsburg civil service, especially in Vienna,[18] and some professors indeed found themselves in financial trouble.[19] This issue was also included in the appointment papers; professors often claimed the need for so-called *Naturalwohnung* (i.e., a residence owned by the university) in institutes so that they could closely supervise their research facilities and experiments.[20]

Salary discrepancies across medical faculties were even more serious. University positions were frequently linked to positions at the university clinics and city hospitals (for example, as chief physicians). This made a transfer to a smaller university unattractive even despite an advance in academic rank. The ministry was also reluctant to offer higher salaries than usual in such cases, limiting the possibility of transfers from Vienna.[21] In addition, some associate professors simultaneously had tenured positions as assistants: in this case even the University of Prague, offering the second-highest regular salaries, could not match the earnings of these scholars, particularly those from Vienna.[22]

Lamenting the financial situation of universities was a kind of ritual throughout the empire. Seeking higher salaries, scholars would apply for positions abroad as a bargaining tool. Professors proposed for a position at a foreign university could better their financial status, and home universities often strove to retain them by offering higher remuneration. In such cases, the Ministry of Education was willing to raise salaries considerably, well beyond the regular ones.[23] Further, privileges for one's institute could be gained in this way, including equipment, assistants, or even additional associate professorships.[24] This worked both ways; not only did universities in the German Empire offer better salaries,[25] but professors at non-Habsburg German universities also used the appointment procedures to secure a better position in salary negotiations at their own universities. This means of augmenting one's income was certainly important, and it seems that scholars frequently used it, entering into negotiations with other universities just to bargain with their own administration, with no intention of actually taking an appointment elsewhere.

The introduction of equal salaries for professors at all universities did not change the appointment pattern considerably. Before and after 1898, appointments had a similar structure, following the above-described hierarchy, although one could assume that a position in Innsbruck, for example, would now be more valuable than one in Prague, given the differences in the cost of living. Yet the structure of nominations remained the same after the salary changes. This persistence of traditional hierarchies, resulting from the appeal of both financial and symbolic capital, was best described, somewhat ironically, by Theodor Mommsen, who commented that Habsburg scholars are "sentenced to Chernivtsi, pardoned to Graz, promoted to Vienna."[26] This symbolic hierarchy was also discernible in appointments from other institutions. From 1898 the technical academies and the Academy of Agriculture in Vienna also offered the same salaries as the universities. This too did not change the appointment structure; universities still appointed scholars from technical and agricultural academies, without significant movement in the other direction, apart from a few scholars who taught simultaneously at both universities and technical academies.

The issue of finances was not restricted to salaries but also included the costs of reorganizing institutes to meet the professors' needs; some rearrangements involved considerable expense. This affected appointments; for example, Ludwig Boltzmann was appointed to the chair of experimental physics in Graz in 1876, even though he was proposed in third place in the terna, because appointing the other two candidates would have required modifications to the institute's infrastructure, something Boltzmann did not desire.[27] Several scholars even rejected nominations because of a lack of infrastructure in an institute or the rejection of higher endowments. The natural sciences were especially disadvantaged because new nominations could mean considerable and expensive modifications. Thus, scholars often remained where they were because their own institutions were better tuned to their needs, and even a considerable increase in salary failed to convince them to move.[28]

Such situations also involved comparison with nonuniversity institutions and became a choice involving both gains and losses, which showed that achieving a professorship at a university was not every scholar's ultimate goal. State institutions were effectively competing for the same scholars, especially because academic appointments as such included neither considerable monetary gain nor a change in status. Better conditions at clinics[29] or better access to research material in medicine or veterinary medicine[30] were some of the reasons scholars chose to remain at nonacademic institutions. In rare cases, some professors actually resigned their positions to pursue a nonacademic career.[31]

Choosing a nonuniversity post instead of a professorship was not only a matter of personal preference. The ministry was also eager to retain the best scholars in the most internationally recognized institutes: to keep Rudolf Heberdey as the head of the Austrian Archaeological Institute at Athens, the ministry proposed to make his salary and rank equal to those of a full professor, instead of agreeing to his appointment to Graz.[32] Salaries were clearly an issue here, since state institutions offered comparable salaries, making university appointments expensive. This was especially true when a nominee working at a nonacademic job was proposed to become an associate professor; this academic position had a low nominal payment and thus was not very attractive. When it was clear that a scholar would demand a higher salary, or at least the same salary as in his previous post, the ministry was often hesitant to even enter into talks.[33] Viennese Privatdozenten who had an additional occupation in the city were particularly hard to convince to move to a smaller institution.[34]

Smaller universities were handicapped not only by their financial situation but also by the ministry's ongoing concern with assuring Vienna's role as the central university in the empire. The faculties of the University of Vienna also saw themselves as central institutions in themselves, and based on the "fixed convention," they were allowed (predestined) to acquire "the best people of all"[35] for their chairs. When a provincial university convinced a full professor from Vienna to agree to be included in a faculty proposal, this did not go down well with the ministry. Indeed, this had less to do with finances than with the need to keep the best people in Vienna. "It is not advisable to allow a professor of the University of Vienna to transfer to a smaller university, because this would create a precedent, which would imply critical consequences for the thriving of . . . the University of Vienna"[36] was the reasoning given in one of the few such cases. Only on special occasions did the ministry allow such appointments despite the institutional hierarchy. When Julius Hann, an associate professor in Vienna, retired from the directorship of the Central Bureau for Meteorology and Terrestrial Magnetism in Vienna, he, guided by medical advice, asked for a transfer to a "smaller university, namely, in Graz, or alternately in Innsbruck," to concentrate on teaching; this petition, approved by consensus in Graz, was also accepted by the ministry.[37] After Hann had recovered physically, a second petition, this time a plea to return to Vienna, was issued and accepted.[38]

From Chernivtsi to Vienna: The Structure of the Academic Space

The hierarchical differences described above are clearly discernible in the types of appointments. As noted above, scholars were generally promoted by one rank or more when transferring universities. At the Innsbruck medical faculty, the appointees had mostly been Privatdozenten (75 percent; equal numbers of them were promoted to full and associate professor positions), whereas at Vienna almost all appointees had been full professors at another university. (For details on transfers of full professors, see table 8.) At Graz, the "in-between" university in the academic hierarchy, appointees were either full professors from Innsbruck or, in approximately equal numbers, Privatdozenten and associate professors from Vienna. The Privatdozenten appointed to Graz were, with three exceptions, promoted by only one academic rank, that is, to associate professors. For most professors transferred to Vienna, it was the last stop in their career, whereas slightly fewer than half of imported scholars stayed in Graz (20 percent moved to Vienna and 12 percent to the German University in Prague), and slightly over 30 percent

TABLE 8 Number of full professors who transfered to another university, 1848–1918

	Transfer to							
Transfer from	**Vienna**	**Graz**	**Prague: German**	**Prague**	**Innsbruck**	**Chernivtsi**	**German Empire universities**	**Total**
				Medical faculty				
Vienna	n/a	0	0	1	0	n/a	1	2
Graz	7	n/a	3	0	1	n/a	2	13
Prague: German	6	1	n/a	n/a	1	n/a	8	16
Prague	2	0	0	n/a	0	n/a	1	3
Innsbruck	7	13	3	1	n/a	n/a	3	27
Chernivtsi	n/a	n/a	n/a	n/a	n/a	n/a	n/a	n/a
German Empire universities	9	1	1	3	0	n/a	n/a	14
Total	31	15	7	5	2	n/a	15	75
				Philosophical faculty				
Vienna	n/a	2	0	0	1	0	9	12
Graz	13	n/a	0	6	0	0	10	29
Prague: German	10	2	n/a	n/a	2	n/a	7	21
Prague	9	0	0	n/a	n/a	n/a	6	15
Innsbruck	17	9	1	1	0	0	1	29
Chernivtsi	0	6	6	2	6	n/a	2	22
German Empire universities	10	4	3	2	1	1	n/a	21
Total	59	23	10	11	10	1	35	149

Note: n/a, not applicable.

remained in Innsbruck (25 percent moved to Graz, 10 percent to Vienna, and 10 percent to the German University in Prague).

The number of professors who had only a short residence at a university reflects this difference (see table 9). The medical faculty at Innsbruck was, for thirty-one scholars, an intermediary station in their career (they left after an average of five years). In Graz the same was true for sixteen scholars, while the German University in Prague and the University of Vienna each had only ten appointees from other universities who later pursued a career elsewhere. With regard to philosophical faculties, however, Graz was an intermediary station for twenty-six scholars, the German University in Prague for twenty-five scholars, Innsbruck for twenty scholars, and Vienna for twelve scholars. Vienna was the main university to which scholars returned (that is, those who had acquired at least their habilitation there)—twenty-three to the medical faculty and twenty-nine to the philosophical faculty—while other universities had only an insignificant number of returning scholars.

As noted before, only Vienna can be regarded as a training university for the medical sciences. Other universities rarely promoted their own students, which meant that only a small number of them were appointed to other universities (see table 5). At the same time, Vienna remained the university with the highest number of Privatdozenten who did not advance in their careers: slightly more than 50 percent, in comparison to 40 percent at Graz, 25 percent at Innsbruck, and 14 percent at the German University in Prague.[39]

At this point, the link between science and practice becomes visible and reinforces the idea of Privatdozentur as a secondary occupation. Scholars who remained Privatdozenten were working in disciplines such as ophthalmology, laryngology, dentistry, and internal medicine, where scholars could earn money with additional practice outside the university, and the title of docent was prestigious. Scholars in disciplines where an extra-university occupation was more unlikely, such as anatomy and pathology, mostly achieved professorships or, at least, the title of professor. This local and practical dimension surrounding Privatdozenten in Vienna can be viewed through the disciplinary nexus as well. For example, the fields of balneology, syphidology, the history of medicine, and dentistry had almost no transfers. In internal medicine only around 10 percent changed university in the course of their careers, while around 40 percent of anatomists and pathological anatomists did so.

Global numbers illustrate the centrality of the Vienna medical faculty (on transfers among German-language institutions in Cisleithania, see tables

TABLE 9 Age and transfer statistics, 1848–1918

	Average age at time of appointment	Years between graduation (PhD) and appointment	Average age at time of appointment to another university	Professors passing through[1]		Returning scholars[2]	
				Number of	Average length of stay (years)	Number of	Average time between departure and return (years)
				Medical faculty			
Vienna	44.7	14.6	36.5	10	5.1	23	8.6
Prague: German	39.6	8.9	41.5	10	5.3	4	4.2
Graz	38.5	9.0	43.0	16	10.0	3	3.0
Innsbruck	36.9	6.1	40.7	31	5.2	0	n/a
				Philosophical faculty			
Vienna	43.1	14.8	37.0	12	11.0	29	11.6
Prague: German	38.2	8.7	40.8	25	8.4	2	4.0
Graz	38.6	9.7	41.0	26	9.6	8	8.2
Innsbruck	38.5	8.8	42.2	20	8.5	3	10.0

Note: n/a, not applicable.

[1]These are professors who moved from another university and then moved back, or to a third university.

[2]These are professors who returned to a given university after working at another one.

6 and 7). In 1848–1918 the Vienna medical faculty exported 102 scholars (72 of them to other German-language institutions in Cisleithania), 77 percent of whom had graduated from Vienna and 87 percent of whom had attained the position of Privatdozent there. At the same time, the faculty appointed eighty instructors (half as professors), of whom 33 percent were its own returning graduates. Twenty-two percent of the appointees came from the German Empire. Nevertheless, not all of the scholars in the latter category were foreigners: nearly half had graduated from Vienna, and in total 72 percent of them had graduated from one of the German-language Habsburg universities; however, only four (20 percent) had habilitated in Vienna. In addition, 23 percent of scholars came from Bohemia, predominantly from the German University in Prague.

In the same period, the Graz faculty appointed forty-seven scholars (32 percent of the overall number of instructors), 44 percent of whom came from Vienna and 38 percent from Innsbruck. While the scholars from Vienna were promoted to a higher rank, the scholars from Innsbruck were mostly already full professors and were appointed with no change in rank, although certainly a change in salary. In total, 44 percent of Graz's faculty members transferred to another university: nine moved to Vienna, eight to Prague, and four to the German Empire. However, young scholars from Graz were appointed to other universities in only nine cases (four of them subsequently returned: one from Prague and three from Innsbruck), and five Privatdozenten moved away from Graz (four to Vienna) and habilitated again. This appointment practice strongly encouraged variety in the top positions in Styria. Among the fifty-six scholars holding the position of full professor in Graz, only 10 percent had graduated from that university, with a high turnover among those positions as well.

The German University in Prague similarly remained a university in flux, especially suffering a loss of prestige after the division in 1882. It exported twenty of its own scholars from 1882 onward (this includes scholars who had graduated from the undivided university); they constituted half of all scholars appointed from this university. Ten of them moved to Vienna (in equal parts by being appointed there and by habilitating again), and six to the German Empire (that is, 30 percent of all Prague graduates appointed at other universities), without being subsequently appointed back to Prague (with one exception). During the same period, the faculty appointed thirty-seven scholars, with the majority (twenty-three, or 62 percent) remaining at the university until their retirement. Most common were appointments from Vienna (33 percent), followed by Graz and the German Empire (each around 20 percent). The structure of appointments was not as consistent as in Vienna, however, as the faculty appointed not only full professors from other universities (25 percent) but also associate professors (38 percent) and Privatdozenten (25 percent). Those in the latter two groups were each promoted by at least one position; appointees who were promoted mostly came from Vienna and German Empire universities but also, to a lesser extent, from other universities.

The smallest and youngest medical faculty in Cisleithania, that in Innsbruck, can exemplify the nonformative, transitional faculty. It appointed more than 50 percent of its total teaching faculty between 1869 and 1918. Most of the instructors came from Vienna. Privatdozenten made up a third of the appointees who were promoted to full professorships and a third of those promoted to associate professorships, but these scholars did not remain in Tyrol for long. Two-thirds of those appointed from Vienna left the university (ten moved to Graz, five back to Vienna) after an average of six years spent in Innsbruck, half in fewer than four years (see also table 11). While seven scholars were appointed from German universities, three of whom were Austrian citizens, only two remained in Innsbruck, both scholars who had been born in the Habsburg Empire. Only four scholars who habilitated in Innsbruck moved to other universities, just one of whom was appointed to a professorship, the Transleithania-born medical chemist Leó Liebermann, who was appointed to Budapest in 1902. The other three left Innsbruck and habilitated at other Habsburg universities. The prevailing pattern was that scholars appointed from Vienna moved on from Innsbruck to Graz (eleven cases, i.e., 25 percent of all mobile scholars), while only three scholars returned from Innsbruck to Vienna; similarly, three scholars appointed from Graz returned to that university, and three appointed from the German University in Prague returned there. As at other provincial universities, Innsbruck's own scholars made up only a small percentage of the full professors in the medical faculty: three scholars who had graduated from Innsbruck and three scholars who had gained their venia in Innsbruck (only one both graduated from and habilitated in Innsbruck). Unsurprisingly, scholars with Viennese pasts were prevalent here as well.

Philosophical faculties show a slightly different picture. Similarly to the situation in the medical faculties, a combination of economics and prestige structured academic mobility. From 1875 on, Chernivtsi replaced the University of Innsbruck at the bottom end of the appointment chain, while

Innsbruck showed a pattern of appointments similar to that at the universities in Graz and Prague. In the philosophical faculty, Innsbruck had a much higher proportion of its own scholars among its professors than in the medical sciences, with forty-six of its own Privatdozenten (32 percent of all professors; see also table 5), of whom seventeen were appointed to other universities, including three to Chernivtsi, four to Graz, and six to Vienna, in most cases after having achieved professorships in Innsbruck. Slightly less than a quarter of all full professors working at the philosophical faculty had habilitated in Tyrol, while 20 percent had gained a Privatdozentur in Vienna. The teaching body of the smallest and youngest Cisleithanian university, in Chernivtsi, consisted of 80 percent scholars appointed from other universities in the empire. Almost no graduates from Chernivtsi became Privatdozenten or professors. This was caused (as at the medical faculty in Innsbruck) by its late foundation (in 1875) and the high turnover of professorships, which hindered the development of research groups and schools around professors. Also, the lack of additional occupational activities in this peripheral provincial capital made an unpaid Privatdozentur unattractive. Vienna, in turn, remained the central faculty, filling half of its professorships with its own graduates and Privatdozenten,[40] while a quarter of the professors appointed to the capital from other Habsburg universities had been educated in Vienna.

Graz and the German University in Prague had similar structures of appointments and promotions, and thus a detailed presentation of the Styrian University of Graz perfectly illustrates the characteristics of these two universities' in-between position. Less then 20 percent of the full professors in Graz had graduated or habilitated there (see table 5). Of the eighty-nine Privatdozenten promoted to professorships in Graz, fourteen had habilitated there (seven of these had also graduated there), while thirty-six (40 percent) had habilitated in Vienna, being promoted to Graz mostly from the position of Privatdozent—fourteen became associate professors, six became full professors, and four scholars left the university and habilitated again elsewhere. Two Viennese scholars, the mineralogist Karl Peters and the physicist Ludwig Boltzmann,[41] moved to Graz as full professors; however, both moved in atypical circumstances. Boltzmann changed his university quite frequently, and Peters had only a provisional professorship in Vienna because of his relocation from Pest.[42] Eight Privatdozenten from Vienna came to Graz via other universities. Nine scholars moved on, five to Vienna, two to Prague, and one each to Innsbruck and Berlin. In total, ninety-eight scholars moved to Graz, twenty-nine as associate professors and forty-seven as full professors. Slightly fewer than a third of them (twenty-six) were subsequently appointed to another university—eleven to Vienna (for eight, this was a return), five to the German University in Prague (three were initially appointed to the undivided university, i.e., before 1882), and ten to German universities.

Of all the scholars appointed from Graz, thirty-six had held their first position (including Privatdozentur) there. Of those, twenty-seven were graduates of the faculty. Of those twenty-seven, seven were appointed to Vienna and five to Chernivtsi; two went to Germany; and three were appointed at the technical academy in Graz. For thirty-two scholars, Graz was only a station in their career; most of these were appointed to a university with a higher standing, either the University of Vienna or a German university outside of the Habsburg monarchy. Five scholars returned to Graz: two from Vienna, one each from Chernivtsi and the German University in Prague, and one from Innsbruck via Freiburg. Twenty-seven scholars from Graz who received other appointments had been full professors in Graz before moving to Germany, Prague, or Vienna, while fewer had been associate professors (fourteen, of whom four went to Chernivtsi and four to Vienna) or Privatdozenten (seven left the university and habilitated elsewhere—three moved to Vienna—and eleven were appointed as professors, especially in Prague, Chernivtsi, or Innsbruck).

Through the dominance of Vienna and its (in)formal privilege of appointing the best scholars, the central institution had a considerably more stable faculty than the other universities. Given the low number of scholars for whom the university was only a transitional station (see table 11), in addition to some who returned there, it differed from Graz and Innsbruck, which were often only rungs on a career ladder. Still, Vienna did not turn into a place for retiring scholars, as had been the case before 1848. Although it had the highest average age for full professors[43] and associate professors,[44] the number of new scholars in the faculty (surveyed every ten years, including newly habilitated scholars and those promoted from other universities) was around 50 percent, similar to that for other universities in the empire.[45] In comparison to other universities, however, the rate of promotions within faculties at the University of Vienna was lower by about half (if the award of a title is not considered a promotion),[46] even if the faculties in Cracow and L'viv are taken into consideration. Although no policy explicitly condemned local appointments, the picture of scholars educated in Vienna pursuing careers at other universities and then being appointed back to the capital (or not) is dominant, especially in the medical faculties. Moreover, this had serious consequences for Jewish scholars, as I outline in chapter 6. Despite the statistics being biased by immobile Privatdozenten, full professorships in particular were linked to mobility, including moves to and from Vienna, where mobility largely meant that Viennese graduates taught at other universities before being nominated for a position in Vienna.

A Protestant Counterpart to the Habsburg Empire? The Empire and Its Big Brother

One of the most contentious issues in the Habsburg appointment policy of the nineteenth century remained, however, the relationship with the German Empire, influenced both by geopolitical changes and by the political imagination. As noted earlier, the ministry treated scholars from the German Empire differently from Habsburg Germans, with a lower rate of acceptance if nominated by the faculties. Further, the exchange of professors between the two empires was not always welcomed, not only in the interest of supporting young Habsburg scholars, but also out of concern that German scholars might introduce unwanted ideas and methodologies. Such arguments can be found not only in ideological areas like historiography but also in medicine.[47] At the same time, one can easily discover a certain snobbishness or even orientalism among scholars who thought to bring academic culture to Austria, such as the neoabsolutist linguists mentioned in chapter 2. The "Godliness" of (non-Habsburg) German professors, a mocking description coined by the governor of Bohemia in 1879,[48] hampered Habsburg scholars from nominating foreigners in some cases.

Nevertheless, political reasoning was as important as cultural cautiousness. Immediately after 1870, the ministry feared that German scholars could "use their position in Austria for secondary aims among the youth, which is already fevered by current events."[49] A few years later, the same argument can be found in the appointment records for the chairs of German language and literature in Prague, where the ministry rejected the proposed appointment of German professors who "gave no guarantees regarding their political beliefs"; the ministry appointed only (local) temporary replacement professors.[50] Ironically, the 1870s were, however, one of the periods in which German scholars were most frequently appointed (apart from the 1850s).

As had been the case during Leo Thun-Hohenstein's term as minister of education, there were not enough qualified teachers to satisfy the demands of the growing faculties (see also table 10). Nonetheless, only those whom the ministry considered politically passive were successful.[51]

Although there is no consistent pattern in the exchanges between the two empires, in no period did appointments from abroad exceed those from within the monarchy. The first peak of appointments from abroad occurred between 1849 and 1854, with around 20 percent of scholars appointed at Habsburg universities coming from institutions in non-Habsburg states of the German Confederation. However, a number of the appointees had been exiled to the Habsburg Empire owing to political and religious persecution; they found sanctuary in the philosophical faculties of universities seen as a Catholic counterpart to Prussia. The second peak, in the 1870s, included professors at the philosophical and medical faculties, owing to the strong expansion of university education and the improved financial situation of the Habsburg Empire. Still, the percentage of scholars appointed from abroad was clearly decreasing at Habsburg German-language universities, making them more autarchic but also more hermetic than in the early years after the reform.

The perception that Habsburg universities tended to be autarchic rather than overpopulated with foreigners is reinforced by statistics showing that nominees from the German Empire included up to 30 percent Habsburg returnees,[52] a third of whom had previously held a professorship at a Habsburg university and more than half of whom had gained their doctoral degree in the Habsburg Empire. Of the eighty-two scholars born in the German Empire who taught in the Habsburg Empire in 1848–1918, twenty-six were appointed to the medical faculties (65 percent of them from 1880 onward) and fifty-six to the philosophical faculties, with the overwhelming majority (around 90 percent) in the humanities. Although 35 percent of such professors generally left for the German Empire after several years, there was a significant discrepancy between Vienna, where more professors remained, and other universities. Although appointments from abroad were almost exactly divided between the three possible options for promotion[53] and appointments from the position of full professor, the status division remained quite clear: while the University of Vienna appointed mostly scholars who were already working as full professors and associate professors (who became full professors), other universities promoted Privatdozenten, 25 percent of whom were appointed directly to full professorships.

Table 10 Number of transfers to and from the German-speaking Habsburg universities, 1848–1918

	With other Habsburg German-language academies	Among German-language Habsburg universities	From non-Habsburg German-language universities to Habsburg German-language universities	From Habsburg German-language universities to non-Habsburg German-language universities
1849	3	4	2	0
1850–1854	7	11	10	1
1855–1859	3	5	3	3
1860–1864	5	15	0	1
1865–1869	9	7	2	2
1870–1874	10	9	8	3
1875–1879	11	18	6	6
1880–1884	10	14	8	7
1885–1889	4	21	3	12
1890–1894	12	30	4	2
1895–1899	6	29	5	4
1900–1904	9	29	7	2
1905–1909	16	47	9	6
1910–1914	1	29	1	11

Appointments to the German Empire had a similar configuration. Of the 109 Habsburg scholars appointed to universities in the German Empire, approximately two-thirds were German speakers, almost evenly split between the medical and philosophical faculties. However, while members of the medical faculty were appointed via promotions, especially from Vienna, the majority of appointees from the philosophical faculties were already full professors, with fewer appointed to full professorships from positions as Privatdozenten or associate professors.

In contrast to the strong relationships with the universities of the German Empire, transfers to and from other countries were limited, primarily owing to language issues. While seventeen scholars were appointed from other countries (including eight from Switzerland and six from Italy), they had mostly been born in the Habsburg Empire and had simply worked abroad for a time, or they had necessary skills, as was the case with professors of Italian and Romance languages. Here, personal connections and traditions were also influential. For example, Habsburg surgeons had three consecutive full professorships in surgery in Utrecht thanks to the private connections of the Viennese surgeon Theodor Billroth.[54] One can also find rare instances of transfers resulting from ideological issues. For example, the professor of botany in Belgrade, Lujo Adamović (also Лујо Адамовић), moved to Vienna in 1906 owing to the problems he encountered in Serbia as a foreigner of a different confession from the majority of the intellectuals in the city.[55]

Overseas appointments did not play a large role for Habsburg scholars. However, in 1914 the University of Vienna initiated an exchange program with the United States, with the philosopher and psychologist George Stewart Fullerton being the first visiting scholar in Vienna,[56] but the program did not continue because of the outbreak of war. Fullerton himself was imprisoned when the war broke out and released only in 1918, in poor health, which certainly did not help with reestablishing the program thereafter.[57]

Habsburg ministers of education almost unanimously supported reappointing Habsburg scholars who had been working at foreign (mostly German) universities, seeing it as a positive cultural advantage and the continuation of certain research traditions. The minister of education Sigmund Eybesfeld, for instance, wrote in 1885 that nominating a former Viennese Privatdozent who was working in Liège was "a duty of the administration of education, in consideration of the splendid tradition of [Friedrich] Arlt's school, which [we] should also find in the future representation at . . . the University of Vienna."[58] In even more enthusiastic tones, the minister of education Wilhelm von Hartel happily announced in 1905 that with the appointment of the physiologist Franz Hofmann from Leipzig, "an Austrian scholar [had been] regained."[59] While nominations from abroad incurred higher costs, this was mostly not considered to be as grave an obstacle for Habsburg returnees as it was for foreigners.[60]

Not all such appointments were successful. The most severe response was the ministry's answer to the Viennese medical faculty's proposal for the successor of Theodor Billroth. While Vincenz Czerny, the Bohemian-born chair of surgery in Heidelberg, was proposed in the first place, he was regarded as too expensive. The second nominee, Jan/Johann Mikulicz-Radecki, was rejected because he had moved from Cracow to Königsberg "without urgent reasons." Further, he had "left a teaching position at a university [that is, a domestic university] only because of momentary gain."[61] That Mikulicz-Radecki, one of Billroth's most talented students, was not appointed may also have resulted from the local situation in Galicia. The minister of education, Stanisław Poray-Madeyski (1893–95), the author of the report, had been a colleague of Mikulicz-Radecki at the Jagiellonian University and had himself witnessed the unsuccessful attempts to persuade him to stay: in Cracow Mikulicz-Radecki had been a kind of celebrity, beloved by the faculty and students, who, reportedly together with several hundred city dwellers, had organized marches and meetings to convince him to stay in Galicia.[62]

The preferences of, and pressure from, the faculty typically determined who would be appointed and from where, although in the case of foreign scholars the ministry pushed heavily for a lower number than the universities would have wished. In general, around 14 percent of nominees came from German Empire universities and 80 percent from Habsburg ones, with the highest rates of foreigners (21 percent) in the humanities.[63] There were considerable disparities, however, in the percentage of German Empire scholars placed first in proposals at the different universities, ranging from 33 percent at the German University in Prague (41 percent in the philosophical faculty) to 20 percent in Innsbruck. In Prague slightly more than half of the proposals in the humanities ranked a scholar living in the German Empire as the first choice. Further, the response to proposals also depended on whether the scholars nominated were from the German or the Habsburg Empire. Appointment of the first-place scholar was considerably more likely if he was from a Habsburg university (true in 56 percent of proposals) than if he was from a German university (in 27 percent of proposals), with 76 percent and 29 percent success rates, respectively. If scholars based in the German Empire were proposed first in the terna, such proposals led to a successful appointment (that is, one of the scholars from the German Empire in the terna was appointed) 40 percent of the time. If an Austrian was proposed primo loco, in only 4 percent of cases was a scholar from outside the monarchy ultimately appointed to that position. Overall, the humanities had the highest rate of successful appointments from the German Empire (51 percent) and the medical sciences the lowest (30 percent). Unsurprisingly, the University of Vienna had the most success in appointing scholars from the German Empire, with its proposals approved 70 percent of the time,[64] while in Prague only 50 percent of such proposals met with a positive response.

Looking at these discrepancies, one should also consider that the Ministry of Education was unwilling to appoint scholars from abroad because they were much more likely to be reappointed to a university outside the empire than were scholars from within the empire. Both the universities and the ministry considered whether candidates for chairs would remain at the university and in the empire, exploring whether the candidates would take the appointment seriously or not. The ministry also often referred to prospective open positions, mentioning that a given scholar should not be appointed because in the near future he might be proposed by another faculty. Usually this meant that he would soon be promoted to Vienna and would thus not be a lasting gain for the original university.[65]

Smaller universities tried to counter this by offering contracts to scholars who would agree to stay for a longer period;[66] the University of Graz included a clause about a five-year renunciation of accepting appointments at other universities, but this practice of including such a clause was rare.[67] Some faculties seeking to convince the ministry to promote a local scholar argued that the new scholar would be a more permanent gain for the university. When the philosophical faculty in Innsbruck proposed Alois Cathrein for the chair of mineralogy and petrology, the commission stated two reasons for his primo loco position, which disregarded both his scholarly qualities and the custom of appointing professors from other universities for a chair. The first was his concentration on Tyrolean geology, and the second the fact that he would not be eager to accept a call from another university, as "*might* be the case with other candidates."[68]

The financial disparities between universities in the Habsburg and German Empires (see table 11) made it especially complicated for smaller universities to appoint foreign scholars. Half of the scholars appointed from

TABLE 11 Salaries of professors in 1900 (in marks)

	Full professors			
	Initial salaries	**Official maximum salary**	***Wohnungsgeld* (housing allowance)**	**Associate professors (initial salaries)**
Habsburg monarchy				
Vienna	5,440	6,800	340	3,060
Other universities	5,440	6,800	n/a	3,060
Prussia				
Berlin	4,800–7,200	9,400	900	2,400–4,800
Other universities	4,000–6,000	7,800	540–660	2,000–4,000
Bavaria	4,560	n/a	540	3,180
Tübingen	4,000	6,000	300	2,400–3,900
Sachsen (varied)	3,000 min.	12,000	n/a	Varied
Baden	3,000	10,600	n/a	Varied
Strasbourg	5,000	n/a	n/a	3,600
Giessen	4,500	6,500	n/a	2,500–4,000
Jena	4,000–6,000	7,800	n/a	2,000–4,000
Rostock	4,200	6,600	n/a	2,400–3,600
Russian Empire	9,720	n/a	n/a	6,480

Sources: For the Habsburg monarchy: "Gesetz von 19. September 1898 betreffend die Regelung der Bezüge der Professoren an Universitäten und denselben gleichgehaltenen Hochschulen und Lehranstalten," *Reichsgesetzblatt*, 20 September 1898, 295–96. For the German states: Wilhelm Lexis, *Das Unterrichtswesen im Deutschen Reich*, vol. 1, *Die Universitäten* (Berlin: A. Asher, 1904), 42–50. For the Russian Empire: Joanna Schiller, *Universitas rossica: Koncepcja rosyjskiego uniwersytetu 1863–1917* (Warsaw: Wydawnictwo Instytutu Historii Nauki PAN, 2007), 254.

Note: The following exchange rates were used for salaries in other currencies: 1 krone (2 guldens) = 0.85 marks; 1 ruble = 3.22 marks.

n/a, not applicable.

the German Empire were Privatdozenten (less often associate professors); most were appointed to the Prague and Vienna faculties, and they came from the universities in Göttingen, Munich, and Freiburg (philosophical faculties) and Heidelberg, Strasbourg, and Leipzig (medical faculties). Appointments of full professors from Germany comprised only around 30 percent of the total transfers. In philosophical faculties, such transfers were, with a few exceptions, concentrated in Prague and Vienna, and up to a third of these involved returning scholars who had been born in the Habsburg Empire. In medical faculties, as many as half of the appointed full professors who had been working in the German Empire had been born in the Habsburg Empire, and all of these appointments were to Vienna or Prague. Appointments of full professors from outside the Habsburg realms were, however, a financial burden and were seen as an affront to local scholarship; they were thus not welcomed by the ministry. Only the University of Vienna was privileged, as the principal university in the empire, whereas Prague was gradually but steadily losing status. The Prague faculties saw this as an increasing depreciation of the Charles-Ferdinand University, and the professors of the German University, in particular after 1882, expressed their discontent with Vienna's privileges.[69] In 1899, when the ministry did not appoint one of the two German Empire professors proposed as the faculty's first and second choices, but instead a young Privatdozent from Vienna, the Bohemian faculty protested loudly, seeing it as a vilification of the status of Bohemian academia.[70] But this was to no avail.

The relationship between the two neighboring empires that shared a language was difficult, however, not only owing to the obvious political complications, but also because the Habsburg Empire from the beginning had understood science as a cultural component of its competition with Prussia. Although higher officials advocated the unity of the two empires at an academic level on several occasions, it was the concept of competition that defined academic relations. Especially in the medical sciences, both scholars and the ministry accentuated the idea that the Vienna Medical School was appreciated at German universities. Ministerial papers mentioned not only the welcome spread of Habsburg traditions but also the fact that many young Habsburg scholars would not easily gain a satisfactory position in the Habsburg Empire,[71] thus addressing financial issues related to scientific transfer.

The idea of the "best possible scholars," which one often finds in appointment proposals for the University of Vienna, referred only to

German-speaking academics, however. While it included those from the German Empire, it left Galician and Czech Prague scholars out of the discussions. Those with a confirmed knowledge of German and a Viennese educational background were exceptions, but only a handful of them found their way into proposals.[72]

The story of the dissolution of the empire, which commonly states that the Magyars and Slavs turned away from Vienna, can thus be told in a different way. Well before German nationalism seriously influenced political and academic discourse, German-speaking Habsburg universities had stopped considering Slavic or Magyar scholars as possible appointees and showed much more interest in exchanges with the German Empire. One could speculate that this was an outcome of stereotyping non-Germans in the Habsburg Empire as underdeveloped in scholarship, and the German Empire as a cluster of excellence. Alternatively, Hungarians and Slavs might have disappeared from the radar by publishing less in German. That faculty commissions were also gradually turning toward their own linguistic networks is likewise indubitable. Scholars born and academically socialized outside of the empire were on the commissions, and they also frequently turned to their networks for advice on future nominations. Thus, ironically, the autonomy of the universities and the right to search freely for professorial candidates promoted the nationalization of the universities. For Privatdozenten, however, the boundaries were much blurrier, especially in Vienna, where scholars from throughout the empire studied and habilitated. Nevertheless, the nationalization processes influenced the universities in Pest and L'viv at the same rate as those in Vienna and Graz: as the first two became Magyar and Polish, respectively, Vienna and, in particular, Graz increasingly became German, with Habsburg culture being replaced in all cases.

At the same time, the few non-German scholars who had made careers in Vienna or Graz were valued, and they participated fully in faculty and academic life, regardless of whether they saw themselves as culturally other or whether they were seen as such in the faculty or by the press. As the number of multilingual scholars increased, ethnic stereotypes had a limited, but growing, influence on academic practice. Often historians exaggerate the influence of such stereotypes, however, using categories not yet valid in the nineteenth century.[73] Some such stories were written by the scholars themselves. When Władysław Natanson failed his habilitation in Graz, he attributed it to provincial German nationalism and anti-Semitism, since he could not openly admit scholarly failure.[74] This is not to say that there was no nationalism and anti-Semitism among Habsburg professors, but in most cases these did not affect professorial nominations.

However, from the 1860s the appointments followed strategies of othering. While before 1848 the ministry had favored Habsburg citizens, in the second half of the century, beginning with Thun-Hohenstein, linguistic-cum-cultural ascriptions mattered more—that is, a scholar's academic socialization and the (first or predominant) language of publications. As I show in the next chapter, this was handled in a different way when the Cracow and L'viv universities sought Polish-speaking professors, and Prague sought Czech speakers. But it would be false to speak of a complete ethnicization of the nomination procedures there, and I will show how the national scholars were made national. But I inquire also into other tendencies shaping these two subsystems, such as internal hierarchies and the dialectics between autarchy and internationalization. These show similarities although the geopolitical situation of the Czech and Polish scholars was diametrically opposite, with the former limited to the Habsburg Empire and the latter present in large numbers in all three central European empires.

Chapter 5

Habsburg Slavs and Their Spaces

While [people] at home consider me as a compatriot who has gone astray . . . in the ministry I am depicted as a national radical!

—Antonín Rezek to František Mareš, 1887[1]

The differentiation of academic space within the empire affected scholars and universities of all national identifications. Scholars who considered themselves to be Czechs, Hungarians, or Poles and who pursued careers at universities were also influenced by the linguistic disintegration. German-speaking scholars had less chance of being appointed at non-German-language universities in the empire, and vice versa for their Slavic and Hungarian counterparts. To a certain extent, the latter's appointment opportunities were worse, since there were no Polish, Czech, or Hungarian universities abroad, while universities in the German Empire offered career opportunities for German-speaking Habsburg scholars. Of course, there existed the possibility of being nominated to a university with another language of instruction, but this was rare. This chapter shows how these changes conditioned the development of the Czech-Bohemian and Galician universities.

In Pest the change in the language of instruction in 1860 from German to Hungarian, began a process in which the linguistic competences of scholars were equated with cultural belonging—both in the public eye and in academic policy. Nevertheless, as I argue, cultural belonging was still a very fluid concept, and often a scholar's own national identification would change depending on which university he was appointed to. When German became the language of instruction, most scholars remained in their positions because they had the necessary linguistic skills, as in Cracow in 1853. In contrast, the replacement of German by the respective local language—Hungarian, Polish, or Czech—from the 1860s onward had far-reaching consequences. Universities for Slavs were the most vital elements of cultural and national policy and were seen as a crucial aspect of societal discourse, in a much deeper way than was the case at German-language institutions. This was especially true in Cracow, where every professor was considered godlike.[2]

The movement of scholars that resulted from the changes in the language of instruction brought about three substantial changes. First, they had to be replaced, opening positions for young Privatdozenten, scholars who had been active outside of the university, and scholars from abroad. This process was neither as swift as often supposed—new scholars had to meet the same quality requirements as the previous German-speaking ones—nor as straightforward, since different groups representing differing ideals of scientific development were present in academic institutions. This led to a discussion about how to ascertain quality within a university that had now chosen a Slavic language over German, which I will present here using a Czech university as an example. Second, Czech and Galician scholars had been underrepresented in several disciplines (or not present at all) in the Habsburg monarchy, and they had to be imported from abroad or newly trained. This opened new spaces of exchange in which the identity of scholars would be discussed anew, reaching far beyond the bipolar German versus Slav distinction or a monocultural national discourse. Third, the autonomy of universities, or of linguistically defined networks of tertiary education (Czech Prague and Brno; Cracow and L'viv), brought intraprovincial schisms to the fore. In the Bohemian case, this meant "intellectual disintegration"[3] into Czech and German academic spaces. Conflicts then arose between the young, pro-internationalist generation of scholars and the conservatives seeking to promote local knowledge. In Galicia the generational question remained less obvious, although because L'viv became a Polish institution in 1871, that is, well after Cracow, a more progressive generation of scholars was appointed there. More prominent was, however, the question of Ruthenian scholars, where both sides' reluctance over cooperation and acceptance developed into entrenchment and open conflict in the early twentieth century.

The linguistic boundary did not simply create barriers but also opened distinct spaces, shifting the orientation of appointments from state to linguistic boundaries. Such boundaries had already been altered by appointments from the non-Habsburg German Confederation after 1848, which, as the previous chapter showed, was the space in which German-language universities were functioning.

These changes influenced Galicia in particular. Before these language reforms, the possibility of appointing Polish- or Ruthenian/Ukrainian-speaking scholars from the German or Russian Empires existed, but it was far from the first choice. With the relaxation of appointment policies and increased involvement by the Galician provincial government, such appointments had more chance of success. From the 1860s the ministry even advised the Jagiellonian University to search for candidates abroad if local scholars could not be found; the University of L'viv also later took this advice seriously.[4] Still, similar to the situation at the Habsburg German-language universities, such appointments often aimed, according to the records, at strengthening local academic quality and educating local scholars to prevent future nominations from abroad.[5] Appointments from abroad were also the last resort for the ministry, which opted for Habsburg scholars in cases of dispute, often proposing them against the will of the faculty.[6] Owing to the strengthening of university autonomy, and a desire not to aggravate the political tensions, the ministry only rarely nominated Habsburg scholars if the Galician university and provincial government opposed them.

Becoming Polish: Galicia

Cracow was the first Cisleithanian university to abandon German as the language of instruction; thus, it was a field for experimentation for the politicians. The first language changes, in 1861, targeted only a few scholars, causing little disturbance in the faculties.[7] To balance these departures, however, the ministry had to appoint, among others, two young scholars from Prussia owing to the lack of qualified scholars in Galicia. While the university was bilingual throughout the 1860s, 1869 witnessed an almost complete change to Polish.[8]

This transition was facilitated by national mobilization among Galician and foreign Polish speakers. Following an open letter by Józef Dietl in 1861, in which the newly chosen rector (see figure 5) invited Polish scholars to habilitate in Cracow,[9] the university received a large number of petitions for habilitations and chairs. These were mostly viewed negatively by the professors, who repeatedly stated that only disciplines not covered by professors should be left free for Privatdozenten, thus limiting the number of appointments. But the ministry was also skeptical about habilitating large numbers of scholars and only hesitantly agreed to a few faculty proposals. This affected scholars from the non-Habsburg areas of the German Confederation most of all; they were rejected because their foreign diplomas were not acknowledged. But political issues could also be a problem. In 1862 Józef Oettinger, an active progressive Jewish politician, was proposed as a Privatdozent for the history of medicine but was rejected by the ministry, which accused him of being a "fanatical Pole" who organized nationalist celebrations as a leading member of the Cracow Reform Synagogue.[10] This was, however, one of the very few habilitations that were accepted by the university but opposed by the ministry during this period. After the liberalization of Habsburg policies, the provincial government had no objections to Oettinger, and, seconding this recommendation, the ministry agreed to his habilitation in 1869.[11]

FIGURE 5 Józef Dietl, elected rector in 1861, became the most important spokesman of the pro-Polish professorate of the Jagiellonian University and, after being prematurely pensioned, became the mayor of Cracow. (Walery Rzewuski Museum of History of Photography, Cracow / Muzeum Historii Fotografii im. Walerego Rzewuskiego w Krakowie, MHF 20099/II. Photographer: Zakład fotograficzny Rzewuski Walery.)

The language issue rarely led to conflicts; if it did, it was mostly shortly before the language changes. German-speaking professors obstructed the appointments of scholars who were not fluent in German, and Polish professors proposed Polish-speaking scholars irrespective of their knowledge of German. The trend here was opposition to the appointments of the other group's candidates, with one side claiming that "Polish" scholars had poor scientific qualifications and the other not only arguing in favor of their scholarliness (Polish-language scholars stressed that the nominees were at least equal to the German-speaking candidates) but also emphasizing the importance of the Polish language for practical reasons. In Cracow such a controversy, around the chair of forensic medicine, led to the ministry serving as a mediator.[12] In 1868 at the University of L'viv, the ministry made an exceptional decision to allow parallel lectures in Polish in philosophy and economics in the law faculty. Two other issues of contention were habilitation procedures (whether they could be conducted in Polish and based on a Polish-language publication) and the use of Polish in history lectures.[13] In these cases, the faculty was divided almost perfectly along linguistic lines, with the exception of one German scholar who voted in favor of Polish.

The language changes of 1871 affected L'viv more seriously than Cracow, not only because all but four of the scholars who had been active in 1870 left L'viv, but also because the faculty encountered severe problems in determining who should propose their successors. That non-Polish-speaking scholars would continue their activities until they were replaced was regarded as unrealistic: the ministry reported that the press and the students were campaigning against these scholars, which hindered their work at the university.[14] Only in three cases did the ministry and the university agree to an exception to the condition of learning Polish within three years and lecturing in this language. Two of these three scholars had been transferred to L'viv from Cracow in 1869. However, only Eduard Buhl, who taught the history of German state and law in the law faculty, remained at the university after a three-year probationary period, knowing Polish but lecturing in German.[15] In 1877 the university vehemently refused to make Buhl's situation the basis for a legal exception that would allow instructors to teach in German.[16] The only chair that used German for instruction remained that in German language and literature, to which the bilingual Catholic priest Eugeniusz Janota was appointed in 1871. He was, in fact, the only Galician scholar affiliated with Polish culture to hold a professorship in this discipline until 1919.

While in Cracow the language question was solved with the introduction of Polish, this issue remained pivotal for several decades in L'viv, where teaching and other activities were conducted in two languages, Polish and Ruthenian. Because the administrative language was Polish from 1879, the obligation for instructors to know Polish was seen as an issue of practice; later this worsened the academic opportunities for Ruthenian- and German-speaking Jewish scholars. Both Ruthenian and German were the language of instruction in some gymnasia, despite growing pressure for assimilation.[17] In general, habilitations of scholars who lectured on Ruthenian topics, or who published in Ruthenian and on Ruthenian topics, were not welcomed before 1890 and were often rejected (officially) owing to language issues.[18] Ruthenians were not the only victims of the dominance of Polish-language purists. For the well-known Jewish neurologist Gustaw Bikeles, who spoke broken Polish (his low level of competency probably caused both by his hearing impairment and by the fact that German was his first language), language was a vital issue. After five years as a Privatdozent, in 1906 he was proposed as an associate professor. The faculty, supported by a medical expert, agreed to award him only the title and character of associate professor, claiming that Bikeles would never gain a full professorship owing to his deafness and thus should not be fully supported.[19]

The Division of the Charles-Ferdinand University and the Disintegration of Bohemia

The language issue was also a major problem in Bohemia: only through ministerial support could Czech scholars acquire chairs at the Charles-Ferdinand University until its division in 1882. Apart from ideologically motivated appointments immediately after 1848, and the nominations of Czech professors by the minister of education Josef Jireček in 1871 (described in chapter 3), characteristic here is the situation in 1870. In this year the gynecologist Jan Streng was promoted to the chair of the Institute of Gynecology because he spoke Czech.[20] The Prague medical faculty had proposed three German Bohemian scholars, claiming that they "had all gained their education at the University of Prague, are completely fluent in the Czech language, and had been appointed to other universities [Graz, Bern, and Tübingen, respectively]. This was because of their scientific achievements during their early careers as young scholars."[21] What the faculty proposal meant by "fluent in the Czech language" was, however, different from what Czech-speaking Prague scholars expected. For the scholars in the proposal, Czech was a second language, while Czech scholars had requested a true native speaker, who would count as a Czech national. As I show below, this nationalist-driven distinction was not always as clear as it seems.

Similar disagreements came to the fore in 1881. The rules for the division of the university stated that each institute would be located at the university (German or Czech) where the head of the institute chose to teach. Thus, the conflict over who would be appointed was particularly meaningful.

Consequently, for the first chair of internal medicine in 1881, the faculty proposed, unsurprisingly, three scholars with a German-language cultural background. The ministry, however, appointed associate professor Bohumil Eiselt, one of the few university scholars in Prague who published in both German and Czech and one of the most prominent organizers of academic medical research in the Czech language, as well as the founder of Časopis lékařův českých.[22]

Even though the ministry supported Czech-speaking scholars, those regarded as too nationalistic were treated differently. The following story of the events leading to the appointment of the third director of a medical institute, the surgeon Vilém/Wilhelm Weiss (who later continued his activity at the Czech University), exemplifies this. Pronounced Czech patriotism was perhaps no obstacle to obtaining a professorship at Habsburg universities, but Prague, seen as the main locus of conflict between German and Czech patriotism, was subject to special consideration in this regard, and the ministry balanced these two opposing groups. In 1878, while teaching as a full professor in Innsbruck, Eduard Albert, a pronounced Czech patriot, was proposed for the chair of surgery in a minority opinion (*Minoritätsvotum*) of the Prague medical faculty. The ministry, however, decided not to appoint him because the "peaceful life of the faculty" might be troubled through the appointment of a scholar who "is not completely objective toward Czech national efforts."[23] In 1880 Albert tried once again to transfer to Prague after the previous incumbent, a Czech-speaking surgeon, retired. This time, the faculty decided overwhelmingly against including him in the proposal, proposing only German-speaking scholars for the position.[24] (Albert received the chair of surgery at Vienna in 1881 only as a kind of compensation for his unsuccessful attempt to gain a position in Prague.) After long deliberations, the ministry decided to appoint a scholar from outside the terna, Weiss, who later taught at the Czech University after 1883. Weiss, like Eiselt, had previously been active in Czech medical organizations and journals and had the support of the Czech public and scholars as well.[25]

There were, in fact, a large number of Czech scholars who were working abroad as well as at universities in the empire. Some of these professors had published in Czech as young scholars but for various reasons ceased to do so. This group included some professors who chose the German University in Prague after 1882, thus identifying with German culture.[26]

While in Prague the choice to publish in a language other than Czech was a conscious decision to reject direct participation in the Czech national project, for scholars outside of Bohemia this was not so straightforward. For example, the forensic pathologist Eduard Hofmann, who worked in Prague and later in Innsbruck, published throughout his career a series of articles in Časopis lékařův českých, took part in several Czech-speaking projects, and influenced the appointment of his close friend Albert to Vienna.[27] In most cases, scholars who published in Czech were considered possible candidates for teaching positions, and some of them indeed gained professorships at the Czech Charles-Ferdinand University in Prague or the technical academy in Brno.[28] Although the number of Bohemian scholars who were bilingual is unknown, as are their fates, one can suppose that the either-or dilemma that scholars in Prague faced in the 1880s was the most intense, whereas scholars elsewhere had other options. Vienna, apparently a privileged place for scholars who did not want to be categorized according to nationalist ideologies, provided, for instance, several professors who were nominated to Czech Bohemian institutions.

In Prague a choice had to be made in the language issue, as Anton Gindely, the Bohemian professor of general history, painfully experienced. Because he had signed a petition by Czech professors for an increase in Czech chairs in 1880, he was marked down as a Czech nationalist. In 1882, when he decided to move to the German University in Prague, the professors (with the exception of Julius Jung) published a memorandum opposing his transfer to the German faculty. However, his choice of the German University was reason enough for Czech scholars to deny him a place at the Czech University. Gindely, who from 1870 had been politically active and had unsuccessfully tried to establish a cross-national conservative party, was, however, supported by the ministry and retained (officially) his position at the university and his directorship of the Bohemian Archives. Later in life, he was also a member of the Franz Joseph Czech Academy for Science, Literature and Arts.[29] He was one of only two active members of the German University in Prague elected to membership in the academy.[30] This story should not be read literally as indicating that the Czech Academy completely ignored German scholars as members. The second member of the Czech Academy who came from the German University, the comparative linguist Alfred Ludwig, had been born in Vienna and had no knowledge of Czech before moving to Prague in 1860. There he became interested in Czech culture, and he even published in both Bohemian tongues. Although clearly not identifying as Czech, in 1882 he was allegedly given the choice to join either the Czech or the German University; he joined the latter to give one of his students the opportunity to teach at the Czech institution.[31]

The fluidity of Bohemian identities begs the question of how many Czech-speaking scholars were working at the University of Prague at the time of its division. This is not easy to answer. For some scholars the decision on which side to support was made in 1882, given that there was no possibility of teaching at both universities. Because of the previous dominance of German in publications, scholars who published only in German could have identified as Czech. In any case, the Czech University, and its medical department in particular, was considerably less prepared for its opening than was the German University, and the fears Czech professors had voiced during the debates on the future of the university proved to be correct. For whatever reason, the number of Czech assistants and Privatdozenten at medical faculties was quite low before 1882. The Prague historian Ludmila Hlaváčková states that from 1872 on no Czech scholars had habilitated and that of the thirty-one assistants at the faculty, only three were Czech.[32]

With only three professors thus choosing the Czech-language medical faculty, its foundation was postponed until 1883, and even then it was opened with only sixteen instructors, while the German faculty numbered thirty-six at the time. This discrepancy, however, did not last forever. In 1910 the two faculties were more or less even, with about sixty instructors each, and the Czech University had a few more professors than the German one (see table 3 in the previous chapter). The second issue aggravating the situation in Prague was the question of clinics, which the Czech medical faculty lacked owing to the regulations governing the division of the university. The faculty soon acquired a new building, which allowed a clinic to operate, but it had to be expanded considerably in subsequent decades.[33]

The issue of the medical faculty was a vital one for Czech scholars because of its practical connections to health and sanitary institutions in the city. While delayed at the beginning, this issue was addressed by the university in December 1882, and by 7 January, Franz Joseph wrote to the minister: "I authorize you to begin the preparations to activate the medical faculty of the University with Bohemian [i.e., Czech] as the language of instruction."[34]

Since only a few Czech physicians were active in the university, a commission established by the governor prepared for the medical faculty's opening. It proposed not only candidates for professorships (ironically relying on the professors of the German faculty as experts) but also additional assistants and institutes, which would help the faculty achieve its goals swiftly. The proposals for personnel at the time show where Czech physicians were situated; many were active in universities outside the monarchy. The proposals included several Bohemian practitioners, scholars, and assistants from Vienna as well as two Czech émigré professors from the Russian Empire. Only for medical chemistry could no qualified Czech-speaking scholar be found, and a young Ruthenian assistant from Vienna, Ivan Horbačevs'kyj (Іван Горбачевський, Jan Horbaczewski), was appointed.[35] Some of the proposed scholars were seen as unready for professorships, and in several cases temporary auxiliary professors were appointed instead of permanent professors.[36] Other appointments did not succeed for financial reasons.[37] But even with a limited number of professors, the faculty was officially inaugurated in 1883.

The situation at the philosophical faculty was much simpler since several of the professors teaching there had already been active at the undivided University of Prague. Of the professors who chose the philosophical faculty at the Czech University, only a few had advanced along a normal route with faculty assistance. Most had been nominated by Leo Thun-Hohenstein and Josef Jireček, who did not really care for the faculty's proposals. The Privatdozenten transferred to the Czech University in 1882 had mostly habilitated in the second half of the 1870s. Further, another group of Czech scholars had been appointed to professorships after the division of the university had already been decided; they had previously taught at other Czech-language institutions in the city.[38]

In addition, the philosophical faculty at the Czech University had to appoint a number of scholars from outside the institution. First, Privatdozenten who were recognized as Czechs but were working outside of Bohemia were proposed for professorships. Second, scholars working at other educational institutions, gymnasia, technical universities, and Czech scholarly organizations were appointed; they were often supported from within the faculty. These nominations included scholars who previously would have had no chance at the university but who were already widely known, having authored well-received publications in their respective fields, mostly in both Czech and German.[39] Since the university did not have a full complement of institutes, as these had mostly gone to the German University, not all chairs were filled immediately. For example, the chair of practical astronomy at the Czech University came into being only in the 1890s, after the observatory issue was dealt with and August Seydler, a professor of theoretical astronomy and practical physics, retired; his chair was then divided into two.[40]

The growing division of the scientific landscape worsened when two parallel nationally defined institutions, the Franz Joseph Czech Academy for Science, Literature and Arts and the Association for the Fostering of German Science, Arts and Literature in Bohemia (Die Gesellschaft zur Förderung Deutscher Wissenschaft, Kunst und Literatur in Böhmen),[41] were established in 1890 and 1891, respectively. The bilingual Royal Bohemian Society of Sciences did not cease to exist, but its work increasingly reflected the Czech-German split.[42]

Dividing institutions along cultural lines or establishing separate Czech- and German-language institutions created a largely dual public sphere, which influenced scholarly contacts and even patient-doctor relations. Some clinics had regulations that on odd-numbered days German was used (for German-speaking patients) and on even-numbered days Czech, resulting in a similar division of patients, who, for the sake of communication as well as legal issues concerning childbirth (especially the spelling of names), would wait for "their" physicians, leading to bizarre and often also perilous situations. Similarly, the distribution of cases and even corpses followed this linguistic division.[43] This led to running jokes that Prague scholars from the opposing cultures could only meet during conferences abroad. A more macabre version was that such a meeting was possible only at the deathbed of a prominent nobleman.[44] Since one finds in the records a number of hardly believable stories (e.g., that in the construction of new institutes, only German workers and craftsmen should be employed),[45] such stories might indeed contain a kernel of truth.

However, the division of the faculties was not as fixed as one might imagine, and in several instances it was either questioned or deliberately violated. The theological faculty remained undivided until 1891 owing to the influence of Prague's Prince-Bishop Friedrich Schwarzenberg; after his death in 1885, Prague theologians hindered the division for six more years by appealing to his legacy.[46] In addition, the university administration, the archives, and the university library were not divided until the interwar period. But the supposed academic disintegration had even more flaws. Recent historiographical research has uncovered many more informal contacts among professors in Prague, even though these were hidden from the public eye in informal places such as the Café Louvre. Indeed, renowned scholars such as Albert Einstein drew Czech students to lecture halls. (Although Czech students were allowed to attend lectures at the German University, they did not have the right to take the course exam; the same regulation applied to German students at the Czech University.)

The legal form of the university's division also met with criticism from scholars teaching in Prague. The prominent historian Jaroslav Goll, for instance, criticized the policy of "one university, one language," pinpointing its dysfunctionality in disciplines that would need German lectures, such as Habsburg history, German language, German literature, and German law.[47] Chairs of languages indeed proved to be problematic, as also in Galicia. The German University strove to enhance Slavic philology, which from 1882 was covered only by none other than Alfred Ludwig.[48] Only in 1909 did the first habilitation for Czech language and literature at the German University take place, and the first associate professorship was awarded only in 1917. Conversely, for some years, the Czech University lacked a full professor and an institute for German language, achieving this only in 1894.[49] However, habilitating in "German" and "Habsburg" disciplines at the Czech University in Prague was quite popular. Five scholars habilitated in German literature, five in Czech literature, and five in Austrian history, compared with only three Privatdozenten for Czech history. As I show below, here the change to Czech in the university had different effects than the change to Polish in Galicia; the Czech University retained much of the undivided university's Habsburg character.

At the German University, scholars from the Czech University were not considered as possible appointees, and vice versa, because the "specific circumstances" in Prague made cultural transgressions unfeasible, leaving a limited number of scholars from other institutions who could be appointed. In several cases, however, younger scholars cooperated with each other, for example, in German literature in a group around August Sauer.[50] Sauer, however, openly pleaded for a "recapturing" of Prague by German students.[51]

The local circumstances of the Czech University in Prague, which had fewer possibilities for academic exchange, were not only a common argument for the creation of a second university but also influenced appointment procedures. Five years after the division of the Charles-Ferdinand University, the Czech faculty stated that given that Czech scholars had no possibility of being promoted to other universities, the only way to ensure sufficient high-quality habilitations was to limit appointments of scholars from outside the university. This argument was used to respond to criticisms that a terna proposed in 1887 for the chair of gynecology did not take into account renowned gynecologists from outside Prague. Since capable scholars were already in place, the proposal noted, "this faculty stands in a special position sadly owing to a still existing animosity, and its members have under these conditions scarce expectations to find employment at other, especially German universities, a circumstance that has discouraged some scholars from obtaining a Privatdozentur at this university. So, the faculty has decided . . . to take only its own forces into consideration."[52] Two local scholars mentioned in the terna, although specialized gynecologists, were seen by the ministry (clearly advised by Eduard Albert) as having been proposed only for this reason and not for their scientific qualifications; this resulted in the appointment of Karel Pawlík, a Viennese Privatdozent, as a full professor.[53]

Against the Chinese Spirit: Exchange and Competition in Czech Bohemia

Appointments of scholars from outside Bohemia remained scarce until 1918 and mostly resulted from personal contacts. Most Viennese scholars, including Pawlík, were nominated thanks to the support of Eduard Albert, who was critical of the ideas of the Czech Prague professors who opposed the appointment of scholars from outside the university.[54] Although the Viennese scholars were not the only appointees from non-Czech institutes, both the medical and philosophical faculties had limited transfers from other universities within and outside the empire, nominating only a handful of scholars, mostly émigré Bohemians and Moravians who had established themselves at other institutions. As an analysis of the place of graduation of Prague scholars demonstrates, the local Prague environment clearly predominated here; only a few graduates from outside the Czech Charles-Ferdinand University were working as Privatdozenten or professors (nine scholars, that is, 6 percent).[55] The number of scholars who had studied at other universities in the Habsburg Empire and abroad was higher: 15 percent at the medical faculty and 35 percent at the philosophical faculty; these stays abroad, however, mostly involved one- to two-semester scholarships.

Similarly, few appointments were made from the Czech University to other institutions. Konstantin Jireček's appointment to the University of Vienna in 1894, where he later founded the Institute of East European History (Institut für Osteuropäische Geschichte), was a notable exception.

Only the Galician-born anatomist Andrzej (also in Czech: Ondřej) Obrzut was appointed to another medical faculty, moving to L'viv in 1896. Scholars from the Czech University were seldom considered for chairs at other institutions, and, if so, this mostly occurred through personal contacts. The most prominent was Tomáš Garrigue Masaryk, who expected to receive the chair of philosophy in Vienna[56] (but was not nominated by the faculty).

While exchanges with other Habsburg institutions rarely took place, a significant number of scholars moved between the technical academies in Bohemia and Moravia and the university. Eleven scholars came from the Czech technical academy in Prague and one from the technical academy in Brno, while nine were appointed to these two institutions (see also table 12). In particular, the academy in Brno profited from Prague's Privatdozenten: the number of Privatdozenten in Prague around 1900 (that is, when the technical academy opened) was exorbitant in comparison with the number of possible professorships, and thus moving to Moravia was a welcome career choice.[57]

The scarcity of opportunities for appointments outside Prague, as well as the limited exchange with other institutes, was criticized, and various solutions were proposed. From the 1890s on, Czech scholars pleaded for the establishment of a second university in either Brno or Olomouc, which, it was hoped, would also improve academic quality through exchanges and competition among scholars. Masaryk wrote on this occasion that "a second university, giving more freedom for the students and also for some professors, would speed up and strengthen scientific development. This moment can be named with a word: scientific competition—students would have a broader choice of teachers, they would be less dependent on individual professors, and the scientific currents and directions of one university would have unmeasured influence on the other university. After all, there is no doubt that if there is no competition, haughtiness and the Chinese spirit appear."[58] Similarly, Goll saw exchange as augmenting scholarly quality and criticized the sacrifice of Czech scientific needs, and thus of the needs of the Habsburg Empire, for political reasons, rebuffing the claims of German nationalists in Moravia who opposed the creation of a Czech-language academy there.[59]

While Czech intellectuals of all political outlooks saw a second Czech university as vital for their culture, some scholars regarded exchanges with German culture, particularly German universities, as integral to maintaining the quality of Czech scholarly culture in particular and intellectual life in

TABLE 12 Transfers between universities (philosophical faculties) and technical and agricultural academies in the Habsburg monarchy, 1848–1918

	Transfer to		
Transfer from	**Technical academies**	**Academy of Agriculture in Vienna**	**Academy of Agriculture in Dubliany/Dublany**
Vienna	15	2	NDA
Graz	3	NDA	NDA
Innsbruck	4	NDA	NDA
Prague	4	1	NDA
Prague: German	8	1	NDA
Chernivtsi	1	1	NDA
Prague: Czech	11	NDA	NDA
Cracow	3	1	4
L'viv	3		1

	Transfer to								
Transfer from	**Vienna**	**Graz**	**Innsbruck**	**Prague**	**Prague: German**	**Chernivtsi**	**Prague: Czech**	**Cracow**	**L'viv**
Technical academies	8	2	1	7	11	4	15	3	2
Academy of Agriculture in Vienna		1			1			1	
Academy of Agriculture in Dubliany/Dublany								3	1

Note: The table excludes scholars who taught as Privatdozenten while simultaneously holding a professorship at another school.
NDA, no data available.

general. The Czech university and technical academies offered more scholarships to study at other universities than did the German university in Prague and German technical academies; thus, the issue of internationality was not merely a rhetorical strategy but led to a search for practical solutions as well.[60]

In the discussions about the need to open Czech culture to exchange, Albert, Masaryk, and Goll took leading roles, questioning the absolute value of Czech culture (as claimed by the older generation) and warning that cultural isolation would hinder scientific productivity. Dependence on, or close interdependence with, German culture was publicly criticized in the conflict over the anonymous publication of Hubert Gordon Schauer's "Naše dvě otázky" ("Our Two Questions," 1886). In this article, the author formulated a provocative thesis, foreseeing a crisis of Czech culture if it enclosed itself in a linguistic ghetto. Bohemian intellectuals strongly opposed this view but only at an emotional level (i.e., without an analytical discussion).[61]

It is clear from the debate surrounding this work that the issue of cultural exchange was a pressing problem for scholars. Masaryk, building on Purkyně's ideas, cautioned against not staying in touch with recent developments in scholarship outside of Bohemia; in the 1880s he envisioned an internationalization of academic institutions that would help achieve this aim. He was, however, severely criticized by the conservatives as a follower of German (i.e., foreign and not native) philosophy.[62] Goll wrote more directly that Czech scholars had a strong tradition of exchange with "German" universities, which they should not abandon because of political tensions. In particular, he felt that historians should spend time at the IAHR: "As we were to prepare for academic careers, our old teachers advised us to visit a German university abroad. . . . At our faculty this tradition is still alive."[63]

While Czech scholars saw interdependence as positive,[64] some German articles claimed that the Czechs' dependency on Germans was responsible for the existence and prospering of Czech culture. The prestigious journal *Hochschulnachrichten* (Higher-education news), which concerned itself with academic issues, wrote, for instance, that, "divided into two universities, this coexistence and thus an always visible competition with German science secures the Czechs from sliding down from the current level, and Czech science and art have the possibility to be seen internationally only through German intermediation."[65] The fierce debate on the interdependence of the two cultures was, however, almost exclusively conducted from the standpoint of asserting cultural hegemony, questioning why Czech scholars were dependent on German science and what the possibilities were for breaking this dependence. If the issue of transfers in the other direction was raised, it was only by Czechs, who questioned the necessity of bilingualism being applied in only one direction.

Interestingly, Czech scholars regarded alternative channels of transfer and exchange as insufficient on their own. Contact with France, although frequent for political reasons, never led to an intensification of student exchanges or long-term fellowships.[66] For fellowships, France was still more popular for Czechs than German-language Habsburg universities. Nevertheless, the German Empire topped the list, which indicates a gradual change from stays at other universities within the state to stays transgressing imperial boundaries in the late nineteenth century.[67]

Slavic reciprocity, after its initial boom up to 1860, now met with increasing skepticism.[68] While they never denounced it as an important source, leading intellectuals saw inter-Slavic communication only as complementary to maintaining and intensifying exchange with the ephemeral "western science."[69] Practical endeavors strengthening this cooperation were also only partially successful; for example, joint meetings of Polish and Czech physicians did not go beyond planning and courtesy visits,[70] although political reasons partially hindered such meetings, such as that planned for Poznań/Posen in 1898.[71] Similarly, the creation of a St. Petersburg–led pan-Slavic Academy of Sciences, supported in Prague, was blocked for political reasons—in this case by Polish elites.[72] Cooperation was more intense among the academies of sciences, with numerous nominations for members (e.g., there were fifteen Czech members in the Cracow Academy, and sixteen Polish members in the Czech Academy)[73] and jointly planned archaeological expeditions.[74]

Galicia and the Cisleithanian Academic Space

While the idea of appointing national scholars also dominated in Galicia, the patterns of scholarly exchange were different there than in Bohemia or at the German-language Habsburg universities. Outside the Habsburg Empire, there were many Polish-speaking scholars but few Polish-language academic institutions. Therefore, contact with other regions was quite one-sided and mostly oriented toward attracting the best Polish-speaking scholars to Galicia. At the same time, support for local students was strong, creating a tension between closeness and openness similar to the situation in Prague. In comparison with German-language Habsburg universities, where turnover in academic positions was the norm, Galician universities primarily promoted their own scientific staff: around 75 percent of scholars spent their entire careers at the university where they had habilitated, and half of these became associate or full professors.[75] Except during the period when Galician universities were German-language institutions, scholars working outside of universities were seldom appointed to chairs. Before the language reforms, such scholars (including litterateurs and gymnasium teachers) had often been directly appointed as professors, comprising around 20 percent of all professors at the time (i.e., before 1861 in Cracow and 1871 in L'viv).

In the nineteenth century, the Polish independence movement, socialism, and the Russophile movement were more developed abroad than in the monarchy, and the provincial government intended to limit the possibility of importing them into Galicia. At the same time, the threat of nationalism as such had diminished in the eyes of political elites in the 1860s; it was no longer seen as a category that excluded candidates from teaching positions so long as it was not linked to independence movements or political radicalism. Still, political supervision was in place, and the L'viv professor of Ruthenian language and literature Jakiv Holovac'kyj, whom the provincial government suspected of being a member of the Russophile movement, was dismissed from that university, although the faculty fought this decision bitterly.[76]

The acceptance of nationalist rhetoric is clearly discernible in both applications for habilitations and professorships at universities and correspondence with the ministry, where the well-being of the Polish nation within (and later also outside of) the Habsburg conglomerate was increasingly accentuated. The case of Oettinger, mentioned above, already signaled a changing attitude toward nationality in the late 1860s. In the same period, when the Cracow philosophical faculty applied for the reinstatement of Wincenty Pol at the university, Pol wrote that the political conditions that had led to his dismissal had "changed constitutionally in the question of national development and education; this could qualify the decision to regain my previous position at . . . the University of Cracow."[77] The faculty welcomed this proposal, and the provincial government emphasized that Pol belonged to "the most acclaimed men of his nation, [and] his restitution would find the approbation and most appreciative gratefulness of the whole country,"[78] referring clearly to the Polish nation and not to Galicia. Even if the application was ultimately unsuccessful, it used a language that had been impossible few years earlier. This change was clearly perceived outside the Galician universities as well, and the following year one of the applicants for habilitation in L'viv referred to his glorious nationalist past and participation in the uprising of 1830.[79]

With the institutionalization of lectures on Polish history, law, and so on, the definitions of scientific patriotism and nationalism blurred, allowing a renegotiation of the boundary between that which was allowed and that which was prohibited. While in the 1850s and into the 1860s nationalism had been rejected in favor of state patriotism, Polish nationalism, in its cultural-patriotic rather than its chauvinistic or openly anti-Habsburg version, was viewed positively from the moment of Galician "autonomy."[80] Thus, it is not surprising that "Poland," as a historical and cultural construct, came to be more clearly referred to as a nation in its own right, by both academics and the ministry.

This change in political discourse clearly influenced the faculties. There were, for instance, no well-qualified young historians to teach Austrian history, and the fact that the young historians who had to hold these lectures specialized in Polish history caused conflicts with the loyalist historians who had been chosen by the ministry in the 1850s and early 1860s.[81] There were also only three habilitations in Austrian history until 1918, as opposed to twelve in Polish history. Similarly, German was defined as a foreign language, and interest in it was seen as merely practical. When Naphtali Sobel applied to habilitate in Old German literature in 1884, the faculty wrote that this was too narrow and that, because German was a foreign language, the university had no interest in accommodating scholars specializing in this subject.[82] That same year, however, Maksymilian Kawczyński habilitated in German philology; he was the sole Polish Privatdozent in this discipline, although only briefly:[83] from 1887 his interest turned toward the philology of the Romance languages, in which he earned first habilitation and then a professorship.[84] He was, apart from Janota, the only Galician-born scholar acknowledged to be teaching German language and literature at the academic level,[85] even though ministerial scholarships for Galician scholars willing to pursue this discipline had been available since 1888.[86] When, in 1913, the Jagiellonian University proposed the creation of a chair of German language and literature with Polish as the medium of instruction, the faculty was unable to suggest any candidates.[87] In comparison, at the Czech University in Prague, habilitation in "Habsburg disciplines" still enjoyed considerable popularity. This difference shows how Galicia detached itself from Habsburg universities, gradually moving toward the creation and analysis of a Polish collective imagination and history.

As in Bohemia, linguistic changes in Galicia brought changes in appointment practice. Transfers between Galician and German-speaking Habsburg universities, however, were more common than in the Czech case. Twenty-five scholars transferred from other Habsburg to Galician institutions, with peaks in 1849–64 (seven) and 1890–1900 (eight), but only a small percentage of scholars transferred in this direction. Most of the mobile instructors who were members of the philosophical faculty had habilitated in Vienna and were promoted from the position of Privatdozent when they moved to Galicia. After the language reforms, the number of such transfers increased in absolute terms, but they made up a smaller proportion of appointments because the number of chairs at Cisleithanian universities grew considerably during this period. The character of such transfers also changed: German-speaking scholars born in Austria and Bohemia had predominated up to 1864, but after that time most appointees from abroad were scholars from the German Empire and Galicia who had habilitated in Vienna, or Polish speakers born outside Galicia. The growing number of Galician civil servants working in Vienna was one of the main reasons for this development. Either they or their children habilitated at the University of Vienna, the German-language Habsburg university from which Galician institutions appointed most scholars.

Moreover, only a few scholars were appointed from universities in the Russian and German Empires, ten and eleven respectively, with the largest number coming from the Warsaw Main School; however, scholars who taught at other universities, such as those in Kazan or St. Petersburg, were also appointed. The number of proposed scholars from abroad who were not appointed was not high, with financial issues being the largest problem in the negotiations.[88] In other cases, the faculty had to withdraw proposals because the candidates did not accept the facilities available.[89] As the nominal salaries in the Habsburg Empire were low compared with those in other empires, appointments from both neighboring empires were limited to Privatdozenten, with a few personally motivated exceptions. Similarly, only a few Galician scholars were appointed to universities in the German and Russian Empires, mostly for disciplines linked with Polish language and history.

An analogous pattern can be found in appointments from Galicia to German-language universities in the Habsburg Empire. These transfers occurred in larger numbers only in 1860–64 (when there were seven) and 1870–74 (nine), which was linked to language reforms and the relocation (*Versetzung*) of German-speaking scholars. Of the scholars leaving Galicia after the 1870s, almost half had occupied chairs with German as the language of instruction, with a negligible number of transfers of scholars who regularly taught in Polish. Polish-speaking scholars teaching in Galicia were also, with four exceptions, not considered as nominees for chairs. Clearly, both the Bohemian and the Galician universities were promoting their own staff, which influenced the relations between these universities. Although Bohemian and Galician scholars cooperated at a personal level, transfers were rare and, when they did occur, were linked with personal connections.

Given the concentration on younger scholars, the facilities of the various university faculties, and the invitation of scholars from abroad (to occupy a chair or habilitate), the low number of transfers between the philosophical faculties of the two Galician universities after 1867 is unsurprising; only nineteen scholars moved from Cracow to L'viv, and eight moved in the opposite direction. These were generally Privatdozenten who were appointed as professors (eight and four, respectively) or who changed their affiliation (five and two), and there were similarly few transfers with other institutions. One cannot speak of returning scholars, as these mobile teachers had graduated either at the university of habilitation or at German-language universities. Transfers remained similarly limited at medical faculties: L'viv acquired its medical faculty only late in the nineteenth century.[90] However, the faculty's most important physician in this period, Ludwik Rydygier, was nominated from Cracow. Further, only a few scholars moved to or from other academic institutions within Galicia (technical and arts academies and the Academy of Agriculture in Dubliany/Dublany; see table 12); nevertheless, a larger number of scholars worked in museums, archives, or libraries in addition to their university positions (e.g., at the Ossolineum, which actively accommodated and supported humanists in L'viv).

Reorienting more and more from the Habsburg system to a "Polish" academia, Galician universities nonetheless remained bound to the Habsburg legal system, which regulated, although with local differences, the number and designations of chairs, remuneration, and habilitation procedures. On all these issues, Cisleithanian universities organized collective efforts, uniting scholars from institutions across the monarchy. The denotations of disciplines were also relatively binding, and the structure of the faculties with regard to the relations among disciplines was rather similar—apart from the existence of chairs for national historiographies and national languages and literatures, as well as the inclusion of agricultural studies in the philosophical faculty at the Jagiellonian University. Some exceptions can be found, though: anthropology was first institutionalized at the Czech University and then in Galicia, before being introduced at other universities (apart from Vienna, which had such a chair early on). In contrast, until 1910 both Galician universities lacked Hebrew and Jewish history, which were taught at other universities in the empire. This is quite surprising, but also very telling, if the cultural statistics of Galicia are taken into consideration (see also chapter 6).

The matter of external experts for habilitations and appointments remained prominent with regard to the unity of the imperial space. These experts were asked not only to assess the qualifications of the candidates but also to help faculties decide whether they had qualified specialists. While these experts were mostly Viennese scholars, the Galician university also asked Czech scholars for expertise in disciplines such as anthropology[91] and oriental studies.[92] Czech scholars could mostly read Polish, which gave them an advantage over specialists from German-language universities. Over the course of the nineteenth century, expert opinions became harder to obtain, especially because scholars' applications were based on publications in their native language, and experts therefore had be found within Galicia. While until the 1880s the ministry had regularly asked Viennese instructors for their opinions on scholars from Bohemia and Galicia,[93] later they could ask only a few who knew Czech or Polish; in this way, such scholars gained political influence over the appointment procedures. Only in formal cases, such as the determination of a habilitation's scope, could the ministry still ask for the participation of specialized scholars.

The issue of expertise also shows the complexity of the Austrian imperial space, as German-language scholars frequently voiced paternalistic opinions of Slavic scholarship. In 1878 such comments on the habilitation of the geographer Karol Benoni led to a clash between the faculties in L'viv and Vienna. The opinions of three Viennese geographers were rather negative, describing the applicant's publication as "cunning compilations" based on outdated theories. More critically, they stated that this would not be adequate for a habilitation in Vienna but would do for L'viv.[94] Unsurprisingly, the Galician faculty took this suggestion as disparaging the standing of the university and accused the Viennese scholars of proposing double standards for scientific quality in the empire, which L'viv's scholars, of course, did not want to accept.[95]

One final feature of the Galician academic exchange within the Habsburg Empire is worth mentioning. By comparing the careers of appointees from the Habsburg Empire, be they Polish-speaking or German-speaking, with those of scholars appointed from the Russian Empire, one sees that after the language change and the first wave of purging the university of "German" professors, new appointees who did not know Polish cooperated peacefully and fruitfully with others in the faculties. (The exception was the notorious German nationalist August Sauer: in 1883, when he was a young man, his contract as a professor in L'viv was not made permanent because he had insulted Polish people in Galicia by criticizing the lack of civilization in the province in a series of articles printed in German-language newspapers in L'viv.[96]) Just about all of the non-Polish Habsburg scholars proposed by the faculties, such as the professors of German literature and language and the professor of animal husbandry Leopold Adametz, learned Polish and took part in the local cultural life of the province.[97]

Ironically, while Vienna- or Graz-educated scholars adapted well to Galicia, most scholars who were educated in the Russian Empire and then moved to Galicia met with conflict, and some even eventually returned to Russia.[98] Several others, including the important Darwinist Benedykt Dybowski, remained after serious clashes.[99] In any case, Habsburg transfers proved much less conflict laden than "intra-Polish" ones, uniting the empire at a nonlinguistic cultural level more than historians have thus far brought to light.

A slightly different and more colorful picture of educational diversity can be obtained by looking at scholars' places of graduation, as the number of scholars who had not graduated from a Galician university was rather high. Because L'viv and Cracow were the only universities with Polish lectures, with the exception of the Warsaw Main School between 1857 and 1863, they attracted Polish-speaking scholars from abroad for habilitation. At the same time, both the universities and the authorities supported young scholars with scholarships to allow them to study outside Galicia; such stays were directed toward the German Empire rather than other Habsburg universities.[100] Some grants included a formal requirement of habilitation within a certain time; these were also limited to provincial universities. Teacher-student relations facilitated this: scholars proposed that their students habilitate in Galicia, or young scholars were sent to German-language universities, following the path of scholars who had studied there before. Here, one can also see that scholarships were awarded predominantly to those from German-speaking countries, although Britain and France became increasingly popular in the late nineteenth century; "German" education was highly valued in all disciplines and mentioned positively in most decisions.

Albeit unsuccessful to begin with, Dietl's invitation to Polish scholars to habilitate in Galicia (see above) bore fruit, and the number of scholars who had been educated abroad soared, especially among new professors. While non-Galician graduates made up around 25 percent of new appointees at both medical faculties in the late nineteenth century, this number was around 50 percent at philosophical faculties; in both cases, graduates from German Empire universities were dominant. In Cracow graduates of German-language Habsburg universities came in third after graduates from the Russian Empire.[101] The scholars who had graduated in the German and Russian Empires included some who were not born in those empires; also, a high number of scholars from the Russian Empire and Galicia had graduated from universities in the German Empire. In contrast, those born in neighboring empires very rarely graduated in Galicia. However, at the individual universities, most nonlocal graduates had earned their PhDs in Vienna, except in the medical faculty in L'viv, where Cracow provided the most young physicians. No university in the German Empire came close to providing as many graduates as Vienna. Of the German Empire universities, Leipzig provided the most graduates, for both the medical and philosophical faculties.

Making National Scholars

As mentioned above, recruiting appropriate scholars was a matter of the utmost importance for Slavic universities. In both Galicia and Bohemia, the question of how to appoint scholars speaking the appropriate language and at the same time sustain scholarly quality was a vital one, not least because German activists closely followed the nomination procedures to find confirmation of the superiority of German scholars. While both universities intended to support local, national scholars, occasionally they had to resort to academics from abroad.

At the Czech University in Prague, scholars identifying as German were clearly not an option. Other Slavic scholars who did not speak Czech were only rarely considered for professorships.[102] The Galician-born anatomist Andrzej/Ondřej Obrzut, who had studied in Galicia but habilitated in Prague, was the only Polish scholar at the Czech University in Prague. The Ruthenian Horbačevs'kyj was nominated in 1882 owing to a lack of qualified scholars in medical chemistry, as mentioned earlier; not only was he unable to speak Czech, but he was also the only Ruthenian appointed.[103]

The case of Josef Rohon shows how cultural and academic appropriateness was a contested issue in Bohemia. Rohon was born in Transleithania to a Slovak Protestant family.[104] After studying and working in Vienna and Munich, he unsuccessfully tried to achieve a tenured position in the Habsburg, German, and Russian Empires. He himself credited his lack of success to the "negative networks" he had in Vienna, which haunted him throughout his life. His frequent changes of workplace resulted primarily from failure to secure a position with longer-term prospects. In the institutions where he did stay, his main roles were either temporarily renewed assistant positions or similarly uncertain scholarships or travel allowances. He also earned additional money as a contract supplier of microscopic preparations and in his later years worked mostly with the Imperial St. Petersburg Mineralogical Society (Императорское Санкт-Петербургское минералогическое общество). Albert was also for several years a financial sponsor for Rohon, who had even asked him directly for support in gaining a position "in his [Rohon's] homeland."[105]

In 1895 Rohon, then fifty years old, was one of the candidates for the chair of embryology and histology in Prague, proposed, unsurprisingly, by Albert's students. The faculty majority, however, clearly favored local scholars, who were, however, not specialists in this discipline. While the minority stressed that Rohon was the only qualified candidate and, as a Slovakian, was able to speak Czech, the majority defended the qualities of the other candidates and voiced concerns about Rohon's capabilities because, in the first place, he had not achieved a Habsburg doctoral degree despite writing his dissertation in Vienna. Second, there were serious concerns about his ability to speak Czech, which was attested only by Albert and not confirmed by his publications. The faculty questioned the authority of Albert in this case, stating that his opinions were not binding in Prague as he was a member of a "foreign faculty" (*Mitglied fremder Fakultät*) and furthermore was a surgeon and not a specialized histologist.[106]

Although this was one of the few cases when a non-Czech was appointed, Rohon exemplifies what scholars working at the Czech University were supposed to do: participate in Czech scholarly life and educate Czech successors. Rohon did both effectively: he was a member of several

Czech organizations and participated in popular projects such as *Ottův slovník naučny* (Otto's encyclopedia).[107] He also published in German in the proceedings of the Royal Bohemian Society of Sciences, much like his Czech colleagues. His educational achievements were even more impressive, and his students shaped histology well into the twentieth century.[108] While Rohon did not publish in Slovakian, Obrzut and Horbačevs'kyj wrote in Polish and Ruthenian (respectively) in addition to Czech and German.[109] As can be observed with other scholars in the empire, there were often three language "types" they could use: that of the institution, that of the linguistic culture they identified with, and German, the scientific lingua franca. In many cases these three converged in one language (i.e., German), and only in rare cases did the three types correspond to three different languages.

Since Galicia often resorted to hiring scholars from the German and Russian Empires, the search for a "Polish" scholar there had a different significance. The language argument was used, as in Bohemia, with national categories in mind, and since the category of linguistic adequacy was flexible, it helped the majority of the faculty force scholars with unwanted cultural affiliations out of the university. When the chair of surgery in Cracow was to be filled in 1882, the faculty clearly favored scholars known for their patriotic Polish engagement.[110] At the same time, it opposed Johann/Jan Mikulicz-Radecki, who was backed by the Viennese star surgeon Theodor Billroth. The commission acknowledged his practical and scientific abilities but could not confirm his language skills.[111] In L'viv the faculties rejected several habilitations by Jewish scholars without any clear reasons.[112] One of the unsuccessful candidates, Naphtali Sobel, was also a victim of Galicia's intellectual life, which was increasingly turning its back on the German language and everything associated with it (see above for details); however, the argument would likely have been different had Sobel not been Jewish.

Candidates' knowledge of Polish was carefully analyzed and discussed during appointment procedures. For non-Galician scholars who published in German, the faculty was often unsure whether the nominees' fluency was sufficient for lecturing. With two personally mediated exceptions, Czech-speaking scholars from Bohemia were not taken into consideration for possible appointments and habilitations owing to their linguistic insufficiency.[113]

Also, Galician universities could not recruit their entire professorship from among their own graduates or scholars identifying as Poles, especially in the initial years after the language change. The problem of a lack of qualified Polish scholars had already arisen in Cracow in the 1860s. When, during the bilingual period (1861–70), a professor of classical philology in the German language was to be appointed, the faculty asked a former member, Bernhard Jülg, then a professor in Innsbruck, if he would agree to return. The request was motivated by the fact that he had learned Polish during his time in Cracow.[114] While Polish speakers were mostly available, in two cases the university decided to propose scholars not entirely fluent in Polish. The first case, which shows that national categories were still fluid, took place in 1873. Two scholars were considered for a chair in Cracow, but they spoke only basic Polish. In this case, for different reasons, neither of them could be appointed; finally (and in fact against the will of the faculty), a local scholar was chosen.[115] The second case took place in 1891, when the Jagiellonian University created the chair of animal husbandry. The faculty proposed Leopold Adametz from the Academy of Agriculture (Hochschule für Bodenkultur) in Vienna; he was required, however, to learn Polish within two years.[116] Such appointments remained exceptions, though, and the faculties were cautious about language issues. In uncertain cases they asked scholars directly whether they were fluent in Polish. Sometimes this led to surprisingly positive answers, although rarely to appointments.[117]

As in Prague, the faculty also consciously used the argument of language to promote local scholars. When, in 1875, both Galician universities were supposed to initiate lectures in geography, a substantial lack of scholars capable of teaching this discipline in Polish was evident. Several German speakers, but also the Ruthenian geographer Anatol' Vachnjanyn (Анатоль Вахнянин),[118] applied to L'viv, but instead of appointing these non-Polish scholars, the university decided to offer scholarships to promising young scholars who identified as Polish; in the meantime, other professors would deliver the lectures in geography.[119] When one of the promising youngsters failed his habilitation in 1878, the University of L'viv still opted for a local Polish Galician instead of Vachnjanyn, waiting for several years until an appropriate candidate habilitated.[120]

As the lectures of the chair of German language and literature were to remain in German, professors proposed for that chair were more valued if they knew at least one Slavic language and thus had a better chance of learning Polish, which also limited potential appointments. In addition, if they had a Polish mother, as Spiridion Wukadinović did, then even the fiercest of Polish nationalists were quieted.[121]

Polish-Ruthenian Schisms

With the Polish nationalists claiming the university in L'viv as the stronghold of Polish interests in Eastern Galicia, the political tensions with Ruthenians were a key issue for university policy there. Unsurprisingly, the main point of confrontation between the university and the government (both the ministry and the provincial government) was the presence and number of Ruthenian scholars and chairs at the University of L'viv. One could definitely say that it was easier for an Austrian German or even a socialist to obtain a position in L'viv than for a Ruthenian, with Jewish scholars having similarly low chances.

The structure of the arguments and the proceedings related to appointments at the university were very similar to when Polish was in the subordinate role and German the hegemonic language. The majority of the scholars at the university (and, if such cases were discussed in Parliament, of Polish nationalists as well) argued consistently that Ruthenian scholars had the possibility of habilitating and that if they conformed to the academic requirements, their advance in academic life would not be obstructed.[122] The idea of equality remained limited to rhetoric, and the practical situation at the university showed how conservative the decision-making was. In the face of a Polish majority in the faculties, Ruthenian professors could only be a minority. When the university pleaded for Polish as the administrative language, only the professor of Ruthenian language and literature, Omelyan Ohonovs'kyj, opposed this.[123] At the University of L'viv—whose self-image as a Polish stronghold intensified after the 1890s, when nationalists started to dominate both the city and the faculties—only political solutions assured a Ruthenian presence at the university.

In many cases, Ruthenian scholars formed a united front against the Polish majority. One such example was their opposition to the Polish nationalist historian Ludwik Finkel. His habilitation in general history in 1884 had already led to controversies in the faculty because it directly challenged the professor of Austrian history, the Ruthenian Isidor Šaranevyč, who had opposed it.[124] The rivalry between Finkel and Šaranevyč over the division of lectures escalated several times thereafter, as the latter complained that the Polish Privatdozent Finkel actually taught Austrian history but, to cover this, added the annotation "against the background of universal history" in the title printed in the lecture catalog.[125] The provincial government elegantly solved this conflict during the subsequent very heated appointment procedures for the chair of universal history, deciding to award Finkel an associate professorship in Austrian history.[126] Most conflicts, however, related to the habilitations of Ruthenian scholars. For example, the geographer Anatol' Vachnjanyn and the historian Volodymyr Myl'kovyč enjoyed the support of Ruthenian professors, but their habilitations were turned down by the Polish majority.[127]

Such cases should not, however, give the impression of persistent and aggressive Polish-Ruthenian conflict at the faculty. At least until the 1890s, that was not the case. In fact, the faculty made most decisions via consensus.[128] To an extent, this resulted from a careful choice of nonnationalist Ruthenian scholars; however, the general prevalence of nonconfrontational conservative cultural Greek Catholic nationalism in Galicia at the time should also be taken into consideration. In fact, older Ruthenian professors did not accept the Ukrainian nationalist ideology of the younger generation of Ruthenian scholars, uniting with the Polish faculty members on this matter.

The most important changes in the cultural division of the L'viv faculty, apart from the language change in the 1870s, took place between 1890 and 1899, a time that also turned Galicia into a powder keg. During the political rapprochement of the 1890s, the so-called New Era (*Nowa Era*, or Нова ера), the Polish-dominated provincial government allowed several concessions for Ruthenians. The most important were the use of Ruthenian as an administrative language, the phonological codification of the language in place of an etymological one, support for Ruthenian educational and cultural organizations, and the strengthening of the Ruthenian presence at the university.[129] These acts seriously strengthened the *narodovtsi* (pro-Ukrainian national populists) against the conservatives and Russophiles, especially in the urban L'viv sphere and among the educated classes. Through a brief glance at this period, I demonstrate the mechanisms of political divisions in L'viv, both those between Poles and Ruthenians and those within Ruthenian culture.[130]

Among the concessions of the New Era, the creation of a chair for Ruthenian history was seen as the most vital. In its designation, the provincial government mentioned not only the scholarly qualifications of the new historian but also the chair's function as a broker between western and eastern cultures: "The professor of the newly created chair should make the university youth acquainted with the historic-literary production of the East, but on the other hand process and use those in the spirit of the West."[131] As a teacher and educator, the new professor had a pronounced political function. Although the faculty also considered Polish scholars for the post, the Ruthenian professors fiercely rejected such proposals.[132] The final terna included the famous Kiev historian Volodymyr Antonovyč (Володимир Антонович, also Włodzimierz Antonowicz), his twenty-seven-year-old student Mychajlo Hruševs'kyj (who had yet to graduate), and Volodymyr Myl'kovyč from the Institute of Austrian Historical Research (a Privatdozent in Chernivtsi). That Antonovyč would get the first place in the terna was in no doubt. More important was the question of whom to grant the second place, as it was clear that the ministry would hardly be able to reach an agreement with the nearly sixty-year-old Antonovyč. He was politically acceptable but would be expensive. Thus, the professorship would go to the second-place scholar. While the majority, along with Šaranevyč, pleaded for Myl'kovyč (Šaranevyč's son-in-law), Ohonovs'kyj and the minority granted Hruševs'kyj the second place. Conflicts also arose because Hruševs'kyj was not a Greek Catholic, as most Ruthenians in Galicia were, but Orthodox.[133] While Antonovyč rejected the call based on his advanced age, he fiercely supported Hruševs'kyj, in his eyes the most skilled of all young Ukrainian historians. This proposition was also approved by the provincial government, which stressed that the young scholar "[belongs] to [the] young-Ruthenian, i.e., Ukrainian party and is an adherent neither of pan-Slavic tendencies nor of an unjustified national chauvinism."[134]

Hruševs'kyj, or Gruszewski, as he was called in the official documents of the university, proved a great deal of trouble for the university, consistently refusing to use Polish and becoming a leader of the Ruthenian nationalists in L'viv. By 1896 Gruszewski had asked to change his name to Hruszewski as this was, in his eyes, the official transliteration of his surname from Cyrillic; the provincial government granted this only after serious deliberations and expert consultations.[135] His conflicts in the faculty were legendary, as he constantly refused to speak Polish. The Polish professors at first asked other professors to translate, but eventually Hruševs'kyj was disciplined.[136] Finally, the dean, Kazimierz Twardowski, refused to acknowledge any statements Hruševs'kyj made in Ruthenian.[137] With these conflicts and his involvement in the Ševčenko Scientific Society, Hruševs'kyj became a spokesperson for Ruthenian demands at the university, which added considerably to his conflicts with the faculty. These demands were publicly discussed by Ruthenian students and were perceived as evidence of Polish oppression, increasing the polarization between the national groups in Galicia. Hruševs'kyj enjoyed immense popularity among Ruthenian cultural and political elites (see figure 6), which finally led him to be chosen as

Figure 6 From the moment of his arrival in L'viv, Mychajlo Hruševs'kyj not only grew to become the political leader of the Ruthenian-Ukrainian movement but also was instrumental in creating and stabilizing its ideological basis. Here he is among participants of the Meeting of Ukrainian Writers (Z'їзд українських письменників) for the hundredth anniversary of the publication of Eneïda by Ivan Kotljarevs'kij. Hruševs'kyj is fourth from the left in the middle row; to his left sits Ivan Franko. (Photographer unknown.)

the first head of the Central Council (Tsentralna Rada, Центральна Рада), the parliament of the short-lived Ukrainian People's Republic (Українська Народня Республіка) in 1918.

The second chair whose filling was influenced by the New Era policies was that for Ruthenian language and literature; it was occupied by Ohonovs'kyj after Holovac'kyj was dismissed for his alleged Russophilism (see above).[138] After Ohonovs'kyj's death in 1894, the question of his successor was raised, but only after Polish-Ruthenian problems had brought an end to the New Era in the autumn of 1894. As the chair was vital for the propagation of the Ruthenian language, conceptions of which differed across political groups, it was also right in the middle of the conflict over the cultural orientation of Ruthenians, massively influenced by New Era policies.

In the early 1890s, the provincial government had decided to introduce a phonetic orthography for Ruthenian schools, legally clarifying an issue that had been discussed throughout the nineteenth century, that of an alphabet for written Ruthenian.[139] The introduction of a phonetic alphabet was a step demarcating "true" Ruthenians from those "who want to bedight Ruthenian with Church Slavonic and Russian ornamentation,"[140] that is, Russophiles, according to the decisive petition, penned by none other than the professor of Romance languages in Chernivtsi, Theodor Gartner, in cooperation with Stepan Smal'-Stoc'kyj, a professor of Ruthenian language and literature at Chernivtsi. The aim of this project was to cleanse Ruthenian orthography of foreign or historical accretions and to establish a codification based purely on folk speech. This issue was highly controversial, leading to opposition by both Greek Catholic church authorities and the Russian movement; both argued that it broke from the historical-religious tradition of Rus' and served as a step toward assimilation with Polish culture. In contrast, this decision strengthened the *narodovtsi*, who not only had initiated this reform but also followed it rigidly in their later publications.

The question of Ohonovs'kyj's successor was thus not merely an academic matter because Smal'-Stoc'kyj was a declared proponent of phonetic orthography, together with Gartner, the author of the first Ruthenian schoolbook that outlined its orthography (1893).[141] The direction the new professor in L'viv would take was of vital interest to both Ruthenian political parties and the church. Directly after Ohonovs'kyj's death, only one person was considered, Oleksandr Kolessa, who had habilitated in 1894 with Smal'-Stoc'kyj in Chernivtsi and habilitated again in L'viv the year after.[142] For a long time, the faculty made no decision on the appointment of future professors, leaving Kolessa as the auxiliary professor for the chair. In the second half of the 1890s, another candidate strove for the chair, Ivan Franko (Іван Франко), a well-known writer and poet, who was supported within the university but rejected by the provincial government, the ministry, and the Ukrainian *narodovtsi*, for whom he was unacceptable because of his political radicalism and his socialist past.[143] In his letters, Franko addressed the issue of the vacant chair, stating that the university would not appoint any of the other candidates and would promote him afterward. After the ministry's rejection, he openly criticized the politicians of the New Era for promoting their own candidate, Kyrylo Studyns'kyj (Studzinski).[144]

Studyns'kyj was the antithesis of Kolessa. While the latter was *narodovets*, the former was a Christian Socialist who had studied in L'viv and Vienna (where he, like Kolessa, had graduated in 1894) and then moved to Berlin to prepare his habilitation. A few months after Ohonovs'kyj's death, Studyns'kyj applied for habilitation in L'viv, which was denied him, officially owing to his low scholarly qualifications.[145] With the support of the provincial government, however, Studyns'kyj was then granted a position at a gymnasium and shortly afterward a fellowship in Cracow. There, he habilitated in the following year and published several articles in Polish.[146] Studyns'kyj resided in L'viv but was granted the possibility of traveling to Cracow once a week, which was clearly against the habilitation laws, which required Privatdozenten to live near the city in which they taught.[147]

After several commissions could find no appropriate candidate for the chair in question,[148] the faculty finally proposed Kolessa as an associate professor.[149] This was countered, however, by the Galician governor, who suggested "another appropriate scholar," Studyns'kyj, based on the credentials supplied by Cracow.[150] The ministry thus requested a new proposal from the faculty that took both their qualifications into account and asked several non-Galician scholars for their expert estimation.[151] Notwithstanding this intervention, the faculty proposed Kolessa once more, supported by the opinions of the specialists, who saw him as a more talented and independent thinker. This time, he succeeded in being appointed as an associate professor (in 1898), after the chair had stood vacant for four years.

The conflict did not end there, however. In the autumn of 1898, the faculty was once more confronted with this issue, as the Greek Catholic Metropolitan-Ordinariate requested a chair of Old Church Slavonic language at the philosophical faculty, which was strongly supported by the provincial government but opposed by the philosophical faculty. The minister of education, Wilhelm von Hartel, proposed instead creating "a second chair of classical philology, alternatively for Ruthenian language and literature with special consideration of Church Slavonic history and literature."[152] The installation of the new chair and nomination of yet another Ruthenian scholar was, unsurprisingly, opposed by the faculty. Polish scholars argued, first, that such a chair would be under church supervision and should be placed at the theological faculty and, second, that a second chair of Ruthenian language was unnecessary, asserting that the ministry should rather create chairs that "relate to the existent needs of the faculty and arise from real scientific needs."[153] Another argument was that since none of the candidates had scholarly qualifications, such a chair should rather be a readership (a lector).[154] This was the official position of the majority of the faculty, including Kolessa, who only wanted to augment the proposal with a sentence that the existing chair already covered the matters of the chair in question.[155]

The remaining Ruthenian professors were not unanimous. Hruševs'kyj argued that the university should rather address a petition for the creation of other chairs that would answer the needs of Ruthenian gymnasia, such as classical philology; if a second philological chair were to be created, he proposed a candidate from the Russian Empire.[156] Only Šaranevyč, a pronounced conservative and the house historian of the Stauropegion Institute, warmly greeted the new chair, proposing Studyns'kyj as the best candidate.[157] Despite an obvious lack of support within the faculty, the ministry appointed Studyns'kyj as an associate professor.[158] A few months later, the faculty successfully proposed Kolessa as a full professor; Studyns'kyj achieved this only in 1908, with the addition *ad personam* (that is, bound to his person and not creating a new chair; this was accepted unanimously in the faculty). In other words, with Studyns'kyj's retirement or death, Church Slavonic would be abandoned at the philosophical faculty.[159]

Studyns'kyj certainly remained an interesting figure in the *narodovtsi*-dominated L'viv. His first major publication after his appointment was an edition of the letters of Holovac'kyj, and he was intensively engaged in the Ruthenian Christian Social Party, later editing its journal, *Ruslan* (Руслан).[160] Still, this politicized appointment should not obscure the fact that the divisions among the Ruthenian faculty members were of secondary importance compared to the issue of nationality; indeed, in subsequent years the Polish versus Ruthenian conflict overshadowed the internal divisions, especially because most Ruthenian scholars (including Franko) were united in the Ševčenko Scientific Society.

Such unity among the Ruthenian scholars can be seen in 1907, as the ministry deliberated the creation of five additional Ruthenian chairs, including in chemistry (Horbačevs'kyj) and the history of literature (Franko). Experts from the university commission, which unsurprisingly had a Polish majority, criticized this approach, listing financial reasons and the violation of university autonomy as crucial. Unsurprisingly, the votum separatum, penned by Hruševs'kyj, Kolessa, and Studyns'kyj, claimed both the need for such chairs and their importance in the future creation of a Ruthenian university.[161] Here, the provincial governor, Andrzej Potocki, took a pro-Polish position, characterizing this decision as the creation of a "university of auxiliary professors,"[162] once more repeating that Ruthenian scholars had the option to habilitate. The issue then dropped off the agenda, partially owing to the intensification of nationalist conflict, which led to the assassination of Potocki in April 1908. The only way for Ruthenian scholars to succeed was to habilitate in disciplines no Polish candidates wanted, such as Austrian history.

Notably, the attitude toward Ruthenian scholars differed between the Cracow and L'viv universities. While polarization was dominant from the 1890s in L'viv, Cracow was more harmonious, allowing pro-Ruthenian demonstrations and accepting, in 1901, the Ruthenian students who had left the University of L'viv in a protest against Polish dominance. Cracow's philosophical faculty also supported Studyns'kyj when he was denied habilitation in L'viv.

Cracow concessions to Ruthenian culture also included a chair in Ruthenian (*ruski*) literature, created in 1893. In the proposal, the faculty accentuated the reciprocity of both nationalities and the importance of knowledge of the Ruthenian language for Poles.[163] The provincial government also supported this claim, with the minister of education Stanisław Poray-Madeyski stating finally that "in the course of centuries one can unmistakably trace Ruthenian influences on the literature, life, and customs of the Polish population; therefore, from a didactic point of view, it is a necessity that at the Jagiellonian University students of Polish nationality should have the opportunity to learn the Ruthenian language and become acquainted with their literature."[164] In practice, Cracow's chair of Ruthenian literature was confined to Polish topics. Józef Tretiak, appointed to the chair in 1893, wanted, in fact, to be transferred to the chair of Polish literature, and he conducted research predominantly on the most famous Polish poet, Adam Mickiewicz, and the influence he had gained in Russia. Nevertheless, Tretiak's professorship and function symbolized his role as a broker dealing with national tensions.

Political Differentiations

With the control over universities in national hands, political conflicts began to affect appointments, replacing previous tensions that had run along linguistic-national lines. These conflicts ran along generational lines as well but also divided scholars with a more internationalist outlook from those wanting the university to concentrate on a national agenda. While many scholars remained aloof from politics, the appointments in Bohemia and Galicia were clearly determined by political strategies.

With a number of conservative Catholic scholars having been appointed in the 1850s, the University of Cracow stood for many years in defense of these values, opening up to liberal and socialist movements only in the

1890s, according to the historian Józef Buszko.[165] With the appointment of some of the most prominent conservative politicians for chairs in humanistic disciplines, the university's link with politics was obvious. Characteristic of this was the appointment proposal for Stanisław Tarnowski, where the faculty did not stress his scholarly achievements but rather his connections with a "noble" family that for more than a hundred years had worked on the field of "motherland" literature.[166]

The direction of these developments in Galicia was fiercely attacked by liberal journals. The Cracow daily *Kraj* (Country), for example, published a series of articles attacking the appointment policy of the university, stating that second-rate scholars from Galicia were being appointed instead of high-class instructors from abroad, and even claiming that some German-speaking scholars should remain at the university as they had proven their scholarly quality.[167] Similarly, the Academy of Arts and Sciences was criticized as being controlled by Cracow conservatives. Ludwik Gumplowicz, then head editor of *Kraj* (1869–74), maintained his negative opinion of the Cracow scholarly environment throughout his life; he constantly rejected any cooperation and publication possibilities there and sent his son to L'viv to study history.[168]

Similar to *Kraj*, the influential fin de siècle left-liberal monthly *Krytyka* (Critics) continually attacked the university for valuing family bonds over scholarly merit and saw a conservative clique consisting of the majority of professors as blocking the appointment of celebrated but liberal scholars. This was particularly evident in the creation of the chair of social sciences at the theological faculty in 1910; as the liberal and socialist press claimed, any number of qualified scholars could have been employed for this subject at the philosophical faculty.[169] In the public sphere, the university acted as a conservative outpost, with strong ties to the conservative journal *Czas*, and participated prominently in state festivities and festive funerals, which were important patriotic manifestations of the formation of collective memory.[170] By 1910 the city and student bodies were already anticlerical, but the faculties were still strongholds of a conservative Catholic outlook.[171]

In L'viv, in contrast, some appointments, especially that of the Darwinian zoologist Benedykt Dybowski, brought the university into conflict with the Catholic clergy.[172] The university, most of whose professoriat had been appointed in the 1870s, when liberal scholars were just beginning their careers, was more open to appointing progressive and socialist intellectuals than the West Galician university. Around the end of the nineteenth century, several pronouncedly nationalist scholars also occupied influential positions,[173] something not possible in Cracow, where contentious scholars were disciplined or removed from faculties.

The difference in the ideological approaches of "progressive" L'viv and "clerical-conservative" Cracow is visible in various subjects such as history and the biological disciplines (the latter owing to the politicization of Darwinism). In historiography two distinct schools emerged, differing in both methodological and political positions, which led to serious conflict at the Second Meeting of Polish Historians (II Zjazd Historyków Polskich) in L'viv in 1890. Cracow historians, according to their L'viv and Warsaw counterparts, concentrated on descriptive political history and criticized the Polish-Lithuanian Commonwealth for its instability, conflicts, moral decay, and general underdevelopment. L'viv historians, especially the most influential of them, Ksawery Liske, propagated a nation-centered historiography, accentuated the positive internal developments of the Commonwealth, saw the impact of imperial and dynastic geopolitics as responsible for the partitions, and, more strongly than scholars from Cracow, argued the need for Polish independence.[174] Although mediating positions were possible,[175] there were almost no transfers between L'viv and Cracow in the historiographical disciplines. The ideological division between the Galician universities should, however, be approached cautiously. In 1959 the Cracow philologist Kazimierz Nitsch, a self-described socialist, anticlerical, and "philoruthenian," claimed in retrospect that his appointment to L'viv in 1908 had failed owing to precisely these three attributes. However, this did not hinder his appointment as an associate professor at the Jagiellonian University in 1910 or his appointment as a full professor in L'viv in 1914.[176]

The situation in Prague was similar, and here visions of the past had also determined current politics. While the German-Czech conflict was most influential until 1882, the creation of a linguistically exclusive university intensified internal conflicts within the Czech faculties. Already in the 1860s, the conflict lines ran between older Czech scholars, who supported the romantic-nationalist Old Czech Party (Staročeši, Národní strana), and the members of the Young Czech Party (Mladočeši, Národní strana svobodomyslná); the latter gained political influence in the Taaffe era (1879–93), allowing it to push through its candidates shortly before the university division in 1882.[177] Discussions on the position of Czech culture and the shape of the "national idea"[178] in particular brought out divisions within the university.

The breach in Czech unity came with a series of publications doubting the authenticity of *Rukopis královédvorský* and *Rukopis zelenohorský* (the Manuscript of Dvůr Králové and Manuscript of Zelená Hora), pivotal documents attesting to old Czech culture and history. The conflict had played out several times from the moment of their discovery/forgery in the late 1810s, with several scholars arguing that they were clearly forgeries, while others, most notably patriotic political celebrities like Palacký and Šafárik, considered them authentic. But in the mid-1880s, the conflict enflamed anew when the young generation of scholars critically analyzed both manuscripts from many sides, which included a chemical examination, in a series of articles in the Masaryk-led journal *Athenaeum*, arguing that they were, indeed, forgeries. The conflict escalated as the older generation of Prague professors criticized their younger colleagues for their doubts, given their national identity, and proposed their own analysis.[179]

Although the political conflict decreased around the turn of the century, the position of Masaryk and his colleagues as outsiders in the faculty was obvious; this was made known to a wider public in articles in *Athenaeum* and caused several serious conflicts during habilitations and professorial appointments. This also led to attempts to remove Masaryk from the university.[180] Masaryk, for his part, opposed appointments of conservative scholars.[181] While the older generation of scholars initially succeeded in appointing their candidates, both the composition of the faculty and the conservative scholars' influence in the ministry changed over time. Although Masaryk, the most polarizing figure, was appointed a full professor only in 1896, younger scholars achieved several gains, supported in Vienna by Eduard Albert and, most important, by the Prague historian Antonín Rezek, whose informal consulting position in Vienna was turned in 1896 into a *Ministerialrat* (secretary of the ministry), and later a Sektionschef, position in the Ministry of Education.[182]

With the Czech past a contentious issue, historical methodology was crucial. Here Jaroslav Goll, a proponent of the German positivist school of Georg Weitz, opposed the philosophical historical creations of the professor of Austrian history Václav Vladivoj Tomek and, later, Masaryk. The struggle had begun to affect the faculty by 1889, when Rezek was appointed as Tomek's successor. Rezek was accused of antinational propaganda owing to his critique of the creation of the Franz Joseph Czech Academy for Science, Literature and Arts. Reflecting on this issue, he noted sarcastically that while he was accused of a lack of patriotism in Prague, in Vienna the ministry saw him as a nationalistic radical (quoted in the epigraph).[183] Rezek was Tomek's student but turned to Goll afterward, and he was influential in supporting Goll's students in Vienna, who faced constant opposition from conservatives in Prague. Such was the case in the appointment of Rezek's replacement: Goll and Rezek secured the appointment of Josef Pekař, a critical-positivist historian of Hussitism, instead of Josef Píč, an archaeologist favoring the view that the manuscripts were authentic, who was supported by Tomek and the conservatives.[184]

These divisions did not run only between old and young politically active professors; from the mid-1890s, they also ran across these boundaries in a fierce conflict between "Masaryk's sect and Goll's school."[185] The trigger was Masaryk's publications in which he described the meaning of Czech history and thus of Czech nationality as a direct outcome of the Hussites, and thus equated Czech nationhood with Protestantism. This socio-philosophical idea met with strong criticism from Goll's students, who accused Masaryk of methodological inconsequence and presentism in which he promoted a political program under the guise of historiosophy. These constant conflicts led Rezek, now a ministerial official, to voice a clear critique in 1899: "What overcomes me is the fight against intrigues from Bohemia and of Czechs against Czechs."[186]

Habsburg Slavic Spaces

This chapter has argued that after the language changes the spatial dynamics of Slavic universities changed significantly. Still part of the legal structure of Cisleithanian universities, they developed their own spaces of recruitment, their own hierarchies, and their own conflicts, although, as I show in the next chapters, they were also heavily influenced by overarching pan-Habsburg phenomena.

Bohemia and Galicia shared several features, such as the idea of finding national scholars, but differed in a few others. Most notably, Galicia opened to scholars from abroad, while the Czech University in Prague could not, although it appointed Czech scholars who had found no place at Habsburg universities before 1882 and had emigrated. In contrast, Galician universities openly invited Polish graduates from the German and Russian Empires to habilitate and thus assured a faculty with a diversified educational background. Second, from the point of view of the Habsburg Empire, Galicia distanced itself, if one considers the disciplines symbolizing imperial unity: Galician faculties considered German as a foreign language and had no habilitations in German literary studies and only a few in Austrian history. To illustrate the effects of this, here is an example from the Commission for the Newer History of Austria (Kommission für Neuere Geschichte Österreichs). Founded in 1896, it included several Bohemian and Moravian scholars, among them the Czechs Rezek and Goll. Although the government insisted on having a "Polish" scholar as well, an appropriate candidate was found and nominated only in 1916.[187]

One should not, however, confuse this change of space with dissolution. The post-neoabsolutist era was characterized by new forms of allegiance and pan-Habsburg loyalty in which diversity-in-unity was the new guiding rule, replacing earlier ideas in which German was the guarantor of cultural unification. In fact, as I will argue in chapter 7, Cisleithanian universities also remained the role model for nationalizing states in the interwar period. In the late nineteenth century, new modes of communication that assured unity emerged, originating not only in the center of the empire, Vienna, but also at Slavic universities, which, for example, began to concentrate on sending young scholars to Vienna or openly promoted scholars with experience at German-language universities. It seems that the relations among scholars within the empire even improved after the centralists in Vienna ceased to prescribe German as the binding cultural element. Although the withdrawal of German scholars from Eastern Galicia led to an intensification of conflicts between Poles and Ruthenians, these followed roughly the same pattern as previous tensions between Germans and Poles.

The growing political tensions at Slavic universities crossed local boundaries and connected with differing visions of nationhood. In Bohemia and Galicia alike, these conflicts were also linked with generational changes as the conservatives who had been promoted under Thun-Hohenstein and in the 1860s began to be challenged by liberal scholars in the 1870s. This trend was clearly pan-Habsburg and affected other universities as well, leading to tensions around 1900. Being an openly Darwinist or anticlerical scholar was a similar experience whether one worked in L'viv, Budapest, or Vienna. Here Thun-Hohenstein's policy showed its long-term influence, especially because university autonomy supported the prevailing ideological positions.

Finally, the norms scholars had to adhere to in order to achieve promotion in the university system remained similar across the empire, with one significant difference. While at the German-language universities mobility was a prerequisite for professorial appointments, this was hardly possible at Slavic universities, simply because of a lack of universities in those regions. Here, short-term stays outside of scholars' home provinces, funded by scholarships, took the place of more permanent relocations. In fact, the universities in Cracow, L'viv, and Prague acted in accordance with Bruno Latour's model of centers of calculation, sending their scholars away to gather knowledge abroad and, later, bringing them back home.[188] In this they were much more international than Vienna and Graz, whose scholars' careers remained limited to German-language universities.

One last detail should reinforce the idea of unity in the Cisleithanian space, namely, the picture of Vienna as the place in which Polish or Czech agitation was indulged. When Eduard Albert was denied a position in Prague but promoted to Vienna, the ministry gave a signal that did not go unnoticed by German nationalists. Albert was not the only nationalist activist promoted to Vienna. Jan Leciejewski, who habilitated in 1884 in Slavic philology, was presented in a report by the Ministry of Foreign Affairs as a well-known and influential Polish nationalist. The Ministry of Education confirmed this in its decision to appoint him, stating, however, that this "does not present an adequate foundation to disqualify Leciejewski from a teaching post, especially as Vienna was not an expedient place for national Polish agitation, and it did not seem clear how Leciejewski could foment national discord as a Privatdozent for Slavic philology at the University of Vienna."[189]

Although the situation was aggravated at the turn of the century, the University of Vienna remained the most open university for scholars of other nationalities, especially owing to its tradition of Slavic philology and a number of nationalist students' organizations that were approved.[190] Vienna was also positively connected to Slavic parts of the empire, compared to Graz or Innsbruck, especially because the number of Polish and Czech scholars who habilitated in Vienna was considerably higher. Scholars depicted Vienna as the most secure place to be during these volatile times, an image that remained powerful after 1918 as well.[191] As I show in the next chapter, Vienna, the imperial capital, was positioned as a safe harbor for many groups, although one has to be cautious: the reality did not always conform to this image, and we must inquire as to what the consequences would have been for the scholars themselves.

CHAPTER 6

Imperial Space and Its Identities

If we [Catholics] would have equal rights, then around 80 percent of the Jews who nowadays frequent the university would have to leave it. . . . Are those equal rights, when just in the recent past among the eight appointed professors we find seven Jews? . . . We want equal rights; we want the university, which once belonged to German Christian people, to belong again to German Christian people.

—KARL LUEGER, 1907[1]

The advance of professionalization and the professional closure of academia to nonacademic scholars during the nineteenth century did not mean that scholarly quality became the only factor influencing academic advancement. Whereas the previous chapters have discussed, among other things, how issues of nationality influenced university careers and dissolved the Habsburg academic space into linguistic subspaces, this chapter concentrates on issues that, until 1918, united Franz Joseph's subjects under one worldview. Catholic identity—one of the founding ideologies of Habsburg statehood, its universities, and most of its peoples—influenced academia across the empire well beyond neoabsolutism. By merging with different nationalisms and conservative ideologies such as Germanness[2] and Polishness,[3] it coproduced pan-imperial cultural othering narratives (*Türcken*, or Turks; *Ostjuden*, or eastern Jews).[4] Scholars did not remain immune to these, both using such stereotypes and coproducing them; for example, Theodor Billroth used the stereotype of low-income Galician *Ostjuden* overcrowding universities, and Adolf Wahrmund and August Rohling wrote openly anti-Semitic pamphlets.

Most important, however, universities, as places of cultural presentation, came to represent not only a national linguistic ideal but also a religious ideal centered around Catholicism, agreed on not only by the public and the ministry but also by the majority of scholars, who were predominantly Catholic.

One cannot, however, say that the universities were exclusively Catholic, or even exclusively Christian. After the Thun-Hohenstein period (1849–60), when the Catholic ideal of a university was virtually prescribed, the number of non-Catholics rose. Nevertheless, the career progress of Jewish and atheist scholars was hindered. In the first place, this occurred in the faculties themselves, since the majority of full professors were conservative Catholics. Also, radical student groups at most universities in the monarchy were increasingly, and also violently, opposed to the appointment of Jewish scholars. All this created difficult questions for faculties and a sense that the appointment of such scholars could cause serious disturbances. This also had significant consequences for the spatial policy of the Cisleithanian universities.

After the liquidation of the Ministry of Religion and Education in 1860, the Unterrichtsrath largely continued the confessional policies of Leo Thun-Hohenstein. Similarly, the church's interest in university matters remained unchanged. In the 1860s controversy arose once more over the religious character of universities, involving the question of the inclusion of the Protestant theological faculty into the university in Vienna, including church officials.[5] This clash intensified the divergences between Catholics and non-Catholics both in the university and in the public sphere, which did not cease until the end of the century. In January 1861 a self-declared majority of scholars, under the leadership of Josef Hyrtl, proposed that a declaration be written that the University of Vienna would become exclusively Catholic, which also found support in the Unterrichtsrath and the Catholic public sphere. The majority of the scientific press, which would have preferred a declaration calling for a clear-cut division between scientific and religious issues, severely criticized this informal assertion of Catholic predominance.[6] But once more the university showed which side of the ideological struggle it favored. When it was the turn of the medical faculty to propose the rector of the university, it chose Hyrtl; his two speeches—his acceptance speech in 1864 on materialism and his speech on church domination over science, given on the occasion of the university's five-hundredth anniversary in 1865—became (in)famous because of their controversial assertion of a conservative Catholic worldview.[7]

Catholic predominance did not mean exclusivity, though. A few other university scholars, such as the notorious liberal outsider Joseph Unger, saw it as a matter of course that "not only Catholics, but also Protestants and Jews, should be appointed not only for professorships, which cannot be challenged from any side, but also for the offices of dean and rector."[8] This can be seen in the case of the unsuccessful nomination of Hermann Bonitz for the office of rector in Vienna in 1852 and the refusal of ministerial confirmation for Friedrich Stein as dean of the philosophical faculty in Prague in 1863.[9] While the conflict over the proposed declaration of Catholic exclusivity was solved by ignoring the demands of the "majority" and keeping the status quo of official pluralism, such a policy de facto kept universities Catholic. This tactic of ministerial silence on ideological issues would be the guiding principle in the coming decades; of course, this silence may have been the ministry's official stance, but when it came to unofficial and semi-official issues, its attitude was quite different. The situation from the 1860s on, however, sheds light on another issue characterizing the universities during this period: the effects of Thun-Hohenstein's personal policy of the 1850s, which turned universities into conservative institutions. One could even be inclined to call the Cisleithanian universities backward, if compared to the zeitgeist represented by public opinion, and the attitudes of full professors in particular confirm this view.

While public opinion in the 1860s and 1870s can be considered to have been more liberal than the views of the majority of scholars for a time, one should not forget that the strengthening of universities' Catholicism after 1848 was a long-term project. Since Thun-Hohenstein had appointed mostly young scholars, they dominated university life as full professors for several decades. One could actually claim that whereas the universities gradually opened up to liberalism toward the end of the century, when scholars who had begun their careers during the liberal period began to achieve full professorships, the majority of the public turned toward racial and cultural nationalism and anti-Semitism. Benedykt Dybowski's inaugural lecture in L'viv in 1885, in which he openly proclaimed Darwinism as the new model of thinking, met with strong critical reactions from high clergy and conservatives alike. However, this failed to influence the university, whose personnel had mostly been recruited in the 1870s.[10] Similarly, in Innsbruck in 1908, the canonical jurist Ludwig Wahrmund harshly accused the Catholic Church of mingling with the academy and violating the division between religion and science. While most scholars, apart from those in the theological faculties and at the University of Cracow,[11] stood behind him, the pressure of public opinion, which accused Wahrmund of religious betrayal and of being a Jew, finally led to Wahrmund's transfer from Innsbruck to Prague.[12]

Because of the Cisleithanian universities' constitution and the dominance of full professors, the universities, as assemblies of scholars, could hardly be progressive, and the strengthening of liberal thinking around 1900 was a belated version of the liberalism of the 1870s rather than a reaction to contemporary developments. (Indeed, most academics remained aloof from the more radical political views commonly held by the public and students around the fin de siècle, particularly ideas of socialism and nationalism.) This was not liberalism in the modern sense but a "fragmented" liberalism,[13] constituting an antithesis to the academic atmosphere following the initial reforms of 1848, which, in turn, at least for the first few years, were constituted as opposing the restrictions of the Vormärz period.

Similarly, as demands for language changes emerged as an internal issue in Cracow, L'viv, and Prague, belatedly in comparison with the demands of the press or public opinion, the question of religion was more an external issue than an internally perceived problem of the universities. Because students assumed the role of pioneers in both the conflict over language and that over religion, professors were increasingly confronted with clashing political positions within academia; at the end of the century, a variety of extreme positions that had broad social and political support contested those of the academics. In comparison with the question of, say, female students and academic teachers, which had been debated in academic senates, declarations on ideological disputes were not officially issued, except that in isolated cases the universities drafted declarations of neutrality. Except during World War I, when the political role of scholars changed, university scholars were far from taking on the pioneering role some had assumed in 1848, and with the exception of a minority of engaged scholars who acted as public intellectuals (who were marginalized in academia),[14] the university was turning into an intellectual ivory tower. Looking at the names of the creators of the *Volksbildung* (folk education, i.e., popular courses for the broader populace) and its most prominent lecturers, one can see that, for these scholars, engagement in the popularization of science went hand in hand with a lack of academic capital in universities.[15]

The unwillingness of professors to accommodate controversy within academia was visible, for example, in the rejection of modern art, not only in the famous conflict over the *Fakultätsbilder* (Faculty paintings) of Gustav

Klimt[16] but also in the appointing of rather antimodernist historians of art and literature. It was also demonstrated through the belated entrance of historical disciplines related to the immediate past and, most directly, the removal of scholars who courted public controversy. The reasons for such removals differed from university to university; they included reviling the memory of the dead,[17] leading spiritual-patriotic organizations,[18] being accused of pedophilia,[19] and supposedly engaging in sacrilege.[20] While most such cases included accusations of acting against Catholic norms, the ministry also occasionally reacted, albeit seldom and belatedly, when scholars openly propagated anti-Semitism.[21]

As different as these examples are, they illustrate that the ministry and the majority of scholars were trying at any cost to lessen the controversy surrounding the university. In many cases, this meant withdrawing support from those who had no influential political and public representation, for example, the Italian minority in Tyrol, Ruthenians in Galicia, and Jewish scholars across the empire. The various forms of nationalism played a substantial role in such conflicts, and a number of scholars publicly presented nationalist views without being seriously threatened in the academic community. One sees, however, an asymmetry here, at both the faculty and the ministerial levels: the involvement of scholars in German or Polish nationalist movements remained largely unpunished, but when Ruthenian or Czech scholars were politically active, conflict resulted.

The differences between the hegemonic and the marginalized discourses appear not only in the press coverage of conflicts but also in the published opinions of the universities. The accounts of universities as antimodernist, conservative, and church-controlled institutions, with politicians and professorial cliques prohibiting all innovation, were countered by critiques that they were a cradle of liberal, socialist, and Jewish scholars propagating their ideas among predominantly Catholic students. One can find this difference in views in accounts written in all the leading languages of Cisleithania. In German one can compare the positive portrayal of the university in the leading daily newspapers of the time (apart from the *Neue Freie Presse* [New free press]) with the negative view voiced in articles in Karl Kraus's journal *Die Fackel* (The torch) or Arthur Schnitzler's drama *Professor Bernhardi* (1912). In Polish the dividing line ran between Cracow's leading journal *Czas* (Time) and the main progressive journals *Kraj* (Country), *Prawda* (Truth), and *Krytyka* (Critics). In the Czech press, the conservative *Národní Listy* (People's papers) contrasted with the liberal *Athenaeum* and *Naše doba* (Our time). These are not even particularly extremist journals; the positions are even more radicalized when one looks further from the center. Similarly, there was no common ground with respect to the national issue, and universities were criticized by nationalists and loyalists from all sorts of positions.

Legal Confessionalization

While countless publications have scrutinized the question of nationalism at universities, the impact of a scholar's religious confession remains an open question; it has mostly been analyzed on a case-by-case basis.[22] Not only are the confessional relations at universities hard to determine, but confession has remained an extremely fluid category and thus requires a flexible methodological approach.

The category "Jew" can be taken as an example of the complexities surrounding one-dimensional descriptions. In the Habsburg Empire, Jews remained officially unacknowledged as a national group but were accepted as a religious community; there were, however, substantial internal conflicts between Orthodoxy, Reform Judaism, and Zionism.[23] With a growing number of conversions, however, this categorization lost some of its explanatory power. In the late nineteenth century, Jewish converts to Catholicism or Protestantism were still referred to as Jews, and many saw themselves as such, despite their change of confession. Likewise, anti-Semites saw ethnicity, which conversion could not change, as the dominant characteristic. And ethnicity could be understood very broadly: in the spring of 1889, anti-Semitic attacks forced Eduard Suess to resign his position as rector of the University of Vienna. Suess had never been Jewish, but ancestors of his mother were.[24] The fluidity of the category of ethnicity in turn influenced political debates, including those concerning universities. Discussing the number of Jewish scholars in a debate on the confessional status of Cisleithanian universities in 1907, the spokesman of the liberals mentioned that the University of Innsbruck had two Jews among professoriat, but his conservative opponents insisted that he should also add two *Judenstämmlinge* (descendants of Jews).[25] In the same debate, similar controversies arose over the number of Jewish scholars teaching at other universities.

Individual accounts present a similar delicate and complex canvas, including regional particularities. In his curriculum vitae in Vienna in 1913, Harry Torczyner (Naftali Herz Tur-Sinai; נפתלי הרץ טור-סיני) described himself as a "German of Jewish nationality and Mosaic confession,"[26] but this combination of terms makes sense only if *German* is not considered a national category. In other words, in the context of the late empire, this combination could hardly be used with *Czech*, *Polish*, or *Ruthenian/Ukrainian* instead of *German*. The designation *Polish Jew* or *Czech Jew* would thus mean something different from *German Jew*. The former were linked to the ominous terms *assimilation* and *acculturation*, incorporating cultural or national transformation,[27] especially the rejection of Haskala (Jewish enlightenment, a cultural intellectual movement affiliated with German culture), whereas *German Jew* usually was not.

Using the ethnic term *Jewish* as a category of analysis accepts an ascription that does not consider cultural identity, leading in this case to less useful results, reminiscent of the categorization of confessions used by the anti-Semitic intellectuals who in 1907 spoke of Austrian universities as being overcrowded with (ethnic) Jews. Whereas anti-Semites spoke of Jews irrespective of conversion or baptism, others distinguished between "Jews" and "people of Jewish origin."[28] Individuals' perspectives on their own identities remain mostly hidden in the official documents, and they can be determined only for some scholars, providing confusing results rather than revealing the situation at the universities. Even a detailed book on Protestant teachers at the University of Vienna was limited to studying professors because of problems with archival sources.[29] Moreover, a recently published detailed monograph on Jewish professors at Prussian universities was made possible not through officially accessible statistics but via the fortunate discovery of intraministerial queries, which hint at a similar privacy of confession in the German Empire.[30]

Religion remained one of the leading issues in the controversies over universities, with the public evidently taking more interest in this than in the scholars themselves. The accusation that universities were liberal, socialist, Jewish, and filled with *Matrikelchristen* ("registered Christians," that is, those who were supposedly Christian in name only) not only caused an extremely serious crisis at the turn of the century, apart from the growing nationalist tensions, but also remained tightly intertwined with nationalisms. The more or less successful recalibration of national self-identification, and of the accompanying cultural rivalries, ran along ethno-religious boundaries: Roman Catholic Poles versus Greek Catholic Ruthenians, Orthodox Russians, and Protestant Germans; Roman Catholic Austrians versus Protestant Prussians; Protestant Czechs versus Catholic "(Bohemian) Germans."[31] These religious nationalities, as stereotypes and auto-stereotypes, had been present since the 1880s in discussions about appointments and the general character of universities, in both the public and private domains.

The question of a confessional-cum-ethnic division was part of a more general question about the Catholic character of Cisleithanian universities. The emancipation of Jews in 1867, the government's denouncement of the Concordat of 1855[32] in 1870, and university reforms three years later solved several of the legal questions concerning the relation between the papacy and the empire. The Constitution of 1867 officially demoted the previously privileged Catholics to just one of many acknowledged religious communities. Although the church's influence on universities was not legally codified, except in the theological faculties, small changes indicated the altering relationship between the two. First, in 1868 the professorial oath concerning religion was slightly modified. While the text of the oath of 1850 included that professors would avoid everything that would threaten "the state, religion, and morality,"[33] the oath from 1868 onward included only passages on legal obedience (*Gesetzestreue*).[34] Nevertheless, until 1918 the oath retained a vow to "God the Almighty" and ended with "so help me God," without legal clarification on what should be done in the case of atheists. Second, after 1873 the influence of the university chancellor (*Universitätskanzler*) was minimized. From 1848 onward, this position had been occupied in Vienna by the auxiliary bishop and the general vicar of the Viennese archbishopric (named in the lecture catalog directly after the rector and pro-rector). Prince-Bishop Friedrich Schwarzenberg, *Kanzler der Universität und Protector Studiorum* (chancellor of the university and protector of studies), had been from 1850 onward the first person named in the University of Prague's lecture catalog, and even after 1873 he continued to influence the university in manifold ways.

While the church's practical influence on the (supposedly) secular faculties of the university was limited, the clarification of the power relations between the state and bishops at the theological faculties remained in limbo. In 1874 the ministry announced preparations for a new policy, but they were never really implemented, leaving the neoabsolutist rules generally unchanged.[35] Since the office of university rector alternated between the faculties, universities would periodically be governed for a four-year term by a professor of theology, who was legally responsible to both the pope, represented by the bishop of the corresponding diocese, and the state. Both the pope and the state influenced appointments and had the right to suspend "unworthy" (*unwürdige*) professors at the theological faculties.[36] While religious scholars of all confessions were responsible to their church authorities and subject to their own festivities and days of rest (which, by the way, is reported to have been taken into account at some universities, for instance, by allowing Jewish assistants a free day on Saturdays), only in the case of Catholics was this inscribed into the academic legislation, influencing all personnel. These were the main points criticized by liberal and socialist politicians. Finally, in 1907 a group of Social Democratic parliamentarians proposed the disassociation of the theological faculties from the universities and their reestablishment as private teaching institutes.[37] This led to parliamentary controversies but not to any change in the law.

Legal issues were not the only area influenced by the Catholic Church. In addition, gradual generational change continued within the universities, although conservative Catholics predominated well into the liberal era, that is, after 1861. The gradual retirement of Thun-Hohenstein's favorites, however, combined with the growth in the number of appointments in the 1870s, gradually liberalized the professorship, although without substantial ideological changes in the most politically sensitive disciplines, such as philosophy and history (see below).

Ludwig Wahrmund and the Culture Wars

To exemplify how tightly intertwined academia and religion were, I turn now to the case of Ludwig Wahrmund, the victim of the most extreme violation of university autonomy in the post-1848 Habsburg state. This example also illustrates the fragmentary unity of the Habsburg Empire and, since the conflict itself was a reaction to events in the German Empire, confirms the Austrian leaning of the German-language community.

The most important aspect of the Wahrmund affair was the papal campaign against modernism. In the late nineteenth and early twentieth centuries, Leo XIII and Pius X had intensified papal interest in scholarship and the sciences, actively promoting the development of Catholic versions of these.[38] In the Habsburg Empire, the most influential act of the new papal policies was the creation in 1892 of the Leo Society, the Association for the Advancement of Science and Art on a Christian Basis (Leo-Gesellschaft, Verein zur Förderung von Wissenschaft und Kunst auf christlicher Grundlage). The papal interest was also evident in the growing frequency of scientific topics in theological periodicals such as the *Zeitschrift für Katholische Theologie* (Journal for Catholic theology) and

Przegląd Powszechny (Universal review).[39] At the theological faculties, new disciplines were established, such as "Christian sociology" (an obligatory subject at the Czech theological faculty in Prague from 1897 on),[40] Christian social science (*Christliche Gesellschaftslehre*), the history of church art, and Christian archaeology.[41]

The Catholic clergy and other interested parties also made requests of universities. The University of Vienna was reminded to maintain its Catholic character, as recorded in its founding charter. More important, however, to counterbalance "secular" academies, the Catholic clergy, supported by the Christian Socialists, proposed the establishment of a Catholic university in Salzburg. This "Free Catholic university in Austria," which was proposed by the episcopacy in 1901 (Catholic organizations had been fighting for it in an organized way since 1884),[42] was to have a slightly different angle than the state universities. It was to be independent of the state, financed by private donors (namely, Habsburg Catholics), and would serve as a training ground for new Catholic cadres rather than as a scientific institution per se. It openly aimed at reforming the state based on the Catholic faith.[43] Although the idea was supported by the bishops of all of the provinces, and a multicultural and multilingual school was proposed, it resonated almost exclusively in the German-speaking parts of the monarchy.[44]

But the idea also had vocal critics. In particular, liberals voiced their concerns,[45] and the *Neue Freie Presse* devoted a long editorial to the impracticability of such a university and the legal problems it would have.[46] Although Pope Leo XIII supported the conference of bishops,[47] the idea was not without critics within the church itself. At virtually the same time, the professor of church history in Vienna, Albert Ehrhard, in his widely discussed book on Catholicism in the twentieth century, warned that founding a Catholic university in Salzburg could be "a retraction from the vast sea of cultural life to an idyllic island, on whose coast the surging waves of the sea will not break." He also saw the mission of the church as lying not in the creation of a ghetto for its needs but rather in the "involvement of the church in all intellectual places of education and culture."[48] Other liberal Catholics, including Ludwig Wahrmund, similarly disliked the idea, fearing that clericalism would dominate over objective research.[49] In Wahrmund's case, this led to severe conflicts within academia.

The direct cause of this struggle over the religious outlook of academia in the Habsburg Empire was, however, of foreign origin: the fight in the German Empire since late 1901 against the appointment of the Catholic historian

Martin Spahn to the (Protestant) University of Strasbourg. In response to the proposed appointment, Theodor Mommsen and Lujo Brentano started a fierce campaign accusing Catholics of representing "science with presuppositions," which differed from the liberal Protestant non-presuppositional approach (*voraussetzungslose Wissenschaft*). Although this term had been used previously, Mommsen stabilized it, marking the German-language discussion on the relation between religion and objectivity, even though the term's philosophical substance (*Voraussetzungslosigkeit*) had already been abandoned in philosophy. In their quest to discredit Spahn, his opponents argued less against Spahn himself and more against the church and papal influence on the matters studied and taught at universities.[50] The point of departure was a critique of ultramontanism, a version of Catholicism in which the pope and the curiae were the highest authorities, in opposition to the liberal and state-led versions popular in the German and Habsburg Empires in various forms, from the *Altkatholiken* (Old Catholics) to the *Los-von-Rom* movement (Away from Rome!, a movement aiming at converting Austrian Catholics to Protestantism).

Whereas in the *Kulturkampf* (culture war) of the 1870s in the German Empire, cultural Protestantism and a secularization movement became strong,[51] papal-led Catholicism remained influential in the Habsburg Empire. Not only was it the basis of the dynasty, but it also achieved additional political representation with the creation of the right-wing Christian Social Party under the direction of Karl Lueger. This was strengthened by Lueger's anti-Semitic rhetoric.[52] In particular, Lueger proposed on several occasions a strategy of re-Catholicization of universities, criticizing the Jewish and socialist presence and the supposed discrimination against Catholic and German students and scholars.

Between 1902 and 1908, Wahrmund, a professor of canonical law in Innsbruck and a member of the Leo Society, became the symbol of the new anticlerical struggle.[53] In 1902, in response to the Spahn affair, Wahrmund published a brochure titled *Religion und Klerikalismus* (Religion and clericalism), proposing a division of state and church and the acceptance of universities as state institutions.[54] Hotly debated, both in academia and in Parliament,[55] this brochure had, however, no serious repercussions.

Shortly afterward, in 1907, Lueger announced at the Sixth Catholic Rally (6. Allgemeiner Katholikentag) a Catholic "reconquista" of the universities,[56] leading to days-long debates in Parliament.[57] Wahrmund answered this with a critique of the Catholic Church, titled *Katholische Weltanschauung und freie Wissenschaft* (Catholic worldview and free scholarship), published later as a brochure.[58] This caused serious disturbances: conservatives protested on the streets of Innsbruck and demanded his dismissal from his professorial post, and university professors criticized Wahrmund in the local media.

What began as a local Innsbruck conflict then expanded owing to the controversial actions of Nuncio Gennaro Granito Pignatelli di Belmonte, who intervened with the ministry, seeking to have Wahrmund dismissed from his duties; even if this intervention was unsuccessful, for many it meant that the nuncio had overstepped his competence. This was seen as a culture war, where *culture* refers not to an ethnic or linguistic affiliation but rather to religious confession. Progressive students of all national affiliations protested against the church's involvement in university matters and against the lack of a ministerial reaction, although for some intellectuals it remained a "German *Volkstheater* [people's theater] in Austria."[59] In Prague, protests in favor of Wahrmund even led to the first joint demonstration by Czech and German students since 1859.[60]

The Wahrmund affair, however, showed not only that some matters in the fin de siècle counted more than nation, language, and ethnicity but also that the government did not really know how to balance the legal autonomy of universities with the growing Catholic and German-national pressure. In this instance, Wahrmund was officially relocated to Prague and even received a higher salary; in other cases, however, the ministry clearly took a pro-Catholic stance.[61] Since future ministers were unwilling to aggravate the confessional ruptures, not unlike the situation with nationalist tensions, Salzburg gained a university only in 1962.[62]

The Ideology of the Empire: Catholicism

While the Wahrmund affair demonstrated that the few religious issues that did arise led to heated debates, the ministry and the faculties discussed the endurance of a Catholic worldview in academia secretly behind closed doors. Catholicism penetrated into Cisleithanian universities throughout the state, and scholars applying for positions were well aware of it; young scholars often mentioned that they were Catholic in their curricula vitae when applying for habilitation. This was a widespread practice, especially at the German University in Prague, where candidates seeking to become Privatdozenten also frequently added the ethnic designation "of German origin." On several occasions, the ministry and the faculties addressed, directly or indirectly, the issue of confession, clearly favoring the Catholic standpoint. In the faculty proposal for the chair of physiology in Innsbruck in 1904, the candidate Franz/Ferenz Tangl was described, first, as the "offspring of a German Catholic family who had come in the eighteenth century from Thuringia to Moravia and from there to Hungary." Next, it was noted that German was his mother tongue, followed by a presentation of his scientific career; only after that did the proposal give a brief description of Tangl's ideas on physiology.[63]

In particular, the chairs of history and philosophy, as constituents of a broadly understood moral and national education, remained seminal in the eyes of the ministry, which did not shrink from making appointments that went against the will of the faculty. Around the turn of the century, the ministry confronted the philosophical faculty in Vienna on several occasions. In 1899 the Innsbruck Privatdozent Joseph Hirn was appointed to the important chair of Austrian history in Vienna, although the faculty had not considered him adequate for the chair and had not included him in their proposal. The minister of education, Arthur Bylandt-Rheidt (March 1898–October 1899), considered this omission a result of Hirn being an exponent of "conservative and Catholic historiography"[64] and proposed his appointment to Franz Joseph. While most Viennese historians were Catholic, Hirn's appointment strengthened the position of ultramontanism, as opposed to liberal Catholicism, in Vienna.

Such appointments followed the pattern of appointing Catholic scholars to the chair of history in Vienna, with the Innsbruck school of Julius Ficker providing a substantial number of scholars who took the desired ideological direction. Apart from Hirn, four of Ficker's students gained full professorships in Vienna and one in Graz.[65] Of the Habsburg German-language universities, only Prague developed an independent school of historiography dominated by local historians. This trend was certainly reinforced by a focus on the development of the auxiliary sciences of history, most successfully among Ficker's students. This was important for political reasons, especially at provincial universities, from the moment Joseph Alexander Helfert proclaimed in the early 1850s the necessity of minute historical research on the Habsburg crown lands (see chapter 2), with the ministers of education adopting this view.[66] However, the overall influence of Catholicism defined the general development of historiography at the universities in which Innsbruck scholars had a say.

The second center of Catholic interest remained philosophy. Its situation was largely an outcome of the teaching of Franz Brentano, appointed a professor of philosophy at the University of Vienna in 1873, who had been recommended by the influential Göttingen philosopher Hermann Lotze, among others, owing to his "professional" philosophical approach.[67] Despite being a Catholic priest and working on liberal Catholic philosophy,[68] Brentano opposed ultramontanism and the newly prescribed papal infallibility. In accordance with these convictions, Brentano withdrew from the priesthood and his position as a professor in Würzburg. Opposition to the all-encompassing papal authority, however, was clearly exactly what Franz Lott (his predecessor, who apparently influenced the faculty's choice through communications with the ministry) and the minister of education Karl Stremayr, a proponent of reducing Catholic influence on the state (precisely at the moment of loosening the Concordat), approved of.[69] In 1880 Brentano moved for a short time to Saxony in order to marry Ida Lieben, which was not legally possible for ordained priests in the empire. Because he had to change his citizenship for the move, this automatically canceled his professorship, which was neither returned to him nor subsequently filled. A proponent of modern philosophy based on the natural sciences and psychology, Brentano remained at the University of Vienna as a Privatdozent, which was unanimously accepted by the faculty without the usual habilitation procedures.[70] He hoped for a future appointment, but over the next few years, the ministers denied him such a chance, which finally led Brentano to resign from his position in 1895.[71]

In his time as a full professor, however, Brentano was able to influence Stremayr to appoint two of his students as professors, Anton Marty (Chernivtsi University, 1875) and Carl Stumpf (University of Prague, 1879). Both had written their dissertations under Lotze's supervision and had previous connections with the Catholic Church. Marty had been a priest but left the priesthood shortly after Brentano; Stumpf had attended the ecclesiastical seminary, leaving it in 1870.[72] Both were something of a rarity in the empire: Marty was Swiss, with no habilitation, and had graduated only shortly before the appointment, which took place (probably) without a terna proposal.[73] Stumpf was not in the Prague faculty terna; the minister consulted with Brentano and decided to appoint Stumpf against the wishes of the faculty, who explicitly wanted a historian of philosophy.[74] Despite the loss of his professorship and his problems with the church, Brentano remained influential. His students achieved high positions at all Cisleithanian universities except Cracow.[75] The ministry also violated university autonomy in some cases, appointing candidates not even named in the ternas, although most of Brentano's students have been acknowledged to have been formidable scholars and were mostly at the head of the faculties' proposals.[76]

Brentano's influence was not seen as entirely positive, and he had opponents in Vienna: Ernst Mach, a professor of philosophy, especially the history and theory of inductive sciences (*Philosophie, insbesondere Geschichte und Theorie der induktiven Wissenschaften*), commented sarcastically on choosing a candidate for his successor: "This school leaves marks on everybody, but they will be shaken off earlier by the most outstanding [scholars]."[77] Nevertheless, Mach acknowledged several of Brentano's students, assessing them as independent scholars but overly influenced by the Viennese philosopher.

Among Brentano's critics was Friedrich Jodl, whose appointment record exemplifies academic and administrative maneuvering between religion and philosophy. In 1885 the chair of philosophy at the German University in Prague was vacated, as Stumpf accepted a call to Halle. The faculty proposed three scholars from the German Empire to succeed him. The minister of education, Sigmund Eybesfeld, decided on Jodl, a liberal Catholic from Munich. The justification for this decision demonstrates that, twenty years after Thun-Hohenstein, the ministry still not only carefully considered the religious views of the scholars in question but even sought the approval of the church with regard to professors of philosophy: "In this concern, it was welcomed by me that the late Archbishop of Prague, Cardinal Schwarzenberg, took the occasion to discuss with me, a short time before his passing away, the question of the appointment for the intended chair, in which he had a lively interest, and in this connection indicated the appointment of Dr. Jodl as particularly recommendable."[78]

Ten years later, in 1896, when deciding on the appointee for the chair of philosophy in Vienna after Robert Zimmermann, the minister of education Paul Gautsch (1879–93, 1895–97) similarly preferred Jodl for religious reasons. In the meantime, however, Jodl had abandoned liberal Catholicism, become a sturdy opponent of ultramontanism, and begun participating in anticlerical organizations. The faculty had proposed three *ex aequo* scholars. Gautsch scrutinized them more with regard to their religion than their philosophical achievements: Benno Erdmann was rejected because he was German and Protestant, while Alois Riehl was a thorn in the flesh of the Catholic authorities, "which [he] seemed to brusquely oppose in Freiburg and which he hurt through his conversion to Protestantism."[79] The third scholar proposed was Jodl, whom Gautsch appointed. While the minister had criticized Riehl for conflicts with religious authorities, he did not use the same argument with respect to Jodl, probably because Mach lobbied the ministry directly in favor of the appointment of the Prague scholar.[80]

However, the ministry and the faculty swiftly balanced Jodl with a philosopher with more conventional confessional ideals. To achieve this, the chair previously held by Brentano, vacant for fifteen years, was filled. Since earlier proposals for this chair had resulted only in the appointment of an associate professor, Franz Hillebrand, to help with the lectures,[81] it seems likely that Jodl's nomination triggered the reactivation of the chair. At the time, Hillebrand was being considered for a professorship at Innsbruck, and thus the potential shortage of teaching staff may have been another reason, although it does not explain the search for a full professor.[82] The faculty committee, with Mach as chair, decided that, to balance Jodl's position, a historian of philosophy should be appointed; it proposed two philosophers from the German Empire. Although this was accepted by an overwhelming majority (forty-one to two), Zimmermann opposed it and proposed Laurenz Müllner, a priest and professor of Christian philosophy at the theological faculty in Vienna. On 18 May 1896, forty days after Jodl's appointment, the ministry presented Franz Joseph with a proposal to move Müllner to the philosophical faculty, with the ultimate aim of teaching Catholic philosophy.[83]

With this decision, two priests had been transferred from theological faculties to teach philosophy within two years, the first being Stefan Pawlicki in Cracow in 1894. In his case, however, the faculty had proposed the transfer, although it was opposed by the only philosopher in Cracow, Maurycy Straszewski, who preferred Wincenty Lutosławski, a young Warsaw-born scholar who was teaching in Kazan. Pawlicki, whose early ideas linked Catholicism with positivism, successfully defended the university against trends in philosophy in later years that were unwelcome to the Catholics. For example, he antagonized Lutosławski, who for a short time taught as a Privatdozent in Cracow, and criticized "materialism," opposing the creation of the Institute of Experimental Psychology (Instytut Psychologii Eksperymentalnej).[84]

The appointments of Jodl, Müllner, and Pawlicki illustrate the general trend of Habsburg philosophy, which constantly sided with Catholicism; scholars opposing the state religion hardly had a chance of being appointed. As an academic discipline, philosophy was connected with pedagogy for most of this period (the chair for the latter was separated from philosophy only in the last decades of the century), leading academic philosophy into a dilemma about how to cope with such a belated change. Finally, the chair of philosophy was divided into one professorship devoted to the natural sciences and logic, and a second focusing on historical aspects, devoted to "social and moral pedagogy."[85] However, as with other disciplines in the empire, the differentiation of chairs depended on the university, leaving smaller institutions, such as Innsbruck and Chernivtsi, disadvantaged.

An Invisible Ghetto Wall: Jewish Scholars

Catholic scholars were not the only people whose mobility was influenced by confessional issues. One of the most pressing questions was that of Jewish scholars. Recently, a number of publications have addressed the issue of latent and open anti-Semitism at Habsburg universities. Also, models of the effects of exclusion have been proposed, underscoring in particular that Jewish scholars who could not find a place at a university were vital to the establishment of independent institutes; in this way they contributed to the cultural thriving of cities, especially the metropolitan capitals. Below I want to delve more into the detailed sphere of negotiations and identity questions and to look beyond the centers; I argue that at the smaller universities, processes took place that enabled the centers, in particular Vienna, to function in the way they did.

To begin with, I want to mention a contradiction between the official view and the public view. While the controversies over the appointments of Jewish scholars were broadly discussed, this issue remained almost completely absent from the official records of the universities and the ministry. These records make precise statements on the confession of professors and Privatdozenten impossible. Since numbers of Jewish scholars converted to Protestantism or Catholicism to facilitate their careers at universities, birth certificates and early life information do not help here either.[86] Conversions remained frequent at least until 1918, and scholars changed religion not only for career reasons but also for the sake of marriage or out of ideological conviction.[87] Hans Kelsen, for instance, born to Jewish parents, was baptized a Catholic in 1905 (for career reasons) and then converted to Lutheranism in 1912 to marry Margarete Bondi.[88] According to the law at least, one's religion could be changed, but not so in the public eye, as I argue later.

Although a religious declaration was not requested in documents on habilitation, an annotation of "Mosaic confession" or (more seldom) Jewish origin can still be found in some papers, such as those of Harry Torczyner, mentioned above.[89] As with Catholic German Bohemians, for some scholars their confession was an important part of their identity, and they did not fear being disadvantaged by openly naming it. (In no cases, however, can one imagine affirmative action as the basis of this practice.) Nevertheless, as noted before, one would have to consider different definitions of Jewishness to draw conclusions about its influence on the appointment policy and thus about the political alignment of the faculty and ministry.

In most cases, it is thus impossible to determine from the official records whether scholars were rejected because of their Jewish confession or origin. However, the historian Urszula Perkowska noted in her analysis of habilitations in Cracow that in many cases she could hardly understand the reasons for declining a habilitation and therefore suspected the presence of conservative Catholic cliques at the university.[90] Indeed, in the case of Szymon Askenazy, members of the Cracow philosophical faculty discouraged him from habilitating because the university already had two Jewish scholars.[91] However, since the young historian never submitted habilitation documents, and this discouragement was articulated in a private letter, it is impossible to tell how the faculty would have reacted if Askenazy had formally applied. One can find cases where habilitations were rejected without no concrete reason given, other than a vague mention of, for example, "personality."[92] Only in a few cases can one find direct statements: anticlericalism and his Jewish confession were the main reasons for the rejection of Ludwik Gumplowicz's habilitation thesis.[93]

In the second half of the nineteenth century, the discursive construction of the Jewish scholar underwent important changes. In the Vormärz period and during the 1850s, it was confession that counted; the ministry saw and treated converted scholars as Catholics, and even promoted them as examples of regained lost sons. Most noteworthy were two converted scholars who worked in the most ideologically important discipline in the empire: philosophy.[94] This situation changed later in the nineteenth century. Cultural and ethnic affiliation, defined and ascribed in different ways across the Habsburg Empire, replaced confession as a marker of Jewishness, especially in the public and political eye. Conversion thus leveled the legal hurdles but not the social and cultural ones.

In the 1850s and 1860s, scholars of Jewish confession had almost no chance of teaching at a university, although Privatdozenten for Hebrew and rabbinic languages were allowed from 1848 on in Vienna, Prague, and L'viv.[95] Of those, only one gained a full professorship, shortly after Thun-Hohenstein resigned, while the other scholars received only associate professorships, even though the Viennese philosophical faculty strove for several years to obtain a full professorship in this discipline.[96]

In other subjects, as well as professorships in general, the effect of a scholar's confession was more complicated. Here, the university was subordinated to more external legal factors, because professors were state officials. Also, the choice of the dean or rector was an issue; this was problematic for Protestants. The discrimination against non-Catholics also applied to schools in general, which were to remain Catholic, according to the Concordat, although the universities were exempted from this.[97]

Until 1867–68 other forms of political discrimination also remained in effect for Jews, including limitations on residency rights and accumulation of property, additional taxes, and so on. The legal confirmation of these discriminatory measures in 1853 caused almost instant protests but also resulted in a falling number of Jewish students at universities, since, given these obstacles, studying constituted a less attractive vehicle for social mobility.[98] The atmosphere of confessional discrimination, especially after the Concordat, was such that, to use Theodor Gomperz's words, the "path to professorships has been closed for the Jews."[99] This discouraged Jews from applying for Privatdozent positions, including in medical studies, the field where a scholar of Jewish confession first attained an associate professorship (in 1861).[100] Anti-Semitic ideas were present in influential media as well. For example, Sebastian Brunner's *Wiener Kirchenzeitung* (Viennese church journal) and the writings of conservative Galician ideologists, which were published more frequently around the 1860s as a reaction to rumors about the legal emancipation of Jews, were ideologically influential in academic and ministerial circles.[101]

Given that rabbinic education took place outside of Habsburg universities,[102] the discipline of Hebrew language and philology entered universities only around 1900; it was separated from the main field of Indo-Germanic oriental languages even later than Sanskrit was, and it also had fewer habilitations than other philological subjects. Consistent with the typical practice for the introduction of new disciplines in the empire, the first full professorship was established in Vienna in 1885, followed by one at the German University in Prague in 1892 and, later, positions at other faculties, except Cracow; most of the universities had associate professorships in this field, but the Czech University in Prague had only a Privatdozent.[103] The professionalization of Semitic philology, which was also taught at theological faculties, meant that it was not covered exclusively by Jewish scholars. Gustav Bickell, a converted Jew and politically involved Catholic, held the chair of Semitic languages in Vienna from 1892.[104] Readers in Semitic languages were similarly rare. While all universities included readers of French, Italian, and English, they rarely offered Czech, Russian/Ruthenian, Hebrew, or Yiddish. In contrast, other, much rarer languages were taught on a regular basis, such as Armenian in L'viv, Lithuanian in Cracow, and Spanish, Modern Greek, and Hungarian in Vienna.

Jewish scholars met with a number of obstacles on their way to acceptance in academia. Although, officially, habilitation did not take confession into account, and in 1867 Jewish emancipation was proclaimed, the general atmosphere of polite hostility in both society and the university certainly inhibited Jews from entering academia, especially Jews who were migrating from the east to the capital, who were victims of a cultural othering by both culturally assimilated Jews and an anti-Semitic public.[105] In Galicia and Bohemia, their options were cultural assimilation or othering.[106] Jewish scholars who assimilated met with fewer obstacles, and most of the national groups of the Habsburg Empire included prominent and influential intellectuals of Jewish faith.

In absolute numbers, the number of Jewish students in Cisleithania grew almost continuously, but since the overall student population was soaring in the Habsburg Empire, relative statistics give a more balanced view of the confessional division at universities. Around 1890 Jewish inhabitants constituted around 9 percent of the population in Vienna (having grown rapidly from 2 percent in 1857), 9 percent in Prague, and around 30 percent in Cracow, L'viv, and Chernivtsi; smaller but growing numbers, especially after 1900, were found in Graz and Innsbruck.[107] At the university in Vienna, young Jews accounted for around a third of all students (peaking in 1885), predominantly in medical and legal studies. Similarly, at the Charles-Ferdinand University in Prague (and later the German Charles-Ferdinand University), Cracow, and L'viv, between 20 and 30 percent of all students were Jewish, similarly concentrated on medical and legal studies; at the law faculty in Chernivtsi, more than 50 percent of the students were Jewish in some semesters.[108] In Vienna and Prague, Jewish students were thus overrepresented relative to the overall population, while in Galicia and Bukovina the proportions were representative of the general population, and in Graz and Innsbruck the numbers were low: in some semesters there were no students of Jewish confession in Tyrol.

At the same time, estimates for Vienna indicate that around 10 percent of those appointed to professorships were Jews, but the number of Jewish Privatdozenten was much higher.[109] Steven Beller, for instance, estimates that the proportion of Jewish scholars in Vienna in 1910 was around 40 percent (between 50 percent and 60 percent in the medical faculty, and 21 percent in the philosophical faculty).[110] While the exact number for Prague is unknown, during the debate of 1907 (see below) it was considered to be disproportionally high, although, as at Vienna, fewer Jewish scholars reached the higher levels of academia. Likewise, statistics for Chernivtsi indicate that 10 percent of professors were Jewish, while the number at other German-language universities was statistically negligible.[111]

This disparity was often discussed in public, and it merged with traditional Catholic anti-Semitism to nourish the popular image of the *Verjudung* (Jewification) of scientific institutions. One must add, however, that Catholic-based anti-Semitism—already of a racial variety—must be considered a public cultural othering that affected, especially in Vienna, assimilated Jews who saw themselves as members of the German bourgeoisie. This was a situation similar to that of the Poles of the Mosaic confession, including those who were clearly aligned with the Polish national groups, such as the above-mentioned Gumplowicz, Natanson, and Askenazy. The issue of assimilation was perceived differently by the different groups involved, ranging from a sign of "civilization" and "progress" (Haskala and Reform Judaism, and the liberal and socialist press) to a signal of racial and cultural decadence (Christian Social parties, radical nationalists), with a nationalist imaginary dominating over the course of the century.

A discussion in the Polish-language journal *Krytyka* in 1914 can help illustrate academic discrimination in the early twentieth century. A letter to the editor described several cases of Jewish assistants at the medical faculty of the University of Cracow who were denied the possibility of habilitation and then emigrated. In response, the anonymous "Doctor K.L.," from the tone of the article neither Jewish himself nor really a pro-Jewish supporter, claimed this to be a loss for Polish science. While the faculty was now closed to Jewish scholars, the author named several Jewish physicians who had previously considerably enriched Polish scholarship.[112] Reactions to this article were of one sort: letter writers argued that there were countless examples of Poles who could not get university positions, and thus one should not criticize the fact that Jews were not being promoted but rather, for the sake of Polish scholarship, should promote Poles, implying that they should, by default, be Catholics.[113] Certainly, several Jewish professors (both converted and not) worked in Galicia, along with a number of Privatdozenten, predominantly in L'viv; their numbers rose only after 1918.[114] One also finds a preponderance of Jewish scholars among Galician-born, German-speaking, university-habilitated scholars,[115] some of whom also began their studies in Cracow and L'viv. This fact points to the trend mentioned by the anonymous Dr. K.L.; however, as there are insufficient data on the situation leading to the migrations and conversions, this statement should be taken with caution.

Felicitas Seebacher has impressively shown, using the example of the medical faculty in Vienna, that such a discourse also occurred in the Austrian capital, although there migration induced by discrimination was not geographic but intra-urban, that is, to other medical institutions in the city.[116] The most prominent issue there was the covert and overt anti-Semitism among scholars and students.[117] The best-known instance of anti-Semitism is Theodor Billroth's 1876 book on the teaching of medical sciences,[118] in which the author used a stereotype of a low-income Jewish student from Galicia to degrade the University of Vienna; the book was heavily criticized, and Billroth ultimately withdrew his statements.[119] Nevertheless, his argument remained influential, being used, for example, in the above-mentioned parliamentary speeches in 1907. Adolf Wahrmund, a professor at the Academy for Oriental Languages (Akademie für Orientalische Sprachen) in Vienna, and August Rohling, a professor of theology in Prague, published a number of widely read and translated pamphlets with anti-Semitic content, supported by their academic authority.[120] After the rise of right-wing parties, not infrequently with reference to Catholicism,[121] and through the consolidation of opposing fronts owing to political affairs,[122] anti-Semitism in its modern racial version gained a firm place in the political landscape of the empire.[123] Harsh commentaries appeared in the press claiming that Wahrmund, Masaryk, and, in 1910, the students who protested in Cracow against pro-Catholic university policy were all Jewish,[124] illustrating how radical parties forged a link between Catholicism and anti-Semitism. Indeed, anti-Semitism was prevalent in the mass media and the public, but one should not forget that it was not the only, or even the most popular, ideology. For example, in 1891 in Vienna, the Association for Defense against Anti-Semitism (Verein zur Abwehr des Antisemitismus) was established, with several University of Vienna professors as both founding fathers and ordinary members. Further, for a large number of scholars, confessional differences played no role in the academic process at all.

The atmosphere surrounding the appointment of Jewish professors remained oppressive throughout the post-1867 period, and the visibility of anti-Semitic views increased after 1890. The university was not only becoming a battleground, as a recent exhibition in Vienna has claimed,[125] but turning the cities into one. In the 1880s the mathematician Seligman Kantor was a victim of street assaults, leading the faculty to consider him an inappropriate candidate for a professorship.[126] Shortly afterward, Kantor moved to Italy. The appointment of Jewish scholars to professorships led to student protests as well. In Vienna the press protested the appointments of Emil Zuckerkandl and Julius Tandler.[127] In Innsbruck in 1900, during the appointment procedure for the ophthalmologist Stephan Bernheimer, the faculty was confronted with a petition for the "purification of the University of Innsbruck from Jewish influence,"[128] along with fierce protests by radical right-wing student organizations. The same university witnessed protests in response to August Haffner's appointment as a professor of Semitic languages (he was transferred from the theological to the philosophical faculty).[129] This tendency was strengthened by the gradual division in student life along religious-national boundaries, resulting in the creation of parallel publics and aggravating potential conflicts.[130]

Divisions based on Christian confessions—Greek Catholic Ruthenians versus Roman Catholic Poles, and Protestant Hussite Czechs versus Roman Catholic Germans—had no obvious influence on appointments and habilitations. For Jews, however, their nationality was defined through their confession, which resulted in their exclusion from other national groups, causing problems. For example, Alfred Přibram's appointment as a full professor of history was blocked several times: in Vienna in 1899, where he was evidently omitted owing to his confession,[131] and in Prague in 1900, when he was proposed primo loco but gained only a titular professorship.[132] He was finally appointed *ad personam* in Vienna in 1913. Samuel Steinherz, a Jewish historian who worked extensively in Rome, acquired a full professorship in Prague owing to the direct support of influential scholars who intervened directly in Vienna, but the ministry rejected his appointment to Vienna, for which he was proposed primo loco, in 1908.[133] When Szymon Askenazy was proposed for the professorship of Polish history in L'viv, the combination of his confession and the prominence of the chair of Polish history was too much for the nationalists to accept—despite Askenazy's writing on the need for Jewish assimilation and his politically engaged assessment of modern history, which put the modern Polish nation at the fore more strenuously than other historians did.[134]

Nonetheless, several Jewish scholars were appointed by the ministry despite these obstacles, especially in the 1870s. For example, Adolf Lieben was promoted twice by the ministry, disregarding the order of candidates in the terna. In 1871 he was promoted from Turin to Prague, while the primo loco proposed candidate went to the technical academy in Brno. In 1875 Lieben was appointed to Vienna, although he was the third choice in the terna; the two other scholars were from the German Empire.[135] Similarly, Theodor Gomperz habilitated in 1867 without first receiving a doctoral degree.[136]

As noted, the discussion on the national question and the increasingly defensive tactics of the ministry (i.e., seeking to avoid igniting conflicts or violating academic autonomy) strengthened the professorial majority, which disadvantaged those groups with less representation at the university. With regard to Jewish scholars, this led to the creation of an "invisible ghetto wall,"[137] leaving few opportunities for promotion. During the appointment of the chair of chemistry in Innsbruck in 1902, Josef Herzig, proposed primo loco (*ex aequo*), was not taken into account because "detrimental events could occur, as they did not long ago [when Stephan Bernheimer was appointed] at the medical faculty."[138] Similarly, in Graz the following year, Josef Jadasson, an associate professor of dermatology at the University of Bern, was rejected with the justification that he, "in consideration of his descent, could lead under present conditions to insalubrities at the university."[139] Six years later, in 1909, when Otto Löwi was proposed for the chair of pharmacology, the ministry voiced the same concerns, stating that his "belonging to the Jewish confession, distinguishable already through the name, [could] impede his activity at the University of Graz and at the worst could lead to insularities."[140] In this case, though, the ministry, having consulted the provincial government, decided to appoint Löwi, who taught in Graz until 1938, winning the Nobel Prize for chemistry in 1936.

Only one place remained unproblematic for Jewish nominees, Vienna, the university with the most Privatdozenten and thus lower chances of appointment in general. This led to rising numbers of Jewish Privatdozenten in Vienna. They were bereft of opportunities to be nominated for professorships at other German-language universities, but experience as a professor at a provincial institution was almost a requirement for nomination as a professor in Vienna or Prague, as shown in chapter 4. However, even in Vienna, the atmosphere was increasingly tense after the late 1890s, and the growing influence of German nationalist scholars and students led to anti-Semitic street protests and, finally, street fights directly before World War I.[141]

Historians often mention that Jewish scholars had to wait longer for professorships because Catholic faculties were promoting Catholics, baiting Jewish Privatdozenten with titles and remunerations but hindering their entrance into faculty positions, which would have bestowed the right to vote on important academic matters. These career inequalities were what the Viennese jurist and politician Josef Redlich meant in a widely discussed speech from 1907.[142] The statistics cited by Karl Lueger in 1907 to substantiate his claim that Cisleithanian universities were turning into Jewish strongholds—that seven of the eight most recent appointees were Jewish—concerned paid and unpaid associate professors,[143] which Lueger neglected to mention. This glass ceiling was most significant in Vienna and Prague; the universities there, hesitating to appoint scholars from within and seeking the best available scholars, tended to look outside their own walls. At the same time, Jewish scholars were generally unwelcome at other universities, which limited their appointment opportunities to universities where they had the most competition, without having any real chance of proving themselves as professors elsewhere. In other words, they had a double burden of work outside the university to improve their financial stability and thus had fewer chances for research and publications. Through this vertical glass ceiling and the horizontal invisible ghetto wall, a large number of Viennese Jewish Privatdozenten were left adrift, leading them to concentrate on other activities, such as *Volkskurse* (people's courses) and semiprivate laboratories, largely contributing to the paradigmatic image of a culturally and scholarly flourishing Vienna in 1900.[144]

CHAPTER 7

Habsburg Legacies

After the murder of Chancellor Dollfuss by a troop of Austrian National Socialists, Schuschnigg had taken over, and had demanded from all persons employed by the state, or by local governments, including all teachers and professors, that they join an organization which he called the Patriotic Front. . . . All university professors signed, (and especially those who were Nazis). There was only one exception: Professor Heinrich Gomperz whose family came from Germany and whose cultural background and Greek scholarship made him partial to a union with Germany where Greek scholars abounded. He himself was of Jewish descent. . . . [H]is failure to sign up with Schuschnigg's Patriotic Front led to the dismissal of Gomperz from his professorship with total loss of his income: and censorship prevented this from ever getting into the papers. Nobody heard of this dismissal. No rumor reached me, until one day he rang me and we met. Then he told me what had happened and that after his dismissal he had decided to emigrate to the United States. But he had not the money to pay for the costly journey. So he went to Prague, to ask his old colleague and friend Masaryk for a loan. Masaryk gave him the money from his own personal savings as a gift, rejecting a loan and explaining to Gomperz that he did not wish to use any kind of official funds for this purpose because the political element in it might make it look as a pro-German act, and even as pro-Hitler. . . . And Gomperz told me how wonderful and moving his meeting with Masaryk had been.

—SIR KARL POPPER, 1994[1]

The events related by Karl Popper took place in 1934, sixteen years after the dissolution of the Habsburg Empire; they united the German Jewish cultural nationalist Heinrich Gomperz with the Czech nationalist Tomáš Garrigue Masaryk, at that time the president of Czechoslovakia. Shortly after most of Europe had swung to the right, the friendship between the two Habsburg philosophers allowed Gomperz to travel to Los Angeles. Popper, who had also been forced to emigrate, propagated an Austria-rooted theory of knowledge in his adopted country of Great Britain.[2] Popper, like many Viennese Jewish intellectuals, may have overvalued the Habsburg legacy and thus the monarchy itself.[3] But Gomperz would have found like-minded scholars with a similar philosophical bent from Warsaw to L'viv to Padua.

This chapter sketches the fate of Cisleithanian universities after World War I, especially focusing on those facets that transformed them from an imperial space to a multistate central European space, defined both by a common intellectual past and by a multitude of weak and strong ties. As I argue, the transformation was less a revolution than a continuation of trends the region had already been experiencing before the Great War, even if new boundaries and legal spaces meant serious changes. However, the habitus, personal networks, a similar ideological orientation, and even the Vienna-centric power structure remained in place, facilitating further contacts and, to a certain degree, unity. Since these new spaces were mostly multicultural, they inherited problems from the empire but also created solutions for dealing with them. The Habsburg system and the universities' experience also proliferated into new regions, both through professorial migration and legal transfer and also because of the political changes during the 1930s, which led to a spread of their influence on a global scale.

Universities at War

World War I seriously disrupted the lives of universities. However, Cisleithanian academic mobility did not change dramatically during the conflict itself, even with central Europe plunged into chaos. Galician Poles even led the ministry in 1917 and 1918, an important sign of Cisleithanian unity. Similarly, nomination procedures continued, although the universities encountered some problems owing to the war, notably the drafting and occasional deaths of young scholars.

From November 1914, just a few months after the beginning of the war, faculty members were dying at the front.[4] The total death toll among professors and Privatdozenten remained low, however, with only a few deaths directly linked to the war.[5] It seems the government was hesitant to draft university members, and when it did, it did not send them directly to the front line; medics, for instance, served in military hospitals, while scholars from the law and philosophical faculties populated the intelligence offices. Some, however, volunteered and joined the soldiers at the front. The famous Viennese physicist Friedrich Hasenörl, for instance, died near Trento/Trient in 1915.[6]

Since universities did not report on their draftees in a consistent manner, and catalogs of lecturers were published irregularly, it is hard to pinpoint the impact on personnel. For example, at the University of Cracow, which closed for some months in 1914/15 owing to the city's reorganization into a fortress (*Festung*), around 30 percent of the staff were drafted into the imperial army and 20 percent joined the Polish Legions (Legiony Polskie).[7] More critically, assistants, adjuncts, demonstrators, and other young academic employees were drafted more frequently, an action that the universities consistently criticized in their reports. The frontline universities in L'viv and Chernivtsi were most affected by the war, with the professors spending most of their time in Vienna.

Still, even by 1918, there was no question for the Habsburg government, and for many of the professors, that a German-language Habsburg university would remain in Bukovina after the war.[8] Among German-Austrian scholars, the idea of imperial unity was widespread, which connected with older patterns of cultural paternalism. The University of Vienna's rector, Emil Reisch, for instance, stated during his inauguration in 1916, with imperialistic and German nationalistic zeal, that after the war the cultural efforts of the state should be intensified through German universities and the German cultural mission (*Kulturmission*), with *German* meaning here a tight, even union-like, cooperation between the Habsburg and German Empires.[9]

Reisch's speech is symptomatic of one other characteristic that began to shape the post-1914 situation. The Great War solidified national categories along linguistic lines: German-Austrian, Hungarian, Polish, or Czech and Slovak. Although these categories had already existed and had already shaped academic practice (as shown in earlier chapters), new power relations meant that they became part of state policies. But they did not remain uncontested, even among nationalist activists. Czech, Slovak, and Czechoslovak identities were debated in Czechoslovakia and partially conflicted with each other, and in Poland a large number of people identified as *tutejsi* (literally, "from here"; that is, they defied national identification). Another question that shaped central Europe was how to deal with nondominant groups within nationalizing states (such as the Jews, Ruthenians, and Germans in Poland).[10] From Tyrol, to Lower Austria, to Bukovina, no post-Habsburg region was monocultural, at least until the ethnic cleansings during and after World War II.[11]

Indeed, academic institutions were unprepared for the final disintegration of the monarchy, and new regulations had to be created swiftly to accommodate the new political realities. Most universities and their faculties readily aligned themselves with the policies of the new states. In Galicia, by 1915 several scholars had already been sent to the newly opened university in Warsaw, and professors there were frequently politically active. In Austria universities readily and apparently happily accepted merging with Germany.[12] The Viennese *Deutsche Hochschul-Zeitung* (German university journal), both pro-German and anti-Semitic, saw German-Austrian reconciliation (*Annäherung*) during the war as the only way forward for German culture.[13] In contrast, Habsburg scholars seemingly regarded the establishment of an Austrian state with uncertainty.[14]

Another area, Tyrol, was also at stake, and there the universities readily participated in continuing an imperial German nationalist discourse. Both the students and the faculties of the University of Innsbruck actively contributed to a propaganda war against "Italian imperialism."[15] Expert reports, memoranda, official participation in marches, and even personal letters to President Woodrow Wilson were used to pressure international politicians.[16] The failure of these efforts (e.g., the loss of South Tyrol to Italy in 1920) and the reality of the new geographies led to an intensification of research, much like phantom limb syndrome, when one loses a limb but has the feeling that it is still there. The Institute of Historical Settlement and Regional Studies of the Alpine Countries (Institut für geschichtliche Siedlungs- und Heimatkunde der Alpenländer, established in 1923) has been described as one of the earliest manifestations of a *völkisch* (folkish, i.e., ideologically populist, ethnic, and racist) historiographical institution, while in Vienna both historians and members of the law faculty proposed the Austrian *Anschluss* (joining) to Germany.[17]

The German University in Prague and Chernivtsi University experienced the most uncertainty at the end of the war, since their teaching staff now belonged to declared national minorities. Scholars from these universities openly opposed the new states and considered changing locations; in Prague the rectorate even proposed the extraterritoriality of the university (that is, the university would remain in Prague but without being subject to the Czechoslovak state), which at first fiercely rejected the Czechoslovak government.[18] Although these efforts failed, both universities remained in the new states, the German University in Czechoslovakia and Chernivtsi University in Romania, but lost some of their faculty members.

Habsburg Multicultural Legacies

As Tara Zahra convincingly argues, schools were one of the places where nationalization processes took place,[19] and universities, which had integrated nation-building and nation-imagining processes even before the schools did, played a major role here, too. Already before World War I, central European multicultural processes had been both shaping and being shaped by universities, and managing the differences in the new states proved at least as problematic for the new political elites as it had been in the late Habsburg Empire. The multiplicity of languages and the issues of multinational coexistence were some of the bequests inherited by the successor states. In the context of universities, several points are of relevance, giving insight into how the post-Habsburg universities experienced these new realities.

First, universities whose faculties had been appointed during the Habsburg Empire were, after 1918, subservient to new state interests, and new curricula were implemented. The reorganization of these universities meant the renegotiation of contracts as well as of long-established scholarly traditions and teacher-student networks. For instance, professors of German language and literature in Galicia and all professors at the German University in Prague were now members of a minority in a foreign state; indeed, particularly in Romania (Chernivtsi) and Czechoslovakia (Bratislava and Prague), many professors faced new and somewhat hostile political realities in 1918–19.

Second, with the transfer of the Russian-language Warsaw Imperial University (Императорский Варшавский университет) to Rostov-on-Don and the dissolution of the short-lived Royal Hungarian Elizabeth University (Magyar Királyi Erzsébet Tudományegyetem) in Bratislava, the universities in Cracow, L'viv, and Prague remained the only fully equipped institutions in Poland and Czechoslovakia. While the universities established across central Europe after 1918 attracted and appointed young scholars who had not previously worked at a university, scholars with academic experience still had the most prestige, becoming central figures at these new institutions. Because patriotism was mobilized to justify an almost mass departure of scholars from the post-Habsburg Slavic universities to the newly opened ones, the universities in Cracow, L'viv, and Prague experienced a severe brain drain. The outcome was a transfer of various types of knowledge—not only academic but also organizational—beyond the former boundaries of the empire, for example, to Warsaw or Vilnius, something that will be developed below.

Third, a number of émigrés, especially Russian and Ukrainian intellectuals, fled the Soviet Union and Eastern Galicia. For Czechoslovakia—which had also inherited Carpathian Ruthenia, making it a multinational state—this meant the creation, in Prague, of the Ukrainian Free University (Український вільний університет), a Russian Law Faculty (Русский юридический факультет в Праге), the Russian People's University (Русский народный университет), and the Ukrainian Academy of Technology and Economics in Poděbrady (Українська господарська академія в Подєбрадах).[20] New forms of cooperation were also implemented to accommodate institutions that were considered foreign, as well as scholars who identified with a different state. This changed Czechoslovakia into a melting pot of Slavic scholarly cultures, with scholars coming from several states that had previously been part of two very different empires.

Finally, the question of students received much attention, especially in Austria, now a small country with postimperial institutions. While the universities in Graz, Innsbruck, and Vienna feared that the new boundaries would mean a plunge in student numbers, this was not the case. The regions from which they had previously recruited their students were now in foreign states, but the liberalization of education and a new intake of women and students from Germany bridged this gap in the 1920s.[21] The fear of low student numbers was, in fact, one of the reasons provincial universities opposed the proposed transfer of Chernivtsi University to Salzburg and the repatriation of professors from Prague.[22] At the same time, especially immediately after the war, returning students and students who had passed the *Abitur* (the final exam in gymnasia) but had been drafted before matriculating made the universities quite crowded. Since universities did not consider this a long-lasting change, ways to cope with this growth without a permanent budget increase were considered. Not only Austrian universities faced this trend: in 1918–19 the Czech University in Prague experienced a rise in students since many who might have studied in Vienna remained in the country.[23]

While the new boundaries were expected to make the studentship more homogeneous, the opposite was true: students from the previously common space of the empire readily followed the path of their predecessors. In Graz, for instance, foreigners, predominantly from southern Europe, who were recorded in the statistics as non–German speakers ("fremdsprachig"), constituted up to 48 percent of students in the 1920s.[24] In Vienna foreign students comprised just under a third of the student population, with a decline during the 1920s; slightly less than half of the foreigners were German citizens.[25] Also, Prague proved to be a multicultural hub among students. It was not, however, the German University[26] that attracted the most foreigners but the Czech University, which had up to 40 percent foreign students at the medical faculty. These students were predominantly Ukrainians from Little Poland, as Galicia was now officially known, in addition to Yugoslav, Russian, and, especially, Jewish students.[27]

The issue of foreigners, in both a civic and a cultural sense, also occupied the universities in another way: the issue of internal others, especially Jews. In Hungary a limitation on the number of Jewish students (numerus clausus) was introduced in 1920 and set at 6 percent. In Poland and Austria, this issue was discussed intensely in 1923, and a numerus clausus was introduced in these countries in 1937 and 1938, respectively. Clearly, anti-Semitism was soaring, and universities had a more or less formally approved means of discrimination. Some universities in Poland had informal quota systems in the 1920s, and in the 1930s witnessed the so-called ghetto benches (*getto ławkowe*), segregated seating of students.[28] In Austria one can also find matriculation hurdles enacted by some universities: in 1923 in Innsbruck, the academic senate, for instance, recommended that the deans should not enroll non-Austrian Jews, especially the "*Ostjuden* . . . and Jews from states that have introduced the numerus clausus."[29] New boundaries and the fear of foreignness also meant pressure to assimilate to the cultures of the new states. While Jewish students had frequently studied at German-language universities in the Habsburg Empire, in the new states they increasingly turned to institutions teaching in the state language. In Prague, for instance, after 1918 the number of students of Jewish nationality or Jewish faith who chose the Czech University grew continually, equaling the numbers at the German University by the 1930s.[30]

Growing nationalism was not the only cause of disturbances. In Vienna between 1923 and 1925, the question whether Karl Horovitz was a communist led to a political scandal over his habilitation. After a right-wing article instigated by university scholars accused the young physicist of being Jewish and a communist,[31] the faculty repeatedly rejected his applications owing to "his personality."[32] After two years, with the conflict having spilled over to the press and the political arena without bringing a decisive solution, the young scholar moved to Purdue University in the United States and made pioneering discoveries in solid-state physics. The problem was that at the University of Vienna an informal clique of eighteen anti-Semitic and antisocialist professors, calling themselves the Bears' Cave (Bärenhöhle), controlled admissions to the university to keep unwanted scholars out. According to Klaus Taschwer, the clique hindered 13 habilitations out of 173, making it clear to Jewish and socialist scholars that their prospects at the university were nonexistent and thus deterring them from pursuing academic careers in Austria.[33] Migration, both abroad and internally, was the result; for example, the sociologist of knowledge Edgar Zilsel, whose habilitation was rejected at the same time as Horovitz's, taught mathematics at a secondary school in Vienna before migrating to the United States in 1938.[34] Thus, even before the surge of right-wing parties in Austria, it was clear that Jews were not welcome at the University of Vienna.[35] This was reminiscent of the period when Leo Thun-Hohenstein called Jews personae non gratae in certain disciplines but went beyond it and extended these restrictions to the universities as a whole.

Not only were the universities unprepared for the new developments after World War I, but both they and their respective governments also reacted to these new realities very slowly, which delayed any changes in how the universities functioned. In fact, except in Chernivtsi, where after a transitional period the university became subject to Romanian laws in 1925, the main rules for university business remained largely unchanged until the 1930s. Then, in a surprisingly parallel development, the governments of Austria, Czechoslovakia, and Poland all strove to reduce university autonomy. This took place mainly in Poland under the authoritarian Lex Jędrzejewicz (Jędrzejewicz's Law, 1933),[36] and to a lesser extent in Austria, which by this time was under an Austrian fascist regime.[37] Only in 1975, 102 years after the last Habsburg reform, did the minister of science and research

Hertha Firnberg (1970–83) substantially reorganize the tertiary education system in Austria.[38]

One reason for the pervasiveness of the Habsburg system, and perhaps the most important one, was that by 1918 the most prominent universities—in Cracow, Prague, and Vienna—had already been acting according to national geographies. Because they were the centers of the three linguistic sections of the Habsburg Empire before the Great War, they simply continued to play this role in the new states. While this is quite clear for Austria and Czechoslovakia, where basically one model predominated, Poland could draw on three imperial experiences. The academic senate of the Jagiellonian University, however, opposed modeling the university laws on Russian or German institutions and succeeded in convincing the government to retain the old Habsburg laws, thereby, they argued, continuing a glorious organizational tradition extending from the times of the university founder Casimir III (1310–70) through Franz Joseph to an independent Poland.[39] Clearly, smaller changes took place, such as the introduction of remuneration for Privatdozenten in most of the new states, increases in specialization (as in the faculties for sciences in Poland and Czechoslovakia), and the distinction between the law curriculum and that for political sciences. However, at the level of metaregulations, the 1849 reforms survived until World War II.

New Spaces

Given the new state boundaries, in 1918 the personnel at post-Habsburg universities encountered the question of loyalties anew. After the Habsburg period had witnessed the dualism of loyalism and nationalism, which were not necessarily conflicting ideals, the new countries demanded that the political and intellectual cadres align themselves with the communities they were to represent. This meant no end of complications, though, and academic migration proved to be a complicated exercise in the geographic amassment of peoples within newly drawn boundaries. Many imperial scholars and politicians had been educated and had lived in structures that no longer existed; hence, for them, moving, say, from Vienna to Bratislava or Warsaw was now more a migration to a new country than a return to the motherland. Indeed, some intellectuals had difficulties finding their own direction after the end of their respective empires.[40] However, some migration was forced because individuals did not adhere to the new rules, did not speak the language of their new country, or rejected oaths to the new states, which were still required of civil servants (a category that included university professors). For others, migration was no longer a case of moving from one state to the other, but occurred within a new state, to the newly opened universities. In all cases, the theories, approaches to education and administration, and implicit and explicit knowledge of the universities that emanated from the old Habsburg Empire brought the academic institutions of central Europe intellectually closer.

The Bukovinian (i.e., Romanian) case was symptomatic of the new boundaries and language changes. Contrary to the impression that the new states employed only professors who matched the new national ideals, and therefore sought to exchange all university personnel from the beginning, most professors had the option of remaining in Chernivtsi, irrespective of the language they spoke and the nationality they saw themselves as belonging to. When efforts to move the university westward failed, not only the Romanian scholars remained, including the whole theological faculty, but also six other German-speaking scholars who had not been born in Bukovina.[41] A seventh, the famous Chernivtsi-born sociologist of law Eugen Ehrlich, moved to Vienna but was later officially reinstated; he died before returning to Bukovina, however.[42] Scholars who did not wish to remain in Bukovina, or who were forced to leave the province, mostly moved to Austrian universities as well as the Ukrainian Free University in Prague and the University of Ljubljana. The Romanian Ferdinand I University in Cluj-Napoca also enticed the famous Vienna-educated philologist Sextil Puşcariu away from Chernivtsi; the government gave Puşcariu the task of organizing the university once the Royal Hungarian Franz Joseph University (Magyar Királyi Ferenc József Tudományegyetem) in Cluj had closed and moved to Szeged.[43]

Thus, the new states profited not only from scientific knowledge of Habsburg origin but also from organizational know-how. The university in Cluj was not the only one importing scholars to fulfill both functions; the same can be said of Ljubljana. Scholars moving there from Chernivtsi were bestowed with academic honors: the positions of dean of the philosophical faculty and rector of the university were filled with scholars with Bukovinian pasts; in fact, all deans except the dean of theology had had Habsburg scholarly careers. Another émigré, Mihajlo Rostohar from the Czech University in Prague, drafted the university statutes based on the Habsburg ones.[44] The Vienna Medical School provided (unsurprisingly, one could add) several scholars for the new universities. Đorđe Joannović (Ђорђе Јоанновић), one of the founders of the University of Belgrade's medical faculty, can be named as one of several prominent examples.[45] Nonetheless, such moves away from Vienna were more the exception than the rule. One has to consider that this faculty was gigantic in comparison to others, with many scholars having low chances at a professorship, especially given the new smaller Austrian reality.

In general, most scholars active at German-language universities during the Habsburg period remained in Austria. Although there is no comprehensive list, one can find only a few scholars per faculty who left for abroad.[46] Indeed, the potential for migration was not high since the imperial, multicultural pan-Habsburg universities had disappeared long before the Great War.

These changes did not affect only Austria, since every Habsburg university entertained scholars with various cultural allegiances. Thus, this also held for the former Galicia, where the new boundaries meant that some professors were now "foreigners." Notably, both Cracow and L'viv professors of German language and literature remained at their universities until they retired; as noted above, there were no really qualified Poles to teach this subject. But most non-Polish scholars left. The goodbyes were not always easy. Two Czech physicians who left during the war, moving to universities in Czechoslovakia, apparently retained no contact with their previous institutions, and one of them was the subject of a local scandal.[47] After the Polish-Ukrainian conflict over L'viv, there was no option for Ukrainian scholars to remain there either, and they turned to both internal and external migration (see below); unresolved issues also remained, such as the issue of pensions for scholars who left Poland.[48]

The German ~~Charles-Ferdinand~~ University in Prague

The case of the German University in Prague can help illustrate how minority institutions functioned in the new states. The new cultural power relations meant major changes and problems for the power and cultural divisions within Prague. Even before the Lex Mareš (Mareš Law, 19 February 1920)[49] declared the Czech University the only successor to the ancient Charles (-Ferdinand) University and named the German University simply Německá univerzita v Praze (German University in Prague), the Czech University had seized the previously joint university buildings, archives, insignia, astronomical observatory, and so on. New rules were enacted, even for the division of cadavers among the medical faculties; instead of the previous fifty-fifty division, the new rules stated that the German University would receive one-sixth of the cadavers.[50] Fearing a loss of influence, the German University seriously considered moving to a region more densely populated by German speakers and started discussions with the city councils of Litoměřice/Leitmeritz, Ústí nad Labem/Aussig, Cheb/Eger, and Teplice/Teplitz. However, although the government and President Masaryk seemed to support this development, it ultimately failed for financial reasons.[51]

While scholars at the German University were certainly not satisfied with the developments there in the immediate aftermath of the war, they hesitantly took the loyalty oath to the new state. Only a few had left Prague after 1918, moving to Vienna, clearly an act of refusal to swear the oath to the new state, since such scholars received only a Privatdozentur in the Austrian capital. Since at least one of the émigrés, Anton Lampa, had been a full professor and could have remained at the German University, political protest is the inevitable conclusion.[52]

The nominations in Prague seemed to continue as usual during and after the war, as if the new boundaries were no problem, although some rectors in the 1920s did challenge the citizenship rules for professors as preventing the German University from recruiting the best people. However, the structure of appointments changed in comparison with the Habsburg period. According to Ota Konrád, of the thirty-eight newly appointed professors in the philosophical faculty of the German University in Prague between 1921 and 1937, nineteen were promoted from within the Prague staff, and sixteen were foreigners. Konrád considers scholars from both Austria and Germany as foreigners, however. While twenty-three of the fifty-nine teaching staff in this period had been born in Bohemia or Moravia, taught there, and mostly spent their whole careers there, those appointed from abroad remained in Bohemia relatively briefly.[53] In addition to several possible cultural reasons, such as foreigner status and the peripheral nature of the German University in Prague, financial reasons were certainly important, since Czechoslovak salaries were not competitive with German ones.[54] Even so, the university successfully retained some professors who received calls from German universities.

For financial reasons, new appointees for full professors had mostly previously held positions as associate professors and Privatdozenten, which was characteristic of in-between universities, as outlined in the previous chapters. While the German University in Prague was certainly not a top-notch German-language institution, it was still one of the largest in terms of the number of students. In 1929 it had the second-largest German-language medical faculty behind Vienna.[55]

Given the constraints in Prague and the relatively low population of German speakers in Czechoslovakia, it is hardly surprising that the German University eagerly participated in "scholarly exchange with the whole German culture circle [*Kulturkreis*]," as Otto Grosser put it.[56] This also brought criticism that foreigners were promoted more readily than the university's own home-educated graduates; several rectors responded with harsh words about the dangers of provincialism and inbreeding (*Inzucht*).[57]

These two metaphors—of a culture circle (*Kulturkreis*) and inbreeding (*Inzucht*)—show that the new rhetoric of official university statements was heavily imbued with *völkisch* elements. Appellations for culture that transgressed state boundaries and referred to a cultural and spiritual community (*Kulturgemeinschaft* and *Geistesgemeinschaft*) were also used. One can observe here a continuity with the situation before 1918: the exchanges among all the German-speaking universities in the (former) empire were aligned with the German cultural space. The element of state, previously often appealed to by university officials, was now absent, however. While Habsburg scholars had always included the needs of the fatherland (*Bedürfnisse des Vaterlandes*), explicitly or implicitly, when talking about academic graduates, Czechoslovak German Bohemians (*Deutschböhmen*) at the university depicted the state rather as an obstacle to their rightful needs.

However, the German University was a state university. Apart from occasional boycotts, throughout the interwar period it took part in meetings of the rectors of Czechoslovak tertiary institutions and thus participated in policy-making processes on par with other universities.[58] As the statistics show, the Czechoslovak government did not try to minimize the teaching staff there. The official statistics for 1929–30 list 117 full professors, 40 associate professors, and 131 Privatdozenten at the Czech University, compared to 66 full professors, 36 associate professors, and 88 Privatdozenten at the German University. Relative to the number of students (9,934 and 4,714, respectively), this gives the German University better ratios, both for the ratio of students to full professors (74:1, compared with 88:1 at the Czech University) and for the ratio of students to all instructors (25:1 versus 34:1).[59] Nonetheless, both universities were underprivileged in comparison with the smaller ones in Brno (55 professors, 117 total teaching faculty, 2,933 students) and Bratislava (34 professors, 72 total teaching faculty, 1,761 students). This well-known established pattern of a low number of Privatdozenten (now called Soukromí docenti) advantaged the small universities, if only as far as the number of professors was concerned.[60]

Bratislava: Becoming (Czecho)Slovakian

The new state of Czechoslovakia inherited three universities from the Habsburg period: the two universities in Prague and the Hungarian-language Royal Hungarian Elizabeth University in Bratislava, which had opened in 1914 with law and medical faculties. Bratislava was now a Slovak city, which meant major changes for the university. Although from the earliest period of the new state Prague politicians had signaled that they were interested in keeping Elizabeth University, it was dissolved in 1919–20, partly because of the lack of a Hungarian minority in the state that the university would serve. Slovak politicians were unanimous that Bratislava should get a Slovak university in exchange. Most wanted to have one built once Slovakian schools were producing students and young scholars, but the pressure of medical scholars in Prague lobbying for a new university moved the ministry to open Comenius University in Bratislava in June 1919.[61] Yet, despite the efforts to find Slovak students and attract Czech ones, in the first few years students identifying as Hungarian or German made up the majority at the medical faculty, while in the law faculty Slovaks and Czechs prevailed.[62]

If we can trust the statistics, the medical and law faculties at Comenius University opened with only one scholar identifying as Slovak, Augustín Ráth. With the opening of the philosophical faculty two years later, three other Slovaks were nominated, although only one of them actively taught at the university.[63] Surprisingly, while a few Slovak scholars had previously worked in Hungary, they were not considered for professorial positions; cultural separatism and an anti-Czech position were probably the reasons for their rejection.[64] It seems that even though the principal aim of the university was the reinforcement of Slovak elites, the Slovak public remained skeptical of the project. Those favoring a national project certainly had reason for skepticism, since behind closed doors Prague politicians had asked the professors appointed to Bratislava to support and habilitate only those who accepted and promoted Czechoslovakism.[65] For the tenth jubilee of the university's establishment, the rector, Albert Pražák, wrote openly in a German-language interview that "the Slovak public is somewhat cautious [about the university]. The reason is new scholarly methods . . . which changed the picture of a once and current Slovakia."[66] That the Czechoslovak activist and historian then criticized the conservatism and traditionalism of Slovak political elites, and employed the arguments of a Czech civilizing mission toward Slovaks, might have contributed to the nonacceptance problem.

Given the lack of Slovakian universities before the creation of the new state and the Czechoslovak policy of the government, it was thus not surprising that Prague was also the faculty at which the vast majority of Bratislava's scholars had been educated. In the academic year 1924–25, nine out of twelve full professors and four out of five associate professors at the medical faculty were Prague graduates.[67] At the philosophical faculty, only two scholars had carried out their studies predominantly abroad, and they taught Ruthenian and Russian history and literature, respectively.[68] The law faculty's scholars had all habilitated in Prague after 1919. In this way, with few exceptions, the implementation of Czech cultural policy in Slovakia meant that the scholarly habitus of the University of Prague was transferred east, with Bratislava thus coming under the influence of the Cisleithanian principles of higher education.

The transfer, or domination, of Cisleithanian knowledge had its drawbacks as well. This was notable especially in the law faculty, which guaranteed that students of Elizabeth University and of the law faculty in Košice/Kassa could take their exams according to the Hungarian rules, which would also include Hungarian-specific subject matter. This resulted in problems, since the new professors were not familiar with Hungarian legislation. In this case, they had expertise in Cisleithanian law and had additionally to learn and teach the Hungarian and Czechoslovak legal systems.[69]

Another drawback was the lack of extensive experience abroad on the part of most of the scholars nominated to Bratislava. Since most nominees were young Privatdozenten, this had a long-lasting effect. While two scholars, Vinzenz Chlumský and Ján Buchtala, had previously worked outside the new state, both were, after 1918, acknowledged as pioneers in their disciplines and recognized as the founders of their respective schools in Czechoslovakia.[70] While Prague scholars also predominated at Brno's Masaryk University (established in 1919), the faculty was more diversified there. Similarly, in Bratislava mostly younger scholars were appointed, as well as, in the medical faculty, several local practitioners. One exception was the faculty of sciences, which employed three professors with substantial experience abroad, two of whom had been active mostly in Switzerland. At other faculties, only shorter stays abroad can be noted. In comparison with

Bratislava, where local scholars were scarce in the early 1920s, the Czech technical academy in Brno constituted a notable source of new professors for Masaryk University: three each in the law faculty and the faculty of sciences, and one each in the philosophical and medical faculties.[71] For instance, the first rector of Masaryk University—one of the scholars most active in the political struggles for its creation—was the professor of economics at the technical academy, Karel Engliš.

Since the new universities drew scholars from Prague, and several professors became political functionaries and ceased teaching, the Czech Charles University faced, after the war, a brief reduction in its teaching staff, which, however, was rapidly compensated for by the growing number of young lecturers.[72] In comparison with the Polish case, where Cracow and L'viv were increasingly turning into local institutions, Prague retained its central status and dominated Czechoslovak education, serving as a nursery for future generations of Czechoslovak scholars and clearly gaining influence after the Habsburg Empire's collapse.

Habsburg Poland

The newly established universities in Poland, such as those in Poznań, Warsaw, and Vilnius, similarly drew on Habsburg cadres. By 1915 the Viennese ministry was more than willing to send L'viv professors there for the founding of the University of Warsaw under the German protectorate. But the final number of seven appointees was seen as a deliberate limitation of the Galician presence in the new institution on the part of the government of the protectorate.[73] For the 1919 Stabilization Commission (Komisja Stabilizacyjna), which was held in Cracow for Galician convenience and was to decide on the final appointments for the University of Warsaw and, a little later, other institutions, the L'viv and Cracow universities sent thirty-four envoys, compared to fifteen from other institutions. This gave them more influence on the decision-making process than scholars from other regions and institutions.[74] Galician professors headed the subcommissions and also the medical and natural sciences, and Kazimierz Twardowski (L'viv) had the deciding vote for the philosophical disciplines.[75] Jan Łukasiewicz, a L'viv logician, was the minister of religion and education from 1919 to 1923, when most of the changes in academia were decided on; three further professors and two graduates from Galician institutions held this position, with only three ministers in the interwar period not being connected to Galician universities.[76]

The dominance of Habsburg scholars among the instructors teaching in interwar Poland is clear. Most came from Galician institutions (44 percent), while 17 percent (186 people) came from the Vistula Land and a further 11 percent from the Russian Empire.[77] For example, thirty-four scholars in the humanities and natural sciences in interwar Poland had previously taught at universities in the Russian Empire, with eleven professors from St. Petersburg and Kiev. In the medical sciences, the number was thirty, with twelve professors and seven docents (predominantly from St. Petersburg), nine professors in law and economics, and thirteen professors in technical disciplines.[78]

While this number hints at the variety of origins of the scholars in the Republic of Poland, the predominance of Habsburg scholars was still visible in most faculties. In Warsaw 30 percent of the instructors up to 1927 had previously taught in Cracow or L'viv; most of these scholars were working in the humanities in the philosophy department. Such dependence on post-Habsburg scholars clearly diminished over time, as the university in the Polish capital, Warsaw, saw more Polish students graduate there.[79] The dominance of Galician scholars at the universities in Poznań and Vilnius, as well as the private Catholic University in Lublin (Katolicki Uniwersytet Lubelski), was even greater, ranging from around 30 percent to 100 percent (the latter at, e.g., the Vilnius philosophical faculty between 1919 and 1920).[80] This dominance is even more evident if one considers that these statistics include scholars in all posts and that Galician scholars were mostly full professors.

Only in the rarest of cases were scholars from German-language Habsburg universities appointed; the universities in the new states were clearly concentrating on local appointments. Also, long-standing ties had weakened, such as those with the Collegium Canisianum in Innsbruck, which before the war had been an important place for Galician clergy and theologians. Five former Canisianum students were instructors at the Cracow theological faculty, twelve in L'viv, and one in Vilnius. Some famous priests of the Second Republic had been students at the Canisianum; of these, Adam Sapieha, archbishop of Cracow and after 1946 a cardinal, was the most prominent.[81]

One important side effect of this wave of appointments at the new universities was the depletion of instructors at the universities in Little Poland:

there were even fewer professors in Cracow a decade after the Great War than in the year before it, while the number of students grew by almost 50 percent. Władysław Natanson, a key figure in organizational matters in Cracow after the war, calculated that the number of professors who had died during the war was lower than the number who left the Jagiellonian University afterward.[82] After a brief period of patriotic zeal for participating in the building of new institutions, Galician faculties had, by 1919, already addressed the ministry with regard to the issue of irreplaceable scholars, claiming that a more balanced appointment policy was needed. In particular, the ministry was requested to list Polish scholars living abroad and to put more effort and money into appointing them in the first place.[83] Given the difficulties all institutions encountered in appointing scholars from abroad, cadres educated in the new independent state could barely fill the vacated positions.

At the same time, Little Poland's universities increasingly became local institutions, especially if one compares them with the prewar situation. At the Jagiellonian University, almost 70 percent of the professors between 1918 and 1939 were from Galicia/Little Poland, and the proportion of local docents exceeded even that number.[84] By the 1930s the university was criticized for having too many overage staff. Aiming for the best scholars and seldom lucky with appointments, the university often left chairs unoccupied or appointed honorary professors.[85] Moreover, the number of scholars appointed from other universities did not compare to the number of instructors that the Jagiellonian University had supplied to other academies, considered to range from 250 to more than 500.[86] The university in L'viv, now renamed the Jan Casimir University, likewise remained locally bound: only seven instructors teaching in 1927–28 had been educated outside Galicia/Little Poland. Four more had returned to L'viv after only a brief period teaching at another institution. The proportion changed only slightly in the 1930s.[87]

Ruthenian Legacies

With the failure to achieve Ukrainian statehood and Galicia's incorporation into the Republic of Poland, Galician Ukrainians retained their position as a subaltern minority deprived of academic institutions. The question of Ukrainian universities proved indeed to be a postimperial legacy that spanned the whole region. The project of creating a Ukrainian university in Galicia, discussed before World War I, was not carried out under the Polish state, although the Ukrainian minority constituted (depending on the method of counting) three to five million people out of slightly more than thirty million people across the large republic.[88] In the early 1920s, conflict erupted. The University of L'viv introduced measures against Ukrainian students, who in turn boycotted the university; this left a substantial number without the possibility of being legally educated in the language they had been promised. In 1921 Ukrainian scholars created the Secret Ukrainian University in L'viv (Таємний український університет), which remained unacknowledged by the Polish state and, after massive arrests of students and professors, was closed in 1925.

While they could not be established in Poland, Ukrainian universities were set up in the short-lived Ukrainian People's Republic, later subsumed into the Soviet Ukraine. The Ukrainian Free University was established in Vienna in 1921, then later that year moved to Prague, where Ukrainian agricultural and pedagogical academies were also subsequently founded. Scholars from Galicia and recent graduates from Habsburg universities constituted a considerable part of the faculty of the Ukrainian Free University in Prague, and students from what was then Little Poland made up a majority.[89] The university's creation and shape in Vienna and its transfer to Prague, where a larger number of émigrés lived, were thanks to Galician scholars and their contacts, especially prewar connections with people such as Jaromír Nečas, at the time Masaryk's secretary; Masaryk had also strongly supported the idea.[90]

Of the other Ukrainian scholars previously active at Habsburg universities, one, Mychajlo Hruševs'kyj, taught in the Soviet Ukraine, after a brief period of exile spent mostly in Vienna. Only three such scholars remained in Poland, teaching at the Secret Ukrainian University, whose faculty consisted mainly of scholars who had been living in L'viv. These scholars were graduates of Habsburg universities (not only Galician but also frequently Viennese universities) and gymnasia teachers. Between 1923 and 1925, the Secret University numbered 1,014 students and 64 professors, making it a substantial institution.

Some of the Habsburg traditions remained in place as part of Ukrainian education in Poland. Teaching at the university followed a slightly adapted Habsburg curriculum, except in the technical faculty (later the Secret Technical University), whose structure was based on the Technical Academy of the Free City of Danzig (Technische Hochschule der Freien

Stadt Danzig).[91] The only Ukrainian postsecondary school in Poland, the Theological Academy (Богословська Академія, now the Ukrainian [Greek] Catholic University in L'viv, founded in 1929), was regarded as a continuation of the Greek Catholic seminary Barbareum (Regium generale Seminarium Graeco-Catholicum Viennae ad Sanctam Barbaram; St. Barbara Royal Greek Catholic Seminary in Vienna), established in 1774 in Vienna and moved in 1783 to L'viv. Its first rector, Josyf Slipyi, had been educated at the Canisianum in Innsbruck, which in the interwar period was of greater importance for Greek Catholic clergy than for their Roman Catholic counterparts.[92] Interestingly, most Ukrainian scholars who gained chairs and docent appointments at universities in Poland had studied for some time in Vienna, notably in Slavic languages and comparative philology.[93]

Polish-language universities also gradually incorporated scholars from the Secret University, but Ukrainian organizations saw these concessions as inadequate. The question of universities and scholarship in general had a significant impact on the collective memory of Ukrainians: the Habsburg government had protected Ukrainian culture, and the Habsburg period therefore still had positive connotations in Western Ukraine (whereas the Second Polish Republic was seen as a period of greater oppression), contributing largely to the myth of Galicia in the collective memory in the eastern borderlands of Ukraine. The Ukrainian émigré historian Ivan Lysiak-Rudnyckyï (also Ivan L. Rudnytsky) used the famous expression "this is worse than a crime; it is a stupidity" to describe the policies of the Polish government toward Ukrainians during this period.[94] Even if one does not agree about the historical accuracy of this statement, it says much about how the period entered the collective memory.

Old Connections

With post-Habsburg scholars dominating academia in the newly founded states, the question remains to what extent this facilitated contacts between these states. The new states by no means ceased pursuing the internationalist goals set before the war, and they made these one of the pillars of their policy of scholarly development.[95] Since the academic exchanges that had taken place during the Habsburg period had weakened during the early twentieth century and were no longer politically prescribed, a geographic reorientation was possible. This was visible, for example, in the case of the

German University in Prague, which gradually turned toward Germany; after World War II, it became a domain of German and not Austrian historians. However, since universities in Czechoslovakia and Poland employed scholars with experience in various provinces of the Habsburg Empire, it is not hard to imagine that contacts that had developed during the imperial period survived and were a substantial factor facilitating future cooperation.

The reorientation of the Polish academic landscape toward the West meant intensified cooperation with scientific centers in France, Great Britain, and the United States, at the cost of sustaining postimperial connections. At the Jagiellonian University, academic exchanges with Austria and Hungary, or even guest lectures, did not play any substantial role in the interwar period.[96] The official statistics on academic travel (for training in a specialization, research, a longer archival trip, or the like) published in 1927 show a clear predominance of visits to France, but Austria was still an important travel destination, although the inclusion of archival research means the statistics are slightly distorted.[97] Similar statistics for Czech universities show a comparable leaning toward France.[98]

Similarly, Polish-Czechoslovak contacts became fewer. The seven-day Polish-Czechoslovak war of 1919, ongoing conflicts over the partition of Silesia, Czech Russophilism, and political support for the Ukrainian cause overshadowed the official relations between the two neighboring states.[99] By the 1930s, even the Polish consul in Prague could hardly present any considerable academic collaborations, except for the then newly established student exchange programs and courses for Czechs in Poland.[100] This does not mean such interactions did not take place, however. Indeed, as far as visiting scholars in Czechoslovakia are concerned, the number of guests from Poland was the second highest, behind those from France.[101] In addition, a chair of Polish language and history, a novelty in comparison with the late Habsburg monarchy, was installed in Prague, and a chair for Czechoslovak in Warsaw.[102]

While the interwar period was not the best time to tighten relations between neighboring countries, the war had not destroyed the entanglements from the Habsburg period. In all post-Habsburg relations, however, it was personal connections that made academic relations possible, rather than state support or exchange policies. For example, the cooperation between Viennese and L'viv-Warsaw neopositivist philosophers—the L'viv-Warsaw school of analytical philosophy and the Vienna Circle (Wiener Kreis)—resulted from Kazimierz Twardowski's contacts with Vienna. Their cooperation included mutual visits and, to a limited extent, joint projects, which extended, after Twardowski's death, into a period of exile in the United States.[103] Władysław Mieczysław Kozłowski, whom the Habsburg ministry had denied a Privatdozentur in L'viv before the war and who was a friend of the Prague professor of philosophy František Drtina, strove in Poznań, where he had held a professorship since 1920, to intensify scientific contact by establishing the Polish-Czechoslovak Society (Towarzystwo Polsko-Czechosłowackie). He also published in Czech and visited Prague several times as a guest lecturer, even living part-time in Czechoslovakia after his retirement from the university.[104]

German studies shows a similarly interesting situation and hints at major changes in university politics in the 1930s. Since no German studies scholars identifying as Poles had gained academic positions before the war, Galician universities retained those from other former Habsburg provinces well into the interwar period. The students of their similarly non-Galician predecessors held most chairs of German studies in Poland after 1918.[105] One exception was the newly created chair in Vilnius in 1927, for which the university nominated the Graz historian of language Franz Doubek, which indicates once more how important the connections with post-Habsburg states were.[106] But the changing geopolitical situation also affected this legacy, and one of the post-Habsburg German-speaking scholars of the German language, Spiridion Wukadinović, had to leave Cracow in 1933 owing to a conflict over a talk he gave in Weimar on Goethe and Poland, during which the scholar referred to several anti-Polish declarations of the poet, as a critique of the independence of Polish culture. An influential diplomat and newly nominated Polish ambassador to Germany, Józef Lipski, denounced Wukadinović's lecture. Wukadinović, who identified as a German and, before his untimely lecture, was a renowned teacher and translator, was a victim not only of growing cultural tensions but also of the tight political oversight of academic institutions introduced in 1933, after which academic autonomy decreased substantially.[107]

Another Austrian scholar, Leopold Adametz, remained highly successful in postwar Poland. After he left Cracow in 1898, returning to Vienna, he maintained intensive contact with his former colleagues and held regular classes in Cracow between 1921 and 1928. Even after his retirement, he often gave guest lectures, and in 1931 his seventieth birthday was commemorated with a special issue of the Polish *Roczniki Nauk Rolniczych i Leśnych* (Agricultural and forest annual).[108]

In Bukovina und Bohemia, German-speaking scholars also remained productive as cultural intermediaries. In Prague one might surmise that the contact between scholars from the Czech and German universities was better than during the Habsburg period, resulting in informal cooperation and formal joint enterprises.[109] The Romance-language scholar Eugen Herzog, in turn, may serve as an example of the new situation in Bukovina. His most important work in philology was an extensive early Romanian grammar, published in 1919 in German, coauthored with Sextil Puşcariu. During his work in Chernivtsi, Herzog also published in *Dacoromania* (Daco Romania), *Codrul Cosminului* (Cosmin forest), and *Revista Filologică* (Philological review) and served as a member of the editorial board of the first journal. Apart from this work on Ruthenian grammar, his research on a glossary of the folk speech of Marginea village (currently in Suceava County, Romania) was also widely and positively reviewed.[110] Not exclusively devoted to Romanian, he published on Old French as well, and his contributions in Chernivtsi were highly valued, as the obituaries published by the most important philologists of the time prove.[111]

With Chernivtsi, the circle of central European exile closes. By around 1900, it was the last place many Habsburg scholars wanted to be, and all were heading to Vienna. Post-Habsburg central Europe, a republic of learning still waiting to be analyzed as a space with a truly transnational intellectual culture amid national boundaries, changed the geography of intellectual relations, but the revolution was still ahead, completed only through the atrocities of World War II and the subsequent Cold War isolationism.

The interwar period instead continued the trends already outlined during the late Habsburg monarchy. The language-based geography of exchange described in previous chapters was realized under the auspices of the new states, putting previously provincial centers in prominent positions under new spatial-political circumstances. Habsburg scholars dominated not only Ljubljana but also Warsaw, thus traversing the boundaries of Habsburg domains. The knowledge that was transferred extended far beyond academic knowledge, for Habsburg scholars were instrumental in devising academic laws and policies, at both the university and the state levels.

The influence of the Habsburg Empire also lived on through personal contacts, for example, when Masaryk helped Gomperz or when, some years earlier, he helped found a Ukrainian university in Prague. One could, however, suggest tentatively that for central Europe, Prague began to play the role Vienna had during the Habsburg period, forming a place of refuge for scholars fleeing nationalism and communism. Vienna, which in the interwar period still had the allure of its former glory, while scholars working there remembered and perpetuated the habitus of one of the foremost German-speaking universities, retained its intellectual capacity, but by reorienting itself toward Germany and the German *Kulturraum* (cultural space), it began progressing toward the self-conscious provincialism that it inflicted on itself after 1945, when the university failed to reappoint scholars who had fled during the 1930s.[112]

Some things did not change at all, however. Catholicism and nationalism, prevailing ideologies before the war, became more radical, even though many intellectuals still clung to prewar tolerance. The Liberal Democrat Hans Kelsen tends to be named as an example of an Austrian liberal scholar, but one often forgets to add that in Vienna it was Othmar Spann, a *völkisch* Austrofascist anti-Semite, and his circle who had more influence at the university.[113] Scholars of the Mosaic faith, Ukrainians in Poland, and, to certain extent, Slovaks all similarly retained their subordinate positions, even given all the uncertainties and ambiguities such subscriptions included.

One final chapter of the Habsburg/central European experience was its globalization, which took place gradually during the 1920s and 1930s. This process began with the emigration from Hungary of scholars escaping the right-wing regime of Miklós Horthy. Austrian scholars followed from the late 1920s, leaving their country in larger numbers after the Austrian fascists seized power. Interwar anti-Semitism also forced scholars from Slavic countries to move away, although in nothing like the same numbers as from Germany and Austria. After 1939 National Socialism and, to a lesser extent, Soviet occupation resulted in another wave of migration. Sonderaktion Krakau, the massacre of L'viv professors, Theresienstadt, and the other atrocities of World War II put an end to the once-blooming Habsburg intellectual landscape.[114]

Conclusion

Paradoxes of the Central European Academic Space

The late nineteenth- and early twentieth-century Habsburg Empire, politically united but culturally drifting further and further apart, was followed after 1918 by a politically divided but still intellectually entangled central Europe. Universities not only were influenced by this political development but also, as the foremost cultural institutions, took a vital part in shaping it. Therefore, in the history of universities, narratives intersect that have thus far been written in parallel, in the sense of both parallel national narratives and also traditions of writing the history of science and scholarship separately from the history of education or of culture in general.

The case study presented in this book—of academic geography as a function of the cultural and political context and practice—thus allows us to draw more general conclusions about the functioning of academia at the interface of scholarship and politics. Although historians have many reservations about discussing the present, an analysis of the processes shaping academia in the nineteenth century can sharpen our understanding of current processes governing scholarly exchange. While I am far from arguing that the Habsburg Empire was a precursor of the European Union, as some scholars and politicians have repeatedly claimed,[1] there are lessons to be learned from its history that would enable a better understanding of the different aspects of mobility that are now shaping both European and global academia. Two areas seem to me especially vital in this respect: the way the requirement of mobility has affected careers, that is, the advantages and disadvantages of careers requiring international mobility; and the issue of internationalism and its relation to politics. Below, I discuss the Habsburg situation in these two areas and its ramifications for contemporary discussions.

Academic Mobility and National Geographies

The characteristics of central Europe, with a demarcation between centers and peripheries that each created their own differentiations and hierarchies, affected academia in many ways. Most important, the nature of the universities was changed by the inclusion of scholars from non-Habsburg universities and, correspondingly, openness to new ideas from outside the empire. This was the result of the 1848–49 reforms that received the most praise from the liberal and progressive scholars of the second half of the nineteenth century. Hailed as an asset, intellectual exchange became entwined in both the praxis and the rhetoric of the faculties, leading to different results depending on the faculties' interests. This helps us to discuss the impact of nationalism and internationalism on Habsburg academia in new terms, refocusing from nations to empires but without overlooking the impact of different nationalisms.

Nationalism influenced academia in many ways, including by changing the geography of academic mobility. The reorientation from empire to nation brought somewhat paradoxical results, as can be seen when we compare the late nineteenth-century German-language universities with the Polish-language ones, that is, the universities of two of the various linguistic groups transgressing Habsburg borders. While from the beginning of the 1870s Galician universities were openly advised to search for candidates abroad and made use of this privilege, German-language Habsburg universities increasingly appointed local scholars. In 1910 a quarter of the instructors at the medical and philosophical faculties in L'viv and Cracow had been appointed from the Russian and German Empires, whereas at the universities in Vienna, Graz, and Innsbruck, the percentage of scholars appointed from abroad fell from around 20 percent in the 1870s to below 10 percent in 1910. This number also includes scholars from the German Empire. With increasing numbers of habilitations, German-language Habsburg universities had a significant number of qualified homegrown scholars striving for positions. This made the appointment of scholars from German Empire universities comparatively less frequent and less popular than in the 1860s. In contrast, the use of Polish as the medium of instruction at Cracow and

L'viv (the only universities using this language at this time) attracted graduates from abroad—especially from German Empire universities—who wished to habilitate in Galicia. The number of instructors who had acquired their doctoral degrees at "foreign" (mostly German) institutions was around 45 percent in 1910 at Cracow and L'viv, while at German-speaking universities in the Habsburg Empire, it dropped to near 10 percent in the same year.

Because nominations in Galicia included mostly Polish scholars, Galician universities were hubs of knowledge from abroad. The resulting trend toward a mixture of research styles was further augmented by Galician scholars who had completed their habilitation process at the universities in Graz, Vienna, Innsbruck, and Chernivtsi. For example, the L'viv-Warsaw school of analytical philosophy originated through cooperation among scholars educated in the Habsburg Empire, the German Empire, and France. The liquefaction of oxygen, the most acclaimed chemistry-related achievement in Galicia, resulted from a combination of knowledge and materials acquired by professors of physics and chemistry during their education abroad.[2] Thus, Galician universities were, in effect, more international than the Habsburg German-language universities, a finding that is particularly striking when contrasted with the stereotypes of nationalist Slavs. But it also demonstrates the shortcomings of our conceptualization of the spaces and boundaries in central Europe, which this book has addressed.

The Czech University can similarly not be regarded only through a national lens. Although it remained geographically bound to Bohemia in its appointments, this geographic enclosure fueled internationalism. In the first years after the inauguration of the Czech Charles-Ferdinand University in 1882, the language change meant that the university had to open itself to scholars from beyond the empire to obtain sufficient teaching staff. This brought together a variety of scholars, who linked the scientific traditions of their respective empires. Later on, the scholarship system facilitated the circulation of students and scholars. One could even venture that Czech internationalization was a direct result of the development of a nationally defined Czech academic system, similar to what happened in Galicia, though under different geopolitical circumstances, which yielded different forms of internationalization. This internationalism led to a number of productive intellectual clashes. The most prominent Czech scholars at the beginning of the twentieth century, Jaroslav Goll and Tomáš Garrigue Masaryk, represented two different traditions they had acquired while students, the first in Göttingen, the latter in Vienna. The explosive mixture of the conservative Ol liberal Young Czech scholars, the latter educated mostly outside of Bohemia, not only proved revolutionary in academia but also led to the revision of the idea of the Czech nation.

International mobility does not, however, necessarily lead to exceeding quality. In fact, German-language Habsburg universities were most successful and influential in precisely those disciplines characterized by continuity and the formation of stable research traditions, such as the medical sciences, biology, art history, Slavic philology, and philosophy.[3] Imported scholars were scarce in these disciplines, albeit in some cases crucial at their beginnings. This does not mean, however, that scholars in these disciplines did not migrate within the empire. Scholars circulated between Innsbruck and Chernivtsi, traversing linguistic boundaries at times, but all within the borders of the Habsburg Empire. In fact, students' networks allowed these schools to thrive, for instance, in Slavic philology, where the Viennese doyens Franz von Miklosich/Franc Miklošič and Vatroslav Jagić enjoyed networks of correspondence that helped them in their comparatist endeavors.

Mobility, Confessional Geography, and the Urban Sphere

The mobility of Habsburg scholars contributed to the intellectual development of the empire, but its impact on individual careers varied. It seems that at German-language universities in the Habsburg Empire, mobility was a synonym for scholarly excellence after 1900, and faculties grew more and more hesitant about home nominations (*Hausberufungen*). But this requirement of mobility strongly disadvantaged a group that was for political reasons prevented from moving, namely, Jewish scholars. Their exclusion was an outcome of the Cisleithanian universities' meandering between liberalism and Catholicism but at the same time of a structural problem in the system of academic career advancement. And it had, unexpectedly, a tremendous impact on the cultural thriving of cities, in particular Vienna.

University policy in the Habsburg Empire remained a political issue throughout the long nineteenth century, from the 1880s falling victim to an extremism-prone "studentocracy"[4] and a general deadlock of reforms in the monarchy. Early appointments by Franz Stadion in 1848 included promotions of liberal and Slavic scholars to both provincial universities and the University of Vienna, clearly an outcome of the 1848 revolution and the demands for reforms, including acceptance of the equality of different national groups. Throughout the 1850s, however, and then under the minister of education Leo Thun-Hohenstein, a pro-state ideological direction was advocated. Conservative Catholic scholarship, promoting conservative nationalism, clearly prevailed. Several disciplines were to be Catholic only, such as philosophy, and positions of academic authority, such as the dean and rector, were similarly reserved for Catholics. While Protestant scholars could be appointed for professorships, this largely resulted from a lack of Catholic scholars in several disciplines, and such appointments remained rare. Most scholars from abroad whom the ministry appointed to teach at Habsburg universities were Catholics, and they had often experienced conflict in their previous environment because of their religious denomination. In the 1870s the anti-ultramontane ministry grew skeptical about nominating Catholic scholars, but this was only a short-term change; most ministers of education preferred Catholics.

In the 1870s scholars of Jewish faith became more widely represented at universities. Before 1868 they were clearly discriminated against by a combination of career discouragement and ministerial policy. This situation changed after liberalization and the enactment of policies requiring equal treatment of denominations. However, the growing numbers of Jewish Privatdozenten and professors at the universities met with strong criticism from right-wing groups like the Christian Socialists. Combined with growing anti-Semitic propaganda, this even led to assaults on individual scholars. By the end of the nineteenth century, the atmosphere in Graz and Innsbruck, cities with few Jewish inhabitants, had grown tense, leading the ministry to consider appointments of Jewish scholars to these cities carefully and mostly to decide on Catholic scholars instead. As a result, the universities in Vienna and Prague had a growing number of Jewish Privatdozenten who had little chance of being appointed to other universities. Owing to the reluctance of the Viennese and Prague faculties to make home nominations, the possibility of promotion there was likewise limited. This meant that these Jewish scholars often worked in private clinics (a widespread practice among physicians), extra-academic institutes (such as the Institute for Radium Research [Institut für Radiumforschung] in Vienna), or Vienna's municipal institutions.[5] Whereas in the German Empire Jewish scholars moved to institutions in smaller cities,[6] in the Habsburg Empire they relocated to Vienna, contributing to the astounding flourishing of extramural research there.

The University of Vienna, the central institution of the empire, offered the most liberal situation not only for Jewish scholars. It was also the place where the government was more lenient toward Polish or Czech nationalist agitation than it was in the respective provinces. At the same time, Vienna was a place in which every conflict could easily be translated into a cultural argument, which led to clashes, especially among the multilingual and multidenominational Privatdozenten, who were competing for fewer and fewer positions and were recruited from groups that after a brief liberal period were becoming more and more disadvantaged. Not only was there cultural uncertainty around 1900,[7] but this also translated into social insecurity for highly educated intellectuals, both in the provinces and in the capital. This uncertainty produced tensions that increased the chance of conflicts. Further, this uncertainty also nurtured ideas of an exclusivist ethnicity. Vienna remained, however, a melting pot of peoples and ideas from which the whole empire profited. Only after the Great War did Vienna lose this dominance and importance; in the 1920s Prague overtook it as the leading light of central European academia.

Mobility and Careering

Career insecurity among Privatdozenten in the Habsburg Empire had both positive and negative effects. Competition soared, and its effects on Habsburg scholars have not yet been scrutinized. Clearly, the scholarly precariat was a problem for the academics themselves and affected both their professional and private lives. At nineteenth-century universities in the German Empire, "the poverty of the Privatdozenten became an almost unquestioned tradition,"[8] and this was equally true for the Habsburg Empire. For universities, however, Privatdozenten were a cheap (mostly free) teaching force, helping the universities cope with rising numbers of students, especially in medicine. This made them particularly attractive for the universities and also produced narratives of competition and precarity as an advantage. Most important, politicians and also professors hailed competition and survival of the fittest in an almost neoliberal manner as a means of increasing productivity among young scholars. And this story is not over, as similar narratives still define current academic discussions.

Privatdozenten were a vital part of the academic system for other reasons as well, connected to their work at private research facilities and their participation in public education (*Volkskurse*).[9] Cities profited from scholars who had to teach outside of universities because they had little chance at a straightforward university career, and this has to be remembered when analyzing the scholarly and cultural productivity of different cities around 1900. At the same time, there was an obvious demand for both nonuniversity research and scholarship in the public sphere: what was and still is debated is what the nonuniversity involvement of university scholars should and can involve and how it relates to an academic career. Universities have long been a privileged place of knowledge, even if historical examples often show that innovation is easier to achieve at smaller, less rigidly organized institutions. And the example of the Habsburg academic system helps us ask what the costs of sustaining scholarly excellence were (and are), as well as what possible strategies for coordinating public and private institutions might be.

Austrian universities are currently facing, in fact, similar dilemmas related to tensions among career opportunities, nonuniversity engagement, and mobility. In the wake of rapid academic internationalization in the past two decades, these involve especially the scholarly exchange with universities abroad. The so-called *Mobilitätszwang*, the necessity to be mobile in order to acquire an academic position, has been strengthened through the construction of contracts at universities. This is true for both young scholars, for whom moving abroad after attaining a doctoral degree is a career prerogative, and also those striving for higher positions; longer stays abroad are regarded as an invaluable asset.

While fin de siècle Jewish scholars were discriminated against since they could not career throughout the empire, the twenty-first-century *Mobilitätszwang* is forcing yet another group out of academia—women—as recent studies convincingly show.[10] The responsibility of childcare still ties women down to a greater extent than men, preventing them from taking either short- or long-term fellowships abroad.[11] While universities have developed more sensitivity to this issue in recent years, this has not resulted in interrogating the predominant idea of excellence, still bound to the assumption that scholars should always be ready and willing to travel abroad.

Discussions about the effects of nominations from abroad are similarly far from over. With internationalism being hailed nowadays as a necessity, ever-higher number of professors have moved to Austria from Germany, leading to tensions and claims that Austria's young scholars should be given priority.[12] The debate over whom to promote is still open, and many avenues are being discussed, such as a gender-sensitive tenure-track model. Looking at the effects of similar discussions 150 years earlier might help us to escape the pitfalls and dangers of educational experiments.[13]

The final point in the discussion of the effects of mobility concerns nationalism. In the Habsburg Empire, scholars with different linguistic and cultural backgrounds than the university majority had dramatically different careers depending on whether they were nominated via political measures or were chosen by the faculties. While in the 1850s foreign scholars were more often rejected than not, by both nationalist scholars in the faculties and the public sphere in certain cities, later they were accepted and could, even after the Great War, make a career for themselves. The examples of German-language scholars appointed to Galicia in the late nineteenth century, such as Leopold Adametz, or of scholars teaching German philology there, show that acceptance and scholarly productivity went hand in hand. But to be successful, scholars had to adapt, at least partially, to the norms of the majority, including in language (if only a passive knowledge) and contacts with the local populace.

The creation of imperial hubs of German-language academics in the 1850s should, however, not be uncritically called a failure. The knowledge they brought with them largely contributed to the thriving of institutions, although it rarely resulted in the creation of local schools and the education of a new generation of local scholars. The relationship between academia and the public sphere further affected developments in both the sciences and the humanities. Engagement in the public sphere was voluntary; some nonlocal scholars did not venture to do so, but most actually did. This affected both their scholarly production and their broader societal knowledge. Adametz, for example, as a professor of domestic animal husbandry and dairy science in Cracow, profited from contact with local farmers while working on new breeds of cattle, which in turn changed Galician farming.[14] Galician professors of the humanities served as reviewers or publicists in the popular press, and most served as translators, thus bridging the ever-growing linguistic divide in the monarchy.[15]

In fact, most scholars whom historians choose to represent nineteenth-century Cisleithanian universities, whether they identified as Germans or Slavs, not only excelled in scholarly matters but were also publicly involved intellectuals. With this historical perspective in mind, one can question the trends in the academic world that reinforce the idea of universities as ivory towers, and scholarship as a practice best done in isolation from the rest of society. Institutes for advanced study, transdisciplinary yet secluded places, would be one example of how these trends manifest themselves at the institutional level. There are, of course, not only losses but also gains from such institutions, but the form of sociability they propagate is having, and will continue to have, a crucial influence on the future shape of knowledge.

Internationalisms and Their Languages

In the first half of the nineteenth century, the understanding of languages and their assumed role in the dissemination of scientific knowledge changed. In the Habsburg Empire, the idea that a national language—at the same time the scholar's mother tongue—was most apt for science prevailed and was picked up by nationalists to substantiate their political claims. The legal support for German as the lingua franca for secondary and tertiary education, a position previously reserved for Latin in the empire, was increasingly perceived as privileging one group and thus devaluing the cultural importance of other languages. As a reaction to this, teaching and publishing science in Czech, Hungarian, and Polish became an issue for local elites, which finally led to the introduction of these languages at all levels of education. Ruthenian elites acted similarly in Galicia, where the Polish language was dominant, although they did not achieve the creation of a Ruthenian university.

Through a combination of political and cultural claims, education—and thus both scholarship and universities—progressively became plurilingual throughout the empire but monolingual within the walls of each university. This meant, however, the codification of a hierarchy of languages, with German as the supralanguage and with culturally defined universities now being able to use their own local language. Of course, this applied not only to Slavic universities: Innsbruck, Graz, Vienna, and the German University in Prague were single-language universities, and the banning of Italian from the University of Innsbruck in 1904 was the final step in this process. The nationalization of universities was thus a complex process involving many parties with vested interests. It was not only the Hungarian and Slavic nationalists who were trying to alter the empire.

By the end of the nineteenth century, institutions of higher education, seen as the most important places for cultural, intellectual, and structural developments, became critical in nationalist propaganda. This led to countless conflicts and even casualties. In 1918 a new political space emerged in which the question of language hegemony did not disappear. German suffered greatly from the dissolution of the monarchy and from sanctions by international scholarly organizations immediately after the war.[16] Still, many scholars teaching in Czechoslovakia and Poland wrote in German, and it remained the language of intellectual communication. At the same time, the persistence and role of micro-imperial languages (Polish in the eastern part of the Republic of Poland and Czech in Czechoslovakia) led to conflicts with Ukrainians and Lithuanians, on one hand, and Slovaks, on the other.

This shift in the understanding of the language/university debate went hand in hand with changes in public perception, historical commemoration, and collective memory concerning universities. It also affected the way in which universities and their scholars participated in the political public sphere. Johannes Feichtinger, a historian of Habsburg scholarship, has called German-speaking Habsburg and Austrian scholars "relatively autonomous," meaning that they were proposing political changes without actively participating in politics; they instead expressed these sentiments in scholarly books and articles.[17] If one wants to apply this to Hungarian or Slavic scholars, one has to distinguish between nationalist politics and politics in a narrower sense.

Scholars working in L'viv, Pest, or Prague took a stance for the nationalist cause in a variety of ways, beginning by signaling national unity through activities in science and culture. This could be as simple as writing in a language other than German, Latin, or French. The staging of culture—its extent and its productivity—was already a political issue, although this politicization had different manifestations and various intensities. In the late nineteenth century, the decision to publish in a particular language of publication was a career choice, and many academics would have simply accepted this as a strategic act and not a political one.

In the historical memory of the new states during the post-1918 period, scholars who had not openly participated in political activities during the Habsburg period were forgotten. Because of this, they are underrepresented in historical writing as well, supporting a narrative that academics jointly and actively supported the national struggle for cultural autonomy. This narrative contains a kernel of truth, albeit a small one: scholars participated in national projects and thus strengthened them, but not through open patriotic support or zeal. It was not really a viable option for scholars to completely back away from national projects and, for instance, write only in German throughout their career, although one finds as many politically silent Slavic or Hungarian scholars as politically active German-language scholars.

Looking at the language changes in the Habsburg Empire brings another facet to the debate on language and scholarship to the fore. It shows, quite clearly, not only that scholarly productivity rose when scholars were allowed to teach and write in their preferred language but also that, at the same time, they published more in international languages and sought out international contacts. The vital difference was whether academics had to write in only one language or had a choice of languages. This was, unsurprisingly, discipline specific. While scholars working in the natural sciences published, with few exceptions, in several languages, those in the humanities chose to write in the languages of their reading public, and this, again unsurprisingly, affected the topics they chose to deal with. This period also saw the foundation of disciplines that pertained to the humanities and that engaged in the processes of nation and empire building. And German-speaking scholars were also involved in these processes, tuning their disciplines to specific needs.

With this observation in mind, one can apply some of the conclusions from this study of Habsburg scholarship to the current debates on the language of science, scholarship, and higher education. This adds neatly to Michael Gordin's history of changing ideas about the principal languages of science by showing the ramifications of nationalism for the German language. In Gordin's narrative, English becomes strengthened as a proxy language in which results by non-English-language scholars are reproduced.[18] In central Europe, German had this role; interestingly, while the motivation to write in it changed—from belonging to the imperial corps to wanting to present national science internationally—its predominance did not. World War I only slightly scratched German's predominance, although it was already, as Gordin also remarks, losing its attraction as the global scientific language by then.

While it is clear that English is currently the language of the natural sciences, the discussion about the language for the humanities is ongoing. In particular, the application of a point system from the natural sciences in the humanities, privileging international peer-reviewed English-language journals, has met with widespread criticism by academics. While I did not analyze in this book the connections among the language of publication, publications' content, and their intended readership in detail, I have attempted to describe the connection between universities and the humanities. One can be sure that the disciplines that supported national claims—and in central Europe nationalism facilitated the humanities' rise to power considerably—will change once the language their findings are presented in changes. One cannot, however, be certain in which direction the trend will go—the recent revival of conservative policies reinforces, for instance, both publishing in one's respective national language, as a means of internal historical politics, and publishing in English, as a means of international propaganda. Also, scholars working on local histories complain about losing readers, and thus the impact they desire, when forced to publish in English. The most probable future form is thus multilingualism for scholars and their publications; academics will most likely publish the same results in one language for local publics and in English for the international forum. With this we are, ironically, back in the late nineteenth century, when scholars at Slavic academies of sciences opted for precisely this solution, with German and French as the languages they published in for readers abroad.

That the interests of scholars and politicians diverge may be a truism, but it connects well to the dynamics of the changes in the academic system. In the Habsburg Empire, centralist politicians' ideas of internationalizing knowledge failed, especially those connected to imperial structures and to German as the imperial language. Rejecting imperial internationalism, scholars opted for a different kind of internationalization and chose different paths to achieve it. One can translate this process into more recent changes in the European and global academic system, in which English became omnipresent at the universities. These changes—often described as a result of the Bologna Process, a process of assuring the compatibility of higher education in Europe that started in 1999—have met with criticism and opposition. One can only assume that these changes might have been accepted more readily if they had been a gradual process led by the academics in their respective institutions rather than being left to politicians.[19] The role of language in international communication becomes even clearer when one looks at the post–Great War discussions on the internationalization of, for instance, Polish science. These discussions of strategies Polish scholars foresaw as guaranteeing that the internationalization process would profit most, and not exclude many, underscore once more that the solutions are numerous and cannot be uncritically imposed.[20]

Empire's Many Spaces

The final question remains what lessons about the Habsburg Empire can be drawn from the story of scholarship meandering between imperial space and national spaces. In this, my findings align well with two recent proposals to conceptualize the nineteenth-century Habsburg state. Pieter M. Judson has recently favored the idea that nationalist movements in the Habsburg

Empire profoundly changed its structure, but not in an either-or relation, as historians writing about nationalisms tearing apart Habsburg central Europe have claimed. Instead of a narrative of the empire's slow demise, Judson speaks of the empire accommodating nationalist demands, and of the ways national movements were shaped by imperial structures and possibilities.[21] In a similar manner, John Deak has described the evolution of imperial statehood into a multinational space.[22] As I have argued throughout this book, the geographic reorganization of the empire similarly reshaped and partially fragmented academia, but most early twentieth-century scholars did not contest the empire as such. They lived it and readily took hold of the opportunities it provided. For instance, when proposing reforms at their universities, they kept the effects these changes would have on the whole empire in mind. This is true of most scholars at the German-language universities but also of most Slavic scholars, like Masaryk or the Cracow scholars who argued for the necessity of mathematical education in 1907. Even nationalist scholars took advantages of the resources the empire provided and bemoaned their lack after 1918.

In contrast with the historiography that has come out of central European scholarship, this work suggests a large number of entanglements that I see as characteristic of the Habsburg Empire: a linguistically divided but still culturally entangled scientific space. Historians in the twentieth century have largely disregarded the productive edge of this multicultural state, the Habsburg Empire, looking at it with a national framework in mind. But during the empire's existence, monoculturalism and trends toward intellectual seclusion were often outweighed by developments and changes favoring interdependence.

Finally, my work suggests the necessity of greater inclusion of peripheral histories in the general narrative of the Habsburg Empire, which also means rethinking it from a spatial perspective. In the particular case of universities, it does not entail rewriting Habsburg history from the viewpoint of the periphery, although that would be a welcome perspective for other research foci.[23] The history of universities, however, helps to demonstrate that the decision-making was imperial; that is, legal documents issued for, say, Chernivtsi, were also binding for Vienna. The legal discrimination against Jewish scholars, illustrated by Hermann Rosenberg's forbidden habilitation in philosophy in L'viv in 1854, is one of many examples. Clearly, the ministry acted in accordance with this particular legal ruling for the next decade or so while making decisions for Vienna or Graz. And the ministry retained this structure of decision-making and legal interdependence until

World War I. One cannot understand processes in Vienna without looking at Cracow or Chernivtsi, just as one cannot understand processes taking place in L'viv without knowing about Graz; similarly, one cannot understand Vienna without taking Berlin into consideration, nor L'viv without Kiev. The Habsburg Empire changed within the sixty years described here, and so did its spaces and the spaces its subjects lived and communicated in. These spatial dynamics and the interdependence of different spaces would seem a rewarding topic for future research.

APPENDIX 1

Disciplines of Habilitation at Austrian Universities

The following tables list the main disciplines of habilitation, in order by date (the year of the first habilitation is in parentheses). The names of the disciplines use the current English terminology.

TABLE A.1 Habilitations at medical faculties

Anatomy (total habilitations: 29)
1. Vienna (year of first habilitation: 1868)
2. Prague (1872)
3. Cracow (1878)

Bacteriology (total: 5)
1. Cracow (1878)
2. L'viv, Prague/Czech (1902)

Balneology (total: 16)
1. Vienna, Prague (1850)
2. Graz (1870)

Chemistry (total: 29)
1. Vienna (1848)
2. Cracow (1862)
3. Prague (1879)

Pediatric medicine (total: 54)
1. Vienna (1848)
2. Prague (1848)
3. Graz (1852)

Dentistry (total: 30)
1. Vienna (1848)
2. Graz (1857)
3. Cracow (1877)

Dermatology/syphidology (total: 49)
1. Cracow, Vienna (1862)
2. Prague (1868)

Illnesses of the ear, nose, and throat (total: 57)
1. Vienna (1861)
2. Prague (1868)
3. Graz (1872)

Forensics (total: 19)
1. Prague (1848)
2. Vienna (1858)
3. Cracow (1869)

Gynecology (total: 100)
1. Prague (1848)
2. Vienna (1849)
3. Cracow (1862)

Histology (together with embryology) (total: 17)
1. Vienna (1849)
2. Prague (1856)
3. Innsbruck (1870)

History of medicine (total: 11)
1. Prague, Vienna (1848)
2. Cracow (1869)

Hygiene (total: 27)
1. Prague (1848)
2. Vienna (1858)
3. Cracow (1874)

Internal medicine (total: 100)
1. Prague (1848)
2. Vienna (1867)
3. Cracow (1881)

Ophthalmology (total: 64)
1. Prague, Vienna (1849)
2. Graz (1881)

Pathological anatomy (total: 41)
1. Vienna (1850)
2. Prague (1872)
3. Cracow (1885)

Experimental pathology (total: 23)
1. Vienna (1862)
2. Graz (1876)
3. Prague (1882)

Pathology and therapy of internal diseases (total: 50)
1. Prague (1856)
2. Vienna (1858)
3. Innsbruck (1878)

Pharmacology (with pharmacognosy) (total: 25)
1. Prague (1848)
2. Vienna (1849)
3. Cracow (1900)

Physiology (total: 36)
1. Prague (1861)
2. Vienna (1869)
3. Cracow (1875)

Psychiatry (total: 63)
1. Prague (1848)
2. Graz (1855)
3. Vienna (1857)

TABLE A.2 Habilitations at philosophical faculties

HUMANITIES (TOTAL: 515)

Historical disciplines (total: 136)

Austrian history (total: 27)
1. Vienna (1850)
2. Graz (1856)
3. Prague (1875)

General history (total: 22)
1. Vienna (1855)
2. L'viv (1872)
3. Graz (1875)

Ancient history (total: 20)
1. Innsbruck, Vienna (1860)
2. Prague/German (1884)

Medieval history (total: 17)
1. Innsbruck (1867)
2. Vienna (1868)
3. Cracow (1875)

Auxiliary sciences of history (total: 14)
1. Vienna (1856)
2. Innsbruck (1878)
3. Cracow (1890)

National histories (total: 12)
1. Prague (1871)[1]
2. L'viv (1871)[2]
3. Cracow (1881)

History of art, including music and aesthetics (total: 45)
1. Vienna (1851)
2. Cracow (1863)
3. Graz (1872)

Archaeology (total: 15)
1. Vienna (1875)
2. Cracow (1893)
3. Prague/German (1902)

Paleontology (total: 17)
1. Vienna (1853)
2. Cracow (1882)
3. Prague Czech (1883)

Anthropology/ethnology (total: 9)
1. Vienna, Prague Czech (1892)
2. L'viv (1911)

Philology (total: 175) and languages (total: 75)

German language and literature, including philology (total: 42)[3]
1. Vienna (1855)
2. Graz (1868)
3. Prague (1875)

Classical philology (total: 47)
1. Vienna (1851)
2. Graz (1857)
3. Prague (1859)

Oriental philology (total: 26)
1. Vienna (1848)
2. Prague (1876)
3. Cracow (1893)

Romance languages and philology (total: 25)
1. Vienna (1871)
2. Prague (1874)
3. Graz (1896)

Slavic philology (total: 19)
1. Prague (1854)
2. Graz (1867)
3. L'viv, Vienna (1878)

Slavic literature (total: 17)
1. Cracow (1868)
2. L'viv (1871)
3. Prague/Czech (1899)

English language and literature, including philology (total: 13)
1. Prague (1881)
2. Vienna (1883)
3. Graz (1888)

Semitic philology (total: 8)
1. Vienna (1848)
2. Prague (1850)
3. L'viv (1851)

Philosophy (total: 60)
1. Prague (1848)
2. Vienna (1849)
3. Graz (1870

Continued

Table A.2 Habilitations at philosophical faculties—cont'd

SCIENCES (TOTAL: 330)

Mathematical and physical sciences (total: 208)

Physics (total: 49)
1. Vienna (1855)
2. Prague (1872)
3. Graz (1875)

Experimental physics (total: 11)
1. Prague (1871)
2. Innsbruck (1879)
3. Graz (1889)

Mathematical physics (total: 8)
1. L'viv (1851)
2. Prague (1863)
3. Vienna (1867)

Chemistry (total: 86)
1. Vienna (1848)
2. Graz (1855)
3. Prague (1861)

Mathematics (total: 54)
1. Vienna (1849)
2. Cracow (1862)
3. Graz (1866)

Life sciences (total: 123)

Botany (total: 51)
1. Prague, Vienna (1857)
2. Graz (1866)

Zoology (total: 39)
1. L'viv (1851)
2. Prague (1852)
3. Graz (1872)

Comparative anatomy (total: 17)
1. Graz (1876)
2. Vienna (1879)
3. Prague/German (1892)

Geosciences (total: 99)

Geology (total: 34)
1. Vienna (1854)
2. Prague (1874)
3. Graz (1880)

Geography (total: 28)
1. Prague (1856)
2. Vienna (1862)
3. Cracow (1876)

Mineralogy (total: 24)
1. Vienna (1861)
2. Prague (1868)
3. Cracow (1870)

Astronomy (total: 20)
1. Vienna (1850)
2. Prague/Czech (1883)
3. Innsbruck (1888)

Meteorology/cosmic physics (total: 10)
1. Vienna (1869)
2. Prague/Czech (1883)
3. Innsbruck (1909)

[1]As "Bohemian history."

[2]As "history of territories of Halych and Vladimir."

[3]The official formula was "German language and literature," with a relatively late distinction between philology and history of literature. At the time when "national" languages became allowed, this distinction was already made, with exception of the first habilitation (Stanisław Tarnowski, Cracow, 1868), who was, however, only a historian of literature.

Appendix 2

Databases of Scholars at Cisleithanian Universities

The databases compiled by the author of this book consist of data on scholars teaching at the medical and philosophical faculties of Habsburg universities in 1848–1918 (with the exception of the Vienna philosophical faculty, where an amended database compiled by Kurt Mühlberger, Archive of the University of Vienna, was used). The databases are based on ministerial documents and catalogs of lecturers from the respective universities. Although compiled with the utmost scrutiny, some bibliographic information is missing, and some was not collected owing to the scope of this project.

The databases are an integral part of this project, and as such the same restrictions apply. If used, please cite them accordingly.

Charles-Ferdinand University in Prague, Medical Faculty, 1848–1882: http://phaidra.univie.ac.at/o:104441.

Charles-Ferdinand University in Prague, Philosophical Faculty, 1848–1881: http://phaidra.univie.ac.at/o:104442.

Czech Charles-Ferdinand University in Prague, Medical Faculty, 1883–1918: http://phaidra.univie.ac.at/o:104437.

Czech Charles-Ferdinand University in Prague, Philosophical Faculty, 1882–1918: http://phaidra.univie.ac.at/o:104438.

German Charles-Ferdinand University in Prague, Medical Faculty, 1883–1918: http://phaidra.univie.ac.at/o:104439.

German Charles-Ferdinand University in Prague, Philosophical Faculty, 1882–1918: http://phaidra.univie.ac.at/o:104440.

Chernivtsi University, Philosophical Faculty, 1875–1918: http://phaidra.univie.ac.at/o:104428.

Jagiellonian University, Medical Faculty, 1848–1918: http://phaidra.univie.ac.at/o:104429.

Jagiellonian University, Philosophical Faculty, 1848–1918: http://phaidra.univie.ac.at/o:104430.

University of Graz, Medical Faculty 1863–1918: http://phaidra.univie.ac.at/o:104431.

University of Graz, Philosophical Faculty, 1848–1918: http://phaidra.univie.ac.at/o:104432.

University of Innsbruck, Medical Faculty, 1869–1918: http://phaidra.univie.ac.at/o:104433.

University of Innsbruck, Philosophical Faculty, 1848–1918: http://phaidra.univie.ac.at/o:104434.

University of L'viv, Medical Faculty, 1848–1918: http://phaidra.univie.ac.at/o:104435.

University of L'viv, Philosophical Faculty, 1848–1918: http://phaidra.univie.ac.at/o:104436.

University of Vienna, Medical Faculty, 1848–1918: http://phaidra.univie.ac.at/o:104443.

Notes

Introduction

1. Fabio Chiocchetti and Roberto Starec, "In Search of the 'Ladin Song': The Project Das Volkslied in Österreich in the Ladin Areas of Tyrol and East Friuli (1904–1914)," *Traditiones* 34, no. 1 (2005): 61–77, esp. 66.
2. See Stepan Smal'-Stoc'kyj, "Fedir Gartner," *Zapysky Naukovoho tovarystva imeni Ševčenka: Praci filolohičnoï sekciï* 86–87 (1925): 239–48.
3. Andriy Zayarnyuk, "Mapping Identities: The Popular Base of Galician *Russophilism* in the 1890s," *Austrian History Yearbook* 41 (2010): 117–42, esp. 121–26.
4. Hans Goebl, "Theodor Gartner und das typologische Denken seiner Zeit," in *Akten der Theodor Gartner-Tagung*, ed. Guntram A. Plangg and Maria Iliescu (Innsbruck: Institut für Romanistik, 1987), 13–23.
5. David Lambert and Alan Lester, eds., *Colonial Lives across the British Empire: Imperial Careering in the Long Nineteenth Century* (Cambridge: Cambridge University Press, 2006). In brief, *imperial careering* denotes a career path throughout a multicultural empire, enabled by the political, economic, cultural, and other structures of the state.
6. Jan Havránek, "Nineteenth Century Universities in Central Europe: Their Dominant Position in the Science and Humanities," in *Bildungswesen und Sozialstruktur in Mitteleuropa im 19. und 20. Jahrhundert = Education and Social Structure in Central Europe in the 19th and 20th Centuries*, ed. Victor Karady and Wolfgang Mitter (Vienna: Böhlau, 1990), 9–26.
7. On the term *pluriculturalism* in contrast to *multiculturalism*, see Moritz Csáky, "Culture as a Space of Communication," in *Understanding Multiculturalism: The Habsburg Central Europe Experience*, ed. Johannes Feichtinger and Gary B. Cohen (New York: Berghahn Books, 2014), 187–208.
8. See also Mitchell G. Ash and Jan Surman, eds., *The Nationalisation of Scientific Knowledge in the Habsburg Empire, 1848–1918* (Basingstoke: Palgrave Macmillan, 2012); and Ralph Jessen and Jacob Vogel, eds., *Wissenschaft und*

Nation in der europäischen Geschichte (Frankfurt am Main: Campus, 2002).

9. See, for example, Barry Smith, *Austrian Philosophy: The Legacy of Franz Brentano* (Chicago: Open Court, 1994).
10. Most recently, Johannes Feichtinger, *Wissenschaft als reflexives Projekt: Von Bolzano über Freud zu Kelsen. Österreichische Wissenschaftsgeschichte 1848–1938* (Bielefeld: Transcript, 2010).
11. See, for example, Martina Bečvářová, "Czech Mathematicians and Their Role in the Development of National Mathematics in the Balkans," in *Mathematics in the Austrian-Hungarian Empire: Proceedings of a Symposium Held in Budapest on August 1, 2009 during the XXIII ICHST*, ed. Martina Bečvářová and Christa Binder (Prague: Matfyzpress, 2010), 9–31.
12. For recent overviews, see Diarmid Finnegan, "The Spatial Turn: Geographical Approaches in the History of Science," *Journal of the History of Biology* 41, no. 2 (2008): 369–88; Steven Shapin and Adi Ophir, "The Place of Knowledge: A Methodological Survey," *Science in Context* 4, no. 1 (1991): 3–22; Peter Meusburger, Michael Welker, and Edgar Wunder, eds., *Clashes of Knowledge: Orthodoxies and Heterodoxies in Science and Religion* (Dordrecht: Springer Science + Business Media, 2008); and Peter Meusburger, David N. Livingstone, and Heike Jöns, eds., *Geographies of Science: Academic Mobilities, Knowledge Spaces, and Public Encounters* (Dordrecht: Springer Science + Business Media, 2010).
13. David N. Livingstone, *Putting Science in Its Place: Geographies of Scientific Knowledge* (Chicago: University of Chicago Press, 2003); and Peter Meusburger, *Bildungsgeographie: Wissen und Ausbildung in der räumlichen Dimension* (Heidelberg: Spektrum Akademischer Verlag, 1998).
14. Kapil Raj, *Relocating Modern Science: Circulation and the Construction of Knowledge in South Asia and Europe, 1650–1900* (Basingstoke: Palgrave Macmillan, 2007).
15. Yuri M. Lotman, *Universe of the Mind: A Semiotic Theory of Culture* (Bloomington: Indiana University Press, 2000).
16. See the discussion of Foucault in Derek Gregory, Peter Meusburger, and Laura Suarsana, "Power, Knowledge, and Space: A Geographical Introduction," in *Geographies of Knowledge and Power*, ed. Peter Meusburger, Derek Gregory, and Laura Suarsana (Dordrecht: Springer Science + Business Media, 2015), 1–18.
17. Martina Löw, "The Constitution of Space: The Structuration of Spaces through the Simultaneity of Effect and Perception," *European Journal of Social Theory* 11, no. 1 (2008): 25–49.
18. Richard White, *Railroaded: The Transcontinentals and the Making of Modern America* (New York: W. W. Norton, 2011); and Frithjof Benjamin Schenk, *Russlands Fahrt in die Moderne: Mobilität und sozialer Raum im Eisenbahnzeitalter* (Stuttgart: Steiner, 2014).
19. Tamson Pietsch, *Empire of Scholars: Universities, Networks and the British Academic World, 1850–1939* (Manchester: Manchester University Press, 2013); and Roger L. Geiger, *The History of American Higher Education: Learning and*

Culture from the Founding to World War II (Princeton, NJ: Princeton University Press, 2014).

20. Heike Jöns, "Grenzüberschreitende Mobilität und Kooperation in den Wissenschaften: Deutschlandaufenthalte US-amerikanischer Humboldt-Forschungspreisträger aus einer erweiterten Akteursnetzwerkperspektive" (PhD diss., Heidelberg University, 2003); and Heike Jöns, "Academic Travel from Cambridge University and the Formation of Centres of Knowledge, 1885–1954," *Journal of Historical Geography* 34, no. 2 (2008): 338–62. I want to thank Nina Wolfeil for referring me to the latter publication.
21. Jean-Louis Guereña, "L'université espagnole vers 1900," in *Sozialer Raum und akademische Kulturen: Studien zur europäischen Hochschul- und Wissenschaftsgeschichte im 19. und 20. Jahrhundert = A la recherche de l'espace universitaire européen*, ed. Jürgen Schriewer, Edwin Keiner, and Christophe Charle (Frankfurt am Main: Peter Lang, 1993), 113–31; and Marita Baumgarten, *Professoren und Universitäten im 19. Jahrhundert: Zur Sozialgeschichte deutscher Geistes- und Naturwissenschaftler* (Göttingen: Vandenhoeck & Ruprecht, 1997).
22. See Gerald Stourzh, *Der Umfang der österreichischen Geschichte: Ausgewählte Studien 1990–2010* (Vienna: Böhlau, 2011), esp. 11–105, 283–322.
23. See Gary B. Cohen, "Nationalist Politics and the Dynamics of State and Civil Society in the Habsburg Monarchy, 1867–1914," *Central European History* 40, no. 2 (2007): 241–78; Pieter M. Judson, "L'Autriche-Hongrie était-elle un empire?," *Annales: Histoire, Sciences sociales* 6, no. 3 (2008): 563–96; and Pieter M. Judson, *The Habsburg Empire: A New History* (Cambridge, MA: Belknap Press of Harvard University Press, 2016).
24. See esp. Pieter M. Judson, *Guardians of the Nation: Activists on the Language Frontiers of Imperial Austria* (Cambridge, MA: Harvard University Press, 2006); Aleksei I. Miller and Alfred J. Rieber, eds., *Imperial Rule* (Budapest: Central European University Press, 2004); and Daniel Unowsky, "'Our Gratitude Has No Limit': Polish Nationalism, Dynastic Patriotism, and the 1880 Imperial Inspection Tour of Galicia," *Austrian History Yearbook* 34 (2003): 145–71.
25. For the latest critical appropriation of Miroslav Hroch's A-B-C schema, see "Twenty-Five Years of A-B-C: Miroslav Hroch's Impact on Nationalism Studies," special issue, *Nationalities Papers: The Journal of Nationalism and Ethnicity* 38, no. 6 (2010). For the Polish case, see Brian A. Porter, *When Nationalism Began to Hate: Imagining Modern Politics in Nineteenth-Century Poland* (New York: Oxford University Press, 2000); for more on the Habsburg versus German Austrian cultural situation, see Pieter M. Judson, *Exclusive Revolutionaries: Liberal Politics, Social Experience, and National Identity in the Austrian Empire, 1848–1914* (Ann Arbor: University of Michigan Press, 1996).
26. Tatjana Buklijas and Emese Lafferton, "Science, Medicine and Nationalism in the Habsburg Empire from the 1840s to 1918," *Studies in History and Philosophy of Biological and Biomedical Sciences* 38, no. 4 (2007): 679–86; Ernest Gellner, *Language and Solitude: Wittgenstein, Malinowski, and the Habsburg Dilemma*

(Cambridge: Cambridge University Press, 1998); and Marianne Klemun, "Wissenschaft und Kolonialismus—Verschränkungen und Figurationen," *Wiener Zeitschrift zur Geschichte der Neuzeit* 9, no. 1 (2009): 3–12.

27. I argued the merits of such an approach more broadly in Jan Surman, "Imperial Knowledge? Die Wissenschaften in der späten Habsburgermonarchie zwischen Kolonialismus, Nationalismus und Imperialismus," *Wiener Zeitschrift zur Geschichte der Neuzeit* 9, no. 2 (2009): 119–33.
28. Benjamin Feyen and Ewa Krzaklewska, eds., *The ERASMUS Phenomenon—Symbol of a New European Generation?* (Bern: Peter Lang, 2013); and Ettore Recchi, *Mobile Europe: The Theory and Practice of Free Movement in the EU* (Basingstoke: Palgrave Macmillan, 2015), esp. chap. 2.
29. See, e.g., the discussion in Voldemar Tomusk, "European Higher Education Considering Gellner, Malinowski und Wittgenstein," in *The Bologna Process in Central and Eastern Europe*, ed. Tamás Kozma, Magdolna Rébay, Andrea Óhidy, and Éva Szolár (Wiesbaden: Springer Fachmedien, 2014), 33–63.
30. See on this point also Stefan Berger and Alexei Miller, eds., *Nationalizing Empires* (Budapest: Central European University Press, 2015).
31. See, e.g., the theoretically refined reactions in Ulrich Ammon, ed., *The Dominance of English as a Language of Science: Effects on Other Languages* (Berlin: Walter de Gruyter, 2001); Charles Durand, *La mise en place des monopoles du savoir* (Paris: Harmattan, 2001); and Franz Nies, ed., *Europa denkt mehrsprachig / L' Europe pense en plusieurs langues* (Tübingen: Narr, 2005).
32. "Garden" was a metaphor used for Vienna owing to its many spacious parks and gardens, while "Workshop" was used to describe Budapest, characterized by dense architecture and a large number of factories. See Péter Hanák, *The Garden and the Workshop: Essays on the Cultural History of Vienna and Budapest*, with a preface by Carl E. Schorske (Princeton, NJ: Princeton University Press, 1998).
33. Markian Prokopovych, "Staging Empires and Nations: Politics in the Public Space of Habsburg Lemberg," in *Die Besetzung des öffentlichen Raumes: Politische Plätze, Denkmäler und Straßennamen im europäischen Vergleich*, ed. Peter Stachel and Rudolf Jaworski (Berlin: Frank & Timme, 2007), 427–53.
34. See esp. Patrice M. Dabrowski, *Commemorations and the Shaping of Modern Poland* (Bloomington: Indiana University Press, 2004); for a general overview, see Maria Bucur and Nancy Meriwether Wingfield, eds., *Staging the Past: The Politics of Commemoration in Habsburg Central Europe, 1848 to the Present* (West Lafayette, IN: Purdue University Press, 2001).
35. Wojciech Bałus, *Krakau zwischen Traditionen und Wegen in die Moderne: Zur Geschichte der Architektur und der öffentlichen Grünanlagen im 19. Jahrhundert* (Stuttgart: Franz Steiner, 2003), 39–43; and Jakub Lewicki, *Między tradycją a nowoczesnością: Architektura Lwowa lat 1893–1918* (Warsaw: Oficyna Wydawnicza Towarzystwa Opieki nad Zabytkami; Neriton, 2005), 173–78.
36. Jan Fellerer, *Mehrsprachigkeit im galizischen Verwaltungswesen (1772–1914): Eine historisch-soziolinguistische Studie zum Polnischen und Ruthenischen (Ukrainischen)* (Cologne: Böhlau, 2005).

37. Danuta Sosnowska, *Inna Galicja* (Warsaw: Dom Wydawniczy "Elipsa," 2008); Stefan Simonek, "Möglichkeiten und Grenzen postkolonialistischer Literaturtheorie aus slawistischer Sicht," in *Habsburg postcolonial: Machtstrukturen und kollektives Gedächtnis*, ed. Ursula Prutsch, Johannes Feichtinger, and Moritz Csáky (Innsbruck: Studien Verlag, 2003), 129–39; and Jan Surman and Klemens Kaps, eds., "Galicia Postcolonial: Prospects and Possibilities," special issue, *Historyka: Studia metodologiczne* 42 (2012), https://www.academia.edu/15087613/_full_issue_Jan_Surman_Klemens_Kaps_Postcolonial_Galicia_Prospects_and_Possibilities_Historyka_42_2012.
38. For the theoretical background of spatial conflict within empires, see Stefan Berger and Alexei Miller, "Nation-Building and Regional Integration, c. 1800–1914: The Role of Empires," *European Review of History—Revue Européene d'histoire* 15, no. 3 (June 2008): 317–30; see also Jeremy King, "The Nationalisation of East Central Europe: Ethnicism, Ethnicity and Beyond," in Bucur and Wingfield, *Staging the Past*, 112–52, esp. 131–33; and Philipp Ther, "Das Europa der Nationalkulturen: Die Nationalisierung und Europäisierung der Oper im 'langen' 19. Jahrhundert," *Journal of Modern European History* 5, no. 1 (2007): 39–66.
39. See the discussion of the different *Ausgleiche* (compromises) in Cisleithania in Lukáš Fasora, ed., *Moravské vyrovnání z roku 1905: Možnosti a limity národnostního smíru ve střední Evropě* (Brno: Matice moravská, 2006).
40. On changing ideas of "culture" in the nineteenth century, see Franz Leander Fillafer, "The 'Imperial Idea' and Civilizing Missions," *Historyka: Studia metodologiczne* 42 (2012): 37–60, https://www.academia.edu/15087613/_full_issue_Jan_Surman_Klemens_Kaps_Postcolonial_Galicia_Prospects_and_Possibilities_Historyka_42_2012.
41. See, e.g., Oskar Kolberg's monumental ethnographic works on the "Polish" regions of the three empires, Romanov, Habsburg, and German: *Lud: Jego zwyczaje, sposób życia, mowa, podania, przysłowia, obrzędy, gusła, zabawy, pieśni, muzyka i tańce*, 86 vols. (various publishers, 1857–).
42. See the historiographies of Volodymyr Antonovyč, Mychajlo Hruševs'kyj, and Oleksandra/Aleksandra Yefymenko as well as Stepan Rudnytsky's geography: Oleksandr Kyjan, *Volodymyr Antonovyč: Istoryk j orhanizator Kyïvs'koï istoryčnoï školy* (Kiev: Instytut istoriï Ukraïny Nacional'noï Akademiï Nauk Ukraïny, 2005); and Stephen Rudnitsky [Stepan Rudnytsky], *Ukraine, the Land and Its People: An Introduction to Its Geography* (1910; New York: Rand McNally, 1918) (analyzed in Guido Hausmann, "Das Territorium der Ukraine: Stepan Rudnyc'kys Beitrag zur Geschichte räumlich-territorialen Denkens über die Ukraine," in *Die Ukraine: Prozesse der Nationsbildung*, ed. Andreas Kappeler [Cologne: Böhlau, 2011], 145–58).
43. Hedwig Kadletz-Schöffel, "Metternich und die Wissenschaften" (PhD diss., University of Vienna, 1989), 299.
44. Eighteen nominees came from Vienna, seven from Lombardy and Veneto, six from Bohemia, four from Hungary and Transylvania, two from Styria, two from

Tyrol, and one from Upper Austria, according to the *Denkschrift der kaiserlichen Akademie der Wissenschaften: Philosophisch-historische Klasse*, vol. 1 (Vienna: Aus der Kaiserlich-Königlichen Hof- und Staatsdruckerei, 1850), vii–ix.

45. In the usual sense of the term, corresponding members, in contrast to full members, may reside outside of a given country and do not have to be present at the meetings. In the Habsburg case, corresponding members could also be from within the monarchy; in such cases the term denoted the preliminary stage to becoming a full member, although most corresponding members never received that honor.
46. Richard Meister, *Geschichte der Akademie der Wissenschaften in Wien 1847–1947* (Vienna: Adolf Holzhausens Nachfolger, 1947), 56.
47. See Christine Ottner, "Zwischen Wiener Localanstalt und Centralpunct der Monarchie: Einzugsbereich und erste Geschichtsforschungsunternehmen der kaiserlichen Akademie der Wissenschaften," *Anzeiger der phil.-hist. Klasse der Österreichischen Akademie der Wissenschaften* 143, no. 1 (2008): 171–96; and František Palacký, *Urkundliche Beiträge zur Geschichte Böhmens und seiner Nachbarländer im Zeitalter Georg's von Podiebrad (1450–1471)* (Vienna: Kaiserlich-Königliche Hof- und Staatsdruckerei, 1860).
48. "Vorbericht," in Joseph Chmel, ed., *Urkunden zur Geschichte von Österreich, Steiermark, Kärnten, Krain, Görz, Triest, Istrien, Tirol: Aus den Jahren 1246–1300; aus den Originalen des Kais. Kön. Haus- Hof- und Staats-Archives (= Fontes rerum Austriacarum. Österreichische Geschichtsquellen, herausgegeben von der Historischen Commission der Kaiserlichen Akademie der Wissenschaften in Wien. Zweite Abtheilung Diplomataria et Acta I. Band. Diplomatatium Miscellum Seculi XIII)* (Vienna: k.k. Hof- und Staatsdruckerei, 1849), v; for a definition of "Austria," see p. xxxi.
49. James E. McClellan, *Science Reorganized: Scientific Societies in the Eighteenth Century* (New York: Columbia University Press, 1985).
50. Hugh LeCaine Agnew, "Czechs, Germans, Bohemians? Images of Self and Other in Bohemia to 1848," in *Creating the Other: Ethnic Conflict and Nationalism in Habsburg Central Europe*, ed. Nancy Wingfield (New York: Berghahn Books, 2003), 56–80, esp. 66 on the question of the name of the society.
51. Antoni Zygmunt Helcel, *Starodawne prawa polskiego pomniki poprzedzone wywodem historyczno krytycznym tak zwanego Prawodawstwa Wiślickiego Kazimiérza Wielkiego w texcie ze starych rękopism krytycznie dobranym* (Cracow and Warsaw: Drukarnia C. K. Uniwersytetu; nakładem Księgarni Gustawa Sennewalda, 1856).
52. See Helcel's introduction, in which he hails the end of nationalistic particularity in science. Antoni Zygmunt Helcel, "O teoretycznéj i praktycznéj oświacie," *Kwartalnik naukowy, wydawany w połączeniu prac miłośników umiejętności* 1, no. 1 (1835): 1–26.
53. Jan Svatopluk Presl, "Oznámenj," *Krok: Weřegný spis wšenaučný pro wzdělance národu Česko-Slowanského* 1, no. 1 (1821): 7–18.
54. Milan Kratochvíl, *Jan Evangelista Purkyně a jeho snahy o reformu české*

školy (Prague: Státní pedagogické nakladatelství, 1987); and Thomas Garrique Masaryk, "Jak zvelebovati naši litaraturu náukovou? Clánek III," *Athenaeum: Listy pro literaturu a kritiku vědeckou* 2, no. 9 (1885): 270–75.

55. Franciszek Martinczak, "Geneza zjazdów Związku Lekarzy Słowiańskich," *Archiwum Historii Medycyny* 43, no. 1 (1980): 37–43.
56. Xawer Liske, *Der angebliche Niedergang der Universität Lemberg: Offenes Sendeschreiben an das Reichsrathsmitglied Herrn Eduard Dr. Suess, Prof. an der Universität Wien* (L'viv: Gubrynowicz & Schmidt, 1876).
57. For the sake of brevity, I will refer to the "Ministry of Education" and "minister of education" throughout this book, given my focus on educational institutions. However, the official name was Ministry of Religion and Education throughout the period studied here, and religion did not become a separate ministry until after World War I.
58. See Hans-Georg Hofer, "Jenseits der männlichen Abwehrfront: Arztberuf und Medizinstudium im Spiegel der Neurastheniedebatten um 1900," in *Medizinerinnen*, ed. Sonia Horn and Ingrid Arias (Vienna: Verlagshaus der Ärzte, 2003), 45–53; see also Alison Rose, *Jewish Women in Fin de Siècle Vienna* (Austin: University of Texas Press, 2008), esp. 87–94. For a theoretically interesting approach, see also Rebecca Friedman, *Masculinity, Autocracy and the Russian University, 1804–1863* (Basingstoke: Palgrave Macmillan, 2005).
59. See the databases of scholars teaching at medical and philosophical faculties in Cisleithania in 1848–1918, listed in Appendix 2.

Chapter 1

1. [Viktor Andrian Werburg], *Österreich und dessen Zukunft* (Hamburg: Hoffmann und Campe, 1843), 56–57. Unless otherwise indicated, all translations are my own.
2. Ministry of Religion and Education, answer to parliamentary inquiry, KUMAkte Z. 2272/KUM ex 1897, quoted in Leo Beck von Mannagetta and Karl von Kelle, *Die österreichischen Universitätsgesetze: Sammlung der für die österreichischen Universitäten giltigen Gesetze, Verordnungen, Erlässe, Studien- und Prüfungsordnungen usw.* (Vienna: Manz'sche k.k. Hofverlags- und Universitäts-Buchhandlung, 1906), 1.
3. See *Die Neugestaltung der österreichischen Universitäten über Allerhöchsten Befehl dargestellt von dem k.k. Ministerium für Kultus und Unterricht* (Vienna: k.k. Hof- und Staatsdruckerei, 1853); and Rudolf Kink, *Geschichte der kaiserlichen Universität zu Wien*, 2 vols. (Vienna: Carl Gerold & Sohn, 1854), esp. 1:432–532 for the more recent history.
4. See the critique in Johannes Feichtinger and Franz Leander Fillafer, "Leo Thun und die Nachwelt: Der Wissenschaftsreformer in der österreichischen Geschichts- und Kulturpolitik des 19. und 20. Jahrhunderts," in *Die Thun-Hohenstein'schen*

Universitätsreformen 1849–1860: Konzeption—Umsetzung—Nachwirkungen, ed. Christof Aichner and Brigitte Mazohl (Vienna: Böhlau, 2017), 347–78.

5. Un diplomate étranger qui a longtemps résidé dans ce pays [Liudwig Tengoborskij], *De l'instruction publique en Autriche* (Paris: Cousin, 1841), 306.
6. See the chapters on the First Vienna Medical School (Wiener Medizinische Schule) in Erna Lesky, *The Vienna Medical School of the 19th Century*, trans. L. Williams and I. S. Levi (Baltimore: Johns Hopkins University Press, 1976); for an example of international acknowledgment, see Marcel Chahrour, "'A Civilizing Mission'? Austrian Medicine and the Reform of Medical Structures in the Ottoman Empire, 1838–1850," *Studies in History and Philosophy of Science, Part C: Studies in History and Philosophy of Biological and Biomedical Sciences* 38, no. 4 (2007): 687–705.
7. Lorenz Oken, foreword to Franz Graf von Kolowrat, "An die vaterländischen Freunde der Wissenschaften," *ISIS oder Encyclopädische Zeitung*, nos. 1–4 (1818): 1103.
8. Oken, foreword to Kolowrat, "An die vaterländischen Freunde," 1103.
9. William Wilde, *Austria, Its Literary, Scientific and Medical Institutions with Notes and a Guide to the Hospitals and Sanatory Establishments of Vienna* (Dublin: Curry, 1843), xxii.
10. Wilde, *Austria*, 84.
11. Franz Sartori, *Historisch-ethnographische Übersicht der wissenschaftlichen Cultur, Geistesthätigkeit und Literatur des österreichischen Kaiserthumes nach seinen mannigfaltigen Sprachen und deren Bildungsstufen in skizzierten Umrissen bearbeitet: Erster Theil* (Vienna: Carl Gerold, 1830), ix, emphasis in original.
12. Yvonne Steif, *Wenn Wissenschaftler feiern: Die Versammlungen deutscher Naturforscher und Ärzte 1822 bis 1913* (Stuttgart: Wissenschaftliche Verlags-Gesellschaft, 2003), esp. 47–57.
13. See, on Britain, Louise Miskell, *Meeting Places: Scientific Congresses and Urban Identity in Victorian Britain* (Farnham: Ashgate, 2013); on Scandinavian scientific congresses, see Dan C. Christensen, *Hans Christian Ørsted: Reading Nature's Mind* (Oxford: Oxford University Press, 2013), 517–18.
14. E.g., Karel Ignac Tham, *Obrana gazyka českého protí zlobiwym geho vtrhačům, též mnohym wlastencům, w cwičenj se w něm liknawym a nedbalym sepsaná* (Prague: J. F. ze Schönfeldu, 1783); see its partial translation by Derek Paton, "Apology of the Czech Language against Slanderers as Well as Many Countrymen Negligent and Indolent in the Practice of the Language," in *Discourses of Collective Identity in Central and Southeast Europe 1770–1945*, vol. 1, *Late Enlightenment—Emergence of the Modern "National Idea,"* ed. Balázs Trencsényi and Michal Kopeček (Budapest: Central European University Press, 2006), 205–9.
15. Antonín Kostlán, Jan Janko, and Ladislav Niklíček, "Prosazování myšlenky akademie věd v českých zemích raného nvověku (16.–18. století)," in *Bohemia*

docta: K historickým kořenům vědy v českých zemích, ed. Alena Míšková, Martin Franc, and Antonín Kostlán (Prague: Academia, 2010), 145–71; and Rita Krueger, *Czech, German, and Noble: Status and National Identity in Habsburg Bohemia* (Oxford: Oxford University Press, 2009).

16. See esp. Marlies Raffler, *Museum—Spiegel der Nation? Zugänge zur historischen Museologie am Beispiel der Genese von Landes- und Nationalmuseen in der Habsburgermonarchie* (Vienna: Böhlau, 2007).
17. Claudia Schweizer, "Migrating Objects: The Bohemian National Museum and Its Scientific Collaborations in the Early Nineteenth Century," *Journal of the History of Collections* 18, no. 2 (2006): 187–99.
18. Quoted in Joseph Kalousek, *Geschichte der Königlichen Böhmischen Gesellschaft der Wissenschaften: Sammt einer kritischen Übersicht ihrer Publicationen aus dem Bereiche der Philosophie, Geschichte und Philologie, aus Anlass des hundertjährigen Jubelfestes der Gesellschaft 1884* (Prague: Königlich Böhmische Gesellschaft der Wissenschaften, 1885), 43.
19. Krueger, *Czech, German, and Noble*, 104–5; and Magdaléna Pokorná, "Královská česká společnost nauk," in Míšková, Franc, and Kostlán, *Bohemia docta*, 58–144.
20. Kalousek, *Geschichte*.
21. On the institute, see Janusz Albin, ed., *Z dziejów Zakładu Narodowego imienia Ossolińskich: Studia i materiały* (Wrocław: Ossolineum, 1978).
22. Harry Bresslau, *Geschichte der Monumenta Germaniae Historica* (Hannover: Hahn, 1921), 99–100. The other two were none other than Jernej Kopitar and Klemens Metternich.
23. See the discussion of the development of professional historiography in Galicia in Burkhard Wöller, *"Europa" als historisches Argument: Nationsbildungsstrategien polnischer und ukrainischer Historiker im habsburgischen Galizien* (Bochum: Verlag Dr. Dieter Winkler, 2014), 29–80.
24. Jan Surman, "Objektivität, Bestandsaufnahme, Territorium: Galizische Quelleneditionen und ihre Verortung zwischen wissenschaftlichen und ideologischen Ansprüchen," in *Geschichtsforschung in Deutschland und Österreich im 19. Jahrhundert: Ideen—Akteure—Institutionen*, ed. Christine Ottner and Klaus Ries (Stuttgart: Franz Steiner, 2014), 202–3.
25. Bogdan Suchodolski, ed., *Historia nauki polskiej*, vol. 3, *1795–1862* (Wrocław: Zakład Narodowy im. Ossolińskich, Polska Akademia Nauk, Zakład Historii Nauki, Oświaty i Techniki, 1977); the freedoms within these regions, as compared to Galicia, were recently reevaluated by Maciej Janowski and Jerzy Jedlicki in Jerzy Jedlicki, ed., *Dzieje inteligencji polskiej do roku 1918*, vol. 1, Maciej Janowski, *Narodziny inteligencji (1750–1831)*, and vol. 2, Jerzy Jedlicki, *Błędne koło 1832–1864* (Warsaw: Neriton, 2012).
26. Daniel Beauvois, *Wilno: Polska stolica kulturalna zaboru rosyjskiego 1803–1832* (Wrocław: Wydawnictwo Uniwersytetu Wrocławskiego, 2010).
27. Bernhard Januszewski, "Geneza Katedry języków i literatur słowiańskich we Wrocławiu," *Rocznik Wrocławski* 7/8 (1963/64): 12–76; and *Pamiętnik*

Towarzystwa Literacko-Słowiańskiego przy Uniwersytecie Wrocławskim wydany w roku złotego jubileuszu (Wrocław: Nakładem Drukarni Polskiej [Jan Szymański], 1886).

28. The Studium Ruthenum was officially named the Provisional Educational Institution in the Ruthenian Language (Provisorische Lehranstalt in Ruthenischer Sprache). See Ol'ha Paljoch, "Ukraïns'ke knyhovydannja u L'vivi XIX sr.: Rol' drukaren' Stavropihijśkoho Instytutu ta Naukovoho tovarystva im. Ševčenka," *Zapysky L'vivs'koï naukovoï biblioteky im. V. Stefanyka: Zbyrnyk naukovych prac'* 16 (2008) 1: 54–72, esp. 54–58; on the institute's influence in the early nineteenth century, see Iryna Vasylivna Orlevyč, "Dijal'nist' L'vivs'koho Stavropihijs'koho Instytutu (kinec' XVIII–60-i rr. XIX st.)" (PhD diss., Ivan Kryp'jakevyč Institute of Ukrainian Studies, Ukrainian National Academy of Science, 2000); on the Dormition Brotherhood's role, see Iaroslav Isaievych, *Voluntary Brotherhood: Confraternities of Laymen in Early Modern Ukraine* (Edmonton: Canadian Institute of Ukrainian Studies, 2006).
29. For details see Michael Moser, "Movnyj svit 'Studium ruthenium,'" in *Ucrainica I. Současná ukrajinistika: Problémy jazyka, literatury a kultury. K 65. narozeninám prof. Josefa Anderše* (Olomouc: Universita Palackého, 2004), 316–25.
30. See [Władysław Zawadzki], *Literatura w Galicji (1772–1848): Ustęp z pamiętników Władysława Zawadzkiego* (L'viv: Nakładem Władysława Webera, 1878), 109–10.
31. *Rusalka Dněstrovaja. Ruthenische Volkslieder* (Buda: Pysmom Korol. Vseučylyšča Peštanskoho, 1837). The authors were Jakiv Holovac'kyj (Яків Головацький), Markiyan Šaškevyč (Маркіян Шашкевич), and Ivan Vahylevyč (Іван Вагилевич).
32. Michael Moser, "Die sprachliche Erneuerung der galizischen Ukrainer zwischen 1772 und 1848/1849 im mitteleuropäischen Kontext," in *Contemporary Cultural Studies in Central Europe*, ed. Ivo Pospíšil and Michael Moser (Brno: Ústav slavistiky Filozofické fakulty Masarykovy univerzity v Brně, 2004), 106–7.
33. See Jan Kozik, *The Ukrainian National Movement in Galicia, 1815–1849* (Edmonton: Canadian Institute of Ukrainian Studies, University of Alberta, 1986); and Iaroslav Hrytsak, "Ruslan, Bohdan and Myron: Three Constructed Identities among Galician Ruthenians/Ukrainians, 1830–1914," in *Extending the Borders of Russian History: Essays in Honor of Alfred J. Rieber*, ed. Marsha Siefert (Budapest: Central European University Press, 2003), 97–112.
34. Gábor Palló, "Scientific Nationalism: A Historical Approach to Nature in Late Nineteenth-Century Hungary," in Ash and Surman, *Nationalisation*, 105–6.
35. R. J. W. Evans, "Széchenyi and Austria," in *History and Biography: Essays in Honour of Derek Beals*, ed. T. C. W. Blanning and David Cannadine (Cambridge: Cambridge University Press, 2002), 123–26.
36. Kolowrat, "An die vaterländischen Freunde," 1100.
37. Editorial, *Časopis Společnosti wlastenského museum w Čechách* 1, no. 1 (1827): 4.
38. Editorial, 4–5.

39. "Nachricht über die Fortsetzung dieser Zeitschrift im J. 1832," *Jahrbuch des Böhmischen Museums für Natur und Länderkunde, Geschichte, Kunst und Literatur* 2, no. 4 (1831).
40. *Wýtah z aučtů českého Museum, týkajících se příjmů a wydání Matice české roku 1847* (Prague: Tiskem c. k. dworní knihtiskárny synů Bohumila Háze, 1848).
41. Meister, *Geschichte der Akademie.*
42. Lore Sexl, "Gottfried Wilhelm Leibniz und die geplante Kaiserlicher Akademie der Wissenschaften in Wien," in *Theoria cum praxi: Aus der Welt des Gottfried Wilhelm Leibniz*, ed. Hermann Hunger (Vienna: Verlag der Österreichischen Akademie der Wissenschaften, 2012), 69–239.
43. Imperial decree establishing the Imperial Academy of Sciences and Arts, §1, published in *Wiener Zeitung*, 17 May 1847.
44. Kadletz-Schöffel, *Metternich und die Wissenschaften*, 299.
45. On Purgstall see Paula Sutter Fichtner, "History, Religion, and Politics in the Austrian *Vormärz*," *History and Theory* 10, no. 1 (1971): 33–48.
46. Alfons Huber, *Geschichte der Gründung und der Wirksamkeit der Kaiserlichen Akademie der Wissenschaften während der ersten fünfzig Jahre ihres Bestandes* (Vienna: C. Gerold's Sohn, 1897), 79.
47. Helmut Schelsky, *Einsamkeit und Freiheit, Idee und Gestalt der deutschen Universität und ihrer Reformen* (Hamburg: Rowohlt, 1963), 18.
48. On universities in the German Confederation, see Andreas C. Hofmann, "Studium, Universität und Staat in Bayern 1825 bis 1848: Eine Skizze der Universitätspolitik Ludwigs I," *aventinus bavaric* 2 (2006), http://www.aventinus-online.de/no_cache/persistent/artikel/7760/; and Andreas C. Hofmann, "Deutsche Universitätspolitik im Vormärz zwischen Zentralismus, 'Transstaatlichkeit' und 'Eigenstaatlichkeitsideologien' (1815/19 bis 1848)" (PhD diss., Ludwig-Maximilians-Universität München, 2014).
49. Aleksandr Andreev, *Universitet v Rossijsksj imperii XVIII—pervoj poloviny XIX veka* (Moscow: Rosspen, 2012).
50. Joseph II, "Resolution zum Vortrag der Studienhofkommission. v. 25.11.1782," quoted in Ernst Wangermann, *Aufklärung und staatsbürgerliche Erziehung: Gottfried van Swieten als Reformator des österreichischen Unterrichtswesens 1781–1791* (Vienna: Verlag für die Geschichte und Politik, 1978), 25–26.
51. Quoted in [Charles Sealsfield], *Austria as It Is, or Sketches of Continental Courts, by an Eye-Witness* (London: Hurst, Chance, 1828), 75.
52. Not all graduates were doctors, as other degrees were awarded for, e.g., surgeons or pharmacists. See Joseph Schneller, *Historische Darstellung der Entwickelung der medicinischen Facultät zu Wien, nebst einer kurzen Uebersicht der wissenschaftlichen Leistungen des medicinischen Doctoren-Collegiums* (Vienna: Zamarski, 1856).
53. Wolfgang-Rüdiger Mell, "Ein rechtsgeschichtlicher Beitrag zur 'Selbstbestimmung und Fremdbestimmung der österreichischen Universitäten,'" in *Selbstbestimmung und Fremdbestimmung der österreichischen Universitäten:*

Ein Beitrag zur Soziologie der Universität, ed. Marina Fischer and Hermann Strasser (Vienna: Abteilung Soziologie des Instituts für Höhere Studien und Wissenschaftliche Forschung, 1973), 4.

54. Waltraud Heindl, "Universitätsreform—Gesellschaftsreform: Bemerkungen zum Plan eines 'Universitätsorganisationsgesetzes' in den Jahren 1854/55," *Mitteilungen des österreichischen Staatsarchivs* 35 (1983): 139.
55. Sadek Brim, *Universitäten und Studentenbewegung in Russland im Zeitalter der Großen Reformen 1855–1881* (Frankfurt am Main: Peter Lang, 1985), esp. 25, 32.
56. *Studiendirektoren* were the state-employed supervisors of faculties, mostly concerned with ensuring the conformity of education with the empire's policies; they were responsible for communication with the government.
57. For the philosophical faculty before 1848, see also Erika Rüdegger, "Die philosophischen Studien an der Wiener Universität 1800 bis 1848" (PhD diss., University of Vienna, 1964); and Peter Stachel, "Das österreichische Bildungssystem zwischen 1749 und 1918," in *Geschichte der österreichischen Humanwissenschaften*, vol. 1, *Historischer Kontext, wissenschaftssoziologische Befunde und methodologische Voraussetzungen*, ed. Karl Acham (Vienna: Passagen, 1999), 115–46; see also Eva S. Widmann, "Vormärzliches Studium im Spiegel autobiographischer Quellen," in *Österreichische Bildungs- und Schulgeschichte von der Aufklärung bis zum Liberalismus*, ed. Gerda Mraz (Eisenstadt: Institut für Österreichische Kulturgeschichte, 1974), 118–37.
58. Richard Meister, *Entwicklung und Reformen des österreichischen Studienwesens*, vol. 2, *Dokumente* (Vienna: Böhlhaus Nachf., Kommissionsverlag der Österreichischen Akademie der Wissenschaften, 1963), 33.
59. The best-known cases were Bernhard Bolzano (Prague), Andreas Benedict Freimoser (Innsbruck), and Leopold Rembold (Vienna). See Eduard Winter, *Der Bolzanoprozess: Dokumente zur Geschichte der Prager Karlsuniversität im Vormärz* (Brno: Rohrer, 1944); see also on the political atmosphere at the universities Karel Litsch, "Zur Rechtsstellung der Prager Universitätsprofessoren in der ersten Hälfte des 19. Jahrhunderts," in *Bildungsgeschichte, Bevölkerungsgeschichte, Gesellschaftsgeschichte in den böhmischen Ländern und in Europa: Festschrift für Jan Havránek zum 60. Geburtstag*, ed. Hans Lemberg, Karel Litsch, Richard Georg Plaschka, and Györgi Ránki (Vienna: Verlag für Geschichte und Politik, 1988), 3–17.
60. Meister, *Entwicklung und Reformen*, 58–68.
61. Rüdegger, "Die philosophischen Studien," 55.
62. Clemens Novak and Martin Haidinger, *Der Anteil der organisierten Studentenschaft an der Märzrevolution 1848 in Wien und die Bedeutung dieses Ereignisses für die Korporationen von Heute* (Vienna: Akademischer Corporations-Club, 1999), 13–14.
63. Leo Thun-Hohenstein, for instance, traveled in 1833–35 to Great Britain and France, where he met, among others, Sarah and John Austin, Adam Mickiewicz, and Alexis de Tocqueville; see Dieter Halbwidl, "Life and Times of Count Leo Thun," *Mensch—Wissenschaft—Magie* 30 (2013): 114–16; and Hana Fořtová

and Doubravka Olšáková, "Úvodní studie," in *Lev Thun—Alexis de Tocqueville (Korespondence 1835–1856)*, ed. Hana Fořtová and Doubravka Olšáková (Prague: Oikúmene, 2011), 21–22.

64. Alfred Arneth, *Aus meinem Leben*, vol. 1, *1819–1849* (Vienna: Adolf Holzhausen, 1891), 200.
65. Ferdinand Grassauer, "Universitätsbibliothek," in *Geschichte der Wiener Universität von 1848 bis 1898: Als Huldigungsfestschrift zum 50-jähr. Regierungsjubiläum seiner k.k. apostolischen Majestät des Kaisers Franz Josef I*, ed. Akademischer Senat der Wiener Universität (Vienna: Alfred Hölder, 1898), 367–80.
66. Salomon Frankfurter, *Graf Leo Thun-Hohenstein, Franz Exner und Hermann Bonitz: Beiträge zur Geschichte der österreichischen Unterrichtsreform* (Vienna: Alfred Hölder, 1893), 10.
67. Ludwik Finkel, "Historya Uniwersytetu Lwowskiego do r. 1869," in *Historya Uniwersytetu Lwowskiego*, ed. Ludwik Finkel and Stanisław Starzyński (L'viv: C. K. Uniwersytet Lwowski, 1894), 243; and Isabel Röskau-Rydel, *Kultur an der Peripherie des Habsburger Reiches: Die Geschichte des Bildungswesens und der kultureller Einrichtungen in Lemberg von 1772 bis 1848* (Vienna: Harrassowitz, 1993), 185.
68. Kamilla Mrozowska, "Okres ucisku i daremnych prób wyzwoleńczych, 1833–1850," in *Dzieje Uniwersytetu Jagiellońskiego w latach 1765–1850*, ed. Mirosława Chamcówna and Kamilla Mrozowska (Cracow: Uniwersytet Jagielloński, 1965), vol. 2, pt. 1, 183.
69. Mrozowska, "Okres ucisku."
70. There are no data on this question, however, except for the University of Innsbruck. Gerhard Oberkofler, *Die Rechtslehre in italienischer Sprache an der Universität Innsbruck* (Innsbruck: Kommisionverlag der Österreichischen Kommisionsbuchhandlung, 1975), 9.
71. Helmut Engelbrecht, *Geschichte des österreichischen Bildungswesens*, vol. 3, *Von der frühen Aufklärung bis zum Vormärz* (Vienna: Österreichischer Bundesverlag, 1984), 283. See also for an exemplary description of the procedure Zoran Konstantinović, "'In der praktischen Philosophie ist ihm die Menschenwürde das Wesentlichste …': Zu Franz Miklošič' Bewerbung um die Innsbrucker Lehrkanzel für Philosophie [1838/39]," *Österreichische Osthefte* 33 (1991): 95–103.
72. The Habsburg army kept these rules well beyond 1848; see István Deák, *Beyond Nationalism: A Social and Political History of the Habsburg Officer Corps 1848–1918* (Oxford: Oxford University Press, 1990).
73. "Hofkanzley-Dekret vom 30. November 1810 an sämmtliche Länderstellen," reprinted in Johann G. Megerle von Mühlfeld, ed., *Handbuch für alle kaiserlich-königlichen, ständischen und städtischen Beamten, deren Wittwen und Waisen: Oder Darstellung aller ihnen durch die neuesten allerhöchsten Gesetze vom Jahre 1806 bis 1822 zustehenden Rechte und obliegenden Verbindlichkeiten* (Vienna: Mösle, 1824), 1:78–80.

74. Franz Johannes Richter, "Über öffentliche Erziehungs- und Unterrichtsanstalten in Europa, mit besonderen Rücksicht auf das Schul- und Studienwesen des österreichischen Kaiserstaates," *Österreichische Blätter für Litteratur und Kunst* 2, no. 142 (1845): 1108.
75. Josef Alois Jüstel, quoted in Waltraud Heindl, "Universitäten und Eliten im österreichischen Vormärz," *Études danubiennes* 3, no. 2 (1987): 123.
76. Herbert H. Egglmaier, *Ernst Freiherr von Feuchtersleben als Bildungspolitiker* (Klagenfurt: Abteilung für Historische und Vergleichende Pädagogik, Universität Klagenfurt, 2000), 27–28.
77. "Studienhofkommissionsdekret vom 20. September 1811, Z. 1641 . . . betreffend die Einrichtung einer Pflanzschule der Lehrer aus der Arzneikunde," reprinted in Beck von Mannagetta and Kelle, *Die österreichischen Universitätsgesetze*, 205–6 (no. 223); and "Auszug aus dem Dekret der Studienhofkommission vom 27. September 1811, Z. 1654 Bd. 37, Nr. 42, S. 136 . . . betreffend die Einrichtung einer Pflanzschule künftiger Lehrer für Gymnasien und höhere Lehranstalten," reprinted in Beck von Mannagetta and Kelle, *Die österreichischen Universitätsgesetze*, 206–8 (no. 224).
78. "Auszug aus dem Dekret," §2.6.
79. *Taschenbuch der Wiener k.k. Universität für das Jahr 1847* (Vienna: Besck'sche Universitätsbuchhandlung, 1847), father's occupation based on data in biographies in *Allgemeine Deutsche Biographie*, 56 vols. (Munich: Duncker & Humblot, 1875–1912), *Das Österreichische Biographische Lexikon 1815–1950*, 14 vols. (Vienna: Österreichische Akademie der Wissenschaften, 1957–2015), and *Neue Deutsche Biographie*, 26 vols. (Berlin: Duncker & Humblot, 1953–2016).
80. Leon Wachholz, "Bryk Antoni," in *Polski Słownik Biograficzny*, ed. Władysław Konopczyński (Cracow: Polska Akademia Umiejętności, Gebethner i Wolff, 1937), 3:27.
81. In comparison to low-ranking state officials, who, according to Isabel Röskau-Rydel, were adopted into the Polish urban majority, professors participated more in the German-language public sphere of the city. Isabel Röskau-Rydel, *Niemiecko-austriackie rodziny urzędnicze w Galicji 1772–1918: Kariery zawodowe—środowisko—akulturacja i asymilacja* (Cracow: Wydawnictwo Naukowe Uniwersytetu Pedagogicznego, 2011). For representative examples, see Burkhard Wöller, "Ludwig Zehnmark (1753–1814)," in *Złota księga historiografii lwowskiej XIX i XX w.*, ed. Jerzy Maternicki, Paweł Sierżęga, and Leonid Zaszkilniak (Rzeszów: Wydawnictwo Uniwersytetu Rzeszowskiego, 2014), 2:11–20; and Burkhard Wöller, "Joseph Mauss (1778–1858)," in Maternicki, Sierżęga, and Zaszkilniak, *Złota księga*, 21–30.
82. Mychajlo Kril', "Pedahohična i naukova dijal'nist' Jozefa Maussa," in *Wielokulturowe środowisko historyczne Lwowa w XIX i XX w. / Bahatokul'turne istoryčne seredovyšče L'vova v XIX i XX stolittjach*, ed. Jerzy Maternicki and Leonid Zaszkilniak (L'viv: L'vivs'kij nacional'nyj un-t im. Ivana Franka, Žešuvs'kyj un-t., 2006), 4:98.
83. On the political participation of students in the revolution, see Jan Havránek,

"Karolinum v revoluci 1848," *AUC-HUCP* 26, no. 2 (1986): 35–75; Thomas Maisel, *Alma Mater auf den Barrikaden: Die Universität Wien im Revolutionsjahr 1848* (Vienna: Wiener Universitätsverlag, 1998); Peter Molisch, ed., *Eduard Neussers studentische Erinnerungen aus dem Jahre 1848* (Vienna: Sonderabdruck aus den Mitteilungen des Vereines für Geschichte der Stadt Wien, Vol. XIII/XIV, 1933); and Finkel, "Historya Uniwersytetu Lwowskiego," 297–306.

84. After a one-day dictatorship, Wiszniewski was deposed by Edward Dembowski. Mrozowska, "Okres ucisku," 208; and Karol Lewicki, "Katedra Literatury Polskiej na Uniwersytecie Jagiellońskim w latach 1803–1848," in *Dzieje Katedry Historii Literatury Polskiej*, ed. Tadeusz Ulewicz (Cracow: Wydawnictwo Literackie, 1966), 68–69.
85. Jiří Štaif, "Palackýs Partei der tschechischen Liberalen und die konservative Variante der böhmischen Politik," in *1848/49 Revolutionen in Ostmitteleuropa: Vorträge der Tagung des Collegiums Carolinum in Bad Wiessee vom 30. November bis 1. Dezember 1990*, ed. Rudolf Jaworski and Robert Luft (Munich: Oldenburg, 1996), 65.
86. It is not clear, though, how much influence the legions had; they were badly armed and undisciplined. See Przemysław Majewski, "Gwardie narodowe w Galicji w czasie Wiosny Ludów," in *Per aspera ad astra: Materiały z XVI Ogólnopolskiego Zjazdu Historyków Studentów w Krakowie 16–20 kwietnia 2008*, vol. 7, *Historia Polski pod zaborami*, ed. Adam Świątek (Cracow: Koło Naukowe Historyków Uniwersytetu Jagiellońskiego, AT Group, 2008), 31–48.
87. On the association, see Wilhelm Brauneder, *Leseverein und Rechtskultur: Der Juridisch-Politische Leseverein zu Wien 1840–1990* (Vienna: Manz, 1992).
88. Hans Heiss and Thomas Götz, *Am Rand der Revolution: Tirol 1848/49* (Bozen: Folio, 1998).
89. Finkel, "Historya Uniwersytetu Lwowskiego," 297.
90. Laszlo Szögi, "Der Ausbau des modernen Universitätsnetzes in der bürgerlichen Gesellschaft Ungarns 1848–1918," *Mensch—Wissenschaft—Magie* 26 (2009): 93–106.
91. Moritz Csáky, "Altes Universitätsviertel: Erinnerungsraum, Gedächtnisort," in *Die Verortung von Gedächtnis*, ed. Moritz Csáky and Peter Stachel (Vienna: Passagen, 2001), 257–77; and Finkel, "Historya Uniwersytetu Lwowskiego," 300–306.
92. On Hungary, see László Molnár, "A pesti egyetemi orvoskar 1848/49-ben," *Semmelweis Egyetem* 9 (15 March 2008), http://www.orvostortenelem.hu/tankonyvek/tk-05/pdf/3.4.3/molnar_pesti_egyetemi_orvoskar.pdf. On the other officials, see Waltraud Heindl, *Josephinische Mandarine: Bürokratie und Beamte in Österreich*, vol. 2, *1848 bis 1914* (Vienna: Böhlau, 2013), esp. 61–63.
93. "Adres do Cesarza. Zgromadzenie Profesorów i uczniów Uniwersytetu Jagiell. i wszystkich Zakładów naukowych w mieście Krakowie i jego Okregu obecnie istniejących, zanosi do Waszéj Cesarsko Królewskiéj Mości nastepujące prośby," in *Ważniejsze dokumenta odnoszące się do swojego udziału w sprawach i losach*

Zakładów naukowych b. W. M. Krakowa i jego Okręgu, ed. Józef Brodowicz (Cracow: Uniwersytet Jagielloński, 1874), 79.

94. "Adres do Cesarza," 79.
95. Quoted in Mrozowska, "Okres ucisku," 215.
96. Mrozowska, "Okres ucisku," 215.
97. Finkel, "Historya Uniwersytetu Lwowskiego," 303; Volodymyr Kačmar, "Problema ukraïns'koho universytetu u Lv'ovi v kinci XIX-na počatku XX st.: Suspil'no-polityčnyj aspekt" (PhD diss., Ivan Franko State University in L'viv, 1999), 36.
98. See the appellation of deputy Alexander Borkowski from L'viv during the Constitutive Imperial Congress in Kroměříž (Kremsier), online in *Officielle stenographische Berichte über die Verhandlungen des österr. Reichstages: Neunundsiebzigste (XXVII.) Sitzung des österreichischen constituirenden Reichstages in Kremster am 26. Jänner 1849*, http://www.psp.cz/eknih/1848urrs/stenprot/079schuz/s079001.htm.
99. Havránek, "Karolinum v revoluci 1848," 38–39.
100. "All doctors are allowed from now on, after a provisional registration by the academic senate, to teach scientific lectures." Count Stadion to the Academic Senate, 2 April 1848, announcing the ministerial decree of 31 March 1848, Z. 252. ÚDAUK, Akademický senát Karlo-Ferdinandovy univerzity, Inv.č. 227, Kart. 149. See also ministerial confirmation in "Cirkular-Verordnung des k.k. böhmischen Guberniums vom 5. April 1848: In Bezug auf die Lehr- und Lernfreiheit, Z. 89," reprinted in *Österreichische Zeitschrift für Rechts- und Staatswissenschaft* 3, no. 4 (1848): 92.
101. Count Stadion to the Academic Senate, 2 April 1848.
102. "Letter sent by František Palacký to Frankfurt" (1848), trans. William Beardmore, *Slavonic and East European Review* 26 (1947/48): 303–8; it was originally published anonymously as "Hlas o připojení-se Rakauska k zemi německé," *Národní nowiny*, no. 10 (1848): 37; no. 14 (1848): 45–46.
103. *Der Humorist*, 11 May 1848, 466.
104. *Allgemeine Österreichische Zeitung*, quoted in Jiří Kořalka, *František Palacký (1798–1876): Der Historiker der Tschechen im österreichischen Vielvölkerstaat* (Vienna: Verlag der österreichischen Akademie der Wissenschaften, 2007), 282. Palacký claimed that Franz Exner informed him later that if he had become the minister, his life would have been seriously threatened.
105. *Der Humorist*, 11 May 1848.
106. [Franz Freiherr von Sommaruga], *Rede des Ministers des öffentlichen Unterrichtes Dr. Franz Freiherrn von Sommaruga gehalten in der Aula der Wiener Universität am 30. März 1848*, 1848.
107. [Ernst Freiherr von Feuchtersleben], anonymous commentary in the *Wiener Zeitung*, 15 August 1848, quoted in Egglmaier, *Ernst Freiherr von Feuchtersleben*, 12–13.
108. Egglmaier, *Ernst Freiherr von Feuchtersleben*, 27–28; and Mitchell G. Ash, "Die Universität Wien in politischen Umbrüchen des 19. und 20. Jahrhunderts,"

in *Universität—Politik—Gesellschaft: 650 Jahre Universität Wien. Aufbruch in das neue Jahrhundert*, ed. Mitchell G. Ash and Josef Ehmer (Vienna: Vienna University Press; Göttingen: V&R Unipress, 2015), 51.

109. "Ministerial-Erlass vom 11. December 1848, Z. 8309 . . . womit zur Normirung des Verfahrens bei Wiederbesetzung erledigter Lehrkanzeln an Universitäten, Gymnasien, technischen Instituten und Realschulen eine provisorische Anordnung getroffen wird," reprinted in Georg Thaa, ed., *Sammlung der für die österreichischen Universitäten giltigen Gesetze und Verordnungen* (Vienna: G. J. Manz'sche Buchhandlung, 1871), 109–10 (no. 63).
110. Der akademische Senat der k.k. deutschen C.-F. Universität, *Denkschrift über die Rechtstellung der Professoren und der Universitäten* (Prague, 1896).
111. "Ministerial-Erlass vom 19. December 1848, Z. 8175 . . . womit eine provisorische Anordnung bezüglich der Habilitierung der Privat-Dozenten kundgemacht wird," reprinted in Thaa, *Sammlung*, 129–31 (no. 91).
112. The faculty could, however, choose not to require the public colloquium and test lecture.
113. What constituted a "subject" (*Fach*), and what its boundaries should be, was subsequently a matter of numerous negotiations, with the ministry having the decisive vote.
114. The regulation in Prague on the *Lehrfreiheit* of Privatdozenten was binding: "the freedom of teaching will be acknowledged in a way that, after preliminary registration by the academic senate, it is allowed for all doctors to hold scientific lectures at the faculties which issue the doctoral degree." "Cirkular-Verordnung," 92, here §1.
115. See the critique of the philosophical faculty in Innsbruck for allowing the unapproved instructors Rudolf Kink and Josef Daum to teach in 1849. UAI, Z. 355, 24.1.1849 (ministerial number Z. 8019).
116. On Miklošič, see Al[eksandr] Kotschubinský [Kočubinskij] [also Александр Кочубинский], "Miklosich und Šafárik: Ein Beitrag zu ihren wechselseitigen Beziehungen," *Archiv für slavische Philologie* 25 (1903): 621–27. On Čelakovský, see Artur Závodský, *František Ladislav Čelakovsky* (Prague: Melantrich, 1982), 481–86. On Holovac'kyj, see DALO, F. 26, Op. 5, Spr. 437, PA Glowacki. Holovac'kyj's nomination as auxiliary professor (*supplent*) was announced on 27 November 1848, and that for full professor on 13 December 1848; on Šafárik's assessment of Holovac'kyj, see Mychajlo Kril', "Z čes'koï korespondenciï Jakova Holovac'koho," *Problemy slov'janoznavstva* 50 (1999): 217–22.
117. AT-OeStA/AVA Unterricht UM allg. Akten 1216, PA Celakovsky, Z. 2388/371, 11 April 1849.
118. Franz Exner, *Die Stellung der Studierenden auf der Universität: Eine Rede gehalten an der k.k. Universität zu Prag* (Prague: Gottlieb Haase Söhne, 1837).
119. Franz Exner, "Entwurf der Grundzüge des öffentlichen Unterrichtswesens in Oesterreich," *Wiener Zeitung*, 21 July 1848. See also §62: "the function of the university is to forge the education in general sciences, prepare for the civil service on the basis of *Fachwissen* [specialized knowledge], and finally help

in the process of betterment of the youth's character through scholarship and discipline."

120. On gymnasia, see Rainer Leitner, "Das Reformerwerk von Exner, Bonitz und Thun: Das österreichische Gymnasium in der zweiten Hälfte des 19. Jahrhunderts—Kaderschmiede der Wiener Moderne," in *Zwischen Orientierung und Krise: Zum Umgang mit Wissen in der Moderne*, ed. Sonja Rinhofner-Kreidl (Vienna: Böhlau, 1998), 17–69.
121. Franz Xaver Krones Ritter von Marchland, *Geschichte der Karl-Franzens Universität in Graz: Festgabe zur Feier ihres dreihundertjährigen Bestandes* (Graz: Leuschner und Lubensky, 1886), 545.
122. Peter Wozniak, "Count Leo Thun: A Conservative Savior of Educational Reform in the Decade of Neoabsolutism," *Austrian History Yearbook* 26 (1995): 61–81.

Chapter 2

1. "To the minister of education. / I have to announce a suicide. / The minister of religion has murdered the minister of education." Franz Grillparzer on the Concordat of 1855, quoted in Friedrich Prinz, "Graf Leo Thun," in *Lebensbilder zur Geschichte der böhmischen Länder*, ed. Karl Bosl (Munich: Oldenbourg, 1976), 162.
2. Quotation from Goethe's *Urfaust* in Georg-Emmanuel Haas, *Ueber den Zustand der österr. Universitäten mit besonderer Beziehung auf die Wiener Hochschule* (Augsburg: Schmid, 1853), 34; the words in brackets are amended from Goethe's text and do not appear in Haas's original. English translation from Johann Wolfgang von Goethe, *Faust: A Tragedy*, trans. Taylor Bayard (London: Ward, Lock, 1889), 250.
3. For detailed information on Thun-Hohenstein and his Catholic background, see Gottfried Mayer, *Österreich als katholische Grossmacht: Ein Traum zwischen Revolution und liberaler Ära* (Vienna: Verlag der Österreichischen Akademie der Wissenschaften, 1989); and Alphons Lhotsky, "Das Ende des Josephinismus: Epilegomena zu Hans Lenzes Werk über die Reformen des Ministers Grafen Thun," *Mitteilungen des österreichischen Staatsarchivs* 15 (1962): 526–49.
4. Christof Aichner, "Franz Exner: Professor für Philosophie, Mitschöpfer der Universitätsreform nach 1848," in Ash and Ehmer, *Universität—Politik—Gesellschaft*, esp. 187–88.
5. See Aichner and Mazohl, *Die Thun-Hohenstein'schen Universitätsreformen.*
6. See *Plan einer zeitgemäßen Reform der österreichischen Universitäten: Entworfen und dem Ministerium des Unterrichts überreicht von dem Collegium der Professoren der medic-chirurg. Studien der k.k. Wiener Hochschule* (Vienna, 1848), 57; and the ideas at the Jagiellonian University, referred to in the previous chapter.

7. *Entwurf des Österreichischen Reichstages welcher in der Zeit vom 22. Juli 1848 bis 4. März 1849 getagt hat, zuerst in Wien, ab dem 22. November 1848 in Kremsier* ("Kremsier Entwurf"), §107, http://www.verfassungen.de/at/kremsier49.htm.
8. Hans Lentze, *Andreas Freiherr von Baumgartner und die Thunsche Studienreform* (Vienna: Rohrer, 1959).
9. Heindl, "Universitätsreform—Gesellschaftsreform," 147.
10. "Ministerial-Erlass vom 30. September 1849, Z. 6798 . . . womit das mit allerhöchster Entschließung vom 27. September 1849 genehmigte provisorische Gesetz über die akademischen Behörden kundgemacht wird, und Massregeln zu dessen Durchführung festgesetzt werden," reprinted in Thaa, *Sammlung*, 69–77 (no. 33); as well as other acts from the time (see Thaa, *Sammlung*).
11. Thaa, *Sammlung*, 76; see also Akademischer Senat der Wiener Universität, ed., *Geschichte der Wiener Universität von 1848 bis 1898: Als Huldigungsfestschrift zum 50-jähr. Regierungsjubiläum seiner k.k. apostolischen Majestät des Kaisers Franz Josef I* (Vienna: Alfred Hölder, 1898), 381. Exceptions to this rule were two retired professors, Hieronymus Josef/Jeroným Zeidler, a professor of dogmatic theology and later abbot of Strahov Closter (Prague, chosen rector for the academic year 1855/56), and the L'viv historian Joseph Mauss, pensioned off in 1849 and then chosen as a rector for 1851/52.
12. "Alleruntertänigster Vortrag des . . . Leo Grafen von Thun, über Studienordnung, Disziplinarordnung und Einführung von Collegiengelder an den Universitäten, 30. September 1849," reprinted in Meister, *Entwicklung*, 277–81.
13. These rules were in place from 1849 onward and were tightened in 1853; "Alleruntertänigster Vortrag des . . . Leo Grafen von Thun, über Studienordnung, Disziplinarordnung und Einführung von Collegiengelder an den Universitäten, 30. September 1849," reprinted in Meister, *Entwicklung*, 279.
14. "Die Doktorencollegien an den österreichischen Universitäten," *Akademische Monatsschrift (Deutsche Universitäts-Zeitung)*, February 1852, 57–62, and April 1852, 163–66, here 58.
15. Mell, "Ein rechtsgeschichtlicher Beitrag," 4.
16. In Vienna and Prague, the academic senate consisted of the deans and pro-deans of the *Professorencollegium*, the deans of the Doktoren-Collegium, the rector, and the prorector (and in Vienna also the provost of St. Stephen's Church). The deans of the *Professorenkollegien* had also seats in the Doktoren-Collegien and vice versa.
17. On the request to send the protocols, see §23 of "Ministerial-Erlass vom 30. September 1849, Z. 6798 . . .," reprinted in Thaa, *Sammlung*, 69–77 (no. 33); and "Ministerial-Erlass vom 28. Oktober 1857, Z. 18280 . . . betreffend die Abfassung der Sitzungs-Protokolle," reprinted in Thaa, *Sammlung*, 93–94 (no. 51). Universities whose language of administration was not German were also obliged to send translations; see comment to "Ministerial-Erlass vom 28. Oktober 1857, Z. 18280," in Beck von Mannagetta and Kelle, *Die österreichischen Universitätsgesetze*, 57n4 (no. 56).

18. See Richard Meister, "Lehr- und Lernfreiheit in der Thunschen Universitätsreform und in der Gegenwart in Österreich," *Anzeiger der Österreichischen Akademie der Wissenschaften, philologisch-historische Klasse* 15 (1957): 207–32.
19. "Erlaß des Ministers des Cultus und Unterrichts vom 13. Oktober 1849 . . ., Z. 416, §8, reprinted in *Allgemeines Reichs-Gesetz und Regierungsblatt für das Kaiserthum Österreich, Jahrgang 1849* (Vienna: Kaiserlich-königliche Hof- und Staatsdruckerei, 1850), 746.
20. "Alleruntertänigster Vortrag des . . . Leo Grafen von Thun, über Studienordnung, Disziplinarordnung und Einführung von Collegiengelder an den Universitäten, 30. September 1849," reprinted in Meister, *Entwicklung*, 279. See also the commentary on the introduction of Collegiengelder and their relation to the freedom of teaching and learning in *Akademische Monatsschrift*, October 1850, 466–68.
21. Specified in 1858 as "analogue and homogeneous subjects." See "Aus dem Erlasse des Ministeriums für K. u. U. vom 16. April 1858. Z. 5164. an das Professorenkollegium der rechts- und staatswissenschaftlichen Fakultät der Universität in Graz, (betreffend die Frage, unter welchen Bedingungen die Ausdehnung der Lehrbefugnisse der angestellten Professoren auf Lehrfächer, welche ihrem Nominalfache nicht homogen sind, ausgesprochen werden kann)," in Beck von Mannagetta and Kelle, *Die österreichischen Universitätsgesetze*, 132 (no. 147).
22. For example, Anton Menger's seminar on sociology at the Viennese law faculty, proposed in 1890, was rejected because the authorities apparently confused sociology and socialism, and socialist content was banned from being taught explicitly at Habsburg universities. See Carl Grünberg, "Anton Menger: Sa vie—son œuvre," *Revue d'histoire des doctrines économiques et socials* 2 (1909): 9.
23. Sommaruga, *Rede des Ministers*.
24. Exner, "Entwurf," 1, 2.
25. "Memorandum Jarckes über die Aufgaben eines Unterrichtsministers in Österreich vom 5. August 1849 (Thun Nachlaß C 133)," reprinted in Hans Lentze, *Die Universitätsreform des Ministers Graf Leo Thun-Hohenstein* (Vienna: Verlag der Österreichischen Akademie der Wissenschaften, 1962), 298.
26. "Memorandum Jarckes," 298.
27. *Die Neugestaltung*, 62.
28. Numbers taken from data printed in short statistical notes in *Akademische Monatsschrift* (*Deutsche Universitäts-Zeitung*), June 1853, 348 (Innsbruck, Graz, Vienna) and *Akademische Monatsschrift*, December 1853, 609–11 (Cracow, L'viv, Prague).
29. The number of students is from Akademischer Senat der Wiener Universität, *Geschichte der Wiener Universität*, 402. The number of instructors is from my own calculations based on the catalogs of lecturers.
30. See Thun-Hohenstein's manuscript on the "Polish question" from around 1860 in SOA Litoměřice/Děčín, Rodinný archiv Thun-Hohensteinů, A 3 XXI, D 637.
31. See also Hans Lentze, "Graf Thun und die deutsche Rechtsgeschichte," *Mitteilungen des Instituts für Österreichische Geschichte* 63 (1955): 500–521;

and Ursula Floßmann, *Österreichische Privatrechtsgeschichte*, 5th ed. (Vienna: Springer, 2005), 15.

32. While Exner was a devoted pupil of Herbart, the official memorandum of 1853 described Herbartianism as "a monstrosity and degeneration of the human intellect [*Geist*]." Lentze, *Die Universitätsreform*, 251.
33. Lentze, *Die Universitätsreform*, 217.
34. *Die Neugestaltung*, 105. Lentze claims that Thun-Hohenstein inserted these words on his own, as they contradict the ideas of Alois Flir. Lentze, *Die Universitätsreform*, 217.
35. See the query by Hermann Rosenberg directed to the University of L'viv in 1854. DALO, F. 26, Op. 7, Spr. 30, N. 208, 25 April 1854; F. 26, Op. 12, Spr. 77, Z. 288, 15 May 1854, Z. 289, 15 May 1854; AGAD, MWiO, Kart. 117u, PA Rosenberg, Z. 9458, 4 December 1854.
36. Carsten Wilke, *"Den Talmud und den Kant": Rabbinerausbildung an der Schwelle zur Moderne* (Hildesheim: Olms, 2003), 595.
37. *Allerhöchste Entschließung from 18. January 1834*, quoted in *Die Neugestaltung*, 61. The named reasons were the obligation to participate in the festive Holy Mass and the need to deal with religious matters during faculty meetings.
38. The consistory was the highest body of the university, consisting of the rector, deans, pro-deans (*Prodekane*), chancellor, and four so-called senators, that is, old and experienced members of the faculty, who were nominated by the faculty and in Vienna were required to be Catholic.
39. Franz Leander Fillafer, "Hermann Bonitz: Philologe, Mitschöpfer der Universitätsreform," in Ash and Ehmer, eds., *Universität—Politik—Gesellschaft*, 192–93.
40. See the articles published in the *Wiener Kirchenzeitung* in 1852, by both Brunner and his compatriots and adversaries. Also, in retrospect Brunner recollected the conflict and his own participation in it; see Sebastian Brunner, *Gesammelte Erzählungen und poetische Schriften*, vol. 1, *Woher? Wohin? Geschichten, Gedanken, Bilder und Leute aus meinem Leben. Neue Folge I*, 3rd ed. (Regensburg: G. J. Manz, 1890), 268–72.
41. Karl Pisa, *Ernst Freiherr von Feuchtersleben: Pionier der Psychosomatik (Literatur und Leben)* (Vienna: Böhlau, 1998), 110–12.
42. See Christoph Thienen-Adlerflycht, *Graf Leo Thun im Vormärz: Grundlagen des böhmischen Konservativismus im Kaisertum Österreich* (Vienna: Böhlau, 1967); Fořtová and Olšáková, "Úvodní studie," esp. 19–48; Radim Palouš, *Česká zkušenost: Příspěvek k dějinám české filosofie. O Komenského škole stáři, o Bolzanově významu v našem duchovním vývoji a o Masarykově filosofickém mládí—se závěrečným odkazem k Patočkovi* (Prague: Academia, 1994), 92–93; and Brigitte Mazohl-Wallnig, "Der Einfluß Bolzanos und der Bolzanisten auf die österreichische Universitätsreform der Jahre 1848/49," in *Bernard Bolzano und die Politik: Staat, Nation und Religion als Herausforderung für die Philosophie im Kontext von Spätaufklärung, Frühnationalismus und Restauration*, ed. Helmut Rumpler (Vienna: Böhlau, 2000), 221–46.

43. Leo Thun-Hohenstein, *Úwahy o nynějších poměrech hledíc zwláště k Čechám* (Prague: J. B. Kalwe, 1849). The pamphlet was published first in Czech and then "translated" into German even though Thun-Hohenstein himself stated in the introduction that he did not know Czech well enough to write in that language. According to Joseph Alexander Helfert's memoirs, Josef Jireček, later employed in Thun-Hohenstein's ministry, helped him translate the pamphlet into Czech: Joseph-Alexander Freiherr von Helfert, "Katastrofa: Vlastní zkušenosti a pamětí. (dokončení)," *Osvěta: Listy pro rozhled v uměne, věde a politice* 27, no. 8 (1897): 682–83.
44. Leo Thun-Hohenstein, *Betrachtungen über die Zeitverhältnisse, insbesondere im Hinblicke auf Böhmen* (Prague: J. G. Calve, 1849), esp. 21–39; the quotation is on p. 39.
45. Štaif, "Palackýs Partei," 65–67. Thun-Hohenstein's most vivid critic was the influential writer and journalist Karel Havlíček Borovský.
46. Thun-Hohenstein, *Betrachtungen*, 52–55. I would like to thank John Michael for helping me with the translation.
47. Pieter Judson, *Wien brennt! Die Revolution von 1848 und ihr liberales Erbe* (Vienna: Böhlau, 1998), 92; Arnošt Klíma, *Češi a Němci v revoluci 1848–1849* (Prague: Univerzita Karlova, 1988), 157–220; and Štaif, "Palackýs Partei," 62–66.
48. [František Palacký], "Adresse oder Petition des Slaven Congresses in Prag an Seine k.k. Majestät, 1848," in *Slovanský sjezd v Praze roku 1848: Sbírka dokumentů*, ed. Václav Žáček (Prague: Nakladatelství Československé akademie věd, 1958), 370–75.
49. [František Alexandr Zach], "Manifest des zu Prag versammelten Slaven-Congresses an alle europäischen Völker, 1848," in Žáček, *Slovanský sjezd*, 367.
50. Thun-Hohenstein, *Betrachtungen*, 52–55.
51. "Bericht von Thun-Hohenstein an Doblhoff, 28. Juli 1848," in Žáček, *Slovanský sjezd*, 490.
52. See, for example, ideas about introducing Czech as the second language at the Charles-Ferdinand University. Pavel Jozef Šafárik, "Myšlénky o prowedení stejného práwa českého i německého jazyka w školách českych," *Časopis českého musea* 22, no. 2 (1848): 171–97.
53. See the appellation of the deputy Alexander Borkowski from L'viv during the Constitutive Imperial Congress in Kroměříž (Kremsier), available online in the *Officielle stenographische Berichte.*
54. See Jarosław Moklak, *W walce o tożsamość Ukraińców: Zagadnienie języka wykładowego w szkołach ludowych i średnich w pracach galicyjskiego Sejmu Krajowego 1866–1892* (Cracow: Historia Iagellonica, 2004), 24–39.
55. E.g., *Německo-český slovník vědeckého názvosloví pro gymnasia a reálné školy: Od komise k ustanovení vědeckého názvosloví pro gymnasia a reálné školy* (Prague: Nákladem Kalve'ského knihkupectví; Bedřich Tempský 1853); and *Rus'ka čytanka dla nyžčoï gymnaziï* (Vienna: Nakladom Pravytels'tva, 1852).
56. See Michael Moser, "Some Viennese Contributions to the Development of Ukrainian Terminologies," in *Ukraine's Re-integration into Europe: A*

Historical, Historiographical and Political Urgent Issue, ed. Giovanna Brogi-Bercoff and Giulia Lami (Alessandria: Edizioni dell'Orso, 2005), esp. 156–75. Voljan (known as Basilius/Basil Wolan) was a Privatdozent for forensic medicine and was later appointed to a professorship at the newly opened Chernivtsi University in 1875.

57. See István Nyomárkay, "Spracherneuerungen in Mitteleuropa im 19. Jahrhundert: Versuch der Herausbildung muttersprachlicher Terminologien in den mitteleuropäischen Sprachen," *Studia Slavica* 53, no. 2 (2008): 425–40; Ihor Datsenko, "Das Wörterbuch der Juridischpolitischen Terminologie (1851) als Index der galizischen Realität," in *Galizien: Fragmente eines diskursiven Raums*, ed. Doktoratskolleg Galizien (Innsbruck: StudienVerlag, 2009), 131–41; and Adolf Kamiš, "Tschechisch-deutsche Beziehungen in der politischen Terminologie zu Beginn der konstitutionellen Zeit," in *Deutsch-tschechische Beziehungen im Bereich der Sprache und Kultur: Aufsätze und Studien*, ed. Bohuslav Havránek and Rudolf Fischer (Berlin: Akademie-Verlag, 1968), 71–83.
58. See, for example, for Hungarian scholarship, Andor Mészáros, "Maďarština na (buda)peštské univerzitě a v uherském školství v 19. století," in *Místo Národnich Jazyku ve Výuce, Vědě a Vzdělání v Habsburské Monarchii 1867–1918 / Position of National Languages in Education, Educational System and Science of the Habsburg Monarchy 1867–1918*, ed. Harald Binder, Luboš Velek, and Barbora Křivohlavá (Prague: Výzkumné centrum pro dějiny vědy, 2003), 148.
59. According to the catalogs of lecturers. It is, however, impossible to determine precisely which languages a particular scholar knew. Josef Petráň writes that in Prague only seven professors could lecture in Czech in 1849 and four in 1852. Josef Petráň, *Nástin dějin filozofické fakulty Univerzity Karlovy v Praze (do roku 1948)* (Prague: Univerzita Karlova, 1984), 151.
60. Jan Janko and Soňa Štrbáňová, "Uplatnění nového českého přírodovědného názvosloví na českých vysokých školách v průběhu 19. století," in Binder, Velek, and Křivohlavá, *Místo*, 303. Jan Helcelet named similar reasons for ceasing lecturing in Czech at the technical academy in Brno; see Jiří Malíř, "Otázka založení druhé české univerzity a její nacionální a politická instrumentalizace," in Binder, Velek, and Křivohlavá, *Místo*, 251, 253. In 2016 this issue still divided scholars concerned with education in Prague (I thank Luboš Velek for his insights in our discussions).
61. "107 Hörer der Medizin an der k. k. Universität zu Prag . . .," in *Stenografická zpráva XXXV. sezení třetího ročního zasedání sněmu českého od roku 1861, dne 11. května 1864 / Stenographischer Bericht über die XXXV. Sitzung der dritten Jahres-Session des böhmischen Landtages vom Jahre 1861, am 11. Mai 1864* (=Stenographische Berichte des böhmischen Landtages. Dritte Jahressession 1864. Stenografické zprávy sněmu Českého. Třetí zasedání 1864) (Prague: Statthalterei-Buchdruckerei, 1864), 21.
62. "107 Hörer."
63. AT-OeStA/AVA Unterricht UM allg. Akten 1189, Z. 200, 21 March 1852; Z. 115, 26 March 1852.

64. The whole situation seems to have been provoked by Antoni Walewski, who after his appointment met with hostility at the faculty and sent denunciations of colleagues who had previously opposed his professorship. The straw that broke the camel's back, though, was an overtly patriotic demonstration by the geographer Pol during an excursion to the Tatra Mountains. See Henryk Barycz, "Wincenty Pol jako profesor geografii na Uniwersytecie Jagiellońskim," *Polska Akademia Umiejętności: Prace komisji historii medycyny i nauk matematyczno-przyrodniczych* 3, no. 2 (1949): 104–10.
65. Waltraud Heindl, "Universitätsreform und politisches Programm: Die Sprachenfrage an der Universität Krakau im Neoabsolutismus," *Österreichische Osthefte* 20 (1978): 83.
66. Josef Batron, *Der vergessene Mähr. Verehrer Goethes, Ph. Dr. P. Thomas Bratranek OSA, Professor an der Universität Krakau* (Olomouc: Historický seminář cyrilometodějské fakulty bohoslovecké, 1937), 11.
67. According to *Czas* (no. 187, 18 August 1853), Piotr Bartynowski (letter to Agenor Gołuchowski, 27 September 1827, in CDIAL, F. 146, Op. 14, Spr. 1, p. 3) and notices in *Dziennnik Podawczy* of the Jagiellonian University for 1853 state that while the law faculty fully supported the petition, the medical and philosophical faculties were divided and refused to make a decision on this issue.
68. [František Bratranek], [Der Entwurf einer Petition an das Ministerium für Cultus und Unterricht], undated, identical with petition Z. 730, 27 July 1853 (see Heindl, "Universitätsreform und politisches Programm," 83n40), quoted here from the transcription in Batron, *Der vergessene Mähr. Verehrer Goethes*, 13. The original can be found in MZA Brno, fond E4 (Augustiniáni Staré Brno), kart. 191, folio 40–42 (1457), here 40.
69. Batron, *Der vergessene Mähr. Verehrer Goethes*, 14.
70. Quoted from Batron, *Der vergessene Mähr. Verehrer Goethes*, 14.
71. This and the following fragments are not included in Batron. The quotations are from archival materials in MZA Brno, folio 41. The last passage was inserted by an unidentified hand.
72. AUJ, Z. 1129, 7 December 1853; Z. 1168, 23 December 1853. Martial law was in force from November 1848 until 1854.
73. AUJ, S II 808, 19 October 1854, Z, 25543 (Kozubowski); AUJ, S II 815, 13 June 1856 (Majer).
74. AUJ, Z. 1129, 7 December 1853; Z. 1168, 23 December 1853.
75. Inquiry of the University of Cracow's Academic Senate, in DALO, F. 26, Op. 7, Spr. 39, Z. 351, 9 August 1854.
76. *Czas*, 18 August 1853, 1. The journalist strongly opposed one of the points raised by the petition, namely, the claim that the Polish language did not have the appropriate terminology and vocabulary; see *Czas*, 29 September 1853.
77. In October 1869 the faculty agreed to divide the chair of German literature into two, with lectures in both languages. AUJ, WF II 157, 14 October 1869. The document establishing Polish as the (almost) exclusive language of instruction was issued several months later on the occasion of the relocation of the philologist

Jan Wrobel/Wróbel from his professorship in Cracow to L'viv. AUJ, WF II 151, Z. 4043, 14 May 1870. Bratranek was also liked by students and remained a positive figure in the historiography; see Olga Dobija-Witczak, "František Tomáš Bratránek (1815–1884)," in *Złota Księga Wydziału Filologicznego*, ed. Jan Michalik and Wacław Walecki (Cracow: Księgarnia Akademicka, 2000), 50–58; and Eugeniusz Klin and Jaromir Loužil, eds., *František Tomáš Bratranek—ein Polonophiler Mittler zwischen den Nationen* (Zielona Góra: Wydawictwo Wyższej Szkoły Pedagogicznej, 1987).

78. *Dziennik Polski*, 15 October 1869. The journal nevertheless pleaded for Bratranek's dismissal.
79. Jan Surman, "Symbolism, Communication and Cultural Hierarchy: Galician Discourses of Language Hegemony at the Beginning of the Second Half of the Nineteenth Century," *Historyka: Studia Metodologiczne* 42 (2012): 151–74, https://www.academia.edu/15087613/_full_issue_Jan_Surman_Klemens_Kaps_Postcolonial_Galicia_Prospects_and_Possibilities_Historyka_42_2012.
80. See Lentze, "Graf Thun"; Herbert H. Egglmaier, "Graf Thun und das Rechtsstudium an den oberitalienischen Universitäten Padua und Pavia," in *Forschungen zur Geschichte des Alpen-Adria-Raumes: Festgabe für em.o. Univ.-Prof. Dr. Othmar Pickl zum 70. Geburtstag*, ed. Herwig Ebner, Paul Roth, and Ingeborg Wiesflecker-Friedhuber (Graz: Selbstverlag des Instituts für Geschichte der Karl-Franzens-Universität, 1997), 107–18; and Alois Kernbauer, "Konzeptionen der Österreich-Geschichtsschreibung 1848–1938," in Ebner et al., *Forschungen*, 255–73.
81. Kurt Mühlberger, "Pflanzstätten der Wissenschaft: Zur Einführung von Seminaren und Instituten in den Universitätsbetrieb in Österreich nach 1848," *Mitteilungen des Instituts für Österreichische Geschichtsforschung* (forthcoming); and Walter Höflechner, *Das Historische Seminar der Karl-Franzens-Universität Graz* (Graz: Elektronische Veröffentlichungen aus dem Zentrum für Wissenschaftsgeschichte der Karl-Franzens-Universität Graz, 2007), http://gams.uni-graz.at:8080/fedora/get/o:wissg-hs/bdef:Navigator.fs/get/. In 1850 Václav Vladivoj Tomek criticized the strong philological focus of the seminar, which was, in his opinion, unfruitful for the forging of patriotism. See Václav Vladivoj Tomek, *Paměti z mého žiwota* (Prague: František Řivnáč, 1905), 1:335–60, esp. 1:353–61.
82. See Kurt Bayertz, Walter Jaeschke, and Myriam Gerhard, eds., *Weltanschauung, Philosophie und Naturwissenschaft im 19. Jahrhundert*, vol. 1, *Der Materialismusstreit* (Hamburg: Meiner, 2007); and Tilman M. Schröder, *Naturwissenschaften und Protestantismus im deutschen Kaiserreich: Die Versammlungen der Gesellschaft Deutscher Naturforscher und Ärzte und ihre Bedeutung für die evangelische Theologie* (Stuttgart: Franz Steiner, 2008).
83. Sebastian Brunner, "Zur Wiener Universitätsfrage," *Wiener Kirchenzeitung*, 1 January 1852, 2–3, quoted in Sander Gliboff, "Evolution, Revolution, and Reform in Vienna: Franz Unger's Ideas on Descent and Their Post-1848 Reception," *Journal of the History of Biology* 31, no. 2 (1998): 201. See also

Werner Michler, *Darwinismus und Literatur: Naturwissenschaftliche und literarische Intelligenz in Österreich, 1859–1914* (Vienna: Böhlau, 1998), 38–39; and [Sebastian Brunner], "Der österreichische Vogt-Büchner-Moleschott," *Wiener Kirchenzeitung*, 4 January 1856, 9–10. "Vogt-Büchner-Moleschott" meant the Viennese paleobiologist Franz Unger, an outspoken liberal and materialist.

84. "The natural science, which denies the existence of man's soul as such and chokes all that is spiritual with materialism, exceeds everything [else] in its harmfulness [to the youth]." *Die Neugestaltung*, 20.
85. This particular case was that of the zoologist Ludwig Karl Schmarda, who was accused of expressing materialist outlooks in textbooks for the gymnasia and removed from his chair in Prague in 1855. Constantin von Wurzbach, "Ludwig Karl Schmarda," in *Biographisches Lexikon des Kaiserthums Oesterreich*, ed. Constantin von Wurzbach (Vienna: Kaiserlich-königliche Hof- und Staatsdruckerei, 1875), 30:156; and Michael Wallaschek, *Ludwig Karl Schmarda (1819–1908): Leben und Werk* (Halle, 2014), 51–55.
86. AT-UAW, Dekanatsakten der Philosophischen Fakultät, Zl. 605 from 1850/51.
87. Wilhelm Kergel to Friedrich Haase, 19 December 1849, reprinted in Alfred Schneider, "Briefe österreichischer Gelehrter aus den Jahren 1849–1862: Beiträge zur Geschichte der österreichischen Unterrichtsreform," *Archiv für österreichischen Geschichte* 113 (1936): 237–43; Kergel's quotation about Jan/Johann Helcelet is on p. 239.
88. Nomination act of chemist Franz Pless to L'viv, AGAD, MWiO, Sygn. 117u, PA Pless, Z. 7716, 6 August 1851. See also the nomination act of the chemist Heinrich Hlasiwetz to Innsbruck, in which the ministry accentuated the importance of chemistry for the crafts. Quoted in Robert Rosner, *Chemie in Österreich 1740–1914: Lehre—Forschung—Industrie* (Vienna: Böhlau, 2004), 165–66.
89. Havránek, "Karolinum v revoluci 1848," 38–39. The issue continued to be publicly discussed; see Wojciech Urbański, "Projekt reoranizacji Uniwersytetów we Lwowie i Krakowie ze wzgledu na techniczne akademie," *Dziennik Polski*, 12 October 1869.
90. The academy was established by Thun-Hohenstein in late 1849 by upgrading the Collegium Nobilium (*Ständische Akademie* or *Stavovská akademie*); see Pavel Šišma, *Matematika na německé technice v Brně* (Prague: Prometheus, 2002), 16–19. See also Christian d'Elvert, *Geschichte der Studien- Schul- und Erziehungs-Anstalten in Mähren und Österr. Schlesien insbesondere der olmützer Universität, in den neueren Zeiten* (Brno: Rohrer, 1857), 381–91; and Vaclav Pumpla, "Snahy o zřízení české university na Moravě v 19. století," *Historická Olomouc* 11 (1998): 145–52.
91. Juliane Mikoletzky, "Vom Polytechnischen Institut zur Technischen Hochschule: Die Reform des technischen Studiums in Wien, 1850–1875," *Mitteilungen der österreichischen Gesellschaft für Wissenschaftsgeschichte* 15 (1995): 92; Chahrour, "'A Civilizing Mission'?"
92. On the unsuccessful appointment of Ferdinand Heßler, a professor of physics at the Viennese technical academy, to the Charles-Ferdinand University in Prague, see Walter Höflechner, *Materialien zur Entwicklung der Physik und*

ihrer "Randfächer" Astronomie und Meteorologie an den österreichischen Universitäten 1752–1938 (Graz: Institut für Geschichte der Karl-Franzens-Universität Graz, 2002), 391–95, http://www-gewi.uni-graz.at/wissg/gesch_der_physik.

93. Kurt Mühlberger, "Das 'Antlitz' der Wiener Philosophischen Fakultät in der zweiten Hälfte des 19. Jahrhunderts: Struktur und personelle Erneuerung," in *Eduard Suess und die Entwicklung der Erdwissenschaften zwischen Biedermeier und Sezession*, ed. Johannes Seidl (Göttingen: V&R Unipress, 2009), 67–104.
94. The Joanneum professors also taught chemistry and botany in Graz, although they were not employed as professors at the university, and the corresponding chairs remained vacant. See the *Akademische Behörden, Personalstand und Ordnung der Vorlesungen an der K.K. Karl-Franzens-Universität zu Graz* (Graz: Leykam's Erben, 1850–) of the University of Graz for the corresponding years.
95. For instance, in 1851 the Graz philosophical faculty proposed the comparative zoological anatomist Oskar Schmidt from Jena for the chair of zoology, but the ministry nominated Ludwik Karl Schmarda. AT-OeStA/AVA Unterricht UM allg. Akten 1219, PA Schmarda.
96. Alois Kernbauer, *Das Fach Chemie an der Philosophischen Fakultät der Universität Graz* (Graz: Akademische Druck- und Verlagsanstalt, 1985), 7–13; and Rosner, *Chemie in Österreich.*
97. E.g., the mineralogist Franz Zippe and the zoologist Rudolf Kner, new professors in Vienna. See Johannes Seidl, Franz Pertlik, and Matthias Svojtka, "Franz Xaver Maximilian Zippe (1791–1863): Ein böhmischer Erdwissenschaftler als Inhaber des ersten Lehrstuhls für Mineralogie an der Philosophischen Fakultät der Universität Wien," in Seidl, *Eduard Suess*, 161–209; and Luitfried Salvini-Plawen and Matthias Svojtka, *Fische, Petrefakten und Gedichte: Rudolf Kner (1810–1869)—ein Streifzug durch sein Leben und Werk* (Linz: Biologiezentrum der Oberösterreichischer Landesmuseen, 2008), 99–101.
98. Seidl, Pertlik, and Svojtka, "Franz Xaver Maximilian Zippe."
99. Wacław Szybalski claims that Zawadzki's active participation in the 1848 revolution caused his removal. Most recently, see Szybalski, "Professor Alexander Zawadzki of Lvov University—Gregor Mendel's Mentor and Inspirer," *Biopolymers and Cell* 26, no. 2 (2010): 83–86. In contrast, Ludwik Finkel states that Zawadzki was removed because he neglected physics and mathematics in his lectures. Finkel, "Historya Uniwersytetu Lwowskiego," 322–23.
100. The society united the scholars of Brno regardless of their cultural alignment. In 1904 the Natural Sciences Club in Brno (Přírodovědecký klub v Brně) was established for Czech scholars.
101. Vítezslav Orel, "Professor Alexander Zawadzki (1798–1868)—Mendel's Superior at the Technical Modern School in Brno," *Folia Mendeliana Musei Moraviae* 7 (1972): 13–20.
102. DALO, F. 26, Op. 7, Protokolle, R. 29, 24 April 1857; Finkel, "Historya Uniwersytetu Lwowskiego," 323–24; and Mirosław Dąbrowski, "Wojciech Urbański—polski badacz zjawisk elektrycznych," *Nauka*, no. 2 (2007): 151–60.
103. Calculations based on the Prague and Vienna catalogs of lecturers for 1860/61.

See *Übersicht der akademischen Behörden, der den einzelnen Fakultäten zugehörenden Decane, Pro-Decane, Professoren, Privatdocenten . . . dann der Kirche, Bibliothek, Kanzlei, Quästur an der Kaiserl. Königl. Universität zu Wien für das Studien-Jahr 1860/61* (Vienna: Kaiserlich-königliche Staats-Druckerei, 1861), and *Personalstand der k.k. Universität zu Prag zu Anfang des Winter-Semesters 1860–61* (Prague: k.k. Normalschulbuchdruckerei, 1860).

104. Krzysztofa Michalewska, "Habilitacje w Uniwersytecie Jagiellońskim 1848–1918," *Zeszyty Naukowe Uniwersytetu Jagiellońskiego: Prace Historyczne* 71, no. 12 (1963): 79–80.

105. It was, once again, Šafárik who mediated between the ministry and the university to achieve the appointment of another Czech patriot. See Otakar Brázda, "Příchod Jana Evangelisty Purkyně na pražskou univerzitu," *AUC-HUCP* 27, no. 1 (1987): 55–89; and AT-OeStA/AVA Unterricht UM allg. Akten 1210, PA Purkyne, proposition of the faculty N. 973, 27 September 1849, ministry's proposal Z. 7164/970, 22 October 1849.

106. AT-OeStA/AVA Unterricht UM allg. Akten 1209, PA Jaksch, Z. 2935/383, 28 May 1849.

107. See, for example, the argument in AT-OeStA/AVA Unterricht UM allg. Akten 1186, PA Dietl, Z. 168, 10 February 1851, in which scholars who did not know Polish were not considered, even if they had greater scholarly achievements.

108. Mrozowska, "Okres ucisku," 219. In 1850 Nikodem Bętkowski, the author of the first Polish textbook for pathological anatomy, strove for the chair and was proposed by the faculty, but without result; the same problem was encountered in the history of medicine, as all candidates were rejected by the ministry. See AUJ, S II 810, 20 March 1850; 22 October 1850 (for pathological anatomy); and AUJ, WL II 156 (for the history of medicine).

109. For Kozubowski, see AUJ, S II 808, 19 October 1854, Z. 25543; for Majer, see AUJ, S II 815, 13 June 1856, Z. 11235; AGAD, MWiO, Sygn. 51u, PA Mayer, Z. 13995, 616, 9 March 1856.

110. Stefan Berger and Christoph Conrad, eds., *The Past as History: National Identities and Historical Consciousness in Modern Europe* (Basingstoke: Palgrave Macmillan, 2015), 163.

111. In Prague the imperial-Catholic school of Václav Vladivoj Tomek competed with the positivist-internationalist school of Jaroslav Goll. See, for instance, Bohumil Jiroušek, "Mimořádná profesura Josefa Pekaře (ve světle vztahů Antonína Rezka a Jaroslava Golla)," in *Proměny elit v moderní době*, ed. Milena Lenderová, Zdeněk Bezecný, and Jiří Kubeš (České Budějovice: Historický ústav Jihočeské univerzity, 2003), 167–78. On the Cracow school of historiography, which is still controversial, see, for instance, Andrzej Wierzbicki, "Wokół 'czarnej legendy' historiografii krakowskich konserwatystów," *Kwartalnik Historyczny* 104, no. 2 (1997): 63–87; and Henryk Słoczyński, "Z dziejów czarnej legendy krakowskiej historiografii konserwatywnej: Józef Szujski w opiniach współczesnych i potomnych," *Kwartalnik Historyczny* 102, nos. 3–4 (1995): 209–44.

112. See Vatroslav Jagić, "A Survey of Slavistic Studies," *Slavonic Review* 1,

no. 1 (1922): 40–58; and Stanislaus Hafner, "Geschichte der österreichischen Slawistik," in *Beiträge zur Geschichte der Slawistik in nichtslawischen Ländern*, ed. Josef Hamm and Günther Wytrzens (Vienna: Österreichische Akademie der Wissenschaften, 1985), 11–88.

113. See the recent discussion in Alexander Wilfing, "Kant and 'Austrian Philosophy': An Introduction," in *Detours: Approaches to Immanuel Kant in Vienna, in Austria, and in Eastern Europe*, ed. Violetta L. Waibel (Göttingen: V&R Unipress, 2015), 22–24.
114. Quoted in Lhotsky, "Das Ende des Josephinismus," 534.
115. Erika Weinzierl, "Helfert, Joseph Freiherr von," in *Neue Deutsche Biographie*, ed. Historische Kommission bei der Bayrischen Akademie der Wissenschaften (Berlin: Duncker & Humblot, 1969), 8:469–70.
116. For discussion of models of the origin and commonality of the Habsburg Empire, see Walter Pohl, "National Origin Narratives in the Austro-Hungarian Monarchy," in *Manufacturing Middle Ages: Entangled History of Medievalism in Nineteenth-Century Europe*, ed. Patrick J. Geary and Gábor Klaniczay (Leiden: Brill, 2013), 13–50.
117. Joseph-Alexander Freiherr von Helfert, *Über Nationalgeschichte und den gegenwärtigen Stand ihrer Pflege in Österreich* (Prague: Calve, 1853), 59.
118. See Österreichische Geschichte für das Volk, 17 vols. (Vienna: Prandel, 1864–69), originating from Helfert's ideas and published under his supervision from 1864 onward.
119. Bohumil Jiroušek, "Historik W. W. Tomek," in *W. W. Tomek, historie a politika (1818–1905): Sborník příspěvků královéhradecké konference k 100. výročí úmrtí W. W. Tomka*, ed. Miloš Řezník (Pardubice: Univerzita Pardubice, 2006), 21.
120. The characteristics Monika Báar has described for nationalist historiography fully apply to imperialist historiography as well. Monika Baár, *Historians and Nationalism: East-Central Europe in the Nineteenth Century* (New York: Oxford University Press, 2010).
121. Hannelore Putz, "König Ludwig I. von Bayern und seine Universität," in *Domus Universitatis: Das Hauptgebäude der Ludwig-Maximilians-Universität 1835–1911–2011*, ed. Claudius Stein (Munich: Herbert Utz, 2015), 44–45. Thun-Hohenstein nominated, for example, the historian Konstantin Höfler (Prague) and the legal historian Georg Phillips (Innsbruck, from 1851 Vienna) and, as Phillips's successor in Innsbruck, Ernst Moy de Sons.
122. AT-OeStA/AVA Unterricht UM allg. Akten 668, PA Grauert, Z. 8791/1285, 2 December 1849.
123. Halbwidl, "Life and Times," 121–22; and Emil von Ottenthal, "Theodor von Sickel," *Mitteilungen des Instituts für Österreichische Geschichtsforschung* 29 (1908): 545–59.
124. Lhotsky, "Das Ende des Josephinismus," 545.
125. Böhmer influenced the appointments of Joseph Aschbach and Julius Ficker, among others. Julius Jung, *Julius Ficker (1826–1902): Ein Beitrag zur deutschen Gelehrtengeschichte* (Innsbruck: Wagner'sche Universitäts-Buchhandlung,

1907), 134–48. See also Christine Ottner-Diesenberger, "Joseph Chmel und Johann Friedrich Böhmer: Die Anfänge der Regesta Imperii im Spannungsfeld von Freundschaft und Wissenschaft," in *Wege zur Urkunde, Wege der Urkunde, Wege der Forschung*, ed. Karel Hruza and Paul Herold (Vienna: Böhlau, 2005), 257–91.

126. Jiří Rak, "Obraz Němce v české historiografii 19. století," in *Obraz Němců, Rakouska a Německa v české společnosti 19. a 20. století*, ed. Jan Křen and Eva Broklová (Prague: Karolinum, 1998), 49–75, esp. 57–59.
127. See the wording in the nomination act for Antoni Wacholz: AGAD, MWiO, Sygn. 117u, PA Wacholz, Z. 1190, 4 January 1850.
128. See the negative description of Walewski by the faculty of the university in AUJ, WF II 135, Bericht der Kommission aus 9.8.1850.
129. For an example of the discourse on Walewski, see Krzysztof Baczkowski, "W służbie dworu Habsburskiego: Antoni Walewski (1805–1876)," *Zeszyty Naukowe Uniwersytetu Jagiellońskiego: Prace Historyczne* 132 (2005): 99–108, where, as in other literature quoted in the article, the idea that Walewski was a Habsburg secret agent is based on allegations of the time.
130. See also AT-OeStA/AVA Unterricht UM allg. Akten 1221, PA Tomek, Z. 7547/838, 25 September 1850.
131. *Großdeutsch* refers to the "Great German" solution of restructuring the German Confederation into a national state, including Habsburg regions that were either populated by Germans or seen as belonging historically to them. Höfler is regarded as one of the leading writers of the first generation; see Thomas Brechenmacher, *Großdeutsche Geschichtsschreibung im neunzehnten Jahrhundert: Die erste Generation (1830–48)* (Berlin: Duncker & Humblot, 1996).
132. AT-OeStA/AVA Unterricht UM allg. Akten 1217, PA Höfler, Z. 9331/884, 17 September 1851. On Tomek's influence, see Blanka Zilynská, "Karl Adolf Constantin Ritter von Höfler jako univerzitní učitel," in *Německá medievistika v českých zemích do roku 1945*, ed. Pavel Soukup and František Šmahel (Prague: Výzkumné centrum pro dějiny vědy, 2004), 201–2.
133. Zilynská, "Karl Adolf Constantin Ritter von Höfler," 202; and Karl Adolf Constantin Höfler and Paul Joseph Šafařík, *Glagolitische Fragmente* (Prague: Gottlieb Haase Söhne, 1857).
134. Peter Morée, "Jan Hus as a Threat to the German Future in Central Europe: The Bohemian Reformer in the Controversy between Constantin Höfler and František Palacký," in *The Bohemian Reformation and Religious Practice*, ed. Zdenek V. David and David R. Holeton (Prague: Main Library, 2002), 4:295–307.
135. See the appointment records of Hermann Bonitz in Vienna, AT-OeStA/AVA Unterricht UM allg. Akten 664, PA Bonitz, Z. 377/72, 16 January 1849.
136. AT-OeStA/AVA Unterricht UM allg. Akten 1218, PA Lange, Z. 17544/13191, 7 January 1855.
137. See Leitner, "Das Reformerwerk."
138. Schneider, "Briefe," for example, 237–38.

139. See August Schleicher, *Die Formenlehre der kirchenslawischen Sprache, erklärend und vergleichend dargestellt* (Bonn: H. B. König, 1852).
140. Georg Curtius, *Die Sprachvergleichung in ihrem Verhältniss zur classischen Philologie*, 2nd ed. (Berlin: Wilhelm Besser, 1848), 9.
141. Ludwig Lange, *Die klassische Philologie in ihrer Stellung zum Gesammtgebiete der Wissenschaften und in ihrer inneren Gliederung: Eine Antrittsvorlesung, gehalten am 24. April 1855 in Prag* (Prague: Calve, 1855), 10.
142. Most notably in the case of Małecki, in his nomination to the L'viv chair of Polish language and literature, according to Finkel. Finkel, "Historya Uniwersytetu Lwowskiego," 334–35.
143. See, e.g., on Mychajlo Lučkay's and Josyp Levyc'kyj's proposals of Old Church Slavic Ruthenian as opposed to vernaculars, Andrii Danylenko, "Myxajlo Luckaj: A Dissident Forerunner of Literary Rusyn?," *Slavonic and East European Review* 87, no. 2 (2009): 201–26; and Michael' Mozer [Michael Moser], "Josyf Levyc'kyj jak borec' za kul'turu 'ruskoï' (ukraïns'koï) movy," in *Confraternitas: Jobilejnyj zbirnyk na počanu Jaroslava Isajevyča*, ed. Mykola Krykun and Ostap Sereda (L'viv: Instytut ukraïnoznavstva im. I. Kryp'jakevyča Nacional'na Akademiya Nauk Ukraïny, 2007), 447–60.
144. See Zdeněk Šamberger, "Časopis Vídeňský Denník a jeho poslání v letech 1850–1851 (Ke ztroskotanému pokusu Leo Thuna o založení české konzervativní strany)," *Slovanský přehled* 71, no. 1 (1985): 26–40; Mile Mamić, "Das deutsch-slawische Wörterbuch der juridisch-politischen Terminologie (seine Konzeption und Realisierung)," in *Balten—Slaven—Deutsche: Aspekte und Perspektiven kultureller Kontakte. Festschrift für Friedrich Scholz zum 70. Geburtstag*, ed. Ulrich Obst and Gerhard Ressel (Münster: Lit, 1999), 131–38; and Kamiš, "Tschechisch-deutsche Beziehungen."
145. Mychajlo Kril', "Nevidomi lysty Ivana Holovac'koho do Antona Petruševyča," *Ukraïna moderna* 2–3 (1999): 350–53. Mychaĭlo Vozniak sees the Ministry of the Interior as the responsible and financing body. Mychajlo Voznjak [Mychaĭlo Vozniak], "Z-za redakcijnych kulis videns'koho Vistnyka ta Zori Halyc'koï," *Zapysky Naukovoho Tovarystva imeny Ševčenka* 107 (1912): 73–109.
146. Kořalka, *František Palacký*, 340–44. Thun-Hohenstein appointed Vocel (Wocel) in early 1850 as associate professor of Bohemian archaeology and history of art in Prague. AT-OeStA/AVA Unterricht UM allg. Akten 1221, PA Wocel, Z. 273/47, 30 January 1850.
147. On Hattala, see Wilhelm Zeil, *Slawistik an der deutschen Universität in Prag (1882–1945)* (Munich: Otto Sagner, 1995), 26.
148. DALO, F. 26, Op. 5, Spr. 437, PA Glowacki. Holovac'kyj's appointment as an auxiliary professor (*supplent*) was announced on 27 November 1848, that as a full professor on 13 December 1848; on his pay, see petitions for adjustments in the personnel records and payment list in DALO, F. 26, Op. 7, Spr. 33, list for 1854/55.
149. For one interpretation of the political contacts and problems of the professor of Polish language and literature Jan Szlachtowski, dismissed from the University

of L'viv in 1851, see Karol Estreicher, *Dr. Jan Kanty Szlachtowski: Rzecz czytana na posiedzeniu c. k. Towarzystwa Naukowego Krakowskiego dnia 5 lutego 1872 roku* (Cracow: Kraj, 1872), esp. 19–32; see also a report on his person from 1852 in AGAD, MWiO, Sygn. 117u, PA Szlachtowski, Z. 12651, 15 March 1852.

150. The most important of these were Adam Mickiewicz (Cracow) and August/Augustyn Bielowski (L'viv).
151. For Cracow, these included Wincenty Pol, Henryk Suchecki, Michał Wiszniewski, and Seweryn Goszczyński. On Pol, see Bielak, "Katedra," 89–92. On Suchecki, see Heinrich [Henryk] Suchecki to an unknown professor, SOA Litoměřice/Děčín, Rodinný archiv Thun-Hohensteinů, A 3 XXI D 64, 30 June 1850; see also Miklošič's recommendation, D 40, 1 April 1850. On Wiszniewski, see Henryk Barycz, *Józef Ignacy Kraszewski czterokrotny kandydat do katedry uniwersyteckiej* (Cracow: Wydawnictwo Literackie, 1979), 23–25; and on Goszczński, see Bielak, "Katedra," 89–92. For L'viv, Thun-Hohenstein sought to appoint Mateusz Szrzeniawa Sartyni; see DALO, F. 26, Op. 7, Spr. 22, Z. 420, 8 June 1852; DALO, F. 26, Op. 7, Spr. 39.
152. Auxiliary professors (*supplenten*) were scholars appointed with temporary contracts to cover teaching in a given subject.
153. Bielak, "Katedra," 98–99.
154. See UAI, PF, Z. 141, 1855/56, Malecki; DALO, F. 26, Op. 7, Spr. 56, N.2, 30 September 1856; and Finkel, "Historya Uniwersytetu Lwowskiego," 334–35.
155. In the 1860s Mecherzyński was opposed by the students, who proposed an eminent historian and writer, Karol Szajnocha, for the chair; see Henryk Barycz, ed., *Korespondencja Karola Szajnochy* (Wrocław: Zakład Narodowy im. Ossolińskich, 1959), 355–57.
156. See Antoni Małecki, *Gramatyka języka polskiego: Większa* (L'viv: Nakładem Autora, 1863).
157. See Karol Mecherzyński, *Historya języka niemieckiego w Polsce* (Cracow, 1845).
158. A notable exception was the famous comparative philologist and translator Wilhelm Wackernagel from Basel, proposed in 1850, who remained, however, in Switzerland. AT-OeStA/AVA Unterricht UM allg. Akten 671, PA Karajan, Z. 113/18, 2 January 1850.
159. Most notably, Pater Wilhelm Gärtner, a writer and political theologian appointed to Pest, and Oskar Redwitz, a short-term professor in Vienna and author of the popular epos *Amarath* (1849), a defense of Christian spirituality against rationalism.
160. AT-OeStA/AVA Unterricht UM allg. Akten 1077, PA Ignaz Zingerle, Z. 20842/1394, 25 May 1858.
161. AT-OeStA/AVA Unterricht UM allg. Akten 1077, PA Ignaz Zingerle, minister's proposal, 9 August 1858 (the emperor's annotation on the proposed appointment and rejection); Z. 1786/96, 25 February 1859 (second proposal).
162. AT-OeStA/AVA Unterricht UM allg. Akten 941, PA Weinhold, Z. 961/68, 18 January 1851.
163. Jaromír Loužil, "Franz Thomas Bratraneks Leben und Philosophie," *Bohemia* 13

(1972): 182–210. Kremer's only notable publications after 1852 were his descriptions of Italian art (six volumes). See Henryk Struve, "Życie i Twórczość Józefa Kremera," in *Wykład systematyczny filozofii obejmujący wszystkie jéj części w zarysie: Dla miłośników téj umiejętności pragnących dokładniéj się z nią obeznać*, vol. 1, *Fenomenologia, logika*, by Józef Kremer, ed. Henryk Struve (Warsaw: S. Lewental, 1877), 1–175.

164. See, e.g., Rudolf Haller, "Gibt es eine österreichische Philosophie?," in *Fragen zu Wittgenstein und Aufsätze zu österreichischer Philosophie*, ed. Rudolf Haller (Amsterdam: Rodopi, 1986), 31–43; and Friedrich Stadler, *The Vienna Circle: Studies in the Origins, Development, and Influence of Logical Empiricism* (Vienna: Springer, 2001).
165. Struve, "Życie i Twórczość Józefa Kremera," esp. 56–58.
166. See the annotation of a commission established to deal with this question at the ministry, consisting of Thun-Hohenstein, Joseph Mozart, and Joseph Feil, stressing the nonimportance of a petitioner's denomination, in AGAD, MWiO, Sygn. 117u, PA Rosenberg, Z. 9458, 4 December 1854. The ministry's final word on the appointment can be found in DALO, F. 26, Op. 7, Spr. 43, Z. 427, 19 September 1854.
167. These were Franz Karl Lott, who was moved from Göttingen to Vienna, and Wilhelm Fridolin Volkmann, appointed to Prague. See AT-OeStA/AVA Unterricht UM allg. Akten 673, PA Lott, Z. 735, 16 January 1849; and Frankfurter, *Graf Leo Thun-Hohenstein*, 73.
168. Martin Seiler, "Das 'Manifest der österreichischen Philosophie': Die Materialien Kurt Blaukopfs über die Berufung Robert Zimmermanns an die Universität Wien (1860–1861), im Spiegel von Philosophiegeschichte, Universitätsreform, Berufungspolitik, staatlicher Religions- und Konfessionsgesetzgebung und Verfassungsgeschichte," in Wissenschaftliche Weltauffassung und Kunst: Kunst, Kunsttheorie und Kunstforschung im wissenschaftlichen Diskurs. Ein Projekt des Instituts Wiener Kreis, n.d., http://www.univie.ac.at/ivc/WWUK/manifest.html (accessed 21 October 2014). The expression *Herzensjunge* is taken from Bolzano's letters: Eduard Winter and Liane Zeil, eds., *Wissenschaft und Religion im Vormärz: Der Briefwechsel Bernard Bolzanos mit Michael Josef Fesl, 1822–1848* (Berlin: Akademie-Verlag, 1965), 348, 352, 254.
169. See the discussion in Venanzio Raspa, "Bolzano e la filosofia austriaca," *Discipline filosofiche* 21, no. 2 (2011): 245–85, esp. 250–65.
170. On Thun-Hohenstein's ideals of philosophy, see Walter Höflechner, "Die Auswirkungen politischer und kultureller Veränderungen auf Forschungsorientierung und Wissenschaftsorganisation," in *Geschichte der österreichischen Humanwissenschaften*, vol. 1, *Historischer Kontext, wissenssoziologische Befunde und methodologische Voraussetzungen*, ed. Karl Acham (Vienna: Passagen, 1999), 149–214, esp. 161–63; Peter Goller, *Naturrecht, Rechtsphilosophie oder Rechtstheorie? Zur Geschichte der Rechtsphilosophie an Österreichs Universitäten (1848–1945)* (Frankfurt am Main: Peter Lang, 1997), 39–80; and Johannes Feichtinger, "Positivismus in der österreichischen

Philosophie: Ein historischer Blick auf die frühe Positivismusrezeption," *Newsletter Moderne: Zeitschrift des Spezialforschungsbereichs Moderne—Wien und Zentraleuropa um 1900* 7, no. 2 (2004), http://www-gewi.kfunigraz.ac.at/moderne/heft13f.htm.

171. Enrique M. Ureña, *El Krausismo alemán: Los congresos de filósofos y el krausofrobelismo (1833–1881)* (Madrid: Universidad Pontificia de Comillas de Madrid, 2002), 101, 239. On two proponents, see Otto Busch, "v. Leonhardi," *Philosophische Monatshefte* 11 (1875): 385–90; and on the philosopher of law Heinrich Ahrens, a disciple of Krause, see Goller, *Naturrecht*, 46–54.

172. Thomas W. Simons Jr., "The Prague Origins of the Guntherian Converts (1800–1850)," *Leo Baeck Institute Yearbook* 22, no. 1 (1977): 245–56; on the differences between Herbartianism and Güntherism, see Ivo Tretera, *J. F. Herbart a jeho stoupenci na pražské univerzitě* (Prague: Univerzita Karlova, 1989), 147–50.

173. Alfons Pichler, quoted in Peter Goller, "Georg Schenach (1843–1853): 'Ein Vermittlungsphilosoph,'" in *Die Lehrkanzeln für Philosophie an der Philosophischen Fakultät der Universität Innsbruck 1848 bis 1945*, ed. Peter Goller (Innsbruck: Wagner, 1989), 17.

174. Lentze, *Die Universitätsreform*, e.g., 84; and Otto Weiß, "Bolzanisten und Güntherianer in Wien 1848–1851," in Rumpler, *Bernard Bolzano*, 247–82.

175. Augustin Smetana, *Geschichte eines Excommunicirten: Eine Selbstbiographie von Augustin Smetana*, ed. Alfred Meißner (Leipzig: Grunow, 1863), 152–53.

176. See Augustin Smetana, *Příběh jedné exkomunikace a doprovodné texty*, ed. and trans. Irena Šnebergová (Prague: Filosofia, 2008); on letters to Exner (in German), see pp. 471–500; on losing Exner's favor, see Smetana's letter to Moritz Wilhelm Drobisch, pp. 503–5.

177. AT-OeStA/AVA Unterricht UM allg. Akten 1217, PA Hanusch, Z. 3553/546, 24 May 1849.

178. Pavel Křivský, *Augustin Smetana* (Prague: Univerzita Karlova, 1990), 316.

179. Křivský, *Augustin Smetana*, 311–26, 317–26.

180. Jaromír Loužil, *Ignác Jan Hanuš: Studie s ukázkami z díla* (Prague: Melantrich, 1971), 78–79.

181. Loužil, *Ignác Jan Hanuš*, 79–80.

182. Two physicians in Cracow were not from Galicia; in Prague, six out of twenty-eight had been born outside Bohemia and Moravia, although being born in Bohemia does not mean they were bilingual.

183. Petráň, *Nástin dějin*, 151.

184. On Hungary, see Jan Surman, "Cisleithanisch und transleithanisch oder habsburgisch? Ungarn und das Universitätssystem der Doppelmonarchie," in *Österreichisch-ungarische Beziehungen auf dem Gebiet des Hochschulwesens*, ed. Zsolt K. Lengyel, József Zsigmond Nagy, and Gábor Ujváry (Székesfehérvár: Kodolányi János Főiskola, Eötvös Loránd Tudományegyetem Könyvtára, 2010), 235–52.

185. See Stefan Malfèr, "Die Sprachenfrage und der verstärkte Reichsrat von 1860," in *A Lajtán innen és túl / Jenseits und diesseits der Leitha: Elektronische Festschrift für Éva Somogyi zum 70. Geburtstag*, ed. Imre Ress and Dániel Szabó (Budapest:

Magyar Tudományos Akadémia Törtenettudományi Intezet, 2007), 93–118.

186. See the letters of August Schleicher to Pavel Jozef Šafárik, reprinted in Ilja Lemeškin and Jolanta Zabarskaitė, eds., *Lituanistinis Augusto Schleicherio palikimas / Das lituanistische Erbe August Schleichers* (Vilnius: Lietuvių kalbos institutas, 2008), 1:103–49; see also Schleicher letters, pp. 1:670–740; records from the police are reprinted on pp. 1:813–96.

187. See Jan Surman, "Leon (sic!) Thun in polnischer Historiographie: Zur Tradition der Geschichte," in Aichner and Mazohl, *Die Thun-Hohenstein'schen Universitätsreformen*, 317–46.

188. Lhotsky, "Das Ende des Josephinismus."

189. For contrasting views, see, e.g., Johann-Heinrich Dumreicher, *Über die Nothwendigkeit von Reformen des Unterrichtes an den medicinischen Facultäten Österreichs* (Vienna: Alfred Hölder, 1878); and Wilhelm Hartel, *Festrede zur Enthüllung des Thun-Exner-Bonitz Denkmals gehalten in der 1. Hauptsitzung der 42. Versammlung deutscher Philologen und Schulmänner am 24. Mai 1893* (Vienna: A. Holzhausen, 1893). A positive opinion of the reforms (but not of the political atmosphere of neoabsolutism) is mentioned in most publications in the Czech and Polish historiography of science and education and in at least some newer Hungarian publications; for example, Gábor Farkas, "A birodalmi felsőoktatás modernizációja: A kultuszminiszter oktatási tervezete (1849–1854)," in Lengyel, Nagy, and Ujváry, *Österreichisch-ungarische Beziehungen*, 163–78.

190. Feichtinger and Fillafer, "Leo Thun."

191. Akademischer Senat der Wiener Universität, *Geschichte der Wiener Universität*, 282.

192. AT-OeStA/AVA Unterricht UM allg. Akten 1210, PA Purkyne, Z. 7164/970, 22 October 1849, and Z. 973, 27 September 1849; and AT-OeStA/AVA Unterricht UM allg. Akten 1216, PA Celakovsky, Z. 2852/431, 19 April 1849.

193. AT-OeStA/AVA Unterricht UM allg. Akten 666, PA Eitelberger, Z. 10898/1376, 14 October 1852.

194. Feichtinger, "Positivismus."

195. Compare Józef Szujski, "Mowa na pogrzebie Antoniego Walewskiego prof. Uniw. Jagiell., Członka Akademii," *Rocznik Zarządu Akademii Umiejętności w Krakowie* (1877): 137–42.

196. Karl Langer, a professor of zoology in Pest, and Johann Chiari, an obstetrician in Prague, were among the most prominent nominees.

197. Carl Rokitansky, *Die Conformität der Universitäten mit Rücksicht auf gegenwärtige österreichische Zustände* (Vienna: Sallmayer, 1863).

Chapter 3

1. Jan Evangelista Purkyně, "Řeč na sjezdu německých přírodovědců a lékařů v Karlových Varech dne 25. září 1862," in *Jan Evangelista Purkyně, Sebrané Spisy, Svazek IX: Věda, vychova, společnost; studie a úvahy*, ed. Vladislav Kruta and Zdeněk Hornof (Prague: Nakladatelství Československé akademie věd, 1965), 131.

2. Theodor Billroth, *Briefe von Theodor Billroth*, 8th ed. (Hannover: Hahnsche Buchhandlung, 1910), 91.
3. Armand Dumreicher, *Die Verwaltung der Universitäten seit dem letzten politischen Systemwechsel in Oesterreich* (Vienna: Alfred Hölder, 1873); and Akademischer Senat, ed., *Die Franz-Josephs Universität in Czernowitz im ersten Vierteljahrhundert ihres Bestandes* (Chernivtsi: Bukowinaer Vereinsdruckerei, 1900), xxiii (argumentation of Tobias Wildauer).
4. Theophila Wassilko, "Der Unterrichtsrat: Ein vergessener Zwischenakt aus der Geschichte des österreichischen Unterrichtswesens," *Mitteilungen des Österreichischen Staatsarchivs* 6 (1953): 312–26.
5. See Felicitas Seebacher, *"Freiheit der Naturforschung!" Carl Freiherr von Rokitansky und die Wiener Medizinische Schule: Wissenschaft und Politik im Konflikt* (Vienna: Verlag der österreichischen Akademie der Wissenschaften, 2006), 97; and Emanuel Hofmann to Friedrich Haase, 10 May 1861, reprinted in Schneider, "Briefe," 232–34. The chair of the Unterrichtsrath was Leopold Hasner, appointed by Thun-Hohenstein to the law faculty in Prague; Franz Miklošič, Franz Karl Lott, and Karl Littrow were responsible for the philosophical faculty; and Josef Hyrtl, Franz Pit'ha, and Karl Schroff addressed issues related to medicine.
6. AT-OeStA/AVA Unterricht UM allg. Akten 1216, PA Bayer, Z. 153/863, 29 April 1865.
7. See AT-OeStA/AVA Unterricht UM allg. Akten 941, PA Wertheim, Z. 4231/228, 31 May 1861; Helga Wittmann and Erich Ziegler, *Die Entwicklung chemischer Wissenschaften an der Universität Graz: Ein Leistungsbericht* (Graz: Akademische Druck- u. Verlagsanstalt, 1985), 45–46; and Surman, "Cisleithanisch," 244–45.
8. See Bernhard Hubmann, "Carl Ferdinand Peters (1825–1881): Beitrag zu seiner Biographie," *Berichte der Geologischen Bundesanstalt* 53 (2001): 35; and AT-OeStA/AVA Unterricht UM allg. Akten 939, PA Nahlovsky, Z. 6024/290, 27 September 1862.
9. See AT-OeStA/AVA Unterricht UM allg. Akten 919, PA Heschl, Z. 9938/337, 18 October 1861.
10. The historian Adam Wolf, from 1852 an associate professor in Pest, was appointed to Graz in 1864. For a negative assessment of him, see AT-OeStA/AVA Unterricht UM allg. Akten 939, PA Pebal, Z. 581, 14 November 1864; for his appointment, see Z. 7430, 20 December 1864.
11. See the reasoning for removing the Galician-born Christian August Voigt from Cracow, in AT-OeStA/AVA Unterricht UM allg. Akten 919, PA Heschl, Z. 9938/337, 18 October 1861.
12. See the dismissal of Johann Czermak from the University of Pest. Czermak even proposed lecturing with a translator or teaching in Latin until he was capable of teaching in Hungarian. Petr Svobodný, "Contacts between Bohemian and Hungarian Medical Faculties (14th–20th Centuries)," in *Universitas Budensis 1395–1995, International Conference for the History of Universities on the*

Occasion of the 600th Anniversary of the Foundation of the University of Buda, ed. László Szögi and Júlia Varga (Budapest: Bak-Fisch KTF, 1997), 259–60.

13. Unless otherwise indicated, the information in the following paragraphs is from *Czas*, 23 October 1860 (Franz Joseph's letter), 31 October 1860, 10 November 1860, 15 November 1860, 22 November 1860, 19 December 1860, 20 December 1860, 24 February 1861, and 20 March 1861.
14. The first delegation (November 1860) included Bartynowski; the head of the Cracow Learned Society, Franciszek Stroński; the dean of the medical faculty, Józef Dietl; and the professors Edward Fierich and Józef Majer. The second delegation (March 1861) included Bartynowski, Dietl, and Fierich.
15. This liberal period was short lived, and by 1864 there was a *numerus clausus* for Polish students (*numerus clausus*, literally "closed number," refers to a limitation on the number of students of a certain nationality or confession). This was especially problematic because the Warsaw Main School was closed in 1869 and in its place a Russian-language university was founded. See, on the numerus clausus, Darius Staliūnas, *Making Russians: Meaning and Practice of Russification in Lithuania and Belarus after 1863* (Amsterdam: Rodopi, 2007), 100–105.
16. *Czas*, 17 January 1861.
17. Antoni Zygmunt Helcel, *Uwagi nad kwestyą językową w szkołach i uniwersytetach Galicyi i Krakowa, osnowane na liście odręcznym Jego C. K. Apostolskiej Mości z dnia 20 października 1860 r.* (Cracow: D. E. Friedlein, 1860); and Józef Dietl, *O reformie szkół krajowych*, vol. 1, *Stanowisko szkoły, rada szkolna krajowa, język* (Cracow: Drukarnia Uniwersytetu Jagiellońskiego, 1865).
18. The arguments in favor of German education were familiar to both scholars. Dietl, for example, countered in his book the arguments of Helfert's *Die sprachliche Gleichberechtigung in der Schule und ihre verfassungsmäßige Behandlung* (Prague: Tempsky, 1861).
19. See for a more detailed discussion Surman, "Symbolism."
20. Moklak, *W walce*, 47, 70.
21. The full text, in Polish, of the decree granting the Jagiellonian University bilingual status is found in *Czas*, 17 February 1861 (part 1), and *Czas*, 19 February 1861 (part 2).
22. Andreas Kappeler, *Kleine Geschichte der Ukraine*, 2nd ed. (Munich: Beck, 2000), 126.
23. Because of L'viv's location, the school had three names: the Cisars'ko-Korolivs'ka Akademična Himnazija (Цісарсько-Королівська Академічна Гімназія), the Cesarsko-Królewskie Akademickie Gimnazjum, and the Akademisches Gymnasium.
24. Response of the ministry to an inquiry of the Supreme Ruthenian Council, 9 January 1849, quoted in Moser, "Some Viennese Contributions," 141; on Ruthenian gymnasia, see Petro Polishchuk and Bohdan Struminsky, "Gymnasium," in *Internet Encyclopedia of Ukraine*, n.d., http://www.encyclopediaofukraine.com/display.asp?linkpath=pages\G\Y\Gymnasium.htm (accessed 1 December 2014).
25. Alexei I. Miller, *The Ukrainian Question: The Russian Empire and Nationalism in*

the Nineteenth Century (Budapest: Central European University Press, 2003), 79.

26. Moklak, *W walce*, 71.
27. Petition of 23 October 1869, reprinted in Moklak, *W walce*, appendix 3.
28. On the petition of 1868, see Svjatoslav Pacholkiv, *Emanzipation durch Bildung: Entwicklung und gesellschaftliche Rolle der ukrainischen Intelligenz im habsburgischen Galizien (1890–1914)* (Munich: Oldenbourg, 2002), 50. See also Moklak, *W walce*, 86–87. The question of universities, though, was beyond the competences of the Galician Diet, so both petitions were merely symbolic.
29. "Staatsgrundgesetz vom 21. Dezember 1867, über die allgemeinen Rechte der Staatsbürger für die im Reichsrate vertretenen Königreiche und Länder (R.G.Bl. 142/1867)," in *Reichs-Gesetz-Blatt für das Kaiserthum Oesterreich* (Vienna: k.k. Hof- und Staatsdruckerei, 1867), §19, p. 396.
30. Dorota Litwin, "Udział Polaków w naczelnych organach państwa austriackiego w okresie dualistycznej monarchii," in *Polacy w austriackim parlamencie w 130. rocznicę Klubu Polskiego*, ed. Władysław Kucharski (Lublin: Multico, 1997), esp. 110–13; and Judson, *Exclusive Revolutionaries*, esp. 102, 140.
31. Dan Gawrecki, "Pokus o založení univerzity v Opavě v roce 1870," *Acta historica et museologica Universitatis Silesianae Opaviensis* 2 (1995): 127–32.
32. Stanisław Starzyński, "Historya Uniwersytetu Lwowskiego 1869–1894," in *Historya Uniwersytetu Lwowskiego*, ed. Ludwik Finkel and Stanisław Starzyński (L'viv: C. K. Uniwersytet Lwowski, 1894), 50–54.
33. Mychajlo Slabošpyc'kyj, *Ukraïns'ki mecenaty: Narysy z istoriï ukraïns'koï kul'tury* (Kiev: Vydavnyctvo M.P. Koc', Jaroslaviv Val, 2001).
34. See, on the first years, Volodymyr Kačkan, "Anatol' Vachnjanyn i joho rol' v rozvytku kul'tury Zachidnoï Ukraïny," *Narodna tvorčist' ta etnohrafija*, nos. 1–2 (2003): 15–24.
35. Andriy Zayarnyuk, *Framing the Ukrainian Peasantry in Habsburg Galicia, 1846–1914* (Edmonton: Canadian Institute of Ukrainian Studies Press, 2013), 332–33.
36. Ivan Bartoševskyj, Mychajlo Hruševskyj, Ivan Dobrjanskyj, Stanislav Dnistrjanskyj, Oleksander Kolessa, Josif Komarnyckyj, Tyt. Myškovśkyj, Petro Stebel'skyk, and Kyrylo Studyn'skyj, "Zajava ruskych profesoriv universytetu u L'vovi," *Ruslan*, 6 (19) March 1907.
37. "Borot'ba o ruśkyj universytet," *Svoboda*, 6 (19) December 1907, quoted in Volodymyr Kačmar, "Problema zasnuvannja ukraïns'koho universytetu u L'vovi n rubeži XIX–XX st. u konteksti nacional'noho žyttja," *Visnyk L'vivskoho universytetu: Seria žurnalistyka* 26 (2004), http://www.lnu.edu.ua/faculty/jur/publications/visnyk26/Statti_Kachmar.htm.
38. Mychajlo Hruševs'kyj, "Do svitu: Z podorožnich vražen'," *Visnyk*, no. 44 (1909): 11–23.
39. Mychajlo Hruševs'kyj, "Sprava ukraïns'kych katedr i nasi naukovi potreby," in *Mychajlo Hruševs'kyj: Tvory v 50 tomach, serija "Suspil'no-polityčni tvory,"* vol. 1, *1894–1907*, ed. Jaroslav Daškevyč, Ihor Hyryč, and Pavlo Sochan' (L'viv: Cvit, 2002), 473–74.

40. See Kazimierz Twardowski, *Die Universität Lemberg: Materialien zur Beurteilung der Universitätsfrage* (Vienna: Selbstverlag, 1907). Much more emotional is the brochure entitled *The Attack on the Polish University in L'viv* by Stanisław Głąbiński, a politician of the National Democratic Party and a professor at the University of L'viv. Głąbiński, *Zamach na Uniwersytet Polski we Lwowie* (L'viv: Nakładem Autora, 1902).
41. See Jan Surman, "Du 'barbarisme' et 'civilisation': Le conflit entre les étudiantes polonais et ruthènes en 1907 et sa construction journalistique," in *La Galicie au temps des Habsbourg (1772–1918): Histoire, société, cultures en contact*, ed. Jacques Le Rider and Heinz Raschel (Tours: Presses universitaires François-Rabelais, 2010), 175–88.
42. Milada Sekyrková, "Rozdělení polytechniky—precedens pro univerzitu?," in Binder, Velek, and Křivohlavá, *Místo*, 231–40. For a broader context of the development of Czech-language scholarship in Bohemia, see Jan Janko and Soňa Štrbáňová, *Věda Purkyňovy doby* (Prague: Academia, 1988).
43. On the Czech academic contacts with the Kingdom of Croatia, see Jiří Frajdl, *Zápas s germanizací v období austroslavismu ve Slovinsku a Chorvatsku: Studijní texty* (Prague: Křesťanskosociální hnutí ve spolupráci s Obvodní radou Klubu českého pohraničí, 2005). See also Bečvářová, "Czech Mathematicians"; and Jan Petr, "Přínos Univerzity Karlovy k rozvoji slovanské jazykovědy na univerzitě Klimenta Ochridského v Sofii od doby jejího založení do současnost," *Práce z dějin slavistiky* 13 (1989): 7–41.
44. Most important, Purkyně argued for this in articles on the Bohemian academy of sciences that he published between 1861 and 1863 in the journal *Živa* (Alive), known under the header *Akademie*. These have been republished, among others, as Jan Evangelista Purkyně, *Akademie* (Prague: Nakladatelství Československé akademie věd, 1962).
45. The journals included *Časopis lékařův českých* (Journal of Czech physicians), *Listy chemické* (Chemical letters), and *Časopis pro pěstování matematiky a fyziky* (Journal for fostering mathematics and physics). For more on the societies and their journals, see Jiří Jindra, "České vědecké společnosti exaktních věd v 19. století," in Binder, Velek, and Křivohlavá, *Místo*, 401–20.
46. Karel Chodounský, "K padesátiletí 'Časopisu lékařův českých,'" Časopis lékařův českých 50, no. 53 (1911): 1602–4; and Ludmila Hlaváčková, "Čeština v medicíně a na pražské lékařské fakultě (1784–1918)," in Binder, Velek, and Křivohlavá, *Místo*, 327–44.
47. Martina Bečvářová, *Z historie Jednoty (1862–1869)* (Prague: Prometheus, 1999).
48. See the discourse of František Ladislav Rieger on the petition of 107 medical students from 1864, in *Stenografická zpráva XXXV. sezení / Stenographischer Bericht über die XXXV. Sitzung* (=Stenographische Berichte 1864 / Stenografické zprávy 1864), 30.
49. See the petition from 1863, "107 Hörer."
50. Karel Litsch, "F. L. Rieger o univerzitní otázce v českém sněmu roku 1866," *AUC-HUCP* 22, no. 1 (1982): 27–33.

51. *Utraqustisch* is a legal term meaning that the university should be perfectly bilingual; that is, each professor teaching in German should have a counterpart teaching in Czech. This is different from, for instance, the bilingualism at the University of L'viv, where Polish and Ruthenian were languages of instruction but there was no legal requirement that the number of professors teaching in each language should be equal.
52. Erich Schmied, "Die altösterreichische Gesetzgebung zur Prager Universität: Ein Beitrag zur Geschichte der Prager Universität bis 1918," in *Die Teilung der Prager Universität 1882 und die intellektuelle Desintegration in den böhmischen Ländern*, ed. Ferdinand Seibt (Munich: Oldenbourg, 1984), 18–19.
53. Petition of 1872, reprinted in Jaroslav Goll, *Rozdělení Pražské university Karlo-Ferdinandovy roku 1882 a počátek samostatné University české* (Prague: Nakl. Klubu historického, 1908), 79.
54. F[rantišek] J[osef] Studnička, "O rozvoji naši literatury fysikální za posledních padesáte let," *Časopis musea království českého* 50, no. 1 (1876): 46. See also Emil Brix, "Mentalität ist gut—die Teilung der Prager Universität 1882," *Österreichische Osthefte* 30, no. 3 (1988): 375–77.
55. See Kratochvíl, *Jan Evangelista Purkyně.*
56. Quoted in Jana Mandlerová, "K boji za zřízeni 2. české university w Brně 1882–1918 (Příspěvek o impulzech a struktuře české vědecké politiky na sklonku Rakousko-uherské monarchie)," *AUC-HUCP* 10, no. 1 (1969): 97. See also Jana Burešová, "Úsilí o obnovení Univerzity v Olomouci ve druhé polovině 19. a na začátku 20. stoleti," *Historická Olomouc* 11 (1998): 153–60.
57. Jaroslav Goll, *Der Hass der Völker und die österreichischen Universitäten* (Prague: Bursík & Kohout, 1902).
58. Memorandum of the Czech professors of the philosophical and medical faculties of Charles-Ferdinand University Prague, 18 February 1880, reprinted in Goll, *Hass der Völker*, 88–93; see also Miloslav Čedík, "Národností otazká a rozdělení pražské univerzity," *AUC-HUCP* 22, no. 1 (1982): esp. 15–21.
59. Goll, *Rozdělení*, 24–25.
60. Memorandum of the German professors of the philosophical and medical faculties of Charles-Ferdinand University Prague, December 1879, reprinted in Goll, *Hass der Völker*, 83–88.
61. Memorandum of the Czech professors, 89.
62. [Leopold Wittelshöfer], "Prager Briefe," *Wiener Medizinische Wochenschrift* 32, no. 44 (1882): 1315–16.
63. [Leopold Wittelshöfer], "Das Attentat auf die Prager Universität," *Wiener Medizinische Wochenschrift* 16, no. 20 (1866): 323–26; see also Seebacher, *"Freiheit der Naturforschung!,"* 115–16.
64. [Leopold Wittelshöfer], "Das Ende der deutschen Universität in Prag," *Wiener Medizinische Wochenschrift* 32, no. 7 (1882): 197–98.
65. [Wittelshöfer], "Das Ende."
66. See also Sosnowska, *Inna Galicja.*
67. On the chair of history (Hruševs'kyj), see Ihor Čornovol, *Pol'sko-ukraïns'ka*

uhoda 1890–1894 rr. (L'viv: L'vivs'ka Akademija Mistectv, 2000), 133–37; on the governor's influence on the establishment of a chair of Ruthenian language and literature (with special consideration of Church Slavonic history and literature), see AGAD, MWiO, Sygn. 121u, PA Studziński, Z. 103567, 25 November 1898; and below, pp. 208–10.

68. See the appointment acts in AT-OeStA/AVA Unterricht UM allg. Akten 1217, PA Fric, Z. 768, 11 October 1871.
69. See Hugo Victor Lane, "State Culture and National Identity in a Multi-ethnic Context: Lemberg, 1772–1914" (PhD diss., University of Michigan, 1999), 226–36. While there are still considerable disparities between Polish and Ukrainian historians' accounts of the university's past, since the communist period both groups have accepted and celebrated 1661 as the year of the founding of the university; see, for example, *Universitati Leopoliensi, Trecentesimum Quinquagesimum Anniversarium Suae Fundationis Celebranti: In Memoriam* (Cracow: Polska Akademia Umiejętności, 2011); and J. K. Lazarenko, *300 rokiv L'vivs'koho uniwersytetu* (L'viv: Vydavnyctvo L'vivs'koho uniwersytetu, 1961), as well as more recent Ukrainian publications on this issue.
70. For this argument see Joseph Kalousek, *Über die Nationalität Karl's IV: Entgegnung auf einen von Prof. Dr. J. Loserth unter demselben Titel in den "Mittheilungen des Vereins für die Geschichte der Deutschen Böhmen" veröffentlichten Aufsatz* (Prague: W. Nagel, Selbstverlag, 1879), 15 (quoting from Loserth's article), 17. The reason for this particular controversy was Kalousek's monograph *Karel IV, otec vlasti: Ku pětistoleté památce jeho umrtí*, where he claimed Charles IV was of "Czech nationality" and had a lively interest in Czech culture. Kalousek, *Karel IV, otec vlasti: Ku pětistoleté památce jeho umrtí* (Prague: Jos. R. Vilímek, 1878).
71. See the discussion in Václav Vaněček, "'V Praze budou . . . dvě university . . .' (Zák. z. 28.2.1882, č.24, Ř.z., § 1)," *AUC-HUCP* 22, no. 1 (1982): 7–14.
72. Jiří Müller, "'Insigniáda' a dobové pojetí nacionalismu na stránkách Přítomnosti," *Marginalia historica: Časopis pro dějiny vzdělanosti a kultury* 4, no. 1 (2013): 9–52.
73. Dietl, *O reformie*, 140.
74. See, e.g., Julian Romanczuk [Юліан Романчук], *Die Ruthenen und ihre Gegner in Galizien* (Vienna: C. W. Stern, 1902), esp. 25–30.
75. For example, Twardowski repeatedly claimed that the number of Ruthenian students at the secular faculties of the University of L'viv diminished in the last years before 1907. Twardowski, *Die Universität Lemberg.*
76. Adolf Gürtler, *Deutsche Hochschulnot in Österreich: Referat erstattet in der Monatsversammlung der Vereinigung deutscher Hochschullehrer in Graz am 12. Februar 1913* (Graz: Im Selbstverlage der Vereinigung, 1913), 10.
77. The official German name of Chernivtsi University was Franz-Josephs-Universität Czernowitz; from 1918 to 1940, it was the Universitatea Regele Carol I din Cernăuţi (Romania), from 1940 Чернівецький Державний Університет, from 1989 Чернівецький Державний університет ім. Юрія Федьковича,

and from 1990 Чернівецький національний ім. Юрія Федьковича (Yuriy Fedkovych Chernivtsi National University).

78. Yiddish remained an unofficial "national language" until the end of the monarchy, although proposals to make Yiddish one of the official languages of the Habsburg monarchy were made in Chernivtsi in 1910. Eleonore Lappin, "Die Czernowitzer Sprachkonferenz (1908) und der Streit um die jüdische Nationalsprache," in *Minikosmos Bukowina—Kulturleistungen eines Landstriches: Beiträge der internationalen wissenschaftlichen Konferenz anlässlich des 130. Jahrestages der Czernowitzer Universität am 3.–4.10.2005 / Mini-kosmos Bukovyny: Material Mižnarodnoï konferencii, pryscjačenoï 130-riččjo Černivec'koho un-tu (3–4 žovtnja 2005 r.)*, ed. Österreich Kooperation, Naukovo-doslidnyj Centr Bukovynoznavstva pry Bukowina Zentrum an der Nationalen Jurij-Fedkowitsch Universität Czernowitz, and Avstrijśka kooperacia Černivećkomu nacional'nomu universyteti im. Jorija Fed'kovyča (Chernivtsi: Selena Bukowina/ Zelena Bukovyna, 2006), 260–70.

79. Karl Emil Franzos, *Aus Halb-Asien*, vol. 2, 3rd rev. and exp. ed. (Stuttgart: Adolf Bonz, 1889).

80. Hannelore Burger, "Das Problem der Unterrichtssprache an der Universität Czernowitz," in *Glanz und Elend der Peripherie: 120 Jahre Universität Czernowitz. Eine Veröffentlichung des Österreichischen Ost- und Südeuropainstituts*, ed. Ilona Slawinski and Joseph P. Strelka (Vienna: Peter Lang, 1998), 66.

81. On Hlávka, see the introduction to Jiři Beran, *Vznik České akademie věd a umění v dokumentech* (Prague: Ústřední Archiv Československé akademie věd, 1989).

82. Quoted in Emanuel Turczynski, "Czernowitz, eine vom Bildungsbürgertum errungene Universität im Dienst staatlicher Bildungs- und Wissenschaftsförderung," in *Universitäten im östlichen Mitteleuropa: Zwischen Kirche, Staat und Nation. Sozialgeschichtliche und politische Entwicklungen*, ed. Peter Wörster (Munich: Oldenbourg, 2008), 215.

83. Quoted in Akademischer Senat, *Die Franz-Josephs Universität*, xxiii.

84. See fragments of Stremayr's argumentation for the Austrian Parliament, 9 December 1874, in Akademischer Senat, *Die Franz-Josephs Universität*, xvii; and his own account in Carl von Stremayr, *Erinnerungen aus dem Leben* (Vienna: Eigenverlag, 1899), 55.

85. AT-OeStA/AVA Unterricht UM allg. Akten 1184, Z. 15700, 1874.

86. AT-OeStA/AVA Unterricht UM allg. Akten 1184, Z. 15700, 1874. See also Irina Livezeanu, *Cultural Politics in Greater Romania: Regionalism, Nation Building, and Ethnic Struggle, 1918–1930* (Ithaca, NY: Cornell University Press, 2000), 227–31, esp. 229.

87. Jeroen van Drunen, "'A Sanguine Bunch': Regional Identification in Habsburg Bukovina, 1774–1919" (PhD diss., University of Amsterdam, 2013), 281.

88. Akademischer Senat, *Die Franz-Josephs Universität*, xxiii–xix.

89. Marianna Hausleitner, *Die Rumänisierung der Bukowina: Die Durchsetzung des nationalstaatlichen Anspruchs Grossrumäniens 1918–1944* (Munich:

Oldenbourg, 2001), esp. 50–82, with presentation of different national organizations. On Hasmonäa, established by Igel, see Marianne Hirsch and Leo Spitzer, *Ghosts of Home: The Afterlife of Czernowitz in Jewish Memory* (Berkeley: University of California Press, 2010), 41.

90. Volodymyr Kačmar, "Suspil'no-polityčnyje vidlunnja secesiï ukraïns'kich studentiv z L'vivskoho Universitetu v grudni 1901 roku," *Visnyk L'vivskoho universytetu: Seria istoryčna* 34 (1999): 289–99; see also Vasyl' Mudryj, *Zmahannja za ukraïns'kyj universytet v Halyčyni* (L'viv: Vydavnyctvo Naukovoho Tovarystva imeny Ševčenka, 1999 [1923/1948]), 54–58.
91. Drunen, "'A Sanguine Bunch,'" 281.
92. Oberkofler, *Die Rechtslehre*, 46. A similar argument was used several times in Graz toward southern Slavs; see, e.g., *Denkschrift für die Vervollständigung der k.k. Karl-Franzens-Universität zu Graz* (Graz: Josef A. Kienreich, 1861), 4.
93. Erika Weinzierl, "Aehrenthal and the Italian University Question," in *Intellectual and Social Developments in the Habsburg Empire from Maria Theresa to World War I: Essays Dedicated to Robert S. Kann*, ed. Stanley B. Winters and Robert S. Kann (Boulder, CO: East European Quarterly, 1975; distributed by Columbia University Press), 241–70.
94. Maria Kostner, "Die Geschichte der italienischen Universitätsfrage in der österreichisch-ungarischen Monarchie von 1864–1914" (PhD diss., University of Innsbruck, 1970), 383; and Angelo Ara, "Le problème de l'université italienne en Autriche (1866–1914)," *Études danubiennes* 3, no. 2 (1987): 157–68.
95. Quoted in Andreas Bösche, *Zwischen Kaiser Franz Joseph I. und Schönerer: Die Innsbrucker Universität und ihre Studentenverbindungen 1859–1918* (Innsbruck: StudienVerlag, 2008), 111.
96. See the documents on the reappointment of the chair after the death of Virgin Mayrhofen, who had taught in both languages: UAI, MED, 1876/77, 29 June 1877. The same claim can be found during the discussion on the disciplinary commission against obstetrician Ludwig Kleinwächler. See the claims of the Tyrolean Diet in AT-OeStA/AVA Unterricht UM allg. Akten 1211, PA Stark, 3 December 1880, Z. 4131; 19 November 1880, Z. 5739.
97. Laurence Cole, *Für Gott, Kaiser und Vaterland: Nationale Identität der deutschsprachigen Bevölkerung Tirols 1860–1914* (Frankfurt am Main: Campus, 2000); and Chiocchetti and Starec, "'Ladin Song,'" 64.
98. Kostner, *Die Geschichte*, 81–83, 338–39.
99. Magda Perricelli, "'O Trieste o nulla!': I 'fatti di Innsbruck' nella stampa quotidiana del Regno d'Italia," in *Università e nazionalismi: Innsbruck 1904 e l'assalto alla Facoltà di giurisprudenza italiana*, ed. Günther Pallaver and Michael Gehler (Trento: Fondazione Museo storico del Trentino, 2010), 176–83.
100. See below on Ludwig Wahrmund, or the case of Carl Laker, who accused the senate of the University of Graz of not promoting him despite his long-term affiliation as a Privatdozent, leading to a political discussion, which met with a fierce answer from the university, also reprinted in the press. Der akademische Senat der k.k. Karl-Franzens-Universität Graz, "Erklärung des akademischen

Senates der k.k. Karl-Franzens-Universität Graz anläßlich der Interpellation des Reichsrathsabgeordneten Malik und Genossen, die Verzichtleistung des Dr. Laker auf die Dozenturbetreffend," *Tagespost* (Graz), 11 April 1901, 2.

101. AGAD, MWiO, Sygn. 65u, PA Creizenach, Z. 19335, 3 December 1882.
102. The data are from my own calculations based on 501 known appointments to the medical and philosophical faculties in Graz, Innsbruck, Vienna, and Prague (excluding the Czech University), including archival materials on Chernivtsi University held by the State Archives in Vienna.
103. Bernhard vom Brocke, *Hochschul- und Wissenschaftspolitik in Preußen und im Deutschen Kaiserreich 1882–1907: Das "System Althoff"* (Stuttgart: Klett-Cotta, 1980).
104. See William Lloyd Mathes, "The Struggle for University Autonomy in the Russian Empire during the First Decade of the Reign of Alexander II (1855–1866)" (PhD diss., Columbia University, 1966); and Alen Sinel, *The Classroom and the Chancellery: State Educational Reform in Russia under Count Dmitry Tolstoi* (Cambridge, MA: Harvard University Press, 1973), 85–129.
105. See, e.g., Stremayr, *Erinnerungen aus dem Leben*, 47.
106. See, for example, the procedures for finding a professor of midwifery in Prague in 1852 (AT-OeStA/AVA Unterricht UM allg. Akten 1209, PA Kiwisch, Z. 6683/546, 28 July 1852) or a professor of surgery at the medical-surgical academy in Innsbruck in 1859 (AT-OeStA/AVA Unterricht UM allg. Akten 1055, PA Joseph Fischer, Z. 14179/519, 17 September 1859).
107. AGAD, MWiO, Sygn. 120u, PA Rehmann, Z. 187, 15 September 1882. The documents referred to were Z. 319, 18 March 1875 (proposal of the faculty to hold a contest), and Z. 5048, 17 September 1875 (rejection by the ministry).
108. See the appointment procedures for botany in Innsbruck: UAI, PF, Z. 249 (letters of recommendation by Julius Wiesner and Heinrich Moritz Wilkomm), and Z. 264, 29 March 1888, in which Otto Stampf, at the time a Privatdozent in Vienna, proposes himself as a suitable candidate.
109. For example, even before the chair of surgery in Innsbruck was advertised in 1903, the Prague professors Rudolf Jaksch-Wartenhorst and Anton Wölfler wrote letters presenting the Prague Privatdozent Hermann Schloffer as the best candidate. See August Jaksch-Wartenhorst to (probably) Moritz Löwith, UAI, MF, 7 March 1903; and the letter of Anton Wölfler, 9 June 1903, ad 24130, med 856 1902/1903.
110. See, for example, UAI, Berufung Zoologie, 4/ Carl Heider 1893/94 (Z. Ph. 386, 16 April 1894; a request for expert opinions was sent to the full professors Carl Claus in Vienna, Berthold Hatschek in Prague, Ernst Ehlers in Göttingen, and Franz Eilhard Schulze in Berlin). This practice also took place in several other appointment procedures in Innsbruck.
111. See the appointment of Franz Meyer, in AT-OeStA/AVA Unterricht UM allg. Akten 628, PA Meyer, Z. 32921, 22 September 1904.
112. See, though, AT-OeStA/AVA Unterricht UM allg. Akten 622, PA Durig, 8 March 1918, where the Viennese dean provided his own very unfavorable opinion on

the scholar proposed primo loco; in Prague in 1885, the dean recommended Max Grünert for the chair of Semitic languages (ÚDAUK, FF NU, Inv.č. 229, K/XIVb, Kart. 11, Obsazování mimořádných profesur: [a] pro německou filologii [Jacob Minor, Hans Lambel, dr. Seuffert], [b] pro ang. Filologii [Alois Brandl]—společný spis, 29 December 1885).

113. The scholar appointed, Fritz Hartmann, a Privatdozent in Graz, had been proposed by the faculty *secundo loco* and *ex aequo* (in the same place). See AT-OeStA/AVA Unterricht UM allg. Akten 919, PA Fritz Hartmann, Z. 35079, 4 September 1907.
114. AT-OeStA/AVA Unterricht UM allg. Akten 1216, PA Ambros, Z. 8252, 2 September 1869.
115. AT-OeStA/AVA Unterricht UM allg. Akten 1211, PA Stark, 3 December 1880, Z. 4131, 19 November 1880, Z. 5739; and AT-OeStA/AVA Unterricht UM allg. Akten 1056, PA Kleinwächter, Z. 14759, 13 September 1882. The church's involvement, although not specified, was mentioned in most obituaries and subsequent encyclopedia articles, although I found no records in the press from this period. A similar conflict over the division of competences was mentioned also by Otto Franqué as a main reason for his leaving Prague in 1907. NA, MKV/R, inv.č. 2, fasz.95, PA Franque, Z. 30072, undated.
116. AT-OeStA/AVA Unterricht UM allg. Akten 1208, PA Brandl, Z. 16454, 18 September 1886, repeated on 22 September 1886, repeated on 1 October 1886.
117. Richard Schober, *Theodor Freiherr von Kathrein (1842–1916), Landeshauptmann von Tirol: Briefe und Dokumente zur katholisch-konservativen Politik um die Jahrhundertwende* (Innsbruck: Wagner, 1992); on a similar influence on the part of Vorarlberg governor Adolf Rhomberg, see Alois Niederstätter, "Feuerstein ist nach wie vor bei den Veilchenblauen: Die Briefe von Richard Wollek an den Vorarlberger Landeshauptmann Adolf Rhomberg (1897/98)," *Alemannia Studens: Mitteilungen des Vereins für Vorarlberger Bildungs- und Studenten-Geschichte* 2 (1992): 13–64.
118. For L'viv, see the writings of the provincial government on the situation: AGAD, MWiO, Sygn. 112u, PA Żmurko, Z. 6292, 21 September 1871. For Prague, see the notes in the nomination act for Ivan Horbačevs'kyj: AT-OeStA/AVA Unterricht UM allg. Akten 1209, PA Horbachevsky, Z. 13251, 15 August 1883.
119. For example, for the chair of histology in L'viv in 1896, the commission consisted of one member of the provincial government, a deputy of the Galician Education Authority (Rada Szkolna Krajowa), a provincial referent for health issues, and two deputies of the University of L'viv. AGAD, MWiO, Sygn. 405u, Z. 467, 16 January 1896.
120. AGAD, MWiO, Sygn. 118u, PA Askenazy, Z. 1838, 27 June 1906. On Askenazy's worldview, see Piotr Wróbel, "Szymon Askenazy," in *Nation and History: Polish Historians from the Enlightenment to the Second World War*, ed. Peter Brock, Piotr Wróbel, and John Stanley (Toronto: University of Toronto Press, 2006), 221–45.
121. On the embryologist Édouard van Beneden (proposed primo et unico loco by the

German University in Prague), see AT-OeStA/AVA Unterricht UM allg. Akten 1217, PA Hatschek, Z. 10408, 27 June 1885; on the historian of the German language Matthias Lexer, see AT-OeStA/AVA Unterricht UM allg. Akten 936; PA Hentzel, Z. 5695, 9 July 1868; on the astronomer Johannes Hartmann, see AT-OeStA/AVA Unterricht UM allg. Akten 675, PA Oppenheim Z. 20193, 26 May 1911; on the ophthalmologist Otto Becker, see AT-OeStA/AVA Unterricht UM allg. Akten 1211, PA Sattler, Z. 1026, 26 February 1886; on the physicist Czesław Białobrzeski, see AGAD, MWiO, Sygn. 393u, Z. 55310, 29 November 1913 (for the proposals of the faculty, see Z. 936, 10 July 1913; Z. 78, 10 November 1913; Z. 703, 28 March 1914; Z. 1242, 22 July 1914; for the rejection of the appointment by the Ministry of Finance, see Z. 25923, 6 April 1914; Z. 42691/14, 4 June 1915); and on the chemist Friedrich Kekulé, see AT-OeStA/AVA Unterricht UM allg. Akten 679, PA Franz Schneider, Z. 6978, 3 August 1870 (all named were proposed primo loco).

122. Pogatscher was appointed full professor. AT-OeStA/AVA Unterricht UM allg. Akten 939, PA Pogatscher, Z. 39641, 28 September 1908.
123. AT-OeStA/AVA Unterricht UM allg. Akten 674, PA Luick, Z. 86442, 14 April 1908.
124. NA, MKV/R, fasc. 110, PA Brotanek, Z. 1023, 11 January 1909.
125. In Innsbruck, Vienna, Graz, and the German University in Prague.
126. ÚDAUK, LF NU, Kart. 2, PA Grosser, Z. 1038, 31 March 1909; Z. 21354, 31 May 1909; unnumbered letter from the medical faculty of the German University in Prague, 8 June 1909; Z. 34041, 13 August 1909 (date of appointment: 11 August 1909).
127. See, for example, AT-OeStA/AVA Unterricht UM allg. Akten 663, PA Arnim, Z. 35973, 11 December 1899; Z. 657, 17 January 1900.
128. In 1909 the academy recommended Wilhelm Trabert from Innsbruck (unico loco in the faculty proposal) for this position and in 1916 supported his successor, Felix Exner from Innsbruck, ranked first in the faculty terna, proposing him unico loco. See AT-OeStA/AVA Unterricht UM allg. Akten 681, PA Trabert, Z. 39635, 18 November 1909; and AT-OeStA/AVA Unterricht UM allg. Akten 666, PA Felix Exner, Z. 37893, 26 November 1916.
129. AT-OeStA/AVA Unterricht UM allg. Akten 665, PA Büdinger, Z. 3698, 20 June 1872, emperor's annotation from 9 August 1872; and AT-OeStA/AVA Unterricht UM allg. Akten 940a, PA Franz Schulze, Z. 6945, 22 [month illegible] 1872, emperor's annotation from 20 June 1872. After a second proposal, which stressed the particular qualities of the scholars from abroad, Franz Schulze from Rostock was appointed.
130. AT-OeStA/AVA Unterricht UM allg. Akten 940a, PA Franz Schulze, Z. 6945, 22 [month illegible] 1872, emperor's annotation from 20 June 1872.
131. For the physicist Władysław Natanson in Cracow, see his letter to Ludwik Gumplowicz, 16 February 1889, Collection of the Manuscripts of the Jagiellonian Library, Cracow, sign. 9007 III, vol. 6, fol. 220–22.
132. See the case of Rudolf Dittmar from Graz in 1913: AT-OeStA/AVA Unterricht UM allg. Akten 934, PA Dittmar.

133. See the documents for Dittmar's case: UAG, PF, Z. 1772 ex. 1912/13, 14 June 1913.
134. See the fifty-one-page text of Jan Kvíčala defending his student Justin Prášek. The German version, entitled *Private und vertrauliche Denkschrift*, was sent to the ministry; see NA, MKV/R, inv.č. 9, fasc. 117, PA Prasek.
135. Pavel Kolář, *Geschichtswissenschaft in Zentraleuropa: Die Universitäten Prag, Wien und Berlin um 1900* (Leipzig: Leipziger Universitätsverlag, 2008); for other examples of such practices, see Armin Teske, *Marian Smoluchowski: Leben und Werk* (1955; Wrocław: Zakład Narodowy im. Ossolińskich, Polska Akademia Nauk, 1977), esp. 118; and Raimund von Klebesberg, *Innsbrucker Erinnerungen, 1902–1952* (Innsbruck: Wagner, 1953), 14–19.
136. Mark Granovetter, "The Strength of Weak Ties: A Network Theory Revisited," *Sociological Theory* 1 (1983): 201–33; for an interesting use of *negative ties*, a term rather seldom used in network theory, see Phillip Bonacich and Paulette Lloyd, "Calculating Status with Negative Relations," *Social Networks* 26, no. 4 (2004): 331–38.
137. See Natanson to Gumplowicz, 16 February 1889, Collection of the Manuscripts of the Jagiellonian Library, Cracow, sign. 9007 III, vol. 6, fol. 220–22; and the protocol of the exam, UAG, PF, Z. 205, [day and month illegible] 1888.
138. AT-OeStA/AVA Unterricht UM allg. Akten 632, PA Leopold Unger.
139. At first the rector wrote "general conditions," and then "general" was crossed out and replaced by "special."
140. Letter from the rector of the German University in Prague, answering a question of Chernivtsi University (Czernowitz, 14 January 1908, Z. 455) concerning the possibility of transferring Mahler's habilitation, Prague, 24 February 1908, ÚDAUK, FF NU, Inv.č. 249, fasz. 12, L/53 PA Mahler.
141. AT-OeStA/AVA Unterricht UM allg. Akten 1057, PA Nagy, Z. 4094, 2 February 1916.
142. In this regard, see unsorted material highlighting the assistant positions of university scholars in Lesky, *Vienna Medical School*.
143. To pick two Bohemian examples: Julius Sachs moved with Purkyně from Wrocław/Breslau to Prague, while Karel Maydl moved to Innsbruck and later to Vienna to serve as an assistant to Eduard Albert before occupying the chair of surgery at the Czech University in Prague.
144. For example, the anthropologist Izydor Kopernicki promoted the young physician Julian Talko-Hryncewicz as his successor, helping him to establish contacts and gain financial support. See Michał Ćwirko-Godycki, *Izydor Kopernicki* (Poznań: Poznańskie Towarzystwo Przyjaciół Nauk, 1948), esp. 192–94. See also, on the social dimension of elite reproduction, Józef Buszko, *Społeczno-polityczne oblicze Uniwersytetu Jagiellońskiego w dobie autonomii galicyjskiej (1869–1914)* (Cracow: Uniwersytet Jagielloński, Państwowe Wydawnictwo Naukowe, 1963).
145. See, e.g., Jiři Suk, "Studie o počátích Gollovy školy," *Acta Universitatis Carolinae—Philosophica et Historica* 3 (1993): 147–69.
146. See the rejection of Franz Torggler in Innsbruck in 1888 in AT-OeStA/AVA Unterricht UM allg. Akten 1058, PA Torggler.

147. See the acceptance of Franz Torggler in Innsbruck in 1890 in AT-OeStA/AVA Unterricht UM allg. Akten 1058, PA Torggler.
148. See the commission statement on the occasion of the search for a professor of Sanskrit in L'viv, DALO, F. 26, Op. 7, Spr. 269, p. 18, 3 March 1884.
149. My own calculations are based on the data in the databases for all universities, 1848–1918; only full years are counted. For references, Appendix 2.
150. To my knowledge, the issue of pre-habilitation scholars has been not researched in detail, although the Czech and Polish bibliographical works on the respective faculties, as well as (most) overviews of developments in specific disciplines in the Graz and Innsbruck faculties, include information on assistants, considering this as the first academic position.
151. "Studien-Hofcomissions-Decret vom 20. September 1811, Z. 1641," reprinted in Friedrich Schweickhardt, ed., *Sammlung der für die österreichischen Universitäten giltigen Gesetze und Verordnungen*, 2nd ed. (Vienna: k.k. Schulbuchverlag in Comission bei Manz'schen k.k. Hofverlags- und Universitäts-Buchhandlung, 1885), 163–64 (point no. 5 and Schweickhardt's commentary).
152. "Studienhofkommissionsdekret vom 12. April 1833, Z. 984, PGS Bd. 61, Nr. 57. S. 104, an sämtliche Länderstellen, betreffend die Verleihung von Lehramtsadjunkten- oder Assistentenstellen nur an ledige Individuen," reprinted in Mannagetta and Kelle, *Die österreichischen Universitätsgesetze*, 208 (no. 225). Longtime assistants were exempted from this rule.
153. See the analysis of scholars (predominantly historians) in Galicia in Stefan Ciara, "Finanzielle Probleme galizischer Wissenschaftler um die Wende des 19. zum 20. Jahrhundert," *Mitteilungen des Österreichischen Staatsarchivs* 53 (2009): 313–33.
154. Although no thorough analysis has been carried out here, the mere fact that fewer and fewer scholars were coming from the nobility, while most scholars in the second half of the nineteenth century were the sons of state officials of lower rank and, in only a few cases, from the countryside, points in this direction.
155. See, e.g., *Hohes Abgeordnetenhaus: Petition der Privatdozenten-Vertreter der österreichischen Hochschulen in Angelegenheit von Standesfragen* (Vienna: Fischer, 1910).
156. In Vienna in 1910, 40 percent of Privatdozenten did not have a professional occupation listed in the staff catalogs (mostly in practical specialties), 20 percent were chiefs of clinics, 15 percent were assistants, and slightly fewer were chief physicians. See *Übersicht der akademischen Behörden, Professoren, Privatdocenten, Lehrer, Beamten etc. an der k.k. Universität zu Wien für das Studien-Jahr 1910/11* (Vienna: Adolf Holzhausen, 1910), 15–25.
157. See, on the appointment of the chair of Polish history, where the University of L'viv decided to promote young local historians, DALO, F. 26, Op. 7, Spr. 327, B. 52, 24 July 1891.
158. For the chair of medical chemistry at the University of Graz, the ministry preferred Karl Hofmann, a Privatdozent in Vienna, over the more highly esteemed Ernst Salkowski, an assistant to Rudolf Virchow in Berlin, arguing that the

appointment of the young Austrian Hofmann would "have the most beneficial and animating impression for the Privatdozenten in Austria." AT-OeStA/AVA Unterricht UM allg. Akten 1072, PA Karl Hofmann, Z. 10534, 16 August 1873.

159. See, for example, the explanations of Edmund Neusser for the appointment of Adolf Strümpell from Leipzig in AT-OeStA/AVA Unterricht UM allg. Akten 632, PA Strümpell, Z. 3913, 19 February 1909.

160. For an overview, see Hans-Heinz Eulner, *Die Entwicklung der medizinischen Spezialfächer an den Universitäten des deutschen Sprachgebietes* (Stuttgart: Enke, 1970).

161. Before 1880, there were nine habilitations for pediatric medicine. With three exceptions, those scholars made, however, no further progress at the universities. In contrast, most scholars who habilitated in this subject after 1880 achieved professorial positions.

162. AT-OeStA/AVA Unterricht UM allg. Akten 633, PA Widerhofer, 25 April 1887.

163. On the forensic doctor Karl Ipsen, see AT-OeStA/AVA Unterricht UM allg. Akten 1073, PA Ipsen, Z. 12865, 18 June 1894; Z. 10128, 13 May 1896. On the astronomer Karl Weineck/Weinek László, see AT-OeStA/AVA Unterricht UM allg. Akten 1211, PA Weineck, Z. 12092, 10 July 1883. On Ferenc/Franz Tangl, see AT-OeStA/AVA Unterricht UM allg. Akten 1058, PA Tangl, Z. 32116, 21 September 1904.

164. Such cases included, among others, those of the hygienist Alois Lode (born in Orăştie/Szászváros/Broos, accredited to Most/Brüx in Bohemia), the internist Julius Mannaberg (Pest, Vienna), and the pathologist Arthur Biedl (Comloşu Mic/Kiskomlos/Ostern [Kleinkomlosch], accredited to Lower Austria). See the respective curricula vitae in AT-OeStA/AVA Unterricht UM allg. Akten 1056, PA Lode; AT-OeStA/AVA Unterricht UM allg. Akten 627, PA Mannaberg; and AT-OeStA/AVA Unterricht UM allg. Akten 620, PA Biedl.

165. AGAD, MWiO, Sygn. 122u, PA Zalewski, acts 1888–92, including an expert report by the Viennese biologist Anton Kerner.

166. AT-OeStA/AVA Unterricht UM allg. Akten 629, PA Gustav Paul, Z. 25592 (habilitation record), esp. Z. 27.175 (ministry's decision) from 1906.

167. See the habilitation of Władysław Kretkowski in mathematics (at first "with exclusion of infinitesimal calculus") in AGAD, MWiO, Sygn. 119u, PA Kretkowski, Z. 409, 10 July 1881, and the letter of Emil Weyr from October 1881, in the same file; the issue of the precise designation of Kretkowski's specialization was also debated in the faculty.

168. "Verordnung des Ministers für K. u. U. vom 11. Februar 1888, Z. 2390, betreffend die Habilitierung der Privatdozenten an Universitäten," reprinted in Mannagetta and Kelle, *Die österreichischen Universitätsgesetze*, 169, §2.

169. On social selection through habilitation, see also Susanne Preglau-Hämmerle, *Die politische und soziale Funktion der österreichischen Universität: Von den Anfängen bis zur Gegenwart* (Innsbruck: Inn-Verlag, 1986), 99.

170. "Verordnung des Ministers für K. u. U. vom 11. Februar 1888, Z. 2390, betreffend die Habilitierung der Privatdozenten an Universitäten," §14.

171. AUJ, WF II 121, PA Studzinski, Cyryl, Z. 503, 2 April 1897; Z. 8176, 8 April 1897. The habilitation of the Ruthenian linguist Kyrilo Studyn'skyj was, however, a political matter and can only be seen as an exception.
172. See the argument for the acceptance of the Privatdozentur of Johann Tollinger, director of the School of Agriculture in Rotholz (Landwirtschaftsanstalt in Rotholz), in AT-OeStA/AVA Unterricht UM allg. Akten 1076, PA Tollinger (1887); and the rejection of Leopold Kann, a teacher in Plzeň/Pilsen, in NA, MKV/R, inv.č. 9, fasc. 114, PA Kann (1905) (for a similar case from 1918, similarly linked to Plzeň/Pilsen, see NA, MKV/R, inv.č. 9, fasc. 118, PA Sokol).
173. See the careers of Erwin Hanslick (AT-OeStA/AVA Unterricht UM allg. Akten 668, PA Erwin Hanslik, 30 June 1910 ad Z. 27757), Eugen Herzog (ÚDAUK, FF NU, PA Eugen Herzog, 16 July 1902, Z. 1186; 11 February 1902; Z. 830, 15 January 1909, Z. 764), and David Herzog, who moved from Prague to Graz owing to obligations in the Jewish community (AT-OeStA/AVA Unterricht UM allg. Akten 936, PA Herzog [1909]).
174. In Galicia, for example, Privatdozenten taught eight hours fewer at gymnasia, while retaining a full salary. However, the administration of primary and secondary education was highly autonomous there. Jerzy Starnawski, "Towarzystwa naukowe z zakresie humanistyki na terenie Galicji," in *Galicja i jej dziedzictwo*, vol. 3, *Nauka i oświata*, ed. Andrzej Meissner and Jerzy Wyrozumski (Rzeszów: Wydawnictwo Wyższej Szkoły Pedagogicznej, 1995), 52.
175. See, e.g., Franz Eulenburg, *Der "Akademische Nachwuchs": Eine Untersuchung über die Lage und die Aufgaben der Extraordinarien und Privatdozenten* (Leipzig: B. G. Teubner, 1908); and *Hohes Abgeordnetenhaus.*
176. NA, fond České místodržitelství v Praze, inv.č. 32, fasc. 198, PA Matiegka, Z. 18570, 31 May 1904 (title and character of associate professor); Z. 33803, 28 June 1908 (associate professor; Moritz Hoernes's appointment of 1907 was explicitly mentioned in the appointment records).
177. AT-OeStA/AVA Unterricht UM allg. Akten 628, PA Josef Novak senior, Z. 6712, 11 May 1875. The chair had been proposed by the University of Vienna already in 1871, with reference to the "sanitary construction of schools, hospitals, . . . prisons, further with facilities of colonies"; see AT-UAW, Med. S. 17, 19 January 1871, Z. 285 and Z. 345.
178. AT-OeStA/AVA Unterricht UM allg. Akten 1211, PA Soyka, Z. 6036, 28 March 1884.
179. See, e.g., the application of the faculty of the German University in Prague for the division of the chairs of chemistry into organic and inorganic. NA, MKV/R, inv.č. 9, fasc. 118, PA Rothmund, Z. 835, 4 March 1913.
180. The following chairs were proposed: (1) number theory and higher algebra, (2) mathematical analysis, and (3) geometry; see UAG, PF, Z. 2302, 4 July 1907 (dated 3 July 1907). The list included universities from the Habsburg Empire, the German Empire, the Russian Empire, France, and Italy and, to my knowledge, considered all universities in these empires and countries.
181. On balneology (Enoch Kisch), see AT-OeStA/AVA Unterricht UM allg. Akten 1209, PA Kisch, Z. 3447, 1 March 1884; on medieval history (Johann Lechner),

see AT-OeStA/AVA Unterricht UM allg. Akten 673, PA Lechner; and on comparative anatomy and plant physiology (František Vejdovský), see NA, MKV/R, inv.č. 9, fasc. 120, PA Vejdovsky, 17 December 1889, Z. 217, 28 October 1892, Z. 24049.

182. These were explicitly called such; see AT-OeStA/AVA Unterricht UM allg. Akten 620, PA Albert, Z. 92, 25 January 1881.

183. Walter Höflechner, "Zum Einfluss des deutschen Hochschulwesens auf Österreich in den Jahren 1875–1914," in *Wissenschaftsgeschichte und Wissenschaftspolitik im Industriezeitalter: Das "System Althoff" in historischer Perspektive*, ed. Bernhard vom Brocke (Hildesheim: Lax, 1991), 166.

184. The numbers are from my own calculations based on the printed *Personalverzeichnisse* for the winter term of 1910/11.

185. As Lotman remains a rather peripheral figure in the current debates, there are almost no comparisons of these two theories, at least in their consequences for macrolevel relationships. See, however, Julie A. Buckler, "Writing in a Polluted Semiosphere: Everyday Life in Lotman, Foucault, and De Certeau," in *Lotman and Cultural Studies: Encounters and Extensions*, ed. Andreas Schönle (Madison: University of Wisconsin Press, 2006), 320–44; and, more related to ideas presented here, Tõnis Saart, "Construction of Peripheries: Foucault vs. Lotman and Potential Peripherization of New Member States in the EU," in *European Peripheries*, ed. György Andrássy, Jyrki Käkönen, and Noémi Nagy (Pécs: Publikon, 2012), 21–38.

186. Urszula Perkowska, "La genèse et la caractéristique de la loi sur les écoles supérieures du 13 juillet 1920," *Zeszyty Naukowe Uniwersytetu Jagiellońskiego: Prace Historyczne* 79 (1985): 95–107; Jan Havránek, "Univerzita Karlova, rozmach a perzekuce," in *Dějiny Univerzity Karlovy IV (1918–1990)*, ed. Jan Havránek and Zdeněk Pousta (Prague: Karolinum, 1997), 19–59; Martin Vietor, "Die Gründung der Comenius-Universität in Bratislava (Pressburg) und die österreichischen Universitätsgesetze," in *Festschrift Hans Lentze: Zum 60. Geburtstage dargebracht von Fachgenossen und Freunden*, ed. Nikolaus Grass and Werner Ogris (Innsbruck: Wagner, 1969), 587–97; and Livezeanu, *Cultural Politics*, 219–31.

187. See, e.g., UAG, Z. 340, 4 March 1873 (on habilitation), Z. 5385 (copy), 6 May 1873.

188. See, for example, the survey on the admission of women to universities carried out in 1897, e.g., in DALO, F. 26, Op. 7, Spr. 387, Z. 2945, 15 February 1897.

189. The correspondence, rich in statistics and depicting the networking of universities, can be found in ÚDAUK, FF NU, Sign. K/a (Profesoři), Inv.č. 186–93, Kart. 9.

190. See the report to the faculties on the committee's activities and the memorandum in UAG, PF, Z. 2302, 4 July 1907 (dated 3 July 1907); and the invitation, UAG, PF, Z. 2163, 13 June 1907 (dated 10 June 1907).

191. See *Hohes Abgeordnetenhaus*; and the memorandum to the ministry by the *Assistenten und Konstrukteure der Hochschulen* (representatives of universities, technical academies, and the Academy of Agriculture [Hochschule für

Bodenkultur] in Vienna) in UAG, Z. 2324, 5 July 1907 (dated 10 July 1907).

192. See, on reforms of medical and juridical studies in the 1890s, *Gutachten und Anträge zur Reform der medicinischen Studien- und Rigorosen-Ordnung* (Vienna: k.k. Universitäts-Buchdruckerei Karl Gorischek, 1894); Napoleon Cybulski, *W sprawie reformy studyów lekarskich (Odbitka z Przeglądu Lekarskiego 1896)* (Cracow: Drukarnia Uniwersytetu Jagiellońskiego, 1896); Napoleon Cybulski, *Uwagi nad ministeryalną reformą studyów i egzaminów lekarskich (Odbitka z Przeglądu Lekarskiego 1899)* (Cracow: Drukarnia Uniwersytetu Jagiellońskiego, 1900); and Thomas Garrique Masaryk, "O reformě právnických studií," *Athenaeum: Listy pro literaturu a kritiku vědeckou* 9, no. 4 (1892): 97–110.

193. See Walter Höflechner, *Die österreichische Rektorenkonferenz 1911–1938, 1945–1969* (Vienna: Verein zur Förderung der Rektorenkonferenz, 1993); and Bernhard vom Brocke and Peter Krüger, *Hochschulpolitik im Föderalismus: Die Protokolle der Hochschulkonferenzen der deutschen Bundesstaaten und Österreichs 1898 bis 1918* (Berlin: AkademieVerlag, 1994).

194. See, for example, Gürtler, *Deutsche Hochschulnot.*

195. The most important organizations of higher-education teachers were the Society of German Higher-Education Teachers (Vereinigung deutscher Hochschullehrer), the Society of Higher-Education Teachers (Towarzystwo Nauczycieli Szkół Wyższych) in Galicia, and the (Czech) Central Society of Higher-Education Teachers ([Český] Ústřední spolek učitelů vysokoškolských) in Bohemia; matters of university education were primarily discussed in their journals, the Galician *Muzeum: Czasopismo Towarzystwa Nauczycieli Szkół Wyższych* (Museum: Journal of the Society of Higher-Education Teachers, appearing from 1885) and the Czech *Věstník Českého ústředního spolku učitelů vysokoškolských se sídlem v Praze* (Bulletin of the Czech Central Society of Higher-Education Teachers in Prague, established in 1908).

196. Moritz Rappaport, *Prolog zur Feier des 100-jährigen Geburtstages Friedrich Schillers* (L'viv: Poremba, 1859). On Rappaport's idea of nationality, see Maria Klanska [Kłańska], "Moritz Rappaport als Brückenbauer zwischen der deutschen, jüdischen und polnischen Kultur," *Trans: Internet-Zeitschrift für Kulturwissenschaften* 15 (2004), http://www.inst.at/trans/15Nr/03_5/klanska15 .htm.

197. Ludwik Gumplowicz to Philipp Mansch, 28 October 1859, printed in Polish translation in Jan Surman and Gerald Mozetič, eds., *Dwa życia Ludwika Gumplowicza: Wybór tekstów* (Warsaw: Oficyna Naukowa, 2010), 122–23; the original is found in the National Library of Israel (Department of Archives, Schwadron Autograph Collection).

198. Tobias Wildauer, *Festrede zu Schillers hundertjährigen Geburtstag bei der von der k.k. Universität zu Innsbruck veranstalteten Feier in der Aula am 10 November 1859* (Innsbruck: Wagner, 1859), esp. 30–31.

199. Thomas Garrique Masaryk, "Jak zvelebovati naši litaraturu náukovou? Článek IV," *Athenaeum: Listy pro literaturu a kritiku vědeckou* 2, no. 12 (1885): 76–78; and Liske, *Der angebliche Niedergang.* See also in this regard the discussion of

Hruševs'kyj in Ivan Kryp'jakevyč, "Istoryčno-filosofična sekcija NTŠ pod kerivnyctvom Mychajla Hruševs'koho u 1894–1913 rokach," *Zapysky Naukovoho tovarystv imeni Ševčenka*, vol. 122, *Praci istoryko-filosofs'koï sekciï* (1991): 392–411.

Chapter 4

1. Billroth, *Briefe*, 72.
2. For "pyramidal" versus "tower" models, see Ebbe K. Graversen, "Human Capital Mobility—a Comparable Knowledge Indicator for the Nordic Countries," in *Science and Technology Indicators for the Nordic Countries 2000: A Collection of Articles*, ed. Kirsten Wille Maus (Copenhagen: Nordic Council of Ministers, 2001), 69–81, http://www.oecd.org/dataoecd/35/25/2100200.pdf.
3. This is based on my calculations for the winter terms in 1870–1910; the data are imprecise because they consider only the year of each scholar's birth and not the month and day of birth, but they do not show significant differences across the universities. There are also no significant differences between the faculties.
4. Those scholars whose place of graduation is unknown (around 10 percent) were excluded from the calculation.
5. This includes positions after 1918 and those who received only the title of professor (and not a salaried position) as the next step in their career. Contingency tables indicate no correlation, or even a negative correlation (at the medical faculty in Graz), between promotion and habilitation at a different university from where one graduated.
6. See the nomination acts for Virgil Mayrhofen to Prague, AT-OeStA/AVA Unterricht UM allg. Akten 1057, PA Mayrhofen, Z. 2408, 13 April 1851.
7. See the nomination acts for Karl Foltanek, favored over two Prague Privatdozenten: AT-OeStA/AVA Unterricht UM allg. Akten 1055, PA Foltanek, Z. 15901, 27 July 1892.
8. See the reasoning for not appointing Gustav Preiswerk (from Basel) to Innsbruck: AT-OeStA/AVA Unterricht UM allg. Akten 1057, PA Mayerhofer, Z. 269, 26 May 1905.
9. See especially the appointment for the chair of applied medical chemistry in Innsbruck in 1878: Richard Maly from the technical academy in Graz, proposed primo loco, was, from the start, not considered by the ministry as his appointment to Tyrol would be too expensive, although he was already a full professor. AT-OeStA/AVA Unterricht UM allg. Akten 1056, PA Loebisch, Z. 15440, 7 October 1878.
10. See, e.g., Lorenz von Stein, *Lehrfreiheit, Wissenschaft und Collegiengeld* (Vienna: Alfred Hölder, 1875); and *Zur Frage der Collegiengelder: Denkschrift verfasst von Mitgliedern des Medizinischen Professoren-Collegiums der Wiener Universität* (Vienna: Bergmann, 1896).
11. See Ivana Čornejová, "Výběr a výkvět národa: Snahy o úpravu platů profesorů

na pražké univerzitě na přelomu 19. a 20. století," in *Magister Noster: Sborník Statí Věnovaných in Memoriam Prof.Phdr. Janu Havránkovi, Csc. / Festschrift in Memoriam Prof.Phdr. Jan Havránek, Csc.*, ed. Michal Svatoš, Luboš Velek, and Alice Velková (Prague: Karolinum, 2003), 175–82; and Hedda Leeb, "Geschichte der Universität Innsbruck von 1898 bis 1908," 2 vols. (PhD diss., University of Innsbruck, 1967), 1:386–96.

12. Rokitansky, *Die Conformität*, 10.
13. For example, Jan Purkyně, "Kritika: Carl Rokitansky, Zeitfragen betreffend die Universität mit bes. Beziehung auf Medizin," *Časopis lékařův českých* 2, no. 20 (1863): 256–58; and Jan Purkyně, "Kritika: Carl Rokitansky, Die Conformität der Universitäten mit Rücksicht auf gegenwärtige österreichische Zustände," *Časopis lékařův českých* 3, no. 3 (1863): 22–24.
14. *Zur Frage der Collegiengelder*, 7. See also, in direct reaction to Rokitansky, Johann-Heinrich Dumreicher, *Zeitfragen betreffend die Universität mit besonderer Beziehung auf Medicin, von Carl Rokitansky* (Vienna: Carl Gerold, 1864), 13–14.
15. Joseph Anton Schöpf, *Denkschrift des Comites für Wiederherstellung der Universität in Salzburg* (Salzburg: Zaunrath'sche Buchdruckerei, 1870).
16. See also Heindl, *Josephinische Mandarine*, 119–21.
17. J.-H. Dumreicher, *Zeitfragen*, 13.
18. Heindl, *Josephinische Mandarine*, 200–209.
19. See, for example, Emil Frida (Jaroslav Vrchlický), who asked for a special allowance owing to a "long illness." NA, MKV/R, inv.č. 9, fasc. 111, PA Frida, Z. 1406, 13 January 1910.
20. While this was often included in the negotiations, the most thorough discussion on why the professor and his assistants should have university residence can be found in ÚDAUK, LF NU, Kart. 2, PA Grosser, 30 September 1909.
21. See, for example, the failed appointment of associate professor Franz Chvostek (from Vienna) to Innsbruck, rejected because he demanded that his professorial salary should balance the pay cut he would receive owing to the loss of his position as chief physician. AT-OeStA/AVA Unterricht UM allg. Akten 920, PA Lorenz, Z. 10751, 16 April 1903.
22. See the comments on the promotion of Alfred Kohn from associate professor in Vienna to professor and chair of histology at the German University in Prague: ÚDAUK, LF NU, Kart 4., PA Kohn, Z. 664, 17 December 1910, 5 December 1910.
23. See, on the raise of 4,000 kronen in the case of Friedrich Becke (Graz), AT-OeStA/AVA Unterricht UM allg. Akten 934. Adolf Bauer (Vienna) received a raise from around 11,000 to 14,000 kronen. PA Adolf Bauer, 30047, 28 August 1906; AT-OeStA/AVA Unterricht UM allg. Akten 663, PA Becke. The regular salary was 6,400 kronen.
24. See, on the ministry's efforts to keep the pharmacologist Hans Horst Meyer in Vienna, leading to the creation of an additional associate professorship at his institute, AT-OeStA/AVA Unterricht UM allg. Akten 633, PA Wiechowski, Z. 8979, 2 April 1910.

25. There was no standard salary at all universities: a professor in Prussia earned yearly between 4,000 (Braunsberg) and 6,000 marks (Berlin), while professors in Bavaria received 4,200 marks. See Ludwig Elster, *Die Gehälter der Universitäts-Professoren und die Vorlesungshonorare unter Berücksichtigung der in Aussicht genommenen Reformen in Preussen und Oesterreich* (Jena: Fischer, 1897). See also table 11 later in this chapter.
26. In the original, the quotation read, "Man wird zu einigen Jahren Czernowitz verurteilt und dann zu Innsbruck begnadigt." Karl Emil Franzos, "Erinnerungen an Mommsen," *Neue Freie Presse*, 22 November 1903. In the secondary sources, this is often quoted as "Sentenced to Czernowitz, pardoned to Graz, promoted to Vienna" ("Verurteilt zu Czernowitz, begnadigt zu Graz, befördert nach Wien"), e.g., in Havránek, "Nineteenth Century Universities," 19.
27. AT-OeStA/AVA Unterricht UM allg. Akten 934, PA Boltzmann, Z. 11897, 11 August 1876.
28. See, for example, on the physiologist Franz/Ferenc Tangl from Budapest, proposed for a position in Innsbruck, AT-OeStA/AVA Unterricht UM allg. Akten 1058, PA Tangl, Z. 32116, 21 September 1904 (proposed appointment); Z. 36674, 10 November 1904 (final rejection by Tangl); on the chemist Leon Marchlewski from Cracow, proposed for a position in L'viv, see DALO, F. 26, Op. 7, Spr. 525, p. 3, 28 December 1904; p. 7, 8 January 1905. A few years later, all prominent candidates rejected calls owing to outdated equipment: AGAD, MWiO, Sygn. 66u, PA Dziewoński, Z. 448, 23 February 1911.
29. Johann/Jan Hofmokl preferred the General Hospital in Vienna to a professorship in Cracow. AGAD, MWiO, Sygn. 52u, PA Rydygier, Z. 117, 2 April 1887.
30. For instance, Hugo Schindelka preferred the Military Veterinary Institute in Vienna (Militär-Tierarzneiinstitut) to a chair in Prague. ÚDAUK, FF NU, PA Dexler, 15 July 1894, Z. 1333.
31. See Jolanta Dańczura-Dynowska, "Alfred Biesiadecki, wybitny polski histopatolog i jego wkład do dermatologii," *Archiwum Historii i Filozofii Medycyny* 58, no. 4 (1995): 385. Biesiadecki became a consultant on health issues in the provincial government.
32. Two further examples were the historians Heinrich Kretschmayr and Julius Hermann: AT-OeStA/AVA Unterricht UM allg. Akten 1075, PA Schrader, Z. 33622, 7 August 1905; and AT-OeStA/AVA Unterricht UM allg. Akten 681, PA Trabert, Z. 16585, 2 June 1915.
33. See the case of the art historian Karel Chytil in Prague: NA, MKV/R, inv.č. 9, fasc. 113, PA Chytil, Z. 1761, 16 January 1911.
34. See the appointment records of Heinrich Wopfner: AT-OeStA/AVA Unterricht UM allg. Akten 1077, PA Wopfner, Z. 36418, 10 December 1908.
35. The quotations are from a proposal for a new full professor of ophthalmology: AT-OeStA/AVA Unterricht UM allg. Akten 621, PA Dimmel, Z. 508, 1 March 1909.
36. From acts about the Viennese full professor Wilhelm Meyer-Lübke, proposed as the new professor of Romance languages by Chernivtsi University. AT-OeStA/AVA Unterricht UM allg. Akten 934, PA Cornu, Z. 3365, 4 April 1901.

37. AT-OeStA/AVA Unterricht UM allg. Akten 935, PA Hann, Z. 10080, 13 July 1897.
38. AT-OeStA/AVA Unterricht UM allg. Akten 668, PA Hann, Z. 34553, 5 January 1900.
39. At the University of Prague, 23 percent had only been Privatdozenten before the university was divided.
40. For the period 1848–1900, see Mühlberger, "Das 'Antlitz,'" 85.
41. Ludwig Boltzmann, who moved several times, was an example of an extremely mobile scholar who falls into several of the categories named here; his career path is, however, atypical for Habsburg scholars.
42. See AT-OeStA/AVA Unterricht UM allg. Akten 1221, PA Zepharovich, Z. 12382, 19 February 1864.
43. On average, full professors were three years older than at other German-speaking universities, averaging from forty-six (1880) to fifty-two (1900) years of age.
44. For some years, Vienna had younger associate professors than the other universities, but this trend changed after 1900.
45. Between 30 and 40 percent of instructors (including Privatdozenten) retained their positions (without being promoted), and 15 to 25 percent were appointed from other universities.
46. The number of scholars promoted within faculties was between 5 percent (the lowest rate for Vienna) and 25 percent (the highest rate at other universities).
47. On medicine, see Tatjana Buklijas, "Dissection, Discipline and Urban Transformation: Anatomy at the University of Vienna, 1845–1914" (PhD diss., University of Cambridge, 2005), esp. 208–9; on historiography, see Kamil Krofta, "Anton Gindely," *Zprávy zemského archivu Království českého* 3 (1910): 213.
48. Philipp Werner von Ebenhof, quoted in Jan Havránek, "Česká univerzita v jednání rakouských uřadů do roku 1881," *AUC-HUCP* 22, no. 1 (1982): 48.
49. Words noted on the proposal of the philosophical faculty in Vienna to nominate only German scholars. AT-OeStA/AVA Unterricht UM allg. Akten 679, PA Schneider, Z. 6978, 3 August 1870.
50. AT-OeStA/AVA Unterricht UM allg. Akten 1184, PA Kaluzniacki, Z. 12099, 3 August 1875.
51. See, for example, the appointment of the art historian Alfred Woltmann. AT-OeStA/AVA Unterricht UM allg. Akten 1221, PA Woltmann, Z. 7471, 12 June 1873. Woltmann's appointment was officially supported by the influential Viennese professor of art history Rudolf Eitelberger.
52. Habsburg returnees included thirteen Bohemians and twenty-six Austrians; four of the Bohemian and thirteen of the Austrian returnees had previously held a professorship in the Habsburg Empire. Ninety-five percent of these cases happened from 1880 on, and they included equal numbers of philosophers and physicians. Nearly twenty-two of these scholars had graduated in the Habsburg Empire, eighteen of them from Vienna (six at the philosophical faculty and twelve at the medical faculty).

53. During appointment, one could be promoted from Privatdozent to associate professor, from Privatdozent to full professor, or from associate professor to full professor.
54. Friedrich Salzer was appointed in 1890. After his death in 1893, the chair was filled by Anton von Eiselberg. When Eiselberg was appointed to Königsberg, Albert Narath took the position. Anton von Eiselberg, *Lebensweg eines Chirurgen: Eine Autobiographie aus der großen Zeit der Wiener Medizin 1860–1937* (Vienna: Christian Brandstätter, 1991), 89–97; AT-OeStA/AVA Unterricht UM allg. Akten 1058, PA Schloffer, Z. 32351, 7 October 1903 (on Narath).
55. AT-OeStA/AVA Unterricht UM allg. Akten 663, PA Adamović. Adamović was born in Rovinj/Rovigno in Dalmatia and was a member of the Zagreb-based Yugoslav Academy of Sciences and Arts (Jugoslavenska akademija znanosti i umjetnosti). The fact that he was Catholic caused him all sorts of trouble in Orthodox Belgrade.
56. AT-OeStA/AVA Unterricht UM allg. Akten 667, PA Fullerton.
57. "Fullerton, George Stuart (1825–1925)," in *Pennsylvania Biographical Dictionary*, ed. Jan Onofrio (St. Clair Shores: Somerset, 1999), 1:424–26.
58. See the wording in the appointment of the ophthalmologist Ernst Fuchs (from Liège) to Vienna: AT-OeStA/AVA Unterricht UM allg. Akten 623, PA Ernst Fuchs, Z. 23518/84, 16 July 1885. The ministry, supported by the expertise of Arlt and Karl Stellwag, mentioned that Fuchs was a specialist in pathological-anatomical studies in Arlt's tradition, while other scholars in the proposal represented the "physical school."
59. AT-OeStA/AVA Unterricht UM allg. Akten 1055, PA Franz Hofmann, Z. 6400, 8 January 1905.
60. See, on the appointment of the pediatrician Clemens Pirquet from Wrocław/Breslau, AT-OeStA/AVA Unterricht UM allg. Akten 629, PA Pirquet, Z. 28469, 8 June 1911.
61. AT-OeStA/AVA Unterricht UM allg. Akten 624, PA Gussenbauer, Z. 12711, 2 June 1894.
62. Władysława Bona, "Zasługi Jana Mikulicza dla rozwoju chirurgii w Polsce," *Archiwum Historii i Filozofii Medycyny* 13, nos. 1–2 (1933): 20–100.
63. At the medical faculty, 10 percent were from the German Empire versus 85 percent from Habsburg universities; in the sciences, it was 11 percent versus 86 percent; and in the humanities 21 percent versus 69 percent.
64. This is the percentage of scholars based at German universities who were appointed, if included in the proposal, regardless of their place in the terna.
65. In deliberations about the chair of histology at the German University in Prague, both the ministry and the faculty decided not to nominate the most appropriate candidate because in two years he would be appointed to Vienna. NA, MKV/R, inv.č. 2, fasz.97, PA Kohn, Z. 2888, 9 February 1911 (ministry); 17 December 1910, ad Z. 2888 (faculty).
66. AT-OeStA/AVA Unterricht UM allg. Akten 1210, PA Langer, Z. 28780, 24 September 1915.

67. AT-OeStA/AVA Unterricht UM allg. Akten 918, PA Escherich (ad. Z. 4418), 12 February 1890.
68. Votum separatum of Leopold Gegenbauer in UAI, Ph 476/1890, emphasis in original; see also AT-OeStA/AVA Unterricht UM allg. Akten 1070, PA Blaas, Z. 14771, 13 August 1890.
69. See already Purkyně's "Kritika: Carl Rokitansky, Die Conformität," clearly written from the position of a Prague scholar opposing Vienna's centrality.
70. NA, MKV/R, inv.č. 9, fasc. 116, PA Pelikan, Z. 8604, 18 October 1898; Z. 855, 11 January 1899.
71. AT-OeStA/AVA Unterricht UM allg. Akten 921, PA Payr, Z. 38748, 22 September 1907; and AT-OeStA/AVA Unterricht UM allg. Akten 630, PA Richter, Z. 43166, 7 November 1909.
72. See, e.g., on the physicist Marian Smoluchowski (appointed from Cracow to Vienna), AT-OeStA/AVA Unterricht UM allg. Akten 670, PA Gustav Jäger, Z. 22103, 15 June 1918; see also Teske, *Marian Smoluchowski*, 251–52. On the histopathologist Alfred Biesiadecki (appointed from Cracow to Vienna), see AT-OeStA/AVA Unterricht UM allg. Akten 625, PA Heschl, Z. 232, 13 May 1875..
73. See, for example, Maria Wakounig, "Wissenschaft und Kariere? Polnische Mediziner an der Wiener Uni zwischen 1870 und 1914," in *Polen im alten Österreich: Kultur und Politik*, ed. Walter Leitsch and Stanisław Trawkowski (Vienna: Böhlau, 1993), 107–15.
74. Natanson to Gumplowicz, 16 February 1889, Collection of the Manuscripts of the Jagiellonian Library, Cracow, sign. 9007 III, vol. 6. According to Gumplowicz, "First, all German (and maybe also Galician) professors make a sign of the cross if they see a candidate for a Privatdozent; Second, in the last few years there prevails here an epidemic fear of Jews; Third, they regard it here as a patriotic duty not to admit any non-German, and especially any Pole, for any function." Gumplowicz to Natanson, 20 February 1889, fol. 215.

Chapter 5

1. Quoted in Bohumil Jiroušek, "Jazyky v životě a díle Antonína Rezka," in Binder, Velek, and Křivohlavá, *Místo*, 534.
2. In Cracow professors were considered of equal status to the aristocracy and often came from that group. This phenomenon was codified in countless jokes; see, e.g., Larry Wolff, "Dynastic Conservatism and Poetic Violence in Fin-de-Siècle Cracow: The Habsburg Matrix of Polish Modernism," *American Historical Review* 106, no. 3 (2001): 741; and Antoni Podraza, "Dobry, Lepszy, Bobińska: Wysłuchała i opracowała Rita Pagacz-Moczarska," *Alma Mater*, no. 63 (2004): esp. 48. See also, for a literary impression, Maryla Szymiczkowa, *Tajemnica domu Helclów* (Cracow: Znak, 2015).
3. Seibt, *Die Teilung.*
4. See the documentation from the ministry on the rejection of the appointment

of the mathematician Wojciech Urbański to Cracow, in which the commission was asked not to restrict the search for candidates to Galicia. AUJ, WF II 163, Z. 8839, 30 November 1864.

5. See the documentation in the appointment records of Zygmunt Wróblewski: AGAD, MWiO, Sygn. 70u, PA Wróblewski, Z. 3630, 18 April 1882.
6. On Johann/Jan Hofmokl, see AGAD, MWiO, Sygn. 52u, PA Rydygier, Z. 117, 2 April 1887. In a search for a new professor in Cracow, the ministry preferred the ophthalmologist Michael/Michał Borysiekiewicz (also Михайло Борисикевич) over the German Empire scholar Bolesław Wicherkiewicz but finally appointed the latter: AGAD, MWiO, Sygn. 52u, PA Wicherkiewicz, Z. 405, 10 June 1895 (Borysiekiewicz's comment on his possible move to Galicia), and Z. 21580, 2 December 1895 (the ministry's commentary on the faculty proposal); later, in L'viv, the ministry intended to appoint Borysiekiewicz but finally yielded for financial reasons: AGAD, MWiO, Sygn. 403u, PA Machek, Z. 20817, 9 August 1898. Borysiekiewicz was not considered in the faculty proposal (Z. 640, 22 May 1898).
7. Three of the eleven professors in the medical faculty were removed, two of the eight in the law faculty, and four of the sixteen in the philosophical faculty. These numbers are from my own calculations based on the Cracow lecturers' catalogs.
8. The case of Bratranek, discussed above on p. 63, was a notable exception.
9. Józef Dietl, "O instytucji docentów w ogóle, a szczególnie na Uniwersytecie Jagiellońskim," *Czas*, 31 October 1861, 1–2.
10. See the discussion in AT-OeStA/AVA Unterricht UM allg. Akten 1186, PA Öttinger, Z. 623, 14 August 1862. Also, the faculty was not unanimous about Oettinger's habilitation, officially because of his age (Z. 9909, 3 September 1862).
11. See the notes on AT-OeStA/AVA Unterricht UM allg. Akten 1186, PA Öttinger, Z. 724, 23 December 1868.
12. See the conflict between Richard Heschl and the medical faculty: AT-OeStA/AVA Unterricht UM allg. Akten 1186, PA Kopczynski, Z. 14960, 13 October 1860; and AUJ, S II 819, Medycyna sądowa, Z. 665, 26 September 1860, 21 September 1860, and other documents in this fascicle, as well as the note from the ministry sent to the deanery in AUJ, WL II 152, 26 October 1860.
13. On the habilitation of Wincenty Zakrzewski, see the initial faculty report: DALO, F. 26, Op. 7, Spr. 148, Z. 50, 20 January 1871 and 11 February 1871. On the scholarly level of Zakrzewski's work, see DALO, F. 26, Op. 7, Spr. 142, N. 245, 19 January 1871; on the ministry's negative decision, see DALO, F. 26, Op. 7, Spr. 148, Z. 1544, 17 March 1871. On lectures in Polish (by Ksawery Liske), see AGAD, MWiO, Sygn. 117u, PA Liske, Z. 460, 24 June 1869.
14. See, for example, the note on unspecified assaults on Karl Barach-Rappaport, mentioned in the appointment records of Euzebiusz Czerkawski: AGAD, MWiO, Sygn. 118u, PA Czerkawski, Z. 8147, 19 June 1871.
15. For the correspondence between the ministry and the faculty on the issue of all three professors, see AGAD, MWiO, Sygn. 121u, PA Schmidt, Z. 8999, 12 October 1875; and DALO F. 26, Op. 7, Spr. 184, Z. 16609, 20 October 1875.
16. Finkel, "Historya Uniwersytetu Lwowskiego," 44–47.

17. According to official Habsburg statistics, the majority of the Jews in Galicia spoke Polish as their common language. See Theodor Haas, "Die sprachlichen Verhältnisse der Juden in Österreich," *Zeitschrift für Demographie und Statistik der Juden* 11, no. 1 (1915): 1–12, tables on p. 3.
18. The habilitation of the historian Volodymyr Myl'kovyč (Володимир Милькович, also Wladimir Milkowicz) was rejected in 1890 because of his "inadequate knowledge of the Polish language": DALO, F. 26, Op. 7, Spr. 321, Z. 262, 4 December 1890. Klymentij/Klym Hankevyč (Климентій/Клим Ганкевич, better known as Klemens/Clemens Hankiewicz) was denied habilitation in L'viv and Cracow, owing to his lack of Polish-language publications: DALO, F. 26, Op. 7, Spr. 146; DALO, F. 26, Op. 7, Spr. 132, 12 February 1869; Z. 419, 15 June 1869; and AUJ, WF II 121, PA Hankiewicz. More on Hankevyč can be found in Vjachaslaw Shal'kjevich (Вячаслаў Шалькевіч, Wiaczesław Szalkiewicz), introduction to *Zarys filozofii słowiańskiej*, by Klemens Hankiewicz (Rzeszów: Wydawnictwo Uniwersytetu Rzeszowskiego, 2011). *Zarys filozofii słowiańskiej* is the Polish translation of *Grundzüge der slavischen Philosophie* (1869).
19. AGAD, MWiO, Sygn. 403u, PA Kośmiński, Z. 35837, 15 December 1906 (the final decision of the faculty), Z. 43794, 17 January 1906 (Jan Prus's expert opinion), and Z. 43794, 11 September 1906 (the provincial government's support for the claim that Bikeles was not fluent in Polish). See also Eufemiusz Józef Herman, *Historia neurologii polskiej* (Wrocław: Zakład narodowy im. Ossolińskich, Polska Akademia Nauk, 1975), 152.
20. In 1852 Streng had been nominated by Thun-Hohenstein for the chair of gynecology for midwives, also because of his knowledge of Czech: AT-OeStA/AVA Unterricht UM allg. Akten 1211, PA Streng, Z. 6683/546, 28 July 1852.
21. AT-OeStA/AVA Unterricht UM allg. Akten 1211, Z. 7731, 30 August 1870.
22. AT-OeStA/AVA Unterricht UM allg. Akten 1208, PA Eiselt, Z. 9990, 18 June 1881. The scholars proposed by the faculty included two Bohemian scholars who later went to the German University. On Eiselt's biography, see Ludmila Hlaváčková, "Čtyřnásobné Jubileum Bohumila Eiselta (1831–1908)," *Časopis lékařův českých* 150, no. 5 (2011): 619–23.
23. AT-OeStA/AVA Unterricht UM allg. Akten 1208, PA Gussenbauer, Z. 21118, 15 April 1878.
24. AT-OeStA/AVA Unterricht UM allg. Akten 1211, PA Weiss, Z. 17116, 4 November 1881.
25. See Helena Kokešová, *Eduard Albert: Příspěvek k životopisu a edice korespondence* (Prague: Scriptorium: Výzkumné centrum pro dějiny vědy, Masarykův ústav Akademie věd České republiky, 2004), 22–27; see also the critical assessment of Weiss in Jan Šváb, "I. česká chirurgická kliniká a její vliv na rozvoj chirugie v Českích zemích," in *120 let 1. české chirurgické kliniky 1. lékařské fakulty Univerzity Karlovy v Praze*, ed. Pavel Kleiner (Prague: Karolinum, 2002), esp. 15–17.
26. Most notably, the forensic physician Josef Maschka published in Czech in his early years but from 1865 on only in German.

27. For the appointment, see AT-OeStA/AVA Unterricht UM allg. Akten 620, PA Albert, Z. 92, 25 January 1881; on the probable influence of Josef Skoda/Škoda, see Arnold Jirásek, *Eduard Albert: Pokus o kroniku a rozbor života, práce i významu E. Alberta, učiněný ke stému výročí jeho narození (20. ledna 1941)* (Prague: Československá chirurgická společnost, 1946), 82–84.
28. On the mathematician Jan Sobotka, a professor at the technical academy in Vienna who had published a single Czech-language article (and fifteen in German) before he was nominated to the Czech technical academy in Brno in 1901, see Jaroslav Folta and Pavel Šišma, "Jan Sobotka, Literatura," in *Významní matematici v českých zemích* (Brno, 2003), http://web.math.muni.cz/biografie/.
29. Krofta, "Anton Gindely"; Brigitte Hamann, "Anton Gindely—ein altösterreichisches Schicksal," in *Nationale Vielfalt und gemeinsames Erbe in Mitteleuropa: Vorträge anläßlich der Verleihung des Anton-Gindely-Preises für Geschichte der Donaumonarchie*, ed. Erhard Busek and Gerald Stourzh (Vienna: Verlag für Geschichte und Politik; Munich: Oldenbourg, 1990), 27–37; Jan Havránek, "Anton Gindely, ein Historiker, der zwischen zwei Nationen stand," *Acta Universitatis Carolinae—Philosophica et Historica* 3 (1993): 101–9; and Josef Polišenský, "Anton Gindely und die böhmische Geschichtswissenschaft," *Acta Universitatis Carolinae—Philosophica et Historica* 3 (1993): 13–21.
30. The number of members from Austria who were elected also remained much lower than, for example, the number from Galicia, exceptions being, e.g., Eduard Suess and Robert Zimmermann. Alena Šlechtová and Josef Levora, *Členové České akademie věd a umění 1890–1952* (Prague: Academia, 2004).
31. According to the oral tradition codified in the lexicon *Kdo byl kdo: Čeští a slovenští orientalisté, afrikanisté a iberoamerikanisté* (Prague: Libri, 1999), here quoted from the online version: http://www.libri.cz/databaze/orient/main.php.
32. Ludmila Hlaváčková, "Budování klinických pracovišť české lékařské fakulty v době rozdělen pražské univerzity: II. Snahy o vybudování českých klinických pracovišť před rozdělením univerzity," *Sbornik lékařský* 85, no. 4 (1983): 110–15.
33. Ludmila Hlaváčková and Petr Svobodný, *Dějiny pražskich lékarských fakult 1348–1990* (Prague: Karolinum, 1993), 80–83.
34. See the emperor's note from 7 January 1883 to AT-OeStA/AVA Unterricht UM allg. Akten 1211, PA Schöbl, Z. 21874, 29 December 1882 (ministerial records, Minister Eybesfeld).
35. AT-OeStA/AVA Unterricht UM allg. Akten 1209, PA Horbachevsky, Z. 13251, 15 August 1883.
36. See AT-OeStA/AVA Unterricht UM allg. Akten 1211, PA Steffal.
37. AT-OeStA/AVA Unterricht UM allg. Akten 1209, PA Horbachevsky, Z. 13251, 15 August 1883.
38. An interesting sociological analysis of the professorship around 1882 can be found in Eva Schmidt-Hartmann, "Die philosophische Fakultät der tschechischen Universität um 1882: Kontinuität und Wandel," in Seibt, *Die Teilung*, esp. 96–102.
39. These included the philosopher and psychologist Gustav Adolf Lindner and the gymnasium teacher Alois Vaníček.

40. ÚDAUK, Filozofická fakulta Karlo-Ferdinandovy Univerzity 1882–2012, Inv.č. 640, Kart. 56, PA Seydler, Z. 299, 23 June 1885; Z. 23033, 18 December 1885; Inv.č. 701, PA Safarik, Bericht der Commission, Prag, 4 November 1891; Bericht an das Ministerium, 16 November 1891; and NA, MKV/R, inv.č. 9, Kart. 112, PA Gruss.
41. See Alena Míšková and Michael Neumüller, *Společnost Pro Podporu Německé Vědy, Umění a Literatury v Čechách (Německá Akademie Věd v Praze): Materiály k Dějinám a Inventář Archivního Fondu = Die Gesellschaft zur Förderung deutscher Wissenschaft, Kunst und Literatur in Böhmen (Deutsche Akademie der Wissenschaften in Prag): Materialen zu ihrer Geschichte und Inventar des Archivbestandes: 1891–1945* (Prague: Archiv Akademie věd České republiky, 1994).
42. Pokorná, "Královská česká společnost nauk"; and Martina Niedhammer, "'Slovozpyt' und 'filologie': Nationale Implikationen der sprachwissenschaftlichen Forschung in der Königlich böhmischen Gesellschaft der Wissenschaften im 19. Jahrhundert," in *Sprache, Gesellschaft und Nation in Ostmitteleuropa: Institutionalisierung und Alltagspraxis*, ed. Martina Niedhammer, Klaas-Hinrich Ehlers, Marek Nekula, and Hermann Scheuringer (Göttingen: V&R Unipress, 2014), 33–49.
43. I am indebted to Ludmila Hlaváčková for this information. Certainly, the aula of the Carolinum alternated days, German one day and Czech the next; see Hans Lemberg, "Universität oder Universitäten in Prag—und der Wandel der Lehrsprache," in *Universitäten in nationaler Konkurrenz: Zur Geschichte der Prager Universitäten im 19. und 20. Jahrhundert*, ed. Hans Lemberg (Munich: Oldenbourg, 2003), 29.
44. I thank Luboš Velek in Prague for telling me this story. Both this and the hospital narrative are probably slightly exaggerated to emphasize the importance of nationality at the time and could be answered with counterexamples.
45. This story is from a document by August Sauer, the rector of the German University in Prague, on the necessary development of the university in 1908. NA, MKV/R, fasc. 136, n.d., no number.
46. Kurt A. Huber, "Die Prager theologischen Fakultäten von 1883/1891 bis 1945," in Seibt, *Die Teilung*, 37–54. See also the anonymously published brochure of the auxiliary bishop Wenzel Frind, *Eine eventuelle Theilung (Verdoppelung) der Prager theologischen Fakultät, ist mit dem Prinzipe und der Aufgabe der Kirche vereinbar, Zugleich als Studie über das Verhältnis der Kirche zur Nationalität* (Prague: Ambr. Opitz in Warnsdorf, 1884).
47. Goll, *Der Hass der Völker*, 17.
48. See ÚDAUK, FF NU, Inv.č. 532, K. 52, K/XVIII, PA Trautmann; see also Zeil, *Slawistik*.
49. Lenka Vodrážková-Pokorná, "Die Anfänge der tschechischen Germanistik und ihre ersten Repräsentanten an der Prager Universität," in Lemberg, *Universitäten in nationaler Konkurrenz*, 115–34. German language and literature were also, however, taught by Privatdozenten. On German studies in general, see Lenka

Vodrážková-Pokorná, *Die Prager Germanistik nach 1882: Mit besonderer Berücksichtigung des Lebenswerkes der bis 1900 an die Universität berufenen Persönlichkeiten* (Frankfurt am Main: Peter Lang, 2006).

50. Milan Tvrdík, "August Sauer und die Prager tschechische Germanistik," in *August Sauer (1855–1926): Ein Intellektueller in Prag im Spannungsfeld von Kultur- und Wissenschaftspolitik*, ed. Steffen Höhne (Vienna: Böhlau, 2008), 133–46; and Vodrážková-Pokorná, *Die Prager Germanistik*, 256–68.
51. See in general Höhne, *August Sauer.*
52. From the proposal of the faculty; see AT-OeStA/AVA Unterricht UM allg. Akten 1210, PA Pawlik, Z. 9411, 19 April 1887 (included in the ministerial proposal from 17 May 1887).
53. AT-OeStA/AVA Unterricht UM allg. Akten 1210, PA Pawlik, Z. 9411, 17 May 1887. Another such case was that of Albert's assistant, Karel/Karl Maydl.
54. Alfréd Kotasék, *Karel Pawlík (1849–1914), Osobnost a dílo* (Prague: Univerzita Karlova, 1994).
55. The percentage was higher in the philosophical faculty (7.5 percent, compared to 4.5 percent at the medical faculty).
56. Stanislav Polák, *T. G. Masaryk: Za ideálem a pravdou*, vol. 4, *1900–1914* (Prague: Masarykův ústav Akademie věd České republiky, 2005), 68.
57. See also Jiří Pernes, *Kapitoly z dějin Vysokého učení technického v Brně (cesta moravské techniky 20. stoletím)* (Brno: Vysoké Učeni Technické, Nakl. VUTIUM, 2009).
58. Jana Mandlerová, "K boji," 97. The idiom *the Chinese spirit* (in the dative in Masaryk's Czech original: čínanství) comes from Friedrich Nietzsche's *Antichrist* (in the German original the word is *Chinesenthum*), which in English is translated with either *the Chinese spirit* or *Chinaism*, and in Czech with čínanstvo (according to Rastislav Škoda's translation in 2003; it was unfortunately impossible to check the 1905 translation by Leopold Pudlač, the pseudonym of Arnošt Procházka). The term is used as such in neither English nor Czech.
59. Goll, *Der Hass der Völker*, esp. 21–23.
60. Jana Mandlerová, "K zahraničním cestám učitelů vysokých škol v českých zemích (1888–1918)," *Dějiny věd a techniky* 2, no. 4 (1969): 232–46.
61. See, e.g., Miloš Havelka, "A Hundred Years of the 'Czech Question' and the Czech Question a Hundred Years On," *Czech Sociological Review* 3, no. 1 (1995): 7–19; and Roman Szporluk, *The Political Thought of Thomas G. Masaryk* (Boulder, CO: East European Monographs, 1981).
62. Stanislav Polák, *T. G. Masaryk: Za ideálem a pravdou*, vol. 2, *1882–1893* (Prague: Masarykův ústav Akademie věd České republiky, 2001), 37.
63. Goll, *Der Hass der Völker*, 13.
64. See, however, reactions to Matija/Matthias Murko's *Deutsche Einflüsse auf die Anfänge der slavischen Romantik* (German influence on the beginnings of Romanticism among the Slavs), 2 vols. (Graz: Styria, 1897); the first volume was concerned with early Romanticism in Bohemia: Murko, *Deutsche Einflüsse auf die Anfänge der böhmischen Romantik* (Graz: Styria, 1897). See also Dalibor

Tureček, "Murkovy 'Deutsche Einflüsse' a jejich české přijetí," in *Matija Murko v myšlenkovém kontextu evropské slavistiky: Sborník studií*, ed. Ivo Pospíšil and Miloš Zelenka (Brno: Masarykova univerzita, 2005), 87–99.

65. Ferdinand Hueppe, *Kulturbedürfnisse und Universitäten in Oesterreich (Sonderdruck aus Heft 221/21 der "Hochschul-Nachrichten")* (Munich: Akademischer Verlag, 1909).
66. Jiři Hnilica, "Kulturní a intelektuální výměna mezi Čechami a Francií 1870–1925," *AUC-HUCP* 45, nos. 1–2 (2005): 110–16. See also Jindřich Dejmek, "Učňovská a vandrovní léta Edvarda Beneše (1904–1913)," *Moderní Dějiny* 11 (2003): esp. 8–28.
67. Mandlerová, "K zahraničním cestám."
68. For a recent overview, see Soňa Štrbáňová, "Turning 'Province' to a 'Centre'? Ambitions to Establish an Institutionalized Network of Slavic Scientists at the Turn of the 19th Century," *Dějiny věd a techniky* 48, no. 4 (2015): 274–305.
69. Kratochvíl, *Jan Evangelista Purkyně*, 110; and Goll, *Der Hass der Völker*.
70. Martinczak, "Geneza"; and Soňa Štrbáňová, "Congresses of the Czech Naturalists and Physicians in the Years 1880–1914 and the Czech-Polish Scientific Collaboration," in *Acta historiae rerum naturalium necnon technicarum. Special Issue 21. Studies of Czechoslovak Historians for the 18th International Congress of the History of Science*, ed. Jan Janko (Prague: Institute for Czechoslovak and General History, Institute for Czechoslovak and General History, 1989), 79–122.
71. The congress was forbidden by the Prussian authorities; see Jarosław Obermajer, "Zabroniony Zjazd Lekarzy i Przyrodników Polskich w roku 1898," *Archiwum Historii Medycyny* 28, nos. 1–2 (1965): 119–23.
72. Danuta Rederowa, "Formy współpracy Polskiej Akademii Umiejętności z zagranicą," *Studia i materiały z dziejów nauki polskiej, Seria A* 10 (1966): 79–80.
73. Statistics from Marek Ďurčanský, "Członkostwo zagraniczne polskich i czeskich uczonych w akademiach narodowych: PAU i ČAVU," *Prace Komisji Historii Nauki PAU* 6 (2004): 177–211; see also Julian Dybiec, "Związki Akademii Umiejętności w Krakowie z nauką czeską i słowacką w latach 1873–1918," in *Z dziejów polsko-czeskich i polsko-słowackich kontaktów naukowych*, ed. Irena Jasiukowa-Stasiewicz and Jan Janko (Warsaw: Polska Akademia Nauk, 1990), 34–61; and Emilie Těšínská, "K česko-polským vědeckým stykům v oblasti matematicko-fyzikálních věd," in *Semináře a studie Výzkumneho Centra pro Dêjiny Vêdy z Let 2002–2003*, ed. Antonín Kostlán (Prague: Výzkumne Centrum pro dêjiny vêdy, 2003), 341–76.
74. Jan Hulewicz, *Akademia Umiejętności w Krakowie 1873–1918: Zarys Dziejów* (Wrocław: Ossolineum, 1958), 117. The expedition was, in the end, organized by the Viennese academy, because the *spiritus movens* of Polish-Czech cooperation in Egypt, the young Cracow Egyptologist Tadeusz Smoleński, died before the negotiations over the expeditions had been finalized.
75. Of the 116 scholars who habilitated at the philosophical faculty in Cracow, around half acquired associate professorships there, and slightly fewer than 40 percent obtained full professorships; 35 percent remained Privatdozenten, and 20

percent were appointed to other universities during their careers (the categories are nonexclusive). In L'viv almost half of the scholars who habilitated there remained Privatdozenten, while 40 percent were appointed associate professors, 25 percent became full professors, and 14 percent moved to other academic institutions (including universities) (again the categories are nonexclusive). Apart from the low number of scholars not progressing above Privatdozenten, medical faculties demonstrated a similar distribution.

76. See the police reports and political decisions: DALO, F. 26, Op. 5, Spr. 437, pp. 42–50, 18 March 1867; Z. 554, 29 January 1867; N. 139, 16 April 1868; and the faculty's claim of Holovac'kyj's innocence: AGAD, MWiO, Sygn. 117u, PA Głowacki, Z. 4473, 1 June 1868. On later repercussions, see DALO, F. 26, Op. 5, Spr. 437, p. 55, 6 August 1869. See also, among recent publications on this widely researched topic, Włodzimierz Osadczy, *Święta Ruś: Rozwój i oddziaływanie idei prawosławia w Galicji* (Lublin: Wydawnictwo Uniwersytetu Marii Curie-Skłodowskiej, 2007), 142–54.
77. Wincenty Pol to the philosophical faculty of the Jagiellonian University, AGAD, MWiO, Sygn. 393u, Próba restytuowania W. Pola na katedrę geografii, 18 December 1869.
78. From the opinion the provincial governor on the question, AGAD, MWiO, Sygn. 393u, Z. 1058, 7 May 1870.
79. See the letter of Henryk Niewęgłowski in DALO, F. 26, Op. 7, Spr. 146, p. 108, 20 August 1871.
80. On the practice and rhetorics of Galician provincial autonomy, notwithstanding the lack of a legal basis, see Harald Binder, "'Galizische Autonomie': Ein streitbarer Begriff und seine Karriere," in *Moravské vyrovnání z roku 1905: Možnosti a limity národnostního smíru ve střední Evropě / Der Mährische Ausgleich von 1905: Möglichkeiten und Grenzen für einen nationalen Ausgleich in Mitteleuropa*, ed. Lukáš Fasora (Brno: Matice Moravská pro Výzkumné Středisko pro Dějiny Střední Evropy: Prameny, Země, Kultura, 2006), 239–66.
81. See, for example, Henryk Barycz, "Docenckie kłopoty Józefa Szujskiego," in *Wśród gawędziarzy, pamiętnikarzy i uczonych galicyjskich: Studia i sylwety z Życia umysłowego Galicji XIX w.*, ed. Henryk Barycz (Cracow: Wydawnictwo Literackie, 1963), 91–111. In this chapter Barycz uses the phrase "older generation" as a pejorative to describe those with etatist and loyal leanings.
82. AGAD, MWiO, Sygn. 121u, PA Sobel, Z. 192, 26 January 1884.
83. AGAD, MWiO, Sygn. 119u, PA Kawczyński, Z. 222, 26 January 1884.
84. Maria Julita Nedza, *Polityka Stypendialna Akademii Umiejętności w Latach 1878–1920: Fundacje Gałęzowskiego, Pileckiego i Osławskiego* (Wrocław: Zakład Narodowy im. Ossolińskich. Wydawnictwo Polskiej Akademii Nauk, 1973), 91.
85. Only a few candidates sought to achieve habilitation in this subject but were rejected: Albert Zipper in 1881 and Napthali Sobel in 1884, both in L'viv. Zipper was a translator and later authored Polish-German dictionaries (*Langenscheidt*, among others); for his unsuccessful habilitation, see AGAD, MWiO, Sygn. 122u,

PA Zipper, Z. 146, 26 November 1881; and DALO, F. 26, Op. 7, Spr. 226; see also Wiesław Bieńkowski, "Konstant von Wurzbach und Albert Zipper: Aus der Geschichte der österreichischpolnischen kulturellen Beziehungen im 19. und 20. Jahrhundert," in *Österreich—Polen: 1000 Jahre Beziehungen*, ed. Józef Buszko (Cracow: Wydawnictwo Uniwersytetu Jagiellońskiego, 1996), 481–507.

86. DALO, F. 26, Op. 7, Spr. 293, Z. 9599, 4 August 1888; the scholarships were, however, aimed primarily at the education of gymnasium teachers.
87. AUJ, WF II 157, 18 March 1913.
88. See AGAD, MWiO, Sygn. 112u, PA Żmurko, Z. 11229, 29 September 1871 (Eduard/Edward Strasburger); AUJ, WF II 173, 14 June 1914 (Władysław Rothert); WF II 168, Z. 20295, 30 January 1888 (Marceli Nencki); and WF II 163, [day and month illegible] 1895 (Jan Ptaszycki).
89. In the preparation of the terna for chemistry in Cracow in 1911, five of the best candidates refused to cooperate because of the lack of adequate laboratory equipment: AGAD, MWiO, Sygn. 66u, PA Dziewoński, Z. 448, 23 February 1911.
90. For details on the faculty, see Wanda Wojtkiewicz-Rok, *Dzieje Wydziału Lekarskiego Uniwersytetu Lwowskiego w latach 1894–1918* (Wrocław: Wydawnictwo Uniwersytetu Wrocławskiego, 1992).
91. See the long correspondence with Lubor Niederle, a Czech anthropologist and professor of prehistoric archaeology at Czech University in Prague, on what anthropology was and how to establish a chair for it: DALO, F. 26, Op. 7, Spr. 435, p. 23, 11 August 1902, p. 26, 2 October 1902.
92. See the habilitation of Moses/Mojżesz Schorr in L'viv for "Semitic language and the history of old Semitic language," where the faculty asked three experts from outside Galicia for their opinion: DALO, F. 26, Op. 5, Spr. 2143, PA Schorr, Z. 492, 9 December 1909; and AGAD, MWiO, Sygn. 121u, PA Schorr, Z. 653, 22 January 1910.
93. The professor of art history at the University of Vienna Rudolf Eitelberger, for example, was consulted on the proposed appointment of the art historian Marian Sokołowski. AGAD, MWiO, Sygn. 69u, PA Sokołowski, Z. 13411 ex 1881, 29 January 1882.
94. AGAD, MWiO, Sygn. 118u, PA Benoni, Z. 383, 6 April 1878.
95. AGAD, MWiO, Sygn. 118u, PA Benoni, Z. 234, 17 May 1878.
96. AGAD, MWiO, Sygn. 122u, PA Werner, Z. 458, 26 June 1882; Z. 18879, 3 June 1882; Z. 2591, 13 April 1883.
97. See, e.g., Krzysztof Lipiński, "Wilhelm Creizenach (1851–1919)," in *Uniwersytet Jagielloński: Złota Księga Wydziału Filologicznego*, ed. Jan Michalik and Wacław Walecki (Cracow: Księgarnia Akademicka, 2000), 107–15; and Jerzy Starnawski, "Sylwetki lwowskich historyków literatury: Richard Maria Werner (14 VIII 1854–13 I 1913)," *Przegląd Wschodni* 2, no. 2 (1993): 485–90.
98. On the psychologist Julian Ochorowicz, see AGAD, MWiO, Sygn. 120u, PA Ochorowicz, L. 3, 22 June 1884; Z. 47457, 19 October 1885. On the linguist Jan Baudouin de Courtenay, see Theodore R. Weeks, "Jan Baudouin de Courtenay: The Linguist as Anti-nationalist and Imperial Citizen," in *Eliten im Vielvölkerreich: Imperiale Biographien in Russland und Österreich-Ungarn*

(1850–1918) / Elites and Empire: Imperial Biographies in Russia and Austria-Hungary (1850–1918), ed. Malte Rolf and Tim Buchen (Berlin: de Gruyter, 2015), 338–54. On the philosopher Wincenty Lutosławski, see Tomasz Mróz, *Wincenty Lutosławski (1863–1954): "Jestem Obywatelem Utopii"* (Cracow: Polskia Akademia Umiejętności, 2008).

99. Gabriel Brzęk, "Recepcja darwinizmu w Polsce," in *Recepcja w Polsce nowych kierunków i teorii naukowych*, ed. Adam Strzałkowski (Cracow: Polska Akademia Umiejętności, 2001), 279–81.
100. See, e.g., Maria Julita Nedza, *Polityka Stypendialna Akademii Umiejętności w Latach 1878–1920: Fundacje Gałęzowskiego, Pileckiego i Osławskiego* (Wrocław: Zakład Narodowy im. Ossolińskich. Wydawnictwo Polskiej Akademii Nauk, 1973).
101. That is, if the candidate of science (кандидат наук) or magister (магистр) is counted as the first degree. In more than half of these cases, the candidates later also earned a PhD, mostly at universities in the German Empire.
102. The L'viv star surgeon Ludwik Rydygier is the most prominent case of a scholar who was considered: NA, MKV/R, inv.č. 2, fasc. 97, PA Kukula.
103. Jiří Kraml and Jiří Duchoň, "110 let české lékařské chemie," in *110 let české lékařské chemie a biochemie*, ed. Marie Balíková (Prague: Galén, 1994), 12–14.
104. If not noted otherwise, the information is from Matthias Svojtka, Johannes Seidl, and Barbara Steininger, "Von Neuroanatomie, Paläontologie und slawischem Patriotismus: Leben und Werk des Josef Victor Rohon (1845–1923)," *Mensch—Wissenschaft—Magie* 26 (2009): 123–59.
105. Rohon to Albert, St. Petersburg, 12 December 1892, reprinted in Svojtka, Seidl, and Steininger, "Von Neuroanatomie," 149–52.
106. NA, MKV/R, inv.č 2, fasc. 100, attachment no. 3 to Z. 12714, 24 July 1895, from 6 December 1894 (*Minoritätsvotum* of Vladimir Tomsa, Josef Hlava, and Jan Janosik).
107. *Ottův slovník naučný*, 28 vols. (Prague: J. Otto, 1888–1909).
108. His students included Otakar Srdínko, a professor in Prague, and Josef Florian Babór, a professor in Bratislava. K. Šula, "Otakar Srdínko," *Almanach České Akademie věd a umění* 41 (1930): 957–75; and Gustáv Čatár, Ján Vojtaššák, and Miloš Tichý, "Profesor MUDr. Jozef Florian Babor—významná osobnosť Ústavu pre všeobecnú biológiu LF UK," *História medicíny, farmácie a veterinárnej medicíny v kontexte vývoja európskej vedy 20. storočia*, ed. Ľudmila Pavlíková (Bratislava: Lekárska fakulta Univerzity Komenského, 2000), 61–64.
109. Ivan Holovac'kyj, *Ivan Horbačevs'kij: 1854–1942. Žyttiepysno-bibliografičnyj narys* (L'viv: Naukove Tovarystvo im. Ševčenka, 1995); and Zygmunt Albert, "Prof. Dr. Andrzej Obrzut," *Archiwum Historii i Filozofii Medycyny* 55, no. 1 (1992): 55–61.
110. Waldemar Kozuschek, *Jan Mikulicz-Radecki 1850–1905: Współtwórca Nowoczesnej Chirurgii / Johann von Mikulicz-Radecki: Mitbegründer der modernen Chirurgie* (Wrocław: Wydawnictwo Uniwersytetu Wrocławskiego, 2003), 78–80.
111. AGAD, MWiO, Sygn. 51u, PA Mikulicz, Z. 13062, 20 [month illegible] 1882. See

also Leon Wachholz, "Dwie obsady katedr lekarskich w Uniwersytecie Jagiellonskim w wieku XIX," *Archiwum Historii i Filozofii Medycyny* 10 (1930): 226–33.

112. On Naphtali Sobel, see AGAD, MWiO, Sygn. 121u PA Sobel, Z. 192, 26 January 1884. On Albert Zipper, see AGAD, MWiO, Sygn. 122u, PA Zipper, Z. 146, 26 November 1881; and DALO, F. 26, Op. 7, Spr. 226; see also Bieńkowski, "Konstant von Wurzbach." On Henryk Biegeleisen, see DALO, F. 26, Op. 7, Spr. 244, 9 May 1883, B. 56.
113. See, e.g., Oldřich Kramář's application for the chair of philosophy in Cracow: AUJ, W II 128, 26 January 1877.
114. AUJ, WF II 151, 3 March 1865. The inquiry was caused by the lack of adequate scholars lecturing in Polish. Jülg, however, had also asked earlier whether the chair was filled; see WF II 2, Z. 179, 17 January 1865 (Jülg's letter), Z. 190, 28 January 1865 (the faculty's answer).
115. Zygmunt Radziejewski (Berlin) committed suicide before his appointment: AUJ, WL II 164, 13 April 1873 and 7 July 1873. Wilhelm Zülzer (Berlin) declined the call at the last moment: AUJ, WL II 164, 4 June 1874, 29 June 1874.
116. AGAD, MWiO, Sygn. 65u, PA Adametz, Z. 591, 27 May 1891.
117. On the chemist Julius Braun from Wrocław/Breslau, see AGAD, MWiO, Sygn. 66u, PA Dziewoński, Z. 448, 23 February 1911; on the dermatologist Ernst Finger from Vienna, see AUJ, WL II 174, 11 January 1897.
118. DALO, F. 26, Op. 7, Spr. 175, pp. 24–25, 10 December 1874, p. 32, 14 December 1874.
119. These scholars were Benoni (see DALO, F. 26, Op. 7, Spr. 178, records covering 1874–78) and Franciszek Czerny-Schwarzenberg (see AUJ, WF II 180 Geografia, 28 June 1874).
120. AGAD, MWiO, Sygn. 120u, PA Rehmann.
121. AUJ, WF II 157, Rudolf Kotula's letter of 8 November 1913. The faculty did, however, also ask several experts about Wukadinović's qualification for the professorship.
122. Krzysztofa Michalewska, "Próby utworzenia uniwersytetu ukraińskiego w Polsce 1919–26" (PhD diss., Jagiellonian University, Cracow, 1974), e.g., 14–15 (on the situation in 1867–68); and Twardowski, *Die Universität Lemberg.*
123. Michalewska, "Próby utworzenia," 22.
124. DALO, F. 26, Op. 7, Spr. 269, p. 86, 27 June 1884; p. 96, 12 July 1884. Finkel's habilitation was accepted by all but three votes: those of Šaranevyč, Ohonovs'kyj, and the physicist Oskar Fabian; the latter was, however, not Ruthenian.
125. DALO, F. 26, Op. 7, Spr. 321, Z. 304, 23 January 1891.
126. AGAD, MWiO, Sygn. 118u, PA Dembiński, Z. 2942, 2 March 1892; Z. 847, 7 February 1892.
127. DALO, F. 26, Op. 7, Spr. 175, pp. 24–25, 10 December 1874; p. 32, 14 December 1874; and DALO, F. 26, Op. 7, Spr. 321, Z. 262, 4 December 1890.
128. See, for instance, the appointment of the chair of Polish language and literature (head of commission: Ohonovs'kyj): DALO, F. 26, Op. 7, Spr. 175, p. 104, 15 June 1875.

129. In general, on the New Era and its resolutions, see Dariusz Maciak, *Próba porozumienia polsko-ukraińskiego w Galicji w latach 1888–1895* (Warsaw: Wydawnictwo Uniwersytetu Warszawskiego, 2006); and Ihor Čornovol, *Pol'sko-ukraïns'ka uhoda 1890–1894 rr.* (L'viv: L'vivs'ka Akademija Mistectv, 2000).
130. See on this issue Hrytsak, "Ruslan, Bohdan and Myron"; John Paul Himka, "The Construction of Nationality in Galician Rus': Icarian Flights in Almost All Directions," in *Intellectuals and the Articulation of the Nation*, ed. Ronald Grigor Suny and Michael D. Kennedy (Ann Arbor: University of Michigan Press, 1999), 109–64; and Zayarnyuk, *Framing the Ukrainian Peasantry*, 317–74.
131. Quoted in Pacholkiv, *Emanzipation durch Bildung*, 179.
132. AGAD, MWiO, Sygn. 118u, PA Gruszewski (Hruszewski), Z. 5265, 27 March 1894.
133. DALO, F. 26, Op. 7, Spr. 346, Z. 6898, 9 April 1892; Z. 141, 20 October 1892; distribution of votes in Z. 641, 7 June 1893 and F. 26, Op. 7, Spr. 510, pp. 20–35, 7 June 1893. On the religious issue, see Aleksandr Barvins'kyj, "Zasnovane katedry istoriï Ukraïny v L'vivs'kim universyteti," *Zapysky Naukovoho tovarystva imeni Ševčenka: Praci filologičnoï sekciï* 141–43 (1925): 1–25; on Myl'kovyč, see Vitalij Tel'vak, "Mychajlo Hruševs'kyj contra Volodymyr Mil'kovyč (do problemy formuvannja dyskusijnoho polja ukraïns'koï istoriohrafiï počatku XX stolittja," *Archivoznavstvo. Archeohrafija. Džereloznavstvo* 9 (2008): 259.
134. AGAD, MWiO, Sygn. 118u, PA Gruszewski (Hruszewski), Z. 5265, 27 March 1894; Z. 9018, 14 January 1894.
135. AGAD, MWiO, Sygn. 119u, PA Hruszewski (Gruszewski), Z. 7332, 20 August 1896.
136. DALO, F. 26, Op. 7, Spr. 410, p. 29, 16 July 1901; and F. 26, Op. 7, Spr. 508, p. 12, 26 July 474; Z. 555, 29 November 1907.
137. DALO, F. 26, Op. 5, Spr. 510, PA Hruszewski, p. 72, 11 March 1905.
138. Ohonovs'kyj was appointed an associate professor in 1865 and was given a full professorship in 1871.
139. For the clearest and most up-to-date information dealing with this issue, see Zayarnyuk, "Mapping Identities," esp. 121–26.
140. Theodor Gartner and Stephan Smal-Stocki [Stepan Smal'-Stoc'kyj], *Minoritätsvotum in der vom k.k. Bukowiner Landesschulrathe behufs Regelung der ruthenischen Schulortographie eingesetzten Commission abgegeben im November 1887* (Chernivtsi: Selbstverlag, 1888), 23.
141. On the details see Smal'-Stoc'kyj, "Fedir Gartner."
142. DALO, F. 26, Op. 5, Spr. 899, PA Kolessa, 12 May 1895.
143. The most thorough analysis is still that found in M[ychajlo] Voznjak [Mychaĭlo Vozniak], "Nedopuščennja Ivana Franka do docentury u L'vivs'komu universyteti," *Ivan Franko: Statti i materialy* 1 (1948): 43–63; see also Volodymyr Kačmar, "Sprava ukraïns'koho universytetu na tli pol's'ko-ukraïns'kych superečnostej u Halyčyni 1901–1908 rr.," *Problemy slov'janoznavstva* 42 (2002): 47–58. In 1907 Franko strove once again for the chair. This time the dean passed his application to the mysterious Commission of Chairs for Humanistic

Disciplines (Komisja dla katedr humanistycznych). From this point I could find no further documents on this issue; see DALO, F. 26, Op. 7, Spr. 554, p. 38, Z. 1732, 15 July 1907.

144. Ivan Vakarčuk and Jaroslav Isaievyč, eds., *Listuvannja Ivana Franka ta Mychajla Drahomanova* (L'viv: Vydavnyčyj centr L'vivs'koho Nacional'noho Universytetu im. Ivana Franka, 2006), 523–24, 535; and Voznjak, "Nedopuščennja Ivana Franka."
145. AGAD, MWiO, Sygn. 121u, PA Studziński, Z. 97675, 30 December 1895; and DALO, F. 26, Op. 7, Spr. 381, Z. 249, 22 November 1895.
146. Pacholkiv, *Emanzipation durch Bildung*, 176; details on Studyns'kyj's scientific career in Cracow are given in AUJ, WF II 121, PA Studzinski, Cyryl, especially 10 July 1896 (expertise of Tretiak), 23 January 1897 (confirmation of the ministry).
147. AUJ, WF II 121, PA Studzinski, Cyryl, Z. 503, 2 April 1897 (petition), Z. 8176, 8 April 1897 (acceptance).
148. ALO, F. 26, Op. 5, Spr. 899, PA Kolessa, p. 12, 31 March 1896.
149. DALO, F. 26, Op. 5, Spr. 899, PA Kolessa, pp. 22–24, 15 July 1897; and DALO, F. 26, Op. 7, Spr. 387, Z. 498, 19 June 1897.
150. AGAD, MWiO, Sygn. 119u, PA Kolessa, Z. 11599, 24 October 1897.
151. AGAD, MWiO, Sygn. 119u, PA Kolessa, Z. 1545, 18 January 1898; and DALO, F. 26, Op. 5, Spr. 899, PA Kolessa, Z. 2731, 9 February 1898. The professors of Slavic philology Vatroslav Jagić (Vienna) and Aleksander Brückner (Berlin) provided two experts in support of Kolessa, and Brückner wrote a letter of recommendation, in which he, however, also stated that Franko would actually be the most suitable candidate. AGAD, MWiO, Sygn. 403u, Z. 1545, 9 January 1898.
152. Quoted in the correspondence of the provincial government: AGAD, MWiO, Sygn. 121u, PA Studziński, Z. 103567, 25 November 1898; see also Z. 18236, 4 October 1899.
153. DALO, F. 26, Op. 7, Spr. 421, p. 19, 22 October 1898.
154. The faculty saw Studyns'kyj or the gymnasium teacher in L'viv, Volodymyr Kocovs'skyj (Володимир Коцовський), as the most suitable applicants for the readership.
155. AGAD, MWiO, Sygn. 121u, PA Studziński, votum separatum, dated 6 June 1899.
156. In the documents this candidate was referred to only by his surname, Radchenko. Probably this meant Konstantin/Kostyantyn Radčenko (Константин/Костянтин Радченко), who had earned permission to teach in Kiev in 1898 but had published several articles on Old Slavic languages in Russian and German starting in 1897.
157. See the faculty discussions and presentation of all positions in DALO, F. 26, Op. 7, Spr. 420, Z. 423, 13 March 1899; and F. 26, Op. 7, Spr. 421, Z. 640, 3 June 1899, as well as AGAD, MWiO, Sygn. 121u, PA Studziński.
158. DALO, F. 26, Op. 7, Spr. 434, Z. 326, 27 November 1899.
159. DALO, F. 26, Op. 7, Spr. 554, Z. 1732, 5 July 1907.
160. Harald Binder, "Das Ruthenische Pressewesen," in *Die Habsburgermonarchie 1848–1918*, vol. 8, *Politische Öffentlichkeit und Zivilgesellschaft*, pt. 2, *Die Presse als Faktor der politischen Mobilisierung*, ed. Helmut Rumpler and Peter

Urbanitsch (Vienna: Verlag der Österreichischen Akademie der Wissenschaften, 2006), 2121–23.

161. Michalewska, *Próby utworzenia*, 44–54.
162. Quoted in Michalewska, *Próby utworzenia*, 47.
163. AGAD, MWiO, Sygn. 70u, PA Tretiak, Z. 836, 5 June 1893.
164. AGAD, MWiO, Sygn. 70u, PA Tretiak, Z. 7800, 11 September 1893.
165. Buszko, *Społeczno-Polityczne oblicze*; and Buszko, "L'université Jagellon de Cracovie (1869–1914)," *Études danubiennes* 4, no. 1 (1988): 21–28.
166. AUJ, WF II 153, 21 February 1871; for Tarnowski's ideology and scientific ideas, see Maria Wyka, "Stanisław Tarnowski jako historyk literatury polskiej," in *Stanisław Tarnowski (1837–1917): Materiały z Posiedzenia Naukowego PAU w dniu 14.XI.1997 r.*, ed. Rita Majkowska (Cracow: Wydawnictwo Polskiej Akademii Umiejętności, 1999), 9–16.
167. See, for example, "Wolność Nauki," *Kraj*, 11 November 1870; and "Reforma Uniwersytetu Lwowskiego," *Kraj*, 5 September 1871.
168. Reinhard Müller, "Maksymilian Ernest Gumplowicz (1861–1897)," in Surman and Mozetič, *Dwa życia*, 95.
169. Buszko, *Społeczno-polityczne oblicze*.
170. Wolff, "Dynastic Conservatism"; and Dabrowski, *Commemorations*.
171. Stanisław Konarski, "'Zimmermanniada' w Uniwersytecie Jagiellońskim (1910–1911)," in *Postępowe tradycje młodzieży akademickiej w Krakowie*, ed. Henryk Dobrowolski, Mirosław Frančić, and Stanisław Konarski (Cracow: Wydawnictwo Literackie, 1962), 135–204.
172. Gabriel Brzęk, "Recepcja darwinizmu."
173. For example, the political leaders of the National Democratic Party, Stanisław Głąbiński and Stanisław Grabski, were members of the University of L'viv.
174. Jerzy Maternicki, "Polskie szkoly historyczne we Lwowie w XIX w.," in *Wielokulturowe środowisko historyczne Lwowa w XIX i XX w.*, ed. Jerzy Maternicki and Leonid Zaszkilniak (Rzeszów: Wydawnictwo Uniwersytetu Rzeszowskiego, 2005), 3:23–45.
175. One example was the law historian Oswald Balzer; see Markus Krzoska, "Ein Wissenschaftler zwischen Elfenbeinturm und Öffentlichkeit: Der Lemberger Rechtshistoriker Oswald Balzer (1859–1933)," in *Beruf und Berufung: Geschichtswissenschaft und Nationsbildung in Ostmittel- und Südosteuropa*, ed. Markus Krzoska and Hans-Christian Maner (Münster: Lit, 2005), 217–38.
176. K[azimierz] Nitsch, "Moje wspomnienia językowe VII," *Język Polski* 39, no. 5 (1959): 355–61.
177. See, for the tensions between the Old Czech Party and the Young Czech Party in 1882, Jaromír Čelakovský, *Moje zápisky, 1871–1914*, ed. Luboš Velek and Alice Velková (Prague: Archiv hlavního města Prahy; Výzkumné centrum pro dějiny vědy: Scriptorium, 2004), 41–42.
178. See Havelka, "Hundred Years."
179. A large collection of original articles and the history of the conflict can be found online on the website of the Czech Manuscript Society (Česká společnost rukopisná), http://www.rukopisy-rkz.cz/rkz/csr/.

180. See the comments of Jaroslav Goll in his letters to Eduard Albert, 7 November 1890, [day unknown] November 1890, and 13 November 1890, reprinted in Jaroslav Jedlička, "Eduard Albert—Jaroslav Goll—50 listů korespondence," *AUC-HUCP* 13, nos. 1–2 (1973): 229–32.
181. On conflicts over Jan Palacký (the son of František Palacký), see ÚDAUK, Fond Filozofická fakulta Karlo-Ferdinandovy Univerzity, 1882–2012, Inv.č. 554, Kart. 46, PA Palacký Jan, Masaryk's votum separatum of 8 January 1885. On Petr (Peter) Durdík, brother of the professor of philosophy Josef, AT-OeStA/AVA Unterricht UM allg. Akten 1216, Peter Durdík, Z. 12750, 14 June 1887/24 June 1887, Z. 19335, 19 September 1887.
182. Michal Svatoš, "Univerzitní Působení Filologa Josefa Krále," *AUC-HUCP* 22, no. 2 (1982): 78; and Theodor Syllaba, *Jan Gebauer na pražské Univerzitě* (Prague: Karlova Univerzita, 1983), 60–76.
183. Jiroušek, "Jazyky v životě," 534.
184. Jiroušek, "Mimořádná profesura Josefa Pekaře."
185. Jan Herben, *Masarykova sekta a Gollova škola* (Prague: Pokrok, 1912). See also Tomáš Hermann, *Emanuel Rádl a české dějepisectví: Kritika českého dějepisectví ve sporu o smysl českých dějin* (Prague: Univerzita Karlova v Praze, Filozofická fakulta, 2002).
186. Rezek to Jan Gebauer, 25 December 1899, quoted in Svatoš, "Univerzitní Působení," 75.
187. Fritz Fellner, *". . . ein wahrhaft patriotisches Werk": Die Kommission für Neuere Geschichte Österreichs 1897–2000* (Vienna: Böhlau, 2001), 45, 72, 84. After 1918 members of the commission who worked outside Austria's borders were excluded (252–54).
188. Bruno Latour, *Science in Action: How to Follow Scientists and Engineers through Society* (Cambridge, MA: Harvard University Press, 1987), esp. 220–30; see also, on a similar practice in the Russian Empire, Trude Maurer, "Der Weg zur Mündigkeit: Auslandsaufenthalte rußländischer Wissenschaftler im 19. und frühen 20. Jahrhundert," *Hyperboreus* 10, nos. 1–2 (2004): 60–77.
189. AT-OeStA/AVA Unterricht UM allg. Akten 673, PA Leciejewski, Z. 1730, 29 May 1885.
190. See, for example, Ewa Nowak, *Polska młodzież w Austrii w XIX i XX wieku: Migracje-Edukacja-Stowarzyszenia* (Lublin: Wydawnictwo Uniwersytetu Marii Curie-Skłodowskiej, 2007); Pacholkiv, *Emanzipation durch Bildung*; Wolfgang Petritsch, "Die slowenischen Studenten an der Universität Wien (1848–1890)" (PhD diss., University of Vienna, 1972); and Vasilij Melik and Peter Vodopivec, "Die slowenische Intelligenz und die österreichischen Hochschulen 1848–1918," in *Wegenetz europäischen Geistes II: Universitäten und Studenten. Die Bedeutung studentischer Migration in Mittel und Südosteuropa vom 18. bis zum 20. Jahrhundert*, ed. Richard Georg Plaschka and Karlheinz Mack (Vienna: Verlag für Geschichte und Politik, 1987), 134–54.
191. Henryk Barycz, "Die Rolle der Wiener Universität im geistigen Leben Polens," *Österreichische Osthefte* 7, no. 3 (1965): 176–94; and Urszula Perkowska,

"Études scientifiques des universitaires de Cracovie à Vienne dans les années 1800–1918," *Zeszyty Naukowe Uniwersytetu Jagiellońskiego: Prace Historyczne* 88 (1989): 305–12.

Chapter 6

1. From the speech of Karl Lueger, 5 December 1907, in *Stenographische Protokolle des Abgeordnetenhauses im Jahre 1907*, XVIII. Session, IV Band, 21. bis 41. Sitzung (S. 1851 bis 3054.) (Vienna: Kaiserlich-königliche Hof- und Staatsdruckerei, 1908).
2. E.g., John W. Boyer, *Political Radicalism in Late Imperial Vienna: Origins of the Christian Social Movement, 1848–1897* (Chicago: University of Chicago Press, 1981).
3. Brian A. Porter, *Faith and Fatherland: Catholicism, Modernity, and Poland* (Oxford: Oxford University Press, 2011).
4. Johannes Feichtinger, "Der erinnerte Feind und nationale Integration: Zentraleuropa im langen 19. Jahrhundert," in *Der erinnerte Feind*, ed. Johannes Feichtinger and Johann Heiss (Vienna: Mandelbaumverlag, 2013), 300–322; and Svjatoslav Pacholkiv, "Die 'Ostjuden' als Begriff in der Geschichte," in *"Ostjuden"—Geschichte und Mythos: Juden in Mitteleuropa 2011*, ed. Institut für jüdische Geschichte Österreichs (St. Pölten: Institut für jüdische Geschichte Österreichs, 2011), 2–11.
5. See the statement of Johann Kutschker, the chancellor of the university and auxiliary bishop of the archdiocese in Vienna: Kutschker, "Erklärung des Kanzlers der k.k. Universität zu Wien über die Bitte der protestantisch-theologischen Fakultät um Einverleibung in die genannte Hochschule, abgegeben in der Sitzung Venerabilis Consistorii am 12. Mai 1863," in *Der katholische Charakter der Wiener Universität: Eine Denkschrift der theologischen Facultät*, ed. Anton Horny (Vienna: Verlag der Mechitharisten-Congregations-Buchhandlung, 1863), 153–63.
6. Seebacher, *"Freiheit der Naturforschung!,"* 97; and Mell, "Ein rechtsgeschichtlicher Beitrag," 5.
7. Michler, *Darwinismus und Literatur*, 34–36.
8. Joseph Unger, *Zur Reform der Wiener Universität: Ein Votum erstattet in der Sitzung des Unterrichtsrathes am 29. Dezember 1865* (Vienna: Manz, 1869).
9. *Die deutsche Karl-Ferdinands-Universität in Prag unter der Regierung Sr. Majestät des Kaisers Franz Josef I* (Prague: J. G. Calve, 1899), 18. Notwithstanding the lack of ministerial confirmation, Stein served as dean for the whole year.
10. Brzęk, "Recepcja darwinizmu."
11. See Konarski, "'Zimmermanniada,'" 137–38.
12. Hermann Kuprian, "'Machen Sie diesem Skandal ein Ende. Ihre Rektoren sind eine nette Gesellschaft.' Modernismusdiskussion, Kulturkampf und Freiheit

der Wissenschaft: Die Wahrmund-Affäre," in *Politische Affären und Skandale in Österreich: Von Mayerling bis Waldheim*, ed. Michael Gehler and Hubert Sickinger, 2nd ed. (Vienna: Thaur, 1996), 99–127.

13. Waltraud Heindl, "Der Liberalismus scheiterte. Scheiterte der Liberalismus?," in *"Dürfen's denn das?" Die fortdauernde Frage zum Jahr 1848*, ed. Sigurd Paul Scheichl and Emil Brix (Vienna: Passagen, 1999), 85–95.
14. Robert Luft, "'Politische Professoren' in Böhmen 1861–1914," in Lemberg et al., *Bildungsgeschichte, Bevölkerungsgeschichte, Gesellschaftsgeschichte*, 286–306; and Helmut Slapnicka, "Die juridischen Fakultäten der Prager Universitäten 1900–1939," in Lemberg, *Universitäten in nationaler Konkurrenz*, 79–80.
15. Klaus Taschwer, "Wissenschaft für viele: Zur Wissensvermittlung in der Wiener Volksbildungsbewegung rund um 1900" (PhD diss., University of Vienna, 2002), 121.
16. Stephan Koja, ed., *Gustav Klimt: Der Beethoven-Fries und die Kontroverse um die Freiheit der Kunst* (Munich: Prestel, 2006).
17. Robert Tichy and Johannes Wallner, "Johannes Frischauf—eine schillernde Persönlichkeit in Mathematik und Alpinismus," *Internationale Mathematische Nachrichten* 63, no. 210 (2009): 21–32.
18. Mróz, *Wincenty Lutosławski*, 112–32.
19. Florian Mildenberger, "'. . . als Conträrsexual und als Päderast verleumdet . . .'—der Prozess um den Naturforscher Theodor Beer (1866–1919) im Jahre 1905," *Zeitschrift für Sexualforschung* 18, no. 4 (2005): 332–51; for the wider sociopolitical context of the affair and reactions to it, see Scott Spector, "Where Personal Fate Turns to Public Affair: Homosexual Scandal and Social Order in Vienna, 1900–1910," *Austrian History Yearbook* 38 (2007): 15–24; and other articles in the 2007 *Austrian History Yearbook* section "Writing the History of Sexuality in Fin-de-Siècle Cisleithania."
20. Siegmund Feilbogen from the Academy of Commerce (Exportakademie) in Vienna was removed from office in 1908 after his sister-in-law, with whom he had visited the Sistine Chapel during Holy Mass on Easter Sunday, took the consecrated wafer (according to some, given by the pope himself) from her mouth, causing an international scandal. Nikolaj Beier, *"Vor allem bin ich ich . . .": Judentum, Akkulturation und Antisemitismus in Arthur Schnitzlers Leben und Werk* (Göttingen: Wallstein, 2008), 312–14.
21. August Rohling, known for militant anti-Semitism from the 1870s, was pensioned off only after the Catholic Church placed one of his books, *Der Zukunftstaat* (The state of the future, 1898), on the List of Prohibited Books. "Warum wurde der Professor für hebräische Altertümer an der theologischen Fakultät der Universität Prag, Kanonikus Dr. August Rohling, von der österreichischen Unterrichtsverwaltung seines Postens enthoben?," *Dr. Bloch's Österreichische Wochenschrift*, 3 July 1908, 480–83.
22. Exceptions are Buszko, *Społeczno-polityczne oblicze*; Felicitas Seebacher, *Das Fremde im "deutschen" Tempel der Wissenschaften* (Vienna: Verlag der Österreichischen Akademie der Wissenschaften, 2011); and several recent works on the University of Vienna.

23. See, e.g., Lappin, "Die Czernowitzer Sprachkonferenz (1908)"; and Marsha L. Rozenblit, "The Assertion of Identity: Jewish Student Nationalism at the University of Vienna before the First World War," *Leo Baeck Institute Yearbook* 27, no. 1 (1982): 171–86. See also Henryk Immeles, *Antysemityzm w Austryi* (L'viv: Jedność, 1910); and Feichtinger, *Wissenschaft als reflexives Projekt*, 391–507.
24. Tillfried Cernajsek, Christoph Mentschl, and Johannes Seidl, "Eduard Sueß (1831–1914): Ein Geologe und Politiker des 19. Jahrhunderts," in *Wissenschaft und Forschung in Österreich: Exemplarische Leistungen österreichischer Naturforscher und Techniker*, ed. Gerhard Heindl (Frankfurt am Main: Peter Lang, 2000), 82.
25. Josef Redlich, speech in the Lower House of Parliament, 4 December 1907, reprinted in *Stenographische Protokolle 1907*, 2941.
26. See his curriculum vitae in AT-OeStA/AVA Unterricht UM allg. Akten 681, PA Harry Torczyner.
27. For an analysis of the consequences of adopting an assimilationist perspective, see Agnieszka Jagodzińska, "Asymilacja, czyli bezradność historyka: O krytyce terminu i pojęcia," in *Wokół akulturacji i asymilacji Żydów na ziemiach polskich*, ed. Konrad Zieliński (Lublin: Uniwersytet Marii Curie-Skłodowskiej, 2010), 15–31.
28. See below; and Eugen Ehrlich, "Der Antisemitismus im Professorenkollegium der österreichischen Universität," *Dr. Bloch's Österreichische Wochenschrift*, 6 December 1907, 811–12.
29. Grete Mecenseffy, *Evangelische Lehrer an der Universität Wien* (Vienna: Böhlau, 1967).
30. Andreas D. Ebert, *Jüdische Hochschullehrer an preußischen Universitäten (1870–1924): Eine quantitative Untersuchung mit biografischen Skizzen* (Frankfurt am Main: Mabuse, 2008), 11–19.
31. See especially Pieter M. Judson and Marsha L. Rozenblit, eds., *Constructing Nationalities in East Central Europe* (New York: Berghahn Books, 2005); Theodore R. Weeks, "Assimilation, Nationalism, Modernization, Antisemitism: Notes on Polish-Jewish Relations, 1855–1905," in *Antisemitism and Its Opponents in Modern Poland*, ed. Robert Blobaum (Ithaca, NY: Cornell University Press, 2005), 20–38; and Porter, *Faith and Fatherland.*
32. The Concordat of 1855 assured the role of Catholic clergy in the Habsburg Empire, including influence on early schooling and the civil service in general. While universities were officially not included, the general rules of the concordat were applied there as well.
33. "Verordnung des Ministers des Cultus und Unterrichts vom 24. Mai 1850, RGBl., LXVIII, 11.6.1850, Z. 219," reprinted in Schweickhardt, *Sammlung*, 133–34 (no. 83).
34. Minister's decision (*Ministerialerlass*) of 24 January 1868, Z. 34, reprinted in Schweickhardt, *Sammlung*, 134.
35. See Beck von Mannagetta and Kelle, *Die österreichischen Universitätsgesetze*, 720–21n1–3.

36. "Verordnung des Ministers für Cultus und Unterricht vom 29. März 1858, Z. 264/KUM, RGBl. 1858 . . . die Durchführung der Artikel VI und XVII des Koncordates bezüglich der theologischen Studien betreffend," in Beck von Mannagetta and Kelle, *Die österreichischen Universitätsgesetze*, 740–45 (no. 607). The basis for this ordinance was articles VI and XVII of the Concordat of 1855, quoted in Beck von Mannagetta and Kelle, *Die österreichischen Universitätsgesetze*, 740–41, n. 3.
37. *Stenographische Protokolle 1907*, 4 December 1907, 2958–59.
38. On the pope's science policies, see Thomas C. McGonigle and James F. Quigley, *A History of the Christian Tradition: From the Reformation to the Present* (Mahwah, NJ: Paulist, 1996), 120–52.
39. The *Zeitschrift für Katholische Theologie* was published by the theological faculty in Innsbruck; *Przegląd Powszechny* was a scholarly journal concerned with religious issues, issued in Cracow by the Jesuits.
40. "Erlass des Ministers für K. u. U vom 31. Mai 1897, Z. 631, an das Dekanat der theologischen Fakultät der böhmischen Universität in Prag, betreffend die Einrichtung der 'christlichen Soziologie' unter die für das theologische Studium obligaten Lehrfächer," reprinted in Beck von Mannagetta and Kelle, *Die österreichischen Universitätsgesetze*, 752–53 (no. 614).
41. See "Beschlüsse der Generalversammlung des österreichischen Episkopates zu Wien am 13. November 1901," reprinted in Beck von Mannagetta and Kelle, *Die österreichischen Universitätsgesetze*, 746–48, here point I.2.
42. *Der katholische Universitäts-Verein zu Salzburg, Dem hochwürdigsten Episkopate und den Katholiken Österreichs zum 25-jähr. Vereins-Jubiläum 1884 bis 1909 gewidmet vom Zentral-Ausschuß* (Salzburg: Verlag der katholischen Universitäts-Verein, 1909).
43. "Hirtenschreiben, betreffend Errichtung einer katholischen Universität in Österreich," reprinted in *Das Vaterland*, 19 December 1901, 1–2.
44. See the lists of financial contributors published in *Universitätsblatt: Zeitschrift des Vereines zur Gründung und Erhaltung einer freien katholischen Universität in Salzburg* (University letters: Journal of the Society for the Establishment and Maintenance of a Free Catholic University in Salzburg, published from 1903 on); on various occasions, the publishers criticized the Slavs for concentrating only on their own issues and not helping to establish a superregional academy.
45. See the proceedings of the Political Association for Enlightenment (Politisches Aufklärungsverein) in *Neue Freie Presse*, 19 December 1901, 8.
46. *Neue Freie Presse*, 20 December 1901, 1–2.
47. Leo XIII, "Quod Votis: Encyclical of Pope Leo XIII on the Proposed Catholic University," 30 April 1902, http://www.vatican.va/holy_father/leo_xiii/encyclicals/documents/hf_l-xiii_enc_30041902_quod-votis_en.html.
48. Albert Ehrhard, *Der Katholizismus und das zwanzigste Jahrhundert im Lichte der kirchlichen Entwicklung der Neuzeit* (1901; Stuttgart: Jos. Roth'sche Verlagsbuchhandlung, 1902), 371 (emphasis omitted); see esp. 361–82.
49. Joseph Maria Pernter, *Voraussetzungslose Forschung, freie Wissenschaft und*

Katholizismus (Vienna: Braumüller, 1902), esp. 31; and Ludwig Wahrmund, *Religion und Klerikalismus: Rede, gehalten bei der Gründung der Ortsgruppe Innsbruck des Vereines "Freie Schule"* (Innsbruck: A. Edlinger / Verlag der Ortsgruppe Innsbruck des Vereines "Freie Schule," [1902]).

50. Stefan Rebenich, *Theodor Mommsen und Adolf Harnack: Wissenschaft und Politik im Berlin des ausgehenden 19. Jahrhunderts* (Berlin: de Gruyter, 1997), 414–85.
51. Helmut Walser Smith, *German Nationalism and Religious Conflict: Culture, Ideology, Politics, 1870–1914* (Princeton, NJ: Princeton University Press, 1995), esp. 206–31.
52. See Otto Weiß, *Der Modernismus in Deutschland: Ein Beitrag zur Theologiegeschichte* (Regensburg: Pustet, 1995), esp. 116–18; and Erika Weinzierl, ed., *Der Modernismus: Beiträge zu seiner Erforschung* (Graz: Styria, 1974).
53. See, in general, Alfred Rinnerthaler, "Der Fall Wahrmund: Politische, rechtliche und diplomatische Turbulenzen im Umfeld von Modernismus und Antimodernismus in Österreich," in *Österreich und der Heilige Stuhl im 19. und 20. Jahrhundert*, ed. Hans Paarhammer and Alfred Rinnerthaler (Frankfurt am Main: Peter Lang, 2001), 187–246. See also, with special consideration of differing reactions within the Habsburg public, Jan Surman, "University Scandals and the Public Sphere of Imperial Austria: The Wahrmund and Zimmermann Affairs," *European Review of History—Revue Européene d'histoire* (forthcoming).
54. Wahrmund, *Religion und Klerikalismus.*
55. Ludwig Wahrmund, *Universität und Kirche: Akten zum Fall Wahrmund* (Frankfurt am Main: Neuer Frankfurter Verlag, 1902); and *Stenographische Protokolle über die Sitzungen des Hauses der Abgeordneten im Jahre 1902*, XVII. Session. XII Band, 106. bis 115. Sitzung. (S. 10039 bis 10958.) (Vienna: Kaiserlich-königliche Hof- und Staatsdruckerei, 1902), 109. Sitzung am 13. März 1902, 110. Sitzung am 14. März 1902, esp. 10408–9.
56. *Ein Stich in's Wespennest oder Der 6. allgemeine österreichische Katholikentag (1907) und die katholische Universitätsfrage* (Salzburg: Katholischer Universitäts-Verein, [1907]).
57. *Stenographische Protokolle 1907*, 39. Sitzung der XVIII. Session am 3. Dezember 1907; 40. Sitzung der XVIII. Session am 4. Dezember 1907; 41. Sitzung der XVIII. Session am 5. Dezember 1907.
58. Ludwig Wahrmund, *Katholische Weltanschauung und freie Wissenschaft: Ein populärwissenschaftlicher Vortrag unter Berücksichtigung des Syllabus Pius X und der Enzyklika "Pascendi Dominici gregis"* (Munich: J. F. Lehmann, 1908).
59. "Wahrmund: Ein Lustspiel," *Dr. Bloch's Österreichische Wochenschrift*, 20 March 1908, 214; see also *Czas* on different occasions between April and June 1908.
60. Roland Hoffmann, *T. G. Masaryk und die tschechische Frage: Nationale Ideologie und politische Tätigkeit bis zum Scheitern des deutsch-tschechischen Ausgleichsversuchs vom Februar 1909* (Munich: Oldenburg, 1988), 381–82.

61. Beier, *"Vor allem bin ich ich . . .,"* 312–14 (on Siegmund Feilbogen of the Academy of Commerce in Vienna); and Konarski, "'Zimmermanniada'" (on Kazimierz Zimmerman, Cracow).
62. Alfred Rinnerthaler, "Von der Benediktiner- zur Staatsuniversität: Vom Werden der Salzburger 'Alma Mater,'" in *Bürgerliche Freiheit und christliche Verantwortung: Festschrift für Christoph Link zum siebzigsten Geburtstag*, ed. Heinrich De Wall and Michael Germann (Tübingen: Mohr, Siebeck, 2003), 805.
63. AT-OeStA/AVA Unterricht UM allg. Akten 1058, PA Tangl, Z. 32116, 21 September 1904.
64. AT-OeStA/AVA Unterricht UM allg. Akten 669, PA Hirn, Z. 20935, 5 August 1899.
65. In Vienna, these were Engelbert Mühlbacher, Emil Ottenthal, Oswald Redlich, and Hans Voltelini (the last at the law faculty); in Graz, Arnold Busson.
66. See the statement of Karl Stremayr during the appointment of Josef Emler for the chair of auxiliary history in Prague in 1879 (AT-OeStA/AVA Unterricht UM allg. Akten 1216, PA Emler, Z. 9709, 13 July 1879): "The representatives of this discipline at the provincial universities should at the same time, and in the first place, turn their attention to the sources of their province and prepare the material for the treatment of [the] special history of [the] respective provinces."
67. Franz Lott to Hermann Lotze, 22 August 1872, reprinted in Hermann Lotze, *Briefe und Dokumente*, ed. Reinhardt Pester and Ernst Wolfgang Orth (Würzburg: Königshausen & Neumann, 2003), 572–73.
68. See, for example, Richard Schaefer, "Infallibility and Intentionality: Franz Brentano's Diagnosis of German Catholicism," *Journal of the History of Ideas* 68, no. 3 (2007): 477–99; see also Wolfdietrich Schmied-Kowarzik, "Vergessene Impulse der Wiener Philosophie um die Jahrhundertwende: Eine philosophische Skizze wider den main stream verdrängenden Erinnerns," in *Die Wiener Jahrhundertwende: Einflüsse—Umwelt—Wirkungen*, ed. Jürgen Nautz and Richard Vahrenkamp (Vienna: Böhlau, 1993), 181–201.
69. Lott to Lotze, reprinted in Lotze, *Briefe und Dokumente*, 572–73; and AT-OeStA/AVA Unterricht UM allg. Akten 664, PA Brentano, Z. 9206, 30 December 1873.
70. AT-UAW, PH PA 1118 Franz Brentano, Z. 662, 20 June 1880.
71. See Franz Clemens Brentano, *Meine letzten Wünsche für Oesterreich* (Stuttgart: Cotta, 1895), first published in *Neue Freie Presse*, 2, 5, and 8 December 1894.
72. Karl Stumpf, "Erinnerungen an Franz Brentano," in *Franz Brentano: Zur Kenntnis seines Lebens und seiner Lehre*, ed. Oskar Kraus (Munich: Beck, 1919), esp. 116–18; Robin Rollinger, "Anton Marty," in *The Stanford Encyclopedia of Philosophy (Winter 2008 Edition)*, ed. Edward N. Zalta (Stanford, CA: Metaphysics Research Lab, 2008), http://plato.stanford.edu/archives/win2008/entries/marty/; and Denis Fisette, "Carl Stumpf," in Zalta, *Stanford Encyclopedia of Philosophy* (Stanford, CA: Metaphysics Research Lab, 2009), http://plato.stanford.edu/archives/spr2009/entries/stumpf/.
73. Chernivtsi University opened the same year and thus had no faculty that could prepare the documents.

74. AT-OeStA/AVA Unterricht UM allg. Akten 1220, PA Stumpf, Z. 5599, 15 April 1879.
75. On Franz Hillebrand (appointed in 1896 to Innsbruck), see AT-OeStA/AVA Unterricht UM allg. Akten 1072, PA Hillebrand, Z. 15333, 22 July 1896; on Anton Marty (who received a full professorship in Chernivtsi in 1879 and was appointed to Prague in 1880), see AT-OeStA/AVA Unterricht UM allg. Akten 1219, PA Marty, Z. 1539, 20 March 1880; and on Kazimierz Twardowski (appointed in 1895 to L'viv), see AGAD, MWiO, Sygn. 121u, PA Twardowski, Z. 686, 16 June 1895. Also, Tomáš Garrigue Masaryk was appointed in 1882 at the Czech University in Prague.
76. On Alexius Meinong in Graz, see AT-OeStA/AVA Unterricht UM allg. Akten 938, PA Meinong, Z. 16982, 11 October 1882; on Christian Ehrenfels, appointed to the German University in Prague, see NA, MKV/R, inv.č 9, fasz. 112, PA Ehrenfels, Z. 15334, 22 June 1896.
77. AT-UAW, Ph S 34.15, Ernst Mach, 1 July 1901.
78. AT-OeStA/AVA Unterricht UM allg. Akten 1217, PA Jodl, Z. 5681, 9 April 1885.
79. AT-OeStA/AVA Unterricht UM allg. Akten 671, PA Jodl, Z. 7305, 8 April 1896.
80. Edith Lanser, "Friedrich Jodl: Von Feuerbach zur Gesellschaft für ethische Kultur," *Newsletter Moderne: Zeitschrift des Spezialforschungsbereichs Moderne—Wien und Zentraleuropa um 1900* 6, no. 2 (2003): 16–20.
81. AT-OeStA/AVA Unterricht UM allg. Akten 669, PA Hillebrand, Z. 12225, 13 June 1894; and AT-OeStA/AVA Unterricht UM allg. Akten 674, PA Mach, Z. 7895, 15 April 1895.
82. AT-OeStA/AVA Unterricht UM allg. Akten 1072, PA Hillebrand, Z. 15333, 22 July 1896.
83. AT-OeStA/AVA Unterricht UM allg. Akten 675, PA Müllner, Z. 11935, 18 May 1896.
84. AUJ, W II 128, Z. 692, 14 June 1891; and Wanda Bobrowska-Nowak, *Początki polskiej psychologii* (Wrocław: Zakład narodowy im. Ossolińskich/PAN, 1973). See also the rather uncritical biography by Mirosław Mylik, *Stefan Pawlicki jeden z prekursorów nauki polskiej* (Warsaw: Wydawnictwo Uniwersytetu Kardynała Stefana Wyszyńskiego, 2005).
85. Quotation from the documents on the successor of Müllner, AT-OeStA/AVA Unterricht UM allg. Akten 672, PA Kraus, Z. 5651, 7 February 1913.
86. Hans Liebeschütz, *Das Judentum im deutschen Geschichtsbild von Hegel bis Max Weber* (Tübingen: J. C. B. Mohr [Paul Siebeck], 1967), 70.
87. See Anna L. Staudacher, *". . . meldet den Austritt aus dem mosaischen Glauben": 18000 Austritte aus dem Judentum in Wien, 1868–1914. Namen—Quellen—Daten* (Frankfurt am Main: Peter Lang, 2009), with frequent names of scholars (marked as such, although not thoroughly, in the footnotes).
88. Axel-Johannes Korb, *Kelsens Kritiker: Ein Beitrag zur Geschichte der Rechts- und Staatstheorie (1911–1934)* (Tübingen: Mohr Siebeck, 2010), 250–51; and Anna L. Staudacher, "Zwischen Emanzipation und Assimilation—jüdische Juristen in Wien des Fin-de-Siècle," in *Hans Kelsen Leben—Werk—Wirksamkeit*, ed. Robert Walter, Werner Ogris, and Thomas Olechowski (Vienna: Manz, 2009), 41–53.

89. AT-OeStA/AVA Unterricht UM allg. Akten 668, PA Hatschek; and AT-OeStA/AVA Unterricht UM allg. Akten 681, PA Harry Torczyner.
90. Urszula Perkowska, *Kształtowanie się zespołu naukowego w Uniwersytecie Jagiellońskim (1860–1920)* (Wrocław: Zakład Narodowy im. Ossolińskich, 1975).
91. Henryk Barycz, "Szymon Askenazy: Wśród przeciwieństw i niepowodzeń życiowych i naukowych," in *Na przełomie dwóch stuleci: Z dziejów polskiej humanistyki w dobie Młodej Polski*, ed. Henryk Barycz (Wrocław: Ossolineum, 1977), 238–308.
92. AT-OeStA/AVA Unterricht UM allg. Akten 625, PA Ludwig Hofbauer. Hofbauer was rejected twice, in 1906 and 1913.
93. See, e.g., Surman and Mozetič, *Dwa życia Ludwika Gumplowicza*, 28–33.
94. These were Johann Heinrich Löwe in Vienna and Karl Barach-Rappaport in L'viv (later Innsbruck). Anna L. Staudacher, *Jüdische Konvertiten in Wien 1782–1868*, pt. 1 (Frankfurt am Main: Peter Lang, 2002), 230–31.
95. Wilke, *"Den Talmud und den Kant,"* 595; and Salo W. Baron, "The Revolution of 1848 and Jewish Scholarship: Part II: Austria," *Proceedings of the American Academy for Jewish Research* 20 (1951): 1–100.
96. Guido Kisch, *Die Prager Universität und die Juden, 1348–1848: Mit Beiträgen zur Geschichte des Medizinstudiums* (Amsterdam: John Benjamins, 1969), 63–67; and AT-OeStA/AVA Unterricht UM allg. Akten 667, PA Goldental, Z. 6398/209, 8 May 1860; Z. 11026, 16 December 1868.
97. The Concordat was, however, problematic for Privatdozenten, who were often also teachers. On the Protestant chemist Vojtěch Šafařík, see Ladislav Niklíček, Irena Manová, and Bohumil Hájek, "Profesor Vojtěch Šafařík a počátky výuky chemie na české univerzité v Praze," *AUC-HUCP* 22, no. 1 (1982): 74–75.
98. Filip Friedmann, *Die galizischen Juden im Kampfe um ihre Gleichberechtigung (1848–1868)* (Frankfurt am Main / Łódź: J. Kaufmann / F. Friedmann, 1929), 34–39, 79–84, 134–41.
99. Theodor Gomperz, *Essays und Erinnerungen* (Stuttgart: Deutsche Verlags-Anstalt, 1905), 24.
100. This was Hermann Zeissl in Vienna, a Privatdozent for "primary and consecutive syphilis"; see Richard Landau, *Geschichte der jüdischen Ärzte* (Berlin: Karger, 1895), 131.
101. See, for example, Jean-Michel Helvig, "Antijudaisme ou antisémitisme le procés Sebastian Brunner—Ignaz Kuranda (10 mai 1860)" (PhD diss., University of Paris IV, Sorbonne, 1996); and Arthur Eisenbach, *Emancypacja Żydów na ziemiach polskich 1785–1870 na tle europejskim* (Warsaw: Państowy Instytut Wydawniczy, 1988), 436–43.
102. Rabbinical educational institutions included the academic Franz Joseph Country Rabbinic School (Franz-Josef Landesrabbinerschule) in Budapest (1877) and the Israelite-Theological Seminary (Israelitisch-theologische Lehranstalt) in Vienna (1893).
103. The appointees included August Haffner (associate professor in Innsbruck, 1906), Nikolaus Rhodokanakis (associate professor in Graz, 1907; full professor, 1917),

and Moses Schorr (associate professor in L'viv, 1910). At the Czech University in Prague, Rudolf Růžička habilitated in 1909.

104. Friedrich Wilhelm Bautz, "Bickell, Gustav, Orientalist," in *Biographisch-Bibliographisches Kirchenlexikon*, ed. Friedrich Wilhelm Bautz (Nordhausen: Traugott Bautz, 1990), 579–80. See also Wolfdieter Bihl, *Orientalistik an der Universität Wien: Forschungen zwischen Maghreb und Ost- und Südasien. Die Professoren und Dozenten* (Vienna: Böhlau, 2009), 60; and Bihl's short biographies of the professor of Semitic languages in Vienna David Heinrich Müller (42–46), the Graz associate professor in this discipline Nikolaus Rhodokanakis (73–74), and the Viennese Privatdozent Harry Torczyner (99).

105. Karlheinz Rossbacher, *Literatur und Bürgertum: Fünf Wiener jüdische Familien von der liberalen Ära zum Fin de Siècle* (Vienna: Böhlau, 2003), esp. 226–321.

106. Tim Buchen, *Antisemitismus in Galizien: Agitation, Gewalt und Politik gegen Juden in der Habsburger Monarchie um 1900* (Berlin: Metropol, 2012); and Marcin Soboń, *Polacy wobec Żydów w Galicji doby autonomicznej w latach 1868–1914* (Cracow: Verso, 2011). See also an interesting view from the postcolonial perspective in Michael John and Albert Lichtblau, "Jewries in Galicia and Bukovina, in Lemberg and Czernowitz: Two Divergent Examples of Jewish Communities in the Far East of the Austro-Hungarian Monarchy," in *Jewries at the Frontier: Accommodation, Identity, Conflict*, ed. Sander L. Gilman and Milton Shain (Urbana: University of Illinois Press, 1999), 29–66.

107. See especially Anton G. Rabinbach, "The Migration of Galician Jews to Vienna, 1857–1880," *Austrian History Yearbook* 11 (1975): 43–54; Marsha L. Rozenblit, "A Note on Galician Jewish Migration to Vienna," *Austrian History Yearbook* 19 (1983): 143–52; Piotr Wróbel, "The Jews of Galicia under Austrian-Polish Rule, 1869–1918," *Austrian History Yearbook* 25 (1994): 97–138; Martin Broszat, "Von der Kulturnation zur Volksgruppe: Die nationale Stellung der Juden in der Bukowina im 19. und 20. Jahrhundert," *Historische Zeitschrift* 200, no. 3 (1965): 572–605; Irmgard Plattner, "La città di Innsbruck alla svolta del secolo," in Pallaver and Gehler, *Università e nazionalismi*, 47–80; and T. Haas, "Die sprachlichen Verhältnisse"; see also the critique of Habsburg statistics in, e.g., T. Haas, "Die sprachlichen Verhältnisse"; and Andreas B. Kilcher, "Sprachendiskurse im jüdischen Prag um 1900," in *Franz Kafka im sprachnationalen Kontext seiner Zeit: Sprache und nationale Identität in öffentlichen Institutionen der böhmischen Länder*, ed. Marek Nekula, Ingrid Fleischmann, and Albrecht Greule (Vienna: Böhlau, 2007), 61–62.

108. Robert S. Wistrich, *Die Juden Wiens im Zeitalter Kaiser Franz Josephs* (Vienna: Böhlau, 1999), 55–56; Jiři Pešek, "Jüdische Studenten an den Prager Universitäten 1882–1939," in Nekula, Fleischmann, and Greule, *Franz Kafka*, 213–27; Mariusz Kulczykowski, *Żydzi—studenci Uniwersytetu Jagiellońskiego w dobie autonomicznej Galicji (1867–1918)* (Cracow: Księgarnia Akademicka, 1995); Jakob Thon, "Anteil der Juden am Hochschulstudium in Oesterreich seit dem Jahre 1851," *Zeitschrift für Demographie und Statistik der Juden*, no. 3 (1907): 33–38; and further issues with statistics for the following years.

109. Gary B. Cohen, *Education and Middle-Class Society in Imperial Austria, 1848–1918* (West Lafayette, IN: Purdue University Press, 1996), 232–33.
110. Steven Beller, *Vienna and the Jews, 1867–1938: A Cultural History* (Cambridge: Cambridge University Press, 1991), 36.
111. For Chernivtsi see Erich Prokopowitsch, *Gründung, Entwicklung und Ende der Franz-Josephs-Universität in Czernowitz* (Clausthal-Zellerfeld: Pieper, 1955), 38 (numbers at the law and philosophical faculties).
112. Dr. K.L., "Żydzi na uniwersytecie," *Krytyka* 16, no. 41 (1914): 389–91.
113. See *Krytyka* 16, no. 42 (1914): 116–19, 179–81, 239–42.
114. The numbers rose in the first few years after 1918 but then, owing to growing anti-Semitism and discussions on the numerus clausus rule, decreased again. See especially Kulczykowski, *Żydzi—studenci*, 329–34.
115. The best known are Friedrich Pineles, Sigmund Fraenkel, Jacob Erdheim, Josef Herzig, Max Margules, Leon Kellner, Sigmund Herzberg-Fränkel, and Cäsar Pomeranz; the last three were also later professors in Chernivtsi, but the overall number of Galician-born scholars at German-language universities was low.
116. Seebacher, *Das Fremde*.
117. See, for example, on latent anti-Semitism in academia Sigurd Paul Scheichl, "The Context and Nuances of Anti-Jewish Language: Were All the 'Antisemites' Antisemites?," in *Jews, Antisemitism, and Culture in Vienna*, ed. Ivar Oxaal, Michael Pollak, and Gerhard Both (London: Routledge & Kegan Paul, 1987), 89–110.
118. Theodor Billroth, *Über das Lehren und Lernen der medicinischen Wissenschaften an den Universitäten der deutschen Nation* (Vienna: Gerold, 1876), translated into English as *The Medical Sciences in the German Universities: A Study in the History of Civilization*, with an introduction by William H. Welch (New York: Macmillan, 1924).
119. Billroth, *Medical Sciences*, e.g., 106–7. More on this issue in Felicitas Seebacher, "'Der operierte Chirurg': Theodor Billroths Deutschnationalismus und akademischer Antisemitismus," *Zeitschrift für Geschichtswissenschaft* 56, no. 4 (2006): 317–38; and Tatjana Buklijas, "Surgery and National Identity in Late Nineteenth-Century Vienna," *Studies in History and Philosophy of Biological and Biomedical Sciences* 38, no. 4 (2007): 756–74.
120. See the discussion in Lisa Kienzl, *Nation, Identität und Antisemitismus: Der deutschsprachige Raum der Donaumonarchie 1866 bis 1914* (Göttingen: V&R Unipress, 2014).
121. The most important right-wing Catholic parties in Cisleithania were the Christian Social Party in Austria, the independent Czech Christian Social Party (established in 1894), and National Democracy in Galicia.
122. The most polarizing affairs were the Tiszaeszlár affair in Hungary (1882–83), the Hilsner affair in Bohemia (1899–1900), and the French Dreyfus affair (1894).
123. See, for example, Porter, *When Nationalism*; Michal Frankl, *"Emancipace od židů" Český antisemitismus na konci 19. Století* (Prague: Paseka, 2007); and John W. Boyer, *Culture and Political Crisis in Vienna: Christian Socialism in Power, 1897–1918* (Chicago: University of Chicago Press, 1995).

124. Jan Pawelski, "Po anarchii uniwersyteckiej," *Przegląd Powszechny* 54, no. 2 (1911): esp. 13–15; and Rinnerthaler, "Der Fall Wahrmund," 199.
125. *Die Universität: Eine Kampfzone*, Jewish Museum Vienna, 3 November 2015–28 March 2016.
126. NA, MKV/R, inv.č 9, fasz. 114, PA Kantor.
127. Karl Sablik, *Julius Tandler, Mediziner und Sozialreformer: Eine Biographie* (Vienna: A. Schendl, 1983), 31–32; and Birgit Nemec and Klaus Taschwer, "Terror gegen Tandler: Kontext und Chronik der antisemitischen Attacken am I. Anatomischen Institut der Universität Wien 1910–1933," in *Der lange Schatten des Antisemitismus: Kritische Auseinandersetzungen mit der Geschichte der Universität Wien im 19. und 20. Jahrhundert*, ed. Oliver Rathkolb (Göttingen: V&R Unipress, 2013), 147–72.
128. Michael Gehler, "Studentischer Antisemitismus an der Universität Innsbruck," in *Die Geschichte der Juden in Tirol von den Anfängen bis in die neueste Zeit*, ed. Günther Pallaver (Bozen: Sturzflüge, 1986), 75.
129. Bösche, *Die Innsbrucker Universität*, 134–35.
130. See, e.g., Josef Čermák, "Das Kulturleben der Prager deutschen Studenten seit der Mitte des 19. Jahrhunderts: Die Lese- und Redehalle der deutschen Studenten in Prag," in *Juden zwischen Deutschen und Tschechen: Sprachliche und kulturelle Identitäten in Böhmen 1800–1945*, ed. Marek Nekula and Walter Koschmal (Munich: Oldenbourg, 2006), 33–64; and Michael John and Albert Lichtblau, "Mythos 'deutsche Kultur': Jüdische Gemeinden in Galizien und der Bukowina. Zur unterschiedlichen Ausformung kultureller Identität," in *Studien zur Geschichte der Juden in Österreich*, ed. Martina Keil and Eleonore Lappin (Bodenheim: Philo, 1997), 81–121.
131. AT-OeStA/AVA Unterricht UM allg. Akten 666, PA Fournier, Z. 8811, 6 August 1903, from the appointment records for the chair after Max Büdinger: the appointed records noted that Přibram "seems . . . owing to his [Jewish] descent less adequate for the chair."
132. AT-OeStA/AVA Unterricht UM allg. Akten 1221, PA Weber, Z. 12466, 26 June 1900.
133. AT-OeStA/AVA Unterricht UM allg. Akten 1077, PA Wopfner, Z. 36418, 10 December 1908. See also Gerhard Oberkofler, *Samuel Steinherz (1857–1942): Biographische Skizze über einen altösterreichischen Juden in Prag* (Innsbruck: StudienVerlag, 2008).
134. Wróbel, "Szymon Askenazy"; Askenazy also met with anti-Semitism at the Academy of Arts and Sciences. After he was announced a winner of the prestigious Probus Barczewski Prize, the commission had to change its verdict, as, according to the statute, the prize was reserved for Catholics only; see Marcin Nurowski, *Szymon Askenazy: Wielki Polak wyznania mojżeszowego* (Warsaw: Marcin Nurowski, 2005), 40.
135. AT-OeStA/AVA Unterricht UM allg. Akten 1218, PA Lieben, Z. 12642, 14 October 1871; and AT-OeStA/AVA Unterricht UM allg. Akten 673, PA Lieben, Z. 1251, 6 February 1875.

136. Albin Lesky, "Gomperz, Theodor," *Neue Deutsche Biographie*, ed. Historische Kommission bei der Bayrischen Akademie der Wissenschaften (Berlin: Duncker & Humblot, 1964), 6:641–42.
137. For the metaphor, see Arthur Mahler, *Rede gehalten in der Universitäts-Debatte zum Dringlichkeitsantrag Prof. Masaryk's im oesterreichischen Abgeordnetenhaus, am 3. Dezember 1907* (Vienna: Unitas, 1908).
138. AT-OeStA/AVA Unterricht UM allg. Akten 1070, PA Brunner, Z. 28381, 12 September 1902.
139. AT-OeStA/AVA Unterricht UM allg. Akten 920, PA Kreiblich, Z. 1518, 13 February 1903.
140. AT-OeStA/AVA Unterricht UM allg. Akten 920, PA Löwi, Z. 4944, 16 October 1909.
141. Oliver Rathkolb, "Gewalt und Antisemitismus an der Universität Wien," in Rathkolb, *Der lange Schatten*, 69–92.
142. Josef Redlich, speech in the Lower House of Parliament, 4 December 1907, reprinted in *Stenographische Protokolle 1907*, 2941. At the Viennese medical faculty, for example, from the 1880s on, more than eighty scholars were titular associate professors, and of these, thirty were later promoted to associate professor, but only six ultimately became full professors (up to 1918). In contrast, in Prague most titular associate professors were later promoted; in Graz seven out of twenty-three were promoted; and in Galicia and at the Czech University in Prague, most scholars having the title of professor were later appointed professors. Confessional disparities in promotions are, however, not known.
143. Eugen Ehrlich, "Der Antisemitismus im Professorenkollegium der österreichischen Universität," *Dr. Bloch's Österreichische Wochenschrift*, 6 December 1907, 811–12.
144. On the mechanism of this, see Shulamit Volkov, "Soziale Ursachen des Erfolgs in der Wissenschaft—Juden im Kaiserreich," *Historische Zeitschrift* 245, no. 2 (1987): 315–42. For the process in Vienna, see Karl Schorske, *Fin-de-Siècle Vienna: Politics and Culture* (New York: Knopf, 1980).

Chapter 7

1. Karl Popper, "Prague Lecture," 25 May 1994, https://www.lf3.cuni.cz/3LFEN-255.html.
2. On the Austria-rootedness of Popper's theories, see Malachi Haim Hacohen, *Karl Popper, the Formative Years, 1902–1945: Politics and Philosophy in Interwar Vienna* (Cambridge: Cambridge University Press, 2002); like most historians of philosophy, Hacohen writes of the "Austrian tradition," by which he actually means the Habsburg Empire.
3. See the overview in Adam Kożuchowski, *The Afterlife of Austria-Hungary: The*

Image of the Habsburg Monarchy in Interwar Europe (Pittsburgh, PA: University of Pittsburgh Press, 2013).

4. Two of the first victims were probably Alfred Grund, a professor of geography at the German University in Prague, killed in Smederevo in Serbia, and Josef Stalzer, a Privatdozent for classical philology in Graz, killed in Galicia. AT-OeStA/AVA Unterricht UM allg. Akten 941, PA Wassmuth, Z. 25951, 25 August 1915; and H. Reitterer, "Stalzer, Josef (1880–1914)," in *Das Österreichische Biographische Lexikon*, ed. Österreichische Akademie der Wissenschaften (Vienna: Verlag der Österreichischen Akademie der Wissenschaften, 2007), 13:84–85.
5. These data are from my own calculations, based on information gathered from the databases. Because of the high percentage of missing information, these data remain statistically insignificant. Reports by the universities themselves show similar numbers; for example, Graz lost only two Privatdozenten. Hans Rabl, "Die Verluste der Grazer Universität im Weltkrieg," in *Beiträge zur Geschichte der Karl-Franzens-Universität-Graz* (Graz: Leuschner & Lubensky, 1927), 52.
6. See the biography in Adolf Menzel, "Bericht über das Studienjahr 1915/16," in *Die feierliche Inauguration des Rektors der Wiener Universität für das Jahr 1916/17 am 6. November 1916* (Vienna: Selbstverlag, 1916).
7. Urszula Perkowska, *Uniwersytet Jagielloński w latach I wojny światowej* (Cracow: Universitas, 1990), 97, 101; some staff were part of both the imperial army and the Polish Legions, so these categories are not mutually exclusive.
8. Johannes Uray, "Czernowitz—Salzburg: Die Idee zum Transfer einer Universität (1916–1920)," in *Universitäten in Zeiten des Umbruchs: Fallstudien über das mittlere und östliche Europa im 20. Jahrhundert*, ed. Elmar Schübl and Harald Heppner (Berlin: Lit, 2011), 69–82.
9. Emil Reisch, "Aufgaben unserer Universitäten nach dem Kriege: Inaugurationsrede, gehalten am 6. November 1916," in *Die feierliche Inauguration des Rektors der Wiener Universität für das Studienjahr 1916/17* (Vienna: Holzhausen, 1916), 87–89.
10. See, e.g., Rogers Brubaker, "Nationalizing States in the Old 'New Europe' and the New," *Ethnic and Racial Studies* 19, no. 2 (1996): 411–37.
11. Pieter M. Judson gives special attention to this phenomenon. Judson, *The Habsburg Empire*.
12. Friedrich Johannes Becke, "Bericht über das Studienjahr 1918/1919," in *Die feierliche Inauguration des Rektors der Wiener Universität für das Jahr 1919/20 am 5. November 1919* (Vienna: Selbstverlag, 1919), 7.
13. For example, Dozent Dr. Slawitschek, "Eine allgemeine Deutsch Hochschule," *Deutsche Hochschul-Zeitung*, 30 March 1918, 1. The journal even adopted old-German month names (*Lenzmond* instead of *März* for March, *Gibhard* instead of *Oktober* for October, etc.) to denote its pan-Germanism.
14. See the opening speeches of the rectors in the immediate postwar period; and Brigitte Lichtenberger-Fenz, *". . . deutscher Abstammung und Muttersprache":*

Österreichische Hochschulpolitik in der Ersten Republik (Vienna: Geyer Edition, 1990), 10–22; see also Malachi Haim Hacohen, "Kosmopoliten in einer ethnonationalen Zeit? Juden und Österreicher in der Ersten Republik," in *Das Werden der Ersten Republik: . . . der Rest ist Österreich*, ed. Helmut Konrad and Wolfgang Maderthaner (Vienna: Gerold, 2008), 1:281–316.

15. Gerhard Oberkofler, "Der Kampf der Universität um die Einheit des Landes Tirol (1918–1920)," *Tiroler Heimatblätter* 55 (1980): 78–89.
16. Oberkofler, "Der Kampf der Universität," 81–82.
17. Michael Wedekind, "The Sword of Science: German Scholars and National Socialist Annexation Policy in Slovenia and Northern Italy," in *German Scholars and Ethnic Cleansing (1920–1945)*, ed. Ingo Haar and Michael Fahlbusch (New York: Berghahn, 2005), 110–38; and Wolfgang Huber, "Zur Geschichte der Wissenschaften," in *Österreich 1918–1938*, ed. Erika Weinziert and Kurt Skalnik (Vienna: Styria 1983), 2:571, 575.
18. Pešek, "Jüdische Studenten," 217.
19. Tara Zahra, *Kidnapped Souls: National Indifference and the Battle for Children in the Bohemian Lands, 1900–1948* (Ithaca, NY: Cornell University Press, 2008).
20. In general, see Antonín Kostlán and Alice Velková, eds., *Wissenschaft im Exil: Tschechoslowakei als Kreuzweg 1918–1989* (Prague: Výzkumné centrum pro dějiny vědy, 2004).
21. Helmut Engelbrecht, *Geschichte des österreichischen Bildungswesens: Erziehung und Unterricht auf dem Boden Österreichs*, vol. 5, *Von 1918 bis zur Gegenwart* (Vienna: Österreichischer Bundesverlag, 1988), 228; Preglau-Hämmerle, *Die politische und soziale Funktion*, 177–81; and Irma Völlmecke, "Österreichische Hochschulstatistik 1829 bis 1979," in *Geschichte und Ergebnisse der zentralen amtlichen Statistik in Österreich 1829–1979*, ed. Österreichisches Statistisches Zentralamt (Vienna: Österreichische Staatsdruckerei, 1979), 493.
22. Uray, "Czernowitz—Salzburg."
23. Karl Hermann-Otavský, "Zpráva o studijním roku 1918–19," in *Inaugurace Rektora dne 3. Prosince 1919* (Prague: Univerzita Karlova, 1919), 20.
24. H[ans] Rabl, "Bericht des Prorektors," in *Die feierliche Inauguration des Rektors der Grazer Universität für das Studienjahr 1925/6* (Graz: Leuschner, 1927), 7–8, 30–31.
25. Herbert Posch, "Studierende und die Universität Wien in der Dauerkrise 1918 bis 1938," in *"Anschluß" und Ausschluss 1938: Vertriebene und verbliebene Studierende der Universität Wien*, ed. Herbert Posch, Doris Ingrisch, and Gert Dressel (Vienna: Lit, 2008), 87–89.
26. The German University in Prague had around 20 percent foreign students, mostly German-speaking Jews. This is from my own calculations on the basis of the official statistics for 1929/30, in Štátni úrad štatistický, *Štatistická príručka republiky československé IV* (Prague: Bursík & Kohout, 1932), 351.
27. Petr Svobodný, "Lékařská fakulta 1918–1945," in *Dějiny Univerzity Karlovy*, vol. 4, *1918–1945*, ed. Jan Havránek and Zdeněk Pousta (Prague: Karolinum, 1998), 102.
28. Natalia Aleksiun, "The Cadaver Affair in the Second Polish Republic: A Case

Study of Political Antisemitism?," in *Alma Mater Antisemitica: Akademisches Milieu, Juden und Antisemitismus an den Universitäten Europas zwischen 1918 und 1939*, ed. Regina Fritz, Grzegorz Rossoliński-Liebe, and Jana Starek (Vienna: New Academic Press, 2016), 203–20.

29. Quoted in Gerhard Oberkofler, "Deutschnationalismus und Antisemitismus in der Innsbrucker Studentenschaft um 1920," *Tiroler Heimatblätter* 56 (1981): 70 (quoting from UAI, R. Präsidiale 20/1924).
30. Pešek, "Jüdische Studenten," 220–21. See also, for a broader analysis, Kateřina Čapková, *Češi, Němci, Židé? Národní identita Židů v Čechách, 1918–1938* (Prague: Paseka, 2005), 58–59.
31. Klaus Taschwer, "Geheimsache Bärenhöhle: Wie ein antisemitisches Professorenkartell der Universität Wien nach 1918 jüdische und linke Forscherinnen und Forscher vertrieb," in Fritz, Rossoliński-Liebe, and Starek, *Alma Mater Antisemitica*, 226–27.
32. Arnold Suppan and Marija Wakounig, "Hans Uebersberger (1877–1962)," in *Osteuropäische Geschichte in Wien: 100 Jahre Forschung und Lehre an der Universität*, ed. Arnold Suppan, Marija Wakounig, and Georg Kastner (Innsbruck: StudienVerlag, 2007), 110–15.
33. Taschwer, "Geheimsache Bärenhöhle," 231.
34. Peter Goller, "Ein starkes Stück: Die versuchte Habilitation eines kommunistischen Juden . . . Universitäten im Lichte politischer u. rechtlicher Willkür am Beispiel des Habilitationsverfahrens von Karl Horovitz <1892–1958> an der Wiener Universität 1923–1925," *Jahrbuch 1998: Dokumentationsarchiv des österreichischen Widerstandes* (1998): 115–18.
35. Heinrich Srbik (a professor of history and the minister of religion and education for a short period between October 1929 and September 1930) to Siegfried August Kaehler, 20 August 1930, reprinted in Srbik, *Heinrich Ritter von Srbik: Die wissenschaftliche Korrespondenz eines Historikers 1912–1945*, ed. Jürgen Kammerer (Boppard am Rhein: Harald Boldt, 1988), 359–60.
36. The Lex Jędrzejewicz, or the Jędrzejewicz reform, was a reform of education in Poland, conducted in 1932 and 1933. It strengthened government supervision of universities, including political influence over the choice of rector and the ability to remove professors for political reasons.
37. Jarosław Jastrzębski, "Reforma Jędrzejewicza w państwowym szkolnictwie akademickim II Rzeczypospolitej: Wzmocnienie prerogatyw władz państwowych," *Prace historyczne* 138 (2011): 159–75; Pavel Mates, Petr Prucha, and Jan Svatoň, *Vývoj organizace a řízení československých vysokých škol v letech 1918–1983* (Prague: Ústav školských informací při Ministerstvu Školství Česko-Slovenskej republiky, 1984), 25–30; and Brigitte Lichtenberger-Fenz, "Österreichs Universitäten und Hochschulen—Opfer oder Wegbereiter der nationalsozialistischen Gewaltherrschaft? Am Beispiel der Universität Wien," in *Willfährige Wissenschaft: Die Universität Wien 1938–1945*, ed. Gernot Heiß, Siegfried Mattl, Sebastian Meissl, Edith Saurer, and Karl Stuhlpfarrer (Vienna: Verlag für Gesellschaftskritik, 1989), 3–15.

38. For an outline of the reforms, see Walter Brunner and Helmut Wohnout, "Hochschulrecht," in *Parlamentarismus und öffentliches Recht in Österreich: Entwicklung und Gegenwartsprobleme*, ed. Herbert Schambeck (Berlin: Duncker & Humblot, 1993), 1105–48.
39. Perkowska, "La genèse."
40. Ulrich Steltner, "Grenzgänger zwischen der deutschen und der polnischen Literatur: Tadeusz Rittner und Stanisław Przybyszewski," in *Auf der Suche nach einer größeren Heimat . . . : Sprachwechsel/Kulturwechsel in der slawischen Welt*, ed. Ulrich Steltner (Jena: Collegium Europaeum Jenense, 1999), 105–15.
41. Johannes Uray speaks of four people: the philosopher Carl Siegel, the Romance studies scholar Eugen Herzog, the pharmacologist Fritz Netolizky, and the forensic physician Friedrich Mayer. Uray, "Czernowitz—Salzburg," 77. The biographies of the geologist Karl Alfons Penecke and the retired professor of political economy Friedrich Kleinwächter confirm that they also remained in Romania. Adolf Meixner, "Professor D. Karl Alfons Penecke zur 100. Wiederkehr seines Geburtstages," *Carinthia* 68, no. 2 (1958): 63–90; and "Kleinwächter, Friedrich von," in *Das Österreichische Biographische Lexikon*, ed. Österreichische Akademie der Wissenschaften (Vienna: Verlag der Österreichischen Akademie der Wissenschaften, 1965), 3:392–93.
42. Protests by anti-Semitic students and nationalists had hindered his earlier lectures, though. Manfred Rehbinder, "Die rechts- und staatswissenschaftliche Fakultät der Franz-Josephs-Universität Czernowitz: Ihr Beitrag zur Erforschung des Rechts in einer multikulturellen Gesellschaft," in *Festschrift Hans Stoll zum 75. Geburtstag*, ed. Gerhard Hohloch, Hans Stoll, Rainer Frank, and Peter Schlechtriem (Tübingen: Mohr Siebeck, 2001), 331–33.
43. Livezeanu, *Cultural Politics*, 225–26.
44. Tatjana Dekleva, "Ustanovitev univerze v Ljubljani," in *Ustanovitev Univerze v Ljubljani v letu 1919*, ed. Jože Ciperle and Tatjana Dekleva (Ljubljana: Univerza, 2009), 36–37.
45. Vladimir Kanjuh, "Đorđe Joannović—First Serbian Oncologist-Scientist (on the Occasion of the 75 Anniversary of His Tragic Death)," *Archive of Oncology* 16, no. 12 (2008): 18–21.
46. Those who left the Vienna philosophical faculty were Lujo Adamović/Лујо Адамовић (who moved to Dubrovnik) and Carlo Battisti (who moved first to Trento, then to Florence). Innsbruck lost Bohumil Spáčil and Anton Prešeren (both in theology, first to Rome, then to Prague). Graz lost Matija Murko and Jan/Johann Peisker (to the Czech University in Prague), the medical chemist Ján Buchtala (to Bratislava), Rajko Nahtigal (to Ljubljana), and Vittorio Benussi (to Padua). See "Anton Prešeren," Revija Ognjišče, n.d., http://revija.ognjisce.si/revija-ognjisce/67-pricevanje/1677-anton-preseren (accessed 24 March 2015); Jan Pavlík, *Vzpomínky na zemřelé jezuity, narozené v Čechách, na Moravě a v Moravském Slezsku od roku 1814* (Olomouc: Refugium Velehrad-Roma s.r.o., 2011), fragments of which are available online at http://www.jesuit.cz/vzpominka.php?id=4; Liliana Albertazzi, "Vittorio Benussi," in *The School*

of Alexius Meinongi, ed. Liliana Albertazzi, Dale Jacquette, and Roberto Poli (Aldershot: Ashgate, 2001), 1–35; and Jan Hálek, "Jan Peisker (1851–1933)," *Akademický bulletin*, no. 5 (2008): 5.

47. Vitezslav Chlumský was conscripted and, according to his own report, was wounded and contracted sepsis in 1914, residing after that on his estate in Silesia. When the Cracow medical faculty rejected his appeal to be freed from lecturing until the end of the war, he wrote an emotional letter criticizing the faculty and commented, in a clearly ironic tone, that "if the Medical Faculty wants to get rid of me so much, I would be ready to resign the honour of being its member." The faculty took this seriously, however, and removed him from his position. See AUJ, S II 619, Akt Personalny Chlumsky, esp. Chlumský's letter of 20 April 1917, the faculty's writing to the ministry of 21 April 1917, and the letter to Chlumský informing him of his removal, dated 22 May 1917. See also Zygmunt Albert, "Prof. Dr. Paweł Ludwig Kučera (1872–1928), kierownik zakładu higieny UJK we Lwowie," *Archiwum Historii i Filozofii Medycyny* 55, no. 1 (1992): 62–70; and Franciszek Martinczak, "Profesor ortopedii Vitezslav Chlumsky (karta z działalności we Wrocławiu i Krakowie)," *Archiwum Historii Medycyny* 37, no. 1 (1974): 93–101.
48. See the controversy about Oleksandr Kolessa, denied a pension in 1930 owing to the uncertainty about his division of competences in 1919. DALO, F. 26, Op. 5, Spr. 899, 22 February 1930, IVSV-1581/30.
49. "Zákon č. 135/1920 Sb. z. a n., o poměru pražských universit," in *Sbírka zákonů a nařízení republiky Československé*, no. 30 (Prague: Státni tiskárna, 1920), 319–20.
50. August Naegerle, "Bericht über die Studienjahre 1918/19 u. 1919/20," in *An Stelle der feierlichen Inauguration des Rektors der Deutschen Universität in Prag für die Studienjahre 1919/20 und 1920/21* (Prague: Selbstverlag, 1921), 36.
51. Naegerle, "Studienjahre 1918/19 u. 1919/20," 44–47; and Alena Míšková, "'Heraus aus Prag!'—Pläne für den Aufbau und Umzüge deutscher Hochschulen in Böhmen," in *Die Suche nach dem Zentrum: Wissenschaftliche Institute und Bildungseinrichtungen der Deutschen in Böhmen (1800–1945)*, ed. Kristina Kaiserová and Miroslav Kunštát (Münster: Waxmann, 2014), 119–44.
52. Andreas Kleinert, "Lampa, Anton," in *Neue Deutsche Biographie*, ed. Historische Kommission bei der Bayrischen Akademie der Wissenschaften (Berlin: Duncker & Humblot, 1982), 13:453–54. One of the émigrés, Eugen Steinach, had already during the war worked in a private laboratory in Vienna. I thank Sonja Walch in Vienna for this information.
53. Ota Konrád, *Dějepisectví, germanistika a slavistika na Německé univerzitě v Praze 1918–1945* (Prague: Karolinum, 2011), 53.
54. Konrád, *Dějepisectví, germanistika a slavistika*, 46–47.
55. Otto Grosser, "Bericht über das Studienjahr 1929/30 [sic!]," in *Die feierliche Inauguration des Rektors der Deutschen Universität in Prag für das Studienjahr 1929/30* (Prague: Selbstverlag, 1930), 9.
56. Grosser, "Bericht," 6.

57. Grosser, "Bericht"; August Naegerle, "Bericht über das Studienjahr 1929/30," in *Die feierliche Inauguration des Rektors der Deutschen Universität in Prag für das Studienjahr 1930/31* (Prague: Selbstverlag, 1931), 7–8; and Gerhard Gesemann, "Bericht über das Studienjahr 1933/34," in *Anstelle der feierlichen Inauguration des Rektors der Deutschen Universität in Prag für das Studienjahr 1934/35* (Prague: Selbstverlag, 1935), 7–14.
58. Ota Konrád, "Die Deutsche Universität in Prag in der Ersten Tschechoslowakischen Republik—zwischen Kooperation und Konfrontation," in Schübl and Heppner, *Universitäten*, 29–42.
59. The numbers are from my own calculations on the basis of the official statistics for 1929/30, in Štátni úrad štatistický, *Štatistická príručka*, 350–51.
60. Štátni úrad štatistický, *Štatistická príručka.*
61. Iris Engemann, *Die Slowakisierung Bratislavas: Universität, Theater und Kultusgemeinden 1918–1948* (Wiesbaden: Harrassowitz, 2012), 81.
62. Owen Johnson, *Slovakia, 1918–1938: Education and the Making of a Nation* (Boulder, CO: East European Monographs, 1985).
63. Milan Hodža (professor of modern history) was mostly absent as he held various ministerial offices in Prague, and Krajči (full name unknown) rejected the call. The only actively teaching professor was the literary historian Jozef Škultéty. See the list of professors in E. Paulinyiová, "Vznik a vývin filozofické fakulty v rokov 1921–1945," in *50 rokov Univerzity Komenského*, ed. Július Bartl (Bratislava: Univerzita Komenského, 1969), 285. See also Johnson, *Slovakia*, 240.
64. See, on the professor of international law Vojtech Tuka at Elizabeth University and the moral theologian František Jehlička at the Theological Institute in Budapest, Arnold Suppan, "Katholische Volksparteien in Ostmitteleuropa in der Zwischenkriegszeit am Beispiel der Tschechen und Slovaken," in *Christdemokratie in Europa im 20. Jahrhundert*, ed. Michael Gehler, Wolfram Kaiser, and Helmut Wohnout (Vienna: Böhlau, 2001), 288; on Jehlička's ideology, see Michaela Miroslav, "Plans for Slovak Autonomy in Hungarian Politics 1918–1920," *Historický časopis: Historického ústavu SAV* 58, suppl. (2010): 53–82.
65. According to the memoirs of Albert Pražák, *Politika a revoluce: Paměti* (Prague: Academia, 2004), 37.
66. Albert Pražák, "Zehn Jahre Universität in Bratislava: Die Entwicklung der Hochschule, ihr heutiger Stand und ihre Zukunftsaussichten (repr. from *Prager Presse* 2/7/1929)," in *Ročenka University Komenského za studijní rok 1928–1929*, ed. Albert Pražák (Bratislava: Universita Komenského, 1930), 244–47.
67. *Soznam osôb a ústavov Univerzity Komenského v Bratislave, taktiež štátnych zkúšobných komisií podla stavu na začiatku štud. roku 1924–25* (Bratislava: Univerzita Komenského, 1924).
68. They were Ievhen Perfec'kyj (Євген Перфецький, history of Carpathian Ruthenia) and Valerij Pogorelov (Валерий Погорелов, Russian literature).
69. Martin Vietor, "Pravnická fakulta Univerzity Komenskeho," in Bartl, *50 rokov Univerzity Komenského*, 187–88.
70. Vinzenz Chlumský (orthopedics) had previously worked in Cracow, Ján Buchtala (biochemistry) in Graz. Martinczak, "Profesor ortopedii Vitezslav Chlumsky";

and Darina Heřmanová, "Vedec prof. MUDr. J. Buchtala," in *Oddelenie histórie medicíny a zdravotníctva Ústavu sociálneho lekárstva LF UK v Bratislave 1990–2000*, ed. Eulália Sedláčková, Ľudmila Pavlíková, Darina Heřmanová, and Helena Rapantová (Bratislava: Juga, 2001), 48–51.

71. Pernes, *Kapitoly z dějin*, 47–48.
72. See Josef Petráň, "Filozofická fakulta v letech 1918–1939 (1945)," in Havránek and Pousta, *Dějiny Univerzity Karlovy*, 4:121.
73. Arkadiusz Stempin, "Die Wiedererrichtung einer polnischen Universität: Warschau unter deutscher Besatzung," in *Kollegen—Kommilitonen—Kämpfer: Europäische Universitäten im Ersten Weltkrieg*, ed. Trude Maurer (Stuttgart: Franz Steiner, 2006), 138.
74. Julian Dybiec, *Uniwersytet Jagielloński 1918–1939* (Cracow: Polska Akademia Umiejętności, 2000), 563.
75. Maria Banasiewicz and Antoni Czubiński, eds., *Organizacja i rozwój uczelni od listopada 1918 roku do inauguracji w maju 1919 roku* (Poznań: Uniwersytet im. Adama Mickiewicza, 1973), 86–95.
76. Ministers were, however, seldom in this position for more than six months.
77. Dorota Mycielska, "Drogi życiowe profesorów przed objęciem katedr akademickich w niepodległej Polsce," in *Inteligencja polska pod zaborami: Studia*, ed. Ryszarda Czepulis-Rastenis (Warsaw: Polskie Wydawnictwo Naukowe, 1979), 243–90.
78. Mycielska, "Drogi," 269.
79. The numbers are from my own calculations based on lecturers' biographies in Tadeusz Manteuffel, *Uniwersytet Warszawski w latach 1915/16–1934/35: Kronika* (Warsaw: Uniwersytet Józefa Piłsudskiego, 1936).
80. For Poznań, see Antoni Czubiński, "Utworzenie Uniwersytetu Poznańskiego i jego rozwój do 1922 roku," in *Alma Mater Poznaniensis: W 80 rocznicę utworzenia Uniwersytetu w Poznaniu*, ed. Przemysław Hauser, Tomasz Jasiński, and Jerzy Topolski (Poznań: Wydawnictwo Naukowe Uniwersytetu im. Adama Mickiewicza w Poznaniu, 1999), 131. For Vilnius, the numbers are from my own calculations on the basis of *Program Wykładów i Skład Uniwersytetu w półroczu zimowem roku akademickiego 1919–1920* (Vilnius: Znicz, 1919). For Lublin, see Grażyna Karolewicz, *Nauczyciele akademiccy Katolickiego Uniwersytetu Lubelskiego w okresie międzywojennym*, vol. 1 (Lublin: Katolicki Uniwersytet Lubelski, 1996).
81. Władysław Ćwik, "Polacy w Innsbrucku do 1939 roku," in *Polonia i przyjaciele Polski w Austrii*, ed. Władysław Stanisław Kucharski (Lublin: Multico, 1995), 63–77.
82. Dybiec, *Uniwersytet Jagielloński 1918–1939*, 568.
83. Dybiec, *Uniwersytet Jagielloński 1918–1939*.
84. Dybiec, *Uniwersytet Jagielloński 1918–1939*, 121.
85. Dybiec, *Uniwersytet Jagielloński 1918–1939*, 109–15.
86. Henryk Barycz, *Uniwersytet Jagielloński w życiu narodu polskiego* (Wrocław: Ossolineum, 1964), 115; and Mycielska, "Drogi."
87. Including all faculties, most scholars who were not from L'viv had previously

been appointed in the Russian Empire (Petersburg, Dorpat). See the biographies in *Skład Uniwersytetu Jana Kazimierza we Lwowie w roku akademickim 1927/1928* (L'viv: Uniwersytet Jana Kazimierza, 1927).

88. See Mirosława Papierzyńska-Turek, *Sprawa ukraińska w Drugiej Rzeczypospolitej 1922–1926* (Cracow: Wydawnictwo Literackie, 1979).

89. Andrzej Pilch speaks of 69 percent of students coming from Eastern Galicia. Pilch, "Ukraińcy na wyższych uczelniach Lwowa 1923–1926," *Annales Universitatis Mariae Curie-Skłodowska* 54/55 (1999/2000): 242–46. For a general overview on the university, see *Ukrajinská svobodná univerzita (1921–1996)* (Prague: Národní knihovna České republiky, Slovanská knihovna, 1998).

90. See Jiří Vacek, "Institucionální základna ukrajinské emigrace v Československu v letech 1919–1945," in *Ruská a ukrajinská emigrace v ČSR v letech 1918–1945*, ed. Václav Veber (Prague: Seminář pro dějiny východní Evropy při Ústavu světových dějin Filozofické fakultě Univerzity Karlovy, 1993), 1:36. On the Czechoslovak government policy toward Russian and Ukrainian émigrés, see Catherine Andreyev and Ivan Savicky, *Russia Abroad: Prague and the Russian Diaspora, 1918–1938* (New Haven, CT: Yale University Press, 2004).

91. Mieczysław Iwanicki, *Oświata i szkolnictwo ukraińskie w Polsce w latach 1918–1939* (Siedlce: Wyższa Szkoła Pedagogiczna, 1975), 178.

92. Iwan Dacko, "Hundert Jahre ukrainische Theologiestudenten in Innsbruck: Vergangenheitsanalyse und Zukunftsperspektiven" (lecture, Innsbruck, 13 November 1999), http://www.canisianum.at/zeitbegrenzt/100-jahre-ukrainer-vortrag.html. Among the pre-1918 graduates were Andriy Ishchak and Nykyta Budka (both beatified). I thank Peter Goller, Archives of the University of Innsbruck, for the information on the international contacts of Innsbruck theologians.

93. Iwanicki, *Oświata i szkolnictwo*, 179–82.

94. Ivan L. Rudnytsky, "Polish-Ukrainian Relations: The Burden of History," in *Essays in Modern Ukrainian History*, ed. Peter L. Rudnytsky (Edmonton: Canadian Institute of Ukrainian Studies, University of Alberta, 1987), 71.

95. See the discussion of this issue in Martin Kohlrausch, Katrin Steffen, and Stefan Wiederkehr, "Expert Cultures in Central Eastern Europe: The Internationalization of Knowledge and the Transformation of Nation States since World War I—Introduction," in *Expert Cultures in Central Eastern Europe: The Internationalization of Knowledge and the Transformation of Nation States since World War I*, ed. Martin Kohlrausch, Katrin Steffen, and Stefan Wiederkehr (Osnabrück: fibre, 2010), 9–30.

96. Dybiec, *Uniwersytet Jagielloński 1918–1939*, 535–48.

97. "Stosunki Naukowe Polski z zagranicą w latach 1919–1925," *Nauka Polska* 6 (1927): 413. Of the seven hundred scholars who traveled abroad, 31 percent went to France, while Germany, Italy, and Austria accounted for around 15 percent each, and the United States and Great Britain only 5 percent each.

98. Jiří Hnilica, "'Missionaries of Science and Spirit': The Topic of Visiting Professors Illustrated by the Example of Czechoslovak-French Cultural Relations in the Years 1918–1948," in *Conference Proceedings: Universities in Central Europe—Crossroads of Scholars from All over the World*, ed. Milada Sekyrková

and Petr Cajthaml (Prague: Institute of the History of Charles University and Archive of Charles University Prague, 2012), 207–13.

99. See, for example, the presentation of the Polish Institute for Foreign Cooperation (Polski Instytut Współpracy z Zagranicą) in Wacław Łypacewicz, *Polnisch-Tschechische Beziehungen* (Warsaw: Polnisches Institut fuer Zusammenarbeit mit dem Auslande, 1936).
100. Marjan Kossov [Kossow], *Úkoly kulturní spolupráce československo-polské* (Mladá Boleslav: Národní Zájmy, [1932]).
101. Hnilica, "'Missionaries.'"
102. Michala Benešová, Renata Rusin Dybalska, and Lucie Zakopalová, *90 let pražské polonistiky—dějiny a současnost* (Prague: Karolinum, 2013), 11–24; and Miloslav Trnka, "Profesor Bohumil Karel Vydra ze Lštění na Šumavě," *Rodopisná revue on-line* 9–10 (2010): 20–21, http://rodopisna-revue-online.tode.cz/9-10-10_soubory/20-21_trnka-vydra.pdf.
103. Jan Woleński, "The Reception of the Lvov-Warsaw School," in *The Lvov-Warsaw School and Contemporary Philosophy*, ed. Katarzyna Kijania-Placek and Jan Woleński (Dordrecht: Kluwer, 1998), 13. See also Anita Burdman Feferman and Solomon Feferman, *Alfred Tarski: Life and Logic* (Cambridge: Cambridge University Press, 2004), 76–84.
104. Marek Ďurčanský, "Władysław Mieczysław Kozłowski a česká filosofie," in *Slovanství a věda v 19. a 20. století*, ed. Hana Barvíková (Prague: Archiv Akademie věd České republiky, 2005), 55–76.
105. See Jan Papiór, "Zur Geschichte und Studienpläne der polnischen Germanistik," in *Daß eine Nation die andere verstehen möge: Festschrift für Marian Szyrocki zu seinem 60. Geburtstag*, ed. Norbert Honsza and Hans-Gert Roloff (Amsterdam: Rodopi, 1988), 549–68.
106. Doubek transferred to Berlin in 1933 and was involved in academic research for the Nazi regime, moving finally to Graz in 1945 as a speech therapist; see Waldemar Grzybowski, "Sprachwissenschaftliche Biographien aus Ostdeutschland: Deutsche und polnische Linguisten im sprachwissenschaftlichen und sprachpolitischen Diskurs der Zwischenkriegszeit," in *Grenzüberschreitende Biographien zwischen Ost- und Mitteleuropa: Wirkung—Interaktion—Rezeption*, ed. Tobias Werner (Vienna: Peter Lang, 2009), 307–9.
107. Michał Cieśla, "Spirydion Wukadinović," in *Z dziejów germanistyki historycznoliterackiej w Polsce: Studia i materiały*, ed. Krzysztof Kuczyński (Łódz: Wydawnictwo Uniwersytetu Łódzkiego, 1991), 93–104.
108. See Zbigniew Staliński, "Leopold Adametz (1861–1941)," in *Złota Księga Akademii Rolniczej*, ed. Zbigniew Staliński (Cracow: Wydawnictwo Akademii Rolniczej, 2000), 62–67.
109. See, e.g., Konrád, "Die Deutsche Universität," 34; and Jiří Pešek, Alena Míšková, Petr Svobodný, and Jan Janko, "Německá univerzita v Praze v letech 1918–1939," in Havránek and Pousta, Dějiny Univerzity Karlovy, e.g., 4:200.
110. Iorgu Iordan, "Besprechungen und Anzeigen: V. Rumänisch," *Zeitschrift für romanische Philologie* 54, no. 2 (1934): 363–79.
111. Sextil Puşcariu, "Eugen Herzog [Article nécrologique]," *Dacoromania* 5 (1929)

882–83; and Alexe Procopovici, "[Eugen Herzog†]," *Revista Filologică* 2 (1928/29): 232–45.

112. Christian Fleck, "Autochthone Provinzialisierung: Universität und Wissenschaftspolitik nach dem Ende der nationalsozialistischen Herrschaft in Österreich," *Österreichische Zeitschrift für Geschichtswissenschaften* 7, no. 1 (1997): 67–92.
113. See Reinhard Müller, "Das Leben Othmar Spanns: Ein Vortrag von Hans Riehl 1954," *Zyklos 1: Jahrbuch für Theorie und Geschichte der Soziologie* 1 (2015): esp. 341–52; and Janek Wasserman, *Black Vienna: The Radical Right in the Red City, 1918–1938* (Ithaca, NY: Cornell University Press, 2014), esp. 74–106.
114. Sonderaktion Krakau involved the arrest of 184 academics of the Jagiellonian University by Nazi German occupation forces in November 1939. The academics were later detained at Sachsenhausen and released in the spring of 1940 only following international actions, with seventeen professors dying in the camp or shortly after their release. The massacre of L'viv professors refers to the executions of twenty-five professors from the university and technical academy in L'viv, as well as their families, by Nazi German occupation forces in July 1941. August Jochen, *Sonderaktion Krakau: Die Verhaftung der Krakauer Wissenschaftler am 6. November 1939* (Hamburg: Hamburger Edition, HIS Verlagsgesellschaft, 1997); and Zygmunt Albert, *Kaźń profesorów lwowskich—lipiec 1941* (Wrocław: Wydawnictwo Uniwersytetu Wrocławskiego, 1989).

Conclusion

1. In particular with regard to dealing with cultural diversity, the Habsburg Empire is presented as an important experiment from which the European Union should learn. See, e.g., Reinhard Johler, "Vielfalt," in *Habsburg Neu Denken: Vielfalt und Ambivalenz in Zentraleuropa. 30 Kulturwissenschaftliche Stichworte*, ed. Johannes Feichtinger and Heidemarie Uhl (Vienna: Böhlau, 2016), 229–36.
2. See Jan Woleński, *Filozoficzna szkoła lwowsko-warszawska* (Warsaw: Państwowe Wydawnictwo Naukowe, 1995); and Zdzisław Wojtaszek, Halina Kuzyk, Alojzy Morzyniec, Jerzy Dubowy, and Krystyna Łopata, *Karol Olszewski* (Warsaw: Państwowe Wydawnictwo Naukowe, 1990).
3. Karl Linsbauer, Ludwig Linsbauer, and Leopold Ritter von Portheim, eds., *Wiesner und seine Schule: Ein Beitrag zur Geschichte der Botanik. Festschrift anläszlich der dreiszigjährigen Bestandes des Pflanzenphysiologischen Institutes der Wiener Universität* (Vienna: A. Hölder, 1903); Matthew Rampley, *The Vienna School of Art History: Empire and the Politics of Scholarship* (University Park: Pennsylvania State University Press, 2013); Stanislaus Hafner, "Geschichte der österreichischen Slawistik," in *Beiträge zur Geschichte der Slawistik in nichtslawischen Ländern*, ed. Josef Hamm and Günther Wytrzens (Vienna: Österreichische Akademie der Wissenschaften, 1985), 11–88; and Smith, *Austrian Philosophy*.

4. Perricelli, "'O Trieste o nulla!,'" 179.
5. See, for example, Maria Rentetzi, *Trafficking Materials and Gendered Research Practices: Radium Research in Early 20th Century Vienna* (New York: Columbia University Press, 2008).
6. Volkov, "Soziale Ursachen."
7. For a recent discussion and a more complex description of this phenomenon, see Veronika Hofer and Michael Stöltzner, "What Is the Legacy of Austrian Academic Liberalism?," *NTM Zeitschrift für Geschichte der Wissenschaften, Technik und Medizin* 20, no. 1 (2012): 31–42. This discussion was initiated by Deborah Coen, *Vienna in the Age of Uncertainty* (Chicago: University of Chicago Press, 2007); and Malachi Haim Hacohen, "The Culture of Viennese Science and the Riddle of Austrian Liberalism," *Modern Intellectual History* 6, no. 2 (2009): 369–96.
8. Alexander Busch, a historian of Privatdozenten, quoted in Eric L. Mills, *Biological Oceanography: An Early History, 1870–1960* (Toronto: University of Toronto Press, 2012), 179.
9. Klaus Taschwer, "Universitätskrise und wissenschaftliche Volksbildung: Über die Lage der Wissenschaften in Wien um 1900 und ihre Wechselwirkungen mit der Universitätsausdehnungsbewegung," *Spurensuche*, n.s., 12, no. 1 (2001): 4–23.
10. Nina-Sophie Fritsch, "Warum Wissenschaftlerinnen die Universität verlassen: Eine biografische Fallanalyse zu Ausstiegsgründen aus dem österreichischen Universitätssystem," *SWS-Rundschau* 54, no. 2 (2014): 159–80.
11. Nina-Sophie Fritsch, "At the Leading Edge—Does Gender Still Matter? A Qualitative Study of Prevailing Obstacles and Successful Coping Strategies in Academia," *Current Sociology* 63, no. 4 (2015): 9–10.
12. E.g., "Neue Professoren kommen öfter aus dem Ausland als aus Österreich," *Der Standard*, 9 July 2015, http://derstandard.at/2000018830710/Neue-Professoren-kommen-oefter-aus-dem-Ausland-als-aus-Oesterreich; with more nationalistic tones, see Otmar Lahodynsky, "Verdrängungsbeschwerden," *Profil*, 29 March 2014, http://www.profil.at/home/universitaeten-verdraengungs-beschwerden-355720.
13. See the interesting discussion of recruitment patterns in Europe in Tatiana Fumasoli, Gaële Goastellec, and Barbara M. Kehm, eds., *Academic Work and Careers in Europe: Trends, Challenges, Perspectives* (Dordrecht: Springer, 2014). For an interesting historical-sociological analysis, see Martin Schmeiser, *Akademischer Hasard: Das Berufsschicksal des Professors und das Schicksal der deutschen Universität 1870–1920. Eine verstehend soziologische Untersuchung* (Stuttgart: Klett-Cotta, 1994).
14. Leopold Adametz, "Nowy gatunek dyluwialny rogatego bydła: Bos (brycheros) europaeus," *Sprawozdania z Czynności i Posiedzeń Akademii Umiejętności w Krakowie* 3, no. 3 (1898): 16–17.
15. Krzysztof Lipiński, "Wilhelm Creizenach (1851–1919)," in *Uniwersytet Jagielloński: Złota Księga Wydziału Filologicznego*, ed. Jan Michalik and

Wacław Walecki (Cracow: Księgarnia Akademicka, 2000), 107–15; Starnawski, "Sylwetki"; and Klin and Loužil, *František Tomáš Bratranek.*

16. Roswitha Reinbothe, *Deutsch als internationale Wissenschaftssprache und der Boykott nach dem Ersten Weltkrieg* (Frankfurt am Main: Peter Lang, 2006).
17. Feichtinger, *Wissenschaft als reflexives Projekt*; Denis A. Sdvižkov, *Das Zeitalter der Intelligenz: Zur vergleichenden Geschichte der Gebildeten in Europa bis zum Ersten Weltkrieg* (Göttingen: Vandenhoeck & Ruprecht, 2006); Luft, "'Politische Professoren'"; and Jedlicki, *Dzieje inteligencji polskiej.*
18. Michael Gordin, *Scientific Babel: How Science Was Done before and after Global English* (Chicago: University of Chicago Press, 2015).
19. See also Scott Montgomery, "Impacts of a Global Language on Science: Are There Disadvantages?," in *Language as a Scientific Tool: Shaping Scientific Language across Time and National Traditions*, ed. Miles MacLeod, Rocío G. Sumillera, Jan Surman, and Ekaterina Smirnova (New York: Routledge, 2016), 199–218.
20. See Jan Surman, "Internationality in *Nauka Polska*: Infrastructure, Translation, Language," in *Science Studies in Interwar Poland*, ed. Friedrich Cain and Bernhard Kleeberg (Tubingen: Mohr Siebeck, 2018).
21. Judson, *The Habsburg Empire.*
22. John Deak, *Forging a Multinational State: State Making in Imperial Austria from the Enlightenment to the First World War* (Stanford, CA: Stanford University Press, 2015).
23. See, for an analogous claim, Moritz Csáky, *Das Gedächtnis der Städte: Kulturelle Verflechtungen—Wien und die urbanen Milieus in Zentraleuropa* (Vienna: Böhlau, 2010); and Andreas Kappeler, *Russland und die Ukraine: Verflochtene Biographien und Geschichten* (Vienna: Böhlau, 2012).

Bibliography

Archives

Archiwum Główne Akt Dawnych, Warsaw (AGAD) (Central Archives of Historical Records in Warsaw)
Fond, C. K. Ministerstwo Wyznań i Oświaty (MWiO); K. k. Ministerium für Kultus und Unterricht (304)
Sygn. 51u, Uniwersytet Jagielloński. Wydział Lekarski. Personalia Profesorów K–O.
Sygn. 52u, Uniwersytet Jagielloński. Wydział Lekarski. Personalia Profesorów. P–Z.
Sygn. 65u, Uniwersytet Jagielloński. Wydział Filozoficzny. Personalia Profesorów. A–C.
Sygn. 66u, Uniwersytet Jagielloński. Wydział Filozoficzny. Personalia Profesorów D–J.
Sygn. 69u, Uniwersytet Jagielloński. Wydział Filozoficzny. Personalia Profesorów, S.
Sygn. 70u, Uniwersytet Jagielloński. Wydział Filozoficzny. Personalia Profesorów, T–Z.
Sygn. 112u, Uniwersytet Lwowski. Wydział Lekarski. Zakłady i Instytuty. Zakłady alfabetycznie A–Z.
Sygn. 117u, Uniwersytet Lwowski. Wydział Filozoficzny. Działy: personalia profesorów, asystenci, demonstratorzy, preparatorzy.
Sygn. 118u, Uniwersytet Lwowski. Wydział Filozoficzny. Działy: personalia profesorów, A–G.
Sygn. 119u, Uniwersytet Lwowski. Wydział Filozoficzny. Działy: personalia profesorów, H–N.
Sygn. 120u, Uniwersytet Lwowski. Wydział Filozoficzny. Działy: personalia profesorów, O–R.

Sygn. 121u, Uniwersytet Lwowski. Wydział Filozoficzny. Działy: personalia profesorów, S–P.

Sygn. 122u, Uniwersytet Lwowski. Wydział Filozoficzny. Działy: personalia profesorów, U–Z.

Sygn. 393u, Uniwersytet Jagielloński w Krakowie. Generalia o charakterze osobowym, personalia zbiorowe, wydziałowe i ogólne, spisy wykładów i wykazy profesorów.

Sygn. 403u, Uniwersytet Lwowski. Działy: katedra historii jezyka i literatury Ruskiej. Personalia. Wydział Lekarski. Sprawy terenowe i budowlane uniwersytetu. Instytucje, organizacje i dotacje.

Sygn. 405u, Uniwersytet Lwowski. Pliki: ogród botaniczny, matematyka, muzea przyrodnicze, mineralogia, mechanicy.

Central'nyj deržavnyj istoryčnyj archiv Ukraïny, L'viv (CDIAU) (Central State Historical Archives of Ukraine in L'viv; Центральний державний історичний архів України, Львів)

F. 146, Hal'yćke namisnyctvo, m. L'viv, 1772–1921

Op. 14: 1819–1886

Sp. 1: Rozporyadzžennya Ministerstva virospovidan' i osvity pro zasnuvannya kafedr ukraïns'koyi movy ta pastorśkoho bohosvol'ya u l'vivs'komu universyteti, kafedry moldavs'koyi movy v Černivec'komu universyteti ta lystuvannya po cych pytannyach.

Archiwum Uniwersytetu Jagiellońskiego (AUJ) (Archive of the Jagiellonian University)

WL II, Akta Wydziału Lekarskiego 1849–1949

152, Katedra i Zakład Medycyny Sądowej; materiały z lat 1850–1946.

174, Katedra i Klinika Dermatologiczna; materiały z lat 1862–1949.

156, Katedra i Zakład Historii Medycyny; materiały z lat 1852–1949.

WF II, Akta Wydziału Filozoficznego Uniwersytetu Jagiellońskiego 1849–1945

121, Teczki akt habilitacyjnych z lat 1862–1945.

128, Katedry filozofii 1855–1945.

135, Katedra Historii Powszechnej 1850–1917.

151, Katedry Filologii Klasycznej 1851–1939.

153, Katedra i Seminarium Historii Literatury Polskiej 1853–1939, 1945.

157, Katedra i Seminarium Filologii Germańskiej 1851–1939.

163, Katedry i Instytut Matematyczny 1851–1945.

164, Katedra Astronomii i Obserwatorium Astronomiczne 1851–1939.

173, Katedra i Zakład Anatomii i Cytologii Roślin 1875–1936.

180, Katedra i Instytut Geograficzny 1849–1853, 1874–1945.

S II, Akta Senatu Akademickiego Uniwersytetu Jagiellońskiego 1849–1939

619, Teki osobowe pracowników naukowych czynnych w latach 1850–1939.

815, Wydział Lekarski—Fizjologia 1849–1937.

808, Wydział Lekarski—Anatomia opisowa—sprawy osobowe 1849–1935.
810, Wydział Lekarski—Anatomia patologiczna 1850–1938.
819, Wydział Lekarski—Medycyna sądowa 1850–1938.

Archiv der Universität Wien (AT-UAW) (Archive of the University of Vienna)
Dekanatsakten der Philosophischen Fakultät
AT-UAW/PH Philosophische Fakultät (Artistenfakultät), 1365–2000 (ca.)
Personalakten der Philosophischen Fakultät (19. Jh.–20. Jh.)
PH PA 1118 Brentano, Franz (1880–1952)
PH S Sonderreihe der Philosophischen Fakultät, ca. 1852–ca. 1966
PH S 34 Besetzung und Errichtung von Lehrkanzeln, 1869.04.06–1957.02.01
PH S 34.15 Wiederbesetzung der Philosophischen Lehrkanzel nach Prof. Ernst Mach, 1901.07.01–1903.03.26
AT-UAW/MED Medizinische Fakultät der Universität Wien, 1399–1980
MED S Sonderreihe der Medizinischen Fakultät, 1821.10.25–1991
MED S 17 Lehrkanzelbesetzungen, 1821–1948

Deržavnyj arhiv L'vivs'koï oblasti (DALO) (State Archive of L'viv Oblast; Державний архів Львівської області)
Fond 26: L'vovskij Korolevskij universitet (1817–1918), Universitet im. Jana Kazimira vo L'vove (1919–1939)
Op. 5: Ličnye i pensionnye dela profesorsko-prepodavatel'skogo sostava i služaščich
Spr. 437, Glovackij Jan.
Spr. 510, Gruševskij Mychail.
Spr. 899, Kolessa Aleksandr.
Spr. 2143, Šorr Mojžeš.
Op. 7: Filosofskij Fakultet /1849–1924/, Humanitarnij Fakultet /1924–1939/
Spr. 22, Protokoly zasedanij soveta professorov za 1851/52 učebnyj god i materialy k nim.
Spr. 30, Protokoly zasedanij soveta professorov i konferencji prepodavatel'skogo sostava fakul'teta za 1853/1854 učebnyj god i materialy k nim.
Spr. 33, Perepiska s Ministerstvom veroispovedanij i prosveščenia, Galicijskim kraevym upravleniem, rektoratom o zapreščenii dopuska ne christian k gabilitacii, vybore dekan, naznačenii i peremeščenii professorov, spiski profesorsko-prepodavatel'skogo sostava i dr. materialy po voprosam kadrov.
Spr. 39, Protokoly zasedanij soveta professorov i konferencij prepodaavatel'skogo sostava za 1954/1855 učebnyj god i materialy k nim.
Spr. 43, Perepiska s Ministerstvom veroispovedanij i prosveščenia, Namestničestvom vo L'vove, Krakowskim universitetom ob. Ukomplektovanii kafedry polskoï filologii, utverždenii professorov, začislenii na rabotu vspomogatnogo naučnogo personala i dr. materialy po voprosam kadrov.

Spr. 56, Perepiska s Ministerstvom veroispovedanij i prosveščenia, Namestničestvom vo L'vove i dr. materialy ob izbranii dekana, naznačenii professorov, suplentov, začislenii služaščich na rabotu, vyplate im voznagraždenij i po dr. voprosam kadrov i zarabornoj platy.

Spr. 132, Protokoly zasedanij soveta professorov za 1868/69 učebnyj god i materialy k nim.

Spr. 142, Protokoly zasedanij soveta professorov za 1870/71 učebnyj god i materialy k nim.

Spr. 146, Perepiska s Ministerstvom veroispovedanij i prosveščenia i Namestničestvom vo L'vove o naznačenii professorov, ich perevode na pensiju, po vyšenii zarabotnoj platy, prepostavlenii otpuskov i dr. materialy po voprosam kadrov.

Spr. 148, Zajavlenija Zakševskogo Vincentija, Frejnda Avgusta i dr., nostrifikacii ich doktorskich diplomov, gabilitacii na docenta i perepiska s Ministerstvom veroispovedanij i prosveščenia po etim voprosam, s priloženiem protokola zasedanija voveta professorov or 19 janvarja 1871 g.

Spr. 175, Protokoly zasedanij soveta professorov za 1874/75 učebnyj god i materialy k nim.

Spr. 178, Perepiska s Ministerstvom veroispovedanij i prosveščenia ob. ukompletovanii kafedr geografii, astronomii i chimii.

Spr. 184, Perepiska s Ministerstvom veroispovedanij i prosveščenia, Namestničestvom vo L'vove, rektoratom ob. otkaze v utverždenii prorektora, naznačenii i voznagraždenii professorov i assistentov i po dr. voprosam kadrov.

Spr. 226, Protokoly zasedanij soveta professorov za 1880/81 učebnyj god i materialy k nim.

Spr. 244, Perepiska s Ministerstvom veroispovedanij i prosveščenia, Namestničestvom vo L'vove, i dr. učreždenijami ob. utverždenii rektora i dekana, naznačenii professorov i služaščich.

Spr. 269, Rasporjaženija Ministerstva veroispovedanij i prosveščenia i perepiska s nim, Namestničestvom vo L'vove i dr. materialy ob. usilenii bor'bi i razglašeniem služebnoj tajny, utverždenii docentov, naznačenii assistentov i po dr. voprosam kadrov.

Spr. 293, Perepiska s Ministerstvom veroispovedanij i prosveščenia, Namestničestvom vo L'vove o predostablenii stipendii i posobii studentam docentu, kandidatam na učitelej nemeckogo jazyka.

Spr. 321, Protokoly zasedanij soveta professorov, profesorsko-prepodavatel'skogo sostava i docentov za 1890/91 učebnyj god.

Spr. 327, Perepiska s Ministerstvom veroispovedanij i prosveščenia, Namestničestvom vo L'vove, vysšimi učebnymi zavedenijami Avstro-Vengrii o naznačenii professorov, členov èkzamenacionnych komisij, prieme na rabotu služaščich i dr. materialy po voprosam kadrov.

Spr. 346, Protokoly zasedanij soveta professorov, profesorsko-prepodavatel'skogo sostava i docentov za 1892/93 učebnyj god.

Spr. 381, Protokoly zasedanij soveta professorov za 1895/96 učebnyj god.
Spr. 387, Protokoly zasedanij soveta professorov i sobranij docentov za 1892/93 učebnyj god.
Spr. 410, Pis'ma Ministerstvu veroispovedanij i prosveščenia o nanzačenii Skorskogo Aleksandra profesorom filosofii, ustav filosofskogo seminaria i protokol zaedanija soveta professorov ot 16 ijulja 1901 g.
Spr. 420, Protokoly zasedanij soveta professorov, profesorsko-prepodavatel'skogo sostava, docentov za 1898/99 učebnyj god.
Spr. 421, Perepiska z Kraevym komitetom vo L'vove i rektoratom o proektach osnovanij Kraevego muzeja istorii kul'tury i kafedry cerkovno-slovjanskogo jazyka.
Spr. 434, Protokoly zasedanij soveta professorov, sobranij prepodavatelej za 1899/1900 učebnyj god.
Spr. 435, Protokoly zasedanij komisij soveta professorov za 1899/1905 gg. i materialy k nim.
Spr. 525, Protokoly zasedanij komisij soveta professorov za 1904/1909 gg. i materialy k nim.
Spr. 554, Protokoly zasedanij soveta professorov i sobranij prepodavatelej za 1906/1907 učebnyj god.
Op. 12 Rektorat
Spr. 77, Protokoly zasedanij senata za 1854 god.

Moravský zemský archiv v Brně (MZA Brno) (Moravian Land Archive in Brno)
Fond E4 (Augustiniáni Staré Brno)
Kart. 191

Národní archiv (NA) (National Archives, Prague)
Fond České místodržitelství v Praze
Inv.č. 32, sg. 5 C 1 Prag, záležitosti, týkající se jmenování soukromých docentů mimořádnými profesory, mimořádných profesorů řádnými, potvrzování profesorů v učitelském úřadě, poskytování dovolených profesorům, disciplinárního řízení proti nim a obsazování vedoucích kateder filosofických a lékařských fakult české a německé university 1888–1918.
Fasc. 198
Fond Ministerstvo kultu a vyučování Vídeň 1888–1918
Inv.č. 9, sg. 5 Česká a Německá univerzita Praha, filozofická fakulta—profesoři a docenti / Prag Philosophie-Professoren
Fasc. 110, A–B
Fasc. 111, C–F
Fasc. 113, H–J
Fasc. 114, K
Fasc. 116, N–P
Fasc. 117, P–R
Fasc. 118, R–S

Fasc. 120, U–Ž
Inv.č. 2, sg. 5, Česká a Německá univerzita Praha, lékařská fakulta—profesoři a docenti
Fasc. 95, F–He
Fasc. 97, K
Fasc. 100, R

Österreichisches Staatsarchiv, Allgemeines Verwaltungsarchiv (AT-OeStA/AVA) (Austrian State Archives, General Archive of Administration)
Fond Education and Religion, General Section, Acts Unterricht und Kultus, Unterrichtsministerium, Allgemeine Reihe, Akten (Unterricht UM allg. Akten)
620, Universität Wien, Medizin: Professoren Abeles–Blum.
621, Universität Wien, Medizin: Professoren Bock–Dlauhy.
622, Universität Wien, Medizin: Professoren Doerr–Fischer.
623, Universität Wien, Medizin: Professoren Fleckseder–Fürth.
624, Universität Wien, Medizin: Professoren Gärtner–Hawelka.
625, Universität Wien, Medizin: Professoren Hebra–Jurie.
627, Universität Wien, Medizin: Professoren Landsteiner–Meixner.
628, Universität Wien, Medizin: Professoren Meller–Nyiri.
629, Universität Wien, Medizin: Professoren Obermayer–Puschmann.
630, Universität Wien, Medizin: Professoren Raab–Schintzel.
632, Universität Wien, Medizin: Professoren Starlinger–Volk.
633, Universität Wien, Medizin: Professoren Wagner–Zweig.
663, Universität Wien, Philosophie: Professoren Abel–Berwerth.
664, Universität Wien, Philosophie: Professoren Bibl–Brunswik.
665, Universität Wien, Philosophie: Professoren Büdinger–Dyboski.
666, Universität Wien, Philosophie: Professoren Ebeling–Fournier.
667, Universität Wien, Philosophie: Professoren Frank–Görgey.
668, Universität Wien, Philosophie: Professoren Grafe–Hartmann.
669, Universität Wien, Philosophie: Professoren Haschek–Hirsch.
670, Universität Wien, Philosophie: Professoren Hirschberg–Jagic.
671, Universität Wien, Philosophie: Professoren Jahn–Kaweck.
672, Universität Wien, Philosophie: Professoren Keil–Krumpholz.
673, Universität Wien, Philosophie: Professoren Kubitschek–Löwy.
674, Universität Wien, Philosophie: Professoren Ludloff–Minor.
675, Universität Wien, Philosophie: Professoren Mojsisovits–Otto.
679, Universität Wien, Philosophie: Professoren Schacherl–Schneider.
681, Universität Wien, Philosophie: Professoren Trabert–Weidl.
918, Universität Graz, Medizin: Direktoren, Privatdozenten, Professoren Albrecht –Escherich.
919, Universität Graz, Medizin: Professoren Finotti–Jaksch.
920, Universität Graz, Medizin: Professoren Kaiser–Lundwall.
921, Universität Graz, Medizin: Professoren Mahnert–Rzehaczek.
934, Universität Graz, Philosophie: Professoren Aigner–Doelter.

935, Universität Graz, Philosophie: Professoren Egger–Hausegger.
936, Universität Graz, Philosophie: Professoren Heberdey–Jeiteles.
938, Universität Graz, Philosophie: Professoren Lämmermayer–Murko.
939, Universität Graz, Philosophie: Professoren Nachtigall–Rumpf.
940a, Universität Graz, Philosophie: Professoren Sacher-Masoch–Schwinner.
941, Universität Graz, Philosophie: Professoren Tangl–Zwierzina.
1055, Universität Innsbruck, Medizin: Professoren Fick–Holl.
1056, Universität Innsbruck, Medizin: Professoren Ipsen–Lukasiewicz.
1057, Universität Innsbruck, Medizin: Professoren Malfatti–Rusch.
1058, Universität Innsbruck, Medizin: Professoren Sachs–Zoth.
1070, Universität Innsbruck, Philosophie: Professoren Arleth–Exner.
1072, Universität Innsbruck, Philosophie: Professoren Härdtl–Huber.
1073, Universität Innsbruck, Philosophie: Professoren Jax–Lieber.
1075, Universität Innsbruck, Philosophie: Professoren Pastor–Schweidler.
1076, Universität Innsbruck, Philosophie: Professoren Seemüller–Voltelini.
1077, Universität Innsbruck, Philosophie: Professoren Wackernell–Zwierzina.
1121, Universität Prag, Philosophie: Professoren Tomek–Zubaty, Adjunkten, Assistenten.
1184, Universität Czernowitz, Philosophie: Professoren Becke-Zingerle, Lehrkanzeln: Chemie, allgemeine Geschichte, klassische Philologie, deutsche Rede- und Vortragskunst, Physik; Assistenten; Jus: Professoren Amonn-Petschek, Lehrkanzeln: in genere, Bürgerkunde, allgemeines und österreichisches Staatsrecht, politische Ökonomie, Römisches Recht, Verwaltungsrecht und Verwaltungslehre, Handels- und Wechselrecht; Universität Czernowitz.
1186, Universität Krakau, Medizin: Professoren Biasiadecki–Treitz; Theologie: Professoren Czertunerzakiewicz–Wadolny.
1208, Universität Prag, Medizin: in genere, Dozenten in genere, Professoren Aebyl–Gussenbauer.
1209, Universität Prag, Medizin: Professoren Haas–Köstl.
1210, Universität Prag, Medizin: Professoren Lambl–Purkinje.
1211, Universität Prag, Medizin: Professoren Quadrat–Zobel.
1216, Universität Prag, Philosophie: Dozenten in genere, Professoren Adler–Emler.
1217, Universität Prag, Philosophie: Professoren Fauvin–Jüthner.
1218, Universität Prag, Philosophie: Professoren Kachler–Ludwig.
1219, Universität Prag, Philosophie: Professoren Mach–Rzach.
1220, Universität Prag, Philosophie: Professoren Sachs–Stumpf.
1221, Universität Prag, Philosophie: Professoren Tomek–Zubaty, Adjunkten, Assistenten.

Státní oblastní archiv v Litoměřicích, pobočka Děčín (SOA Litoměřice/Děčín) (State Regional Archives Litoměřice, Děčín Branch)
Fond Rodinný archiv Thun-Hohenstei

Universitätsarchiv Innsbruck (UAI) (Archives of the University of Innsbruck)
Fond Philosophische Fakultät (PF)
Fond Medizinische Fakultät (MF)
Fond Berufung Zoologie

Universitätsarchiv Karl-Franzens-Universität Graz (UAG) (University Archives, University of Graz)
Philosophische Fakultät
Rektorat
Protokolle
Akten

Ústav dějin Univerzity Karlovy a archiv Univerzity Karlovy (ÚDAUK) (Institute of the History of Charles University and Archive of Charles University)
Fond Filozofická fakulta Karlo-Ferdinandovy Univerzity, 1882–2012
Inv.č. 640, Kart. 56, Seydler August.
Inv.č. 554, Kart. 46, PA Palacký Jan.
Filozofická fakulta Německé univerzity (FF NU) v Praze (1857) 1877–1945
Inv.č. 186–193, Sign. K/a, Kart. 9, Záležitosti profesorů.
Inv.č. 229, Sign. K/XIVb, Kart. 11, Obsazování mimořádných profesur: (a) pro německou filologii (Jacob Minor, Hans Lambel, dr, Seuffert), (b) pro ang. Filologii (Alois Brandl)—společný spis, 29.12.1885.
Inv.č. 238, Sign. L/45, Kart. 12, Herzog Eugen.
Inv.č. 249, Sign. L/53, Kart. 12, Mahler Arthur.
Inv.č. 532, Sign. K/XVIII, Kart. 52, Trautmann Reinhold.
Lékařská fakulta Německé univerzity (LF NU) v Praze 1881–1945
Kart. 2, Grosser Otto.
Kart. 2, Dexler Hermann.
Kart. 4, Kohn Alfred.
Akademický senát Karlo-Ferdinandovy univerzity (1791) 1796–1883
Inv.č. 227, Kart. 149, Výkazy dr. J. Witáka.

Zbiory Sekcji Rękopisów, Biblioteka Jagiellońska (Collection of the Manuscripts of the Jagiellonian Library, Cracow)
Sign. 9007 III, Korespondencja Władysława Natansona, vol. 6.

Legal Digests and Stenographic Protocols

Allgemeines Reichs-Gesetz und Regierungsblatt für das Kaiserthum Österreich. Jahrgang 1849. Vienna: Kaiserlich-königliche Hof- und Staatsdruckerei, 1850.
Beck von Mannagetta, Leo, and Karl von Kelle, eds. *Die österreichischen Universitätsgesetze: Sammlung der für die österreichischen Universitäten giltigen*

Gesetze, Verordnungen, Erlässe, Studien- und Prüfungsordnungen usw. Vienna: Manz'sche k.k. Hofverlags- und Universitäts-Buchhandlung, 1906.

Officielle stenographische Berichte über die Verhandlungen des österr. Reichstages: Neunundsiebzigste (XXVII.) Sitzung des österreichischen constituirenden Reichstages in Kremster am 26. Jänner 1849. http://www.psp.cz/eknih/1848urrs/stenprot/079schuz/s079001.htm.

Reichs-Gesetz-Blatt für das Kaiserthum Oesterreich. Vienna: k.k. Hof- und Staatsdruckerei, 1867.

Sbírka zákonů a nařízení republiky Československé. No. 30. Prague: Státni tiskárna, 1920.

Schweickhardt, Friedrich, ed. *Sammlung der für die österreichischen Universitäten giltigen Gesetze und Verordnungen.* 2nd ed. Vienna: k.k. Schulbuchverlag in Comission bei Manz'schen k.k. Hofverlags- und Universitäts-Buchhandlung, 1885.

Štátni úrad štatistický. *Štatistická príručka republiky československé IV.* Prague: Bursík & Kohout, 1932.

Stenografická zpráva XXXV. sezení třetího ročního zasedání sněmu českého od roku 1861, dne 11. května 1864 / Stenographischer Bericht über die XXXV. Sitzung der dritten Jahres-Session des böhmischen Landtages vom Jahre 1861, am 11. Mai 1864 (=Stenographische Berichte des böhmischen Landtages. Dritte Jahressession 1864. Stenografické zprávy sněmu Českého. Třetí zasedání 1864). Prague: Statthalterei-Buchdruckerei, 1864.

Stenographische Protokolle des Abgeordnetenhauses im Jahre 1907. XVIII. Session. IV Band, 21. bis 41. Sitzung (S. 1851 bis 3054.). Vienna: Kaiserlich-königliche Hof- und Staatsdruckerei, 1908.

Stenographische Protokolle über die Sitzungen des Hauses der Abgeordneten im Jahre 1902. XVII. Session. XII Band, 106. bis 115. Sitzung. (S. 10039 bis 10958.). Vienna: Kaiserlich-königliche Hof- und Staatsdruckerei, 1902.

Thaa, Georg, ed. *Sammlung der für die österreichischen Universitäten giltigen Gesetze und Verordnungen.* Vienna: G. J. Manz'sche Buchhandlung, 1871.

Lecture Catalogs

Akademische Behörden, Personalstand und Ordnung der Vorlesungen an der K.K. Karl-Franzens-Universität zu Graz. Graz: Leykam's Erben, 1850–.

Personalstand der k.k. Universität zu Prag zu Anfang des Winter-Semesters 1860–61. Prague: k.k. Normalschulbuchdruckerei, 1860.

Program Wykładów i Skład Uniwersytetu w półroczu zimowem roku akademickiego 1919–1920. Vilnius: Znicz, 1919.

Skład Uniwersytetu Jana Kazimierza we Lwowie w roku akademickim 1927/1928. L'viv: Uniwersytet Jana Kazimierza, 1927.

Soznam osôb a ústavov Univerzity Komenského v Bratislave, taktiež štátnych zkúšobných komisií podla stavu na začiatku štud. roku 1924–25. Bratislava: Univerzita Komenského, 1924.

Übersicht der akademischen Behörden, der den einzelnen Fakultäten zugehörenden Decane, Pro-Decane, Professoren, Privatdocenten . . . dann der Kirche, Bibliothek, Kanzlei, Quästur an der Kaiserl. Königl. Universität zu Wien für das Studien-Jahr 1860/61. Vienna: Kaiserlich-königliche Staats-Druckerei, 1861.

Übersicht der akademischen Behörden, Professoren, *Privatdocenten, Lehrer, Beamten etc. an der k.k. Universität zu Wien für das Studien-Jahr 1910/11.* Vienna: Adolf Holzhausen, 1910.

Encyclopedias

Allgemeine Deutsche Biographie. 56 vols. Munich: Duncker & Humblot, 1875–1912.

Historische Kommission bei der Bayrischen Akademie der Wissenschaften, ed. *Neue Deutsche Biographie.* 26 vols. Berlin: Duncker & Humblot, 1953–2016.

Kadlec, Oskár, ed. *Encyklopédia medicíny.* Vol. 3. Bratislava: Vydavateľstvo Asklepios, 1993–2004.

Kdo byl kdo: Čeští a slovenští orientalisté, afrikanisté a iberoamerikanisté. Prague: Libri, 1999.

Das Österreichische Biographische Lexikon 1815–1950. 14 vols. Vienna: Österreichische Akademie der Wissenschaften, 1957–2015.

Ottův slovník naučný. 28 vols. Prague: J. Otto, 1888–1909.

General Bibliography

Adametz, Leopold. "Nowy gatunek dyluwialny rogatego bydła: Bos (brycheros) europaeus." *Sprawozdania z Czynności i Posiedzeń Akademii Umiejętności w Krakowie* 3, no. 3 (1898): 16–17.

"Adres do Cesarza. Zgromadzenie Profesorów i uczniów Uniwersytetu Jagiell. i wszystkich Zakładów naukowych w mieście Krakowie i jego Okregu obecnie istniejących, zanosi do Waszéj Cesarsko Królewskiéj Mości nastepujące prośby." In *Ważniejsze dokumenta odnoszące się do swojego udziału w sprawach i losach Zakładów naukowych b. W. M. Krakowa i jego Okręgu*, edited by Józef Brodowicz, 77–80. Cracow: Nakładem autora, w drukarni Uniwersytetu Jagiellońskiego, 1874.

Agnew, Hugh LeCaine. "Czechs, Germans, Bohemians? Images of Self and Other in Bohemia to 1848." In *Creating the Other: Ethnic Conflict and Nationalism in Habsburg Central Europe*, edited by Nancy Wingfield, 56–80. New York: Berghahn Books, 2003.

Aichner, Christof. "Franz Exner: Professor für Philosophie, Mitschöpfer der Universitätsreform nach 1848." In *Universität—Politik—Gesellschaft: 650 Jahre Universität Wien—Aufbruch in das neue Jahrhundert*, edited by Mitchell G. Ash and Josef Ehmer, 183–88. Vienna: Vienna University Press; Göttingen: V&R Unipress, 2015.

Akademischer Senat, ed. *Die Franz-Josephs Universität in Czernowitz im ersten Vierteljahrhundert ihres Bestandes*. Chernivtsi: Bukowinaer Vereinsdruckerei, 1900.

Akademischer Senat der k.k. deutschen C.-F. Universität. *Denkschrift über die Rechtstellung der Professoren und der Universitäten*. Prague, 1896.

Akademischer Senat der Wiener Universität, ed. *Geschichte der Wiener Universität von 1848 bis 1898: Als Huldigungsfestschrift zum 50-jähr. Regierungsjubiläum seiner k.k. apostolischen Majestät des Kaisers Franz Josef I.* Vienna: Alfred Hölder, 1898.

Albert, Zygmunt. *Kaźń profesorów lwowskich—lipiec 1941*. Wrocław: Wydawnictwo Uniwersytetu Wrocławskiego, 1989.

———. "Prof. Dr. Andrzej Obrzut." *Archiwum Historii i Filozofii Medycyny* 55, no. 1 (1992): 55–61.

———. "Prof. Dr. Paweł Ludwig Kučera (1872–1928), kierownik zakładu higieny UJK we Lwowie." *Archiwum Historii i Filozofii Medycyny* 55, no. 1 (1992): 62–70.

Albertazzi, Liliana. "Vittorio Benussi." In *The School of Alexius Meinong*, edited by Liliana Albertazzi, Dale Jacquette, and Roberto Poli, 1–35. Aldershot: Ashgate, 2001.

Albin, Janusz, ed. *Z dziejów Zakładu Narodowego imienia Ossolińskich: Studia i materiały*. Wrocław: Ossolineum, 1978.

Aleksiun, Natalia. "The Cadaver Affair in the Second Polish Republic: A Case Study of Political Antisemitism?" In *Alma Mater Antisemitica: Akademisches Milieu, Juden und Antisemitismus an den Universitäten Europas zwischen 1918 und 1939*, edited by Regina Fritz, Grzegorz Rossoliński-Liebe, and Jana Starek, 203–20. Vienna: New Academic Press, 2016.

Ammon, Ulrich, ed. *The Dominance of English as a Language of Science: Effects on Other Languages*. Berlin: Walter de Gruyter, 2001.

Andreev, Aleksandr. *Universitet v Rossijsksj imperii XVIII—pervoj poloviny XIX veka*. Moscow: Rosspèn, 2012.

Andreyev, Catherine, and Ivan Savicky. *Russia Abroad: Prague and the Russian Diaspora, 1918–1938*. New Haven, CT: Yale University Press, 2004.

"Anton Prešeren." Revija Ognjišče, n.d. http://revija.ognjisce.si/revija-ognjisce/67-pricevanje/1677-anton-preseren (accessed 24 March 2015).

Ara, Angelo. "Le problème de l'université italienne en Autriche (1866–1914)." *Études danubiennes* 3, no. 2 (1987): 157–68.

Arneth, Alfred. *Aus meinem Leben*. Vol. 1, *1819–1849*. Vienna: Adolf Holzhausen, 1891.

Ash, Mitchell G. "Die Universität Wien in politischen Umbrüchen des 19. und 20. Jahrhunderts." In *Universität—Politik—Gesellschaft: 650 Jahre Universität Wien—Aufbruch in das neue Jahrhundert*, edited by Mitchell G. Ash and Josef Ehmer, 29–172. Vienna: Vienna University Press; Göttingen: V&R Unipress, 2015.

Ash, Mitchell G., and Jan Surman, eds. *The Nationalisation of Scientific Knowledge in the Habsburg Empire, 1848–1918*. Basingstoke: Palgrave Macmillan, 2012.

Baár, Monika. *Historians and Nationalism: East-Central Europe in the Nineteenth Century*. New York: Oxford University Press, 2010.

Baczkowski, Krzysztof. "W służbie dworu Habsburskiego: Antoni Walewski (1805–1876)." *Zeszyty Naukowe Uniwersytetu Jagiellońskiego: Prace Historyczne* 132 (2005): 99–108.

Bałus, Wojciech. *Krakau zwischen Traditionen und Wegen in die Moderne: Zur Geschichte der Architektur und der öffentlichen Grünanlagen im 19. Jahrhundert.* Stuttgart: Franz Steiner, 2003.

Banasiewicz, Maria, and Antoni Czubiński, eds. *Organizacja i rozwój uczelni od listopada 1918 roku do inauguracji w maju 1919 roku.* Poznań: Uniwersytet im. Adama Mickiewicza, 1973.

Baron, Salo W. "The Revolution of 1848 and Jewish Scholarship: Part II: Austria." *Proceedings of the American Academy for Jewish Research* 20 (1951): 1–100.

Barvins'kyj, Aleksandr. "Zasnovane katedry istoriï Ukraïny v L'vivs'kim universyteti." *Zapysky Naukovoho tovarystva imeni Ševčenka: Praci filologičnoï sekciï* 141–43 (1925): 1–25.

Barycz, Henryk. "Docenckie kłopoty Józefa Szujskiego." In *Wśród gawędziarzy, pamiętnikarzy i uczonych galicyjskich: Studia i sylwety z Życia umysłowego Galicji XIX w.*, edited by Henryk Barycz, 91–111. Cracow: Wydawnictwo Literackie, 1963.

———. *Józef Ignacy Kraszewski czterokrotny kandydat do katedry uniwersyteckiej.* Cracow: Wydawnictwo Literackie, 1979.

———, ed. *Korespondencja Karola Szajnochy.* Wrocław: Zakład Narodowy im. Ossolińskich, 1959.

———. "Die Rolle der Wiener Universität im geistigen Leben Polens." *Österreichische Osthefte* 7, no. 3 (1965): 176–94.

———. "Szymon Askenazy: Wśród przeciwieństw i niepowodzeń życiowych i naukowych." In *Na przełomie dwóch stuleci: Z dziejów polskiej humanistyki w dobie Młodej Polski*, edited by Henryk Barycz, 238–308. Wrocław: Ossolineum, 1977.

———. *Uniwersytet Jagielloński w życiu narodu polskiego.* Wrocław: Ossolineum, 1964.

———. "Wincenty Pol jako profesor geografii na Uniwersytecie Jagiellońskim." *Polska Akademia Umiejętności: Prace komisji historii medycyny i nauk matematyczno-przyrodniczych* 3, no. 2 (1949): 43–128.

Batron, Josef. *Der vergessene Mähr. Verehrer Goethes, Ph. Dr. P. Thomas Bratranek OSA, Professor an der Universität Krakau.* Olomouc: Historický seminář cyrilometodějské fakulty bohoslovecké, 1937.

Baumgarten, Marita. *Professoren und Universitäten im 19. Jahrhundert: Zur Sozialgeschichte deutscher Geistes- und Naturwissenschaftler.* Göttingen: Vandenhoeck & Ruprecht, 1997.

Bautz, Friedrich Wilhelm. "Bickell, Gustav, Orientalist." In *Biographisch-Bibliographisches Kirchenlexikon*, edited by Friedrich Wilhelm Bautz, 579–80. Nordhausen: Traugott Bautz, 1990.

Bayertz, Kurt, Walter Jaeschke, and Myriam Gerhard, eds. *Weltanschauung, Philosophie und Naturwissenschaft im 19. Jahrhundert.* Vol. 1, *Der Materialismusstreit.* Hamburg: Meiner, 2007.

Beauvois, Daniel. *Wilno: Polska stolica kulturalna zaboru rosyjskiego 1803–1832*. Wrocław: Wydawnictwo Uniwersytetu Wrocławskiego, 2010.

Becke, Friedrich Johnanes. "Bericht über das Studienjar 1918/1919." In *Die feierliche Inauguration des Rektors der Wiener Universität für das Jahr 1919/20 am 5. November 1919*. Vienna: Selbstverlag, 1919.

Bečvářová, Martina. "Czech Mathematicians and Their Role in the Development of National Mathematics in the Balkans." In *Mathematics in the Austrian-Hungarian Empire: Proceedings of a Symposium Held in Budapest on August 1, 2009 during the XXIII ICHST*, edited by Martina Bečvářová and Christa Binder, 9–31. Prague: Matfyzpress, 2010.

———. *Z historie Jednoty (1862–1869)*. Prague: Prometheus, 1999.

Beier, Nikolaj. *"Vor allem bin ich ich . . .": Judentum, Akkulturation und Antisemitismus in Arthur Schnitzlers Leben und Werk*. Göttingen: Wallstein, 2008.

Beller, Steven. *Vienna and the Jews, 1867–1938: A Cultural History*. Cambridge: Cambridge University Press, 1991.

Benešová, Michala, Renata Rusin Dybalska, and Lucie Zakopalová. *90 let pražské polonistiky—dějiny a současnost*. Prague: Karolinum, 2013.

Beran, Jiři. *Vznik České akademie věd a umění v dokumentech*. Prague: Ústřední Archiv Československé akademie věd, 1989.

Berger, Stefan, and Christoph Conrad, eds. *The Past as History: National Identities and Historical Consciousness in Modern Europe*. Basingstoke: Palgrave Macmillan, 2015.

Berger, Stefan, and Alexei Miller, eds. *Nationalizing Empires*. Budapest: Central European University Press, 2015.

———. "Nation-Building and Regional Integration, c. 1800–1914: The Role of Empires." *European Review of History—Revue Européene d'histoire* 15, no. 3 (June 2008): 317–30.

Bieńkowski, Wiesław. "Konstant von Wurzbach und Albert Zipper: Aus der Geschichte der österreichischpolnischen kulturellen Beziehungen im 19. und 20. Jahrhundert." In *Österreich—Polen: 1000 Jahre Beziehungen*, edited by Józef Buszko, 481–507. Cracow: Wydawnictwo Uniwersytetu Jagiellońskiego, 1996.

Bihl, Wolfdieter. *Orientalistik an der Universität Wien: Forschungen zwischen Maghreb und Ost- und Südasien. Die Professoren und Dozenten*. Vienna: Böhlau, 2009.

Billroth, Theodor. *Briefe von Theodor Billroth*. 8th ed. Hannover: Hahnsche Buchhandlung, 1910.

———. *The Medical Sciences in the German Universities: A Study in the History of Civilization*. With an introduction by William H. Welch. New York: Macmillan, 1924.

———. Über das Lehren und Lernen der *medicinischen Wissenschaften an den Universitäten der deutschen Nation*. Vienna: Gerold, 1876.

Binder, Harald. "'Galizische Autonomie': Ein streitbarer Begriff und seine Karriere." In *Moravské vyrovnání z roku 1905: Možnosti a limity národnostního smíru ve střední Evropě / Der Mährische Ausgleich von 1905: Möglichkeiten und Grenzen *für einen nationalen Ausgleich in Mitteleuropa*, edited by Lukáš Fasora, 239–66. Brno: Matice Moravská pro Výzkumné Středisko pro Dějiny Střední Evropy: Prameny, Země, Kultura, 2006.

———. "Das Ruthenische Pressewesen." In *Die Habsburgermonarchie 1848–1918*, vol. 8, *Politische Öffentlichkeit und Zivilgesellschaft*, pt. 2, *Die Presse als Faktor der politischen Mobilisierung*, edited by Helmut Rumpler and Peter Urbanitsch, 2091–126. Vienna: Verlag der Österreichischen Akademie der Wissenschaften, 2006.

Bobrowska-Nowak, Wanda. *Początki polskiej psychologii*. Wrocław: Zakład narodowy im. Ossolińskich/PAN, 1973.

Bona, Władysława. "Zasługi Jana Mikulicza dla rozwoju chirurgii w Polsce." *Archiwum Historii i Filozofii Medycyny* 13, nos. 1–2 (1933): 20–100.

Bonacich, Phillip, and Paulette Lloyd. "Calculating Status with Negative Relations." *Social Networks* 26, no. 4 (2004): 331–38.

Bösche, Andreas. *Zwischen Kaiser Franz Joseph I. und Schönerer: Die Innsbrucker Universität und ihre Studentenverbindungen 1859–1918*. Innsbruck: StudienVerlag, 2008.

Boyer, John W. *Culture and Political Crisis in Vienna: Christian Socialism in Power, 1897–1918*. Chicago: University of Chicago Press, 1995.

———. *Political Radicalism in Late Imperial Vienna: Origins of the Christian Social Movement, 1848–1897*. Chicago: University of Chicago Press, 1981.

Brauneder, Wilhelm. *Leseverein und Rechtskultur: Der Juridisch-Politische Leseverein zu Wien 1840–1990*. Vienna: Manz, 1992.

Brázda, Otaka. "Přichod Jana Evangelisty Purkyně na pražskou univerzitu." *AUC-HUCP* 27, no. 1 (1987): 55–89.

Brechenmacher, Thomas. *Großdeutsche Geschichtsschreibung im neunzehnten Jahrhundert: Die erste Generation (1830–48)*. Berlin: Duncker & Humblot, 1996.

Brentano, Franz Clemens. *Meine letzten Wünsche für Oesterreich*. Stuttgart: Cotta, 1895.

Bresslau, Harry. *Geschichte der Monumenta Germaniae Historica*. Hannover: Hahn, 1921.

Brim, Sadek. *Universitäten und Studentenbewegung in Russland im Zeitalter der Großen Reformen 1855–1881*. Frankfurt am Main: Peter Lang, 1985.

Brix, Emil. "Mentalität ist gut—die Teilung der Prager Universität 1882." *Österreichische Osthefte* 30, no. 3 (1988): 371–82.

Brocke, Bernhard vom. *Hochschul- und Wissenschaftspolitik in Preußen und im Deutschen Kaiserreich 1882–1907: Das "System Althoff."* Stuttgart: Klett-Cotta, 1980.

Brocke, Bernhard vom, and Peter Krüger. *Hochschulpolitik im Föderalismus: Die Protokolle der Hochschulkonferenzen der deutschen Bundesstaaten und Österreichs 1898 bis 1918*. Berlin: AkademieVerlag, 1994.

Broszat, Martin. "Von der Kulturnation zur Volksgruppe: Die nationale Stellung der Juden in der Bukowina im 19. und 20. Jahrhundert." *Historische Zeitschrift* 200, no. 3 (1965): 572–605.

Brubaker, Rogers. "Nationalizing States in the Old 'New Europe' and the New." *Ethnic and Racial Studies* 19, no. 2 (1996): 411–37.

Brunner, Sebastian. *Gesammelte Erzählungen und poetische Schriften*. Vol. 1, *Woher? Wohin? Geschichten, Gedanken, Bilder und Leute aus meinem Leben. Neue Folge I*. 3rd ed. Regensburg: G. J. Manz, 1890.

Brunner, Walter, and Helmut Wohnout. "Hochschulrecht." In *Parlamentarismus und öffentliches Recht in Österreich: Entwicklung und Gegenwartsprobleme*, edited by Herbert Schambeck, 1105–48. Berlin: Duncker & Humblot, 1993.

Brzęk, Gabriel. "Recepcja darwinizmu w Polsce." In *Recepcja w Polsce nowych kierunków i teorii naukowych*, edited by Adam Strzałkowski, 273–91. Cracow: Polska Akademia Umiejętności, 2001.

Buchen, Tim. *Antisemitismus in Galizien: Agitation, Gewalt und Politik gegen Juden in der Habsburger Monarchie um 1900*. Berlin: Metropol, 2012.

Buckler, Julie A. "Writing in a Polluted Semiosphere: Everyday Life in Lotman, Foucault, and De Certeau." In *Lotman and Cultural Studies: Encounters and Extensions*, edited by Andreas Schönle, 320–44. Madison: University of Wisconsin Press, 2006.

Bucur, Maria, and Nancy Meriwether Wingfield, eds. *Staging the Past: The Politics of Commemoration in Habsburg Central Europe, 1848 to the Present*. West Lafayette, IN: Purdue University Press, 2001.

Buklijas, Tatjana. "Dissection, Discipline and Urban Transformation: Anatomy at the University of Vienna, 1845–1914." PhD diss., University of Cambridge, 2005.

———. "Surgery and National Identity in Late Nineteenth-Century Vienna." *Studies in History and Philosophy of Biological and Biomedical Sciences* 38, no. 4 (2007): 756–74.

Buklijas, Tatjana, and Emese Lafferton. "Science, Medicine and Nationalism in the Habsburg Empire from the 1840s to 1918." *Studies in History and Philosophy of Biological and Biomedical Sciences* 38, no. 4 (2007): 679–86.

Burešová, Jana. "Úsilí o obnovení Univerzity v Olomouci ve druhé polovině 19. a na začátku 20. stoleti." *Historická Olomouc* 11 (1998): 153–60.

Burger, Hannelore. "Das Problem der Unterrichtssprache an der Universität Czernowitz." In *Glanz und Elend der Peripherie: 120 Jahre Universität Czernowitz. Eine Veröffentlichung des Österreichischen Ost- und Südeuropainstituts*, edited by Ilona Slawinski and Joseph P. Strelka, 65–82. Vienna: Peter Lang, 1998.

Busch, Otto. "v. Leonhardi." *Philosophische Monatshefte* 11 (1875): 385–90.

Buszko, Józef. *Społeczno-polityczne oblicze Uniwersytetu Jagiellońskiego w dobie autonomii galicyjskiej (1869–1914)*. Cracow: Uniwersytet Jagielloński, Państwowe Wydawnictwo Naukowe, 1963.

———. "L'université Jagellon de Cracovie (1869–1914)." *Études danubiennes* 4, no. 1 (1988): 21–28.

Čapková, Kateřina. *Češi, Němci, Židé? Národní identita Židů v Čechách, 1918–1938*. Prague: Paseka, 2005.

Čatár, Gustáv, Ján Vojtaššák, and Miloš Tichý. "Profesor MUDr. Jozef Florian Babor—významná osobnosť Ústavu pre všeobecnú biológiu LF UK." In *História mediciny, farmácie a veterinárnej mediciny v kontexte vývoja európskej vedy 20. storočia*, edited by Ľudmila Pavlíková, 61–64. Bratislava: Lekárska fakulta Univerzity Komenského, 2000.

Čedík, Miloslav. "Národností otazká a rozdělení pražské univerzity." *AUC-HUCP* 22, no. 1 (1982): 15–25.

Čelakovský, Jaromír. *Moje zápisky, 1871–1914.* Edited by Luboš Velek and Alice Velková. Prague: Archiv hlavního města Prahy; Výzkumné centrum pro dějiny vědy: Scriptorium, 2004.

Čermák, Josef. "Das Kulturleben der Prager deutschen Studenten seit der Mitte des 19. Jahrhunderts: Die Lese- und Redehalle der deutschen Studenten in Prag." In *Juden zwischen Deutschen und Tschechen: Sprachliche und kulturelle Identitäten in Böhmen 1800–1945*, edited by Marek Nekula and Walter Koschmal, 33–64. Munich: Oldenbourg, 2006.

Cernajsek, Tillfried, Christoph Mentschl, and Johannes Seidl. "Eduard Sueß (1831–1914): Ein Geologe und Politiker des 19. Jahrhunderts." In *Wissenschaft und Forschung in Österreich: Exemplarische Leistungen österreichischer Naturforscher und Techniker*, edited by Gerhard Heindl, 59–84. Frankfurt am Main: Peter Lang, 2000.

Chahrour, Marcel. "'A Civilizing Mission'? Austrian Medicine and the Reform of Medical Structures in the Ottoman Empire, 1838–1850." *Studies in History and Philosophy of Science, Part C: Studies in History and Philosophy of Biological and Biomedical Sciences* 38, no. 4 (2007): 687–705.

Chiocchetti, Fabio, and Roberto Starec. "In Search of the 'Ladin Song': The Project Das Volkslied in Österreich in the Ladin Areas of Tyrol and East Friuli (1904–1914)." *Traditiones* 34, no. 1 (2005): 61–77.

Chodounský, Karel. "K padesátiletí 'Časopisu lékařův českých.'" *Časopis lékařův českých* 50, no. 53 (1911): 1602–4.

Christensen, Dan C. *Hans Christian Ørsted: Reading Nature's Mind.* Oxford: Oxford University Press, 2013.

Ciara, Stefan. "Finanzielle Probleme galizischer Wissenschaftler um die Wende des 19. zum 20. Jahrhundert." *Mitteilungen des Österreichischen Staatsarchivs* 53 (2009): 313–33.

Cieśla, Michał. "Spirydion Wukadinović." In *Z dziejów germanistyki historycznoliterackiej w Polsce: Studia i materiały*, edited by Krzysztof Kuczyński, 93–104. Łódz: Wydawnictwo Uniwersytetu Łódzkiego, 1991.

"Cirkular-Verordnung des k.k. böhmischen Guberniums vom 5. April 1848: In Bezug auf die Lehr- und Lernfreiheit, Z. 89." Reprinted in *Österreichische Zeitschrift für Rechts- und Staatswissenschaft* 3, no. 4 (1848): 92.

Coen, Deborah. *Vienna in the Age of Uncertainty.* Chicago: University of Chicago Press, 2007.

Cohen, Gary B. *Education and Middle-Class Society in Imperial Austria, 1848–1918.* West Lafayette, IN: Purdue University Press, 1996.

———. "Nationalist Politics and the Dynamics of State and Civil Society in the Habsburg Monarchy, 1867–1914." *Central European History* 40, no. 2 (2007): 241–78.

Cole, Laurence. *Für Gott, Kaiser und Vaterland: Nationale Identität der deutschsprachigen Bevölkerung Tirols 1860–1914.* Frankfurt am Main: Campus, 2000.

Čornejová, Ivana. "Výběr a výkvět národa: Snahy o úpravu platů profesorů na pražké univerzitě na přelomu 19. a 20. století." In *Magister Noster: Sborník Statí Věnovaných in Memoriam Prof.Phdr. Janu Havránkovi, Csc. / Festschrift in Memoriam Prof. Phdr. Jan Havránek, Csc*, edited by Michal Svatoš, Luboš Velek, and Alice Velková, 175–82. Prague: Karolinum, 2003.

Čornovol, Ihor. *Pol'sko-ukraïns'ka uhoda 1890–1894 rr.* L'viv: L'vivs'ka Akademija Mistectv, 2000.

Csáky, Moritz. "Altes Universitätsviertel: Erinnerungsraum, Gedächtnisort." In *Die Verortung von Gedächtnis*, edited by Moritz Csáky and Peter Stachel, 257–77. Vienna: Passagen, 2001.

———. "Culture as a Space of Communication." In *Understanding Multiculturalism: The Habsburg Central Europe Experience*, edited by Johannes Feichtinger and Gary B. Cohen, 187–208. New York: Berghahn Books, 2014.

———. *Das Gedächtnis der Städte: Kulturelle Verflechtungen—Wien und die urbanen Milieus in Zentraleuropa*. Vienna: Böhlau, 2010.

Curtius, Georg. *Die Sprachvergleichung in ihrem Verhältniss zur classischen Philologie*. 2nd ed. Berlin: Wilhelm Besser, 1848.

Ćwik, Władysław. "Polacy w Innsbrucku do 1939 roku." In *Polonia i przyjaciele Polski w Austrii*, edited by Władysław Stanisław Kucharski, 63–77. Lublin: Multico, 1995.

Ćwirko-Godycki, Michał. *Izydor Kopernicki*. Poznań: Poznańskie Towarzystwo Przyjaciół Nauk, 1948.

Cybulski, Napoleon. *Uwagi nad ministeryalną reformą studyów i egzaminów lekarskich (Odbitka z Przeglądu Lekarskiego 1899)*. Cracow: Drukarnia Uniwersytetu Jagiellońskiego, 1900.

———. *W sprawie reformy studyów lekarskich (Odbitka z Przeglądu Lekarskiego 1896)*. Cracow: Drukarnia Uniwersytetu Jagiellońskiego, 1896.

Czubiński, Antoni. "Utworzenie Uniwersytetu Poznańskiego i jego rozwój do 1922 roku." In *Alma Mater Poznaniensis: W 80 rocznicę utworzenia Uniwersytetu w Poznaniu*, edited by Przemysław Hauser, Tomasz Jasiński, and Jerzy Topolski, 123–34. Poznań: Wydawnictwo Naukowe Uniwersytetu im. Adama Mickiewicza w Poznaniu, 1999.

Dąbrowski, Mirosław. "Wojciech Urbański—polski badacz zjawisk elektrycznych." *Nauka*, no. 2 (2007): 151–60.

Dabrowski, Patrice M. *Commemorations and the Shaping of Modern Poland*. Bloomington: Indiana University Press, 2004.

Dacko, Iwan. "Hundert Jahre ukrainische Theologiestudenten in Innsbruck: Vergangenheitsanalyse und Zukunftsperspektiven." Lecture held in Innsbruck, 13 November 1999. http://www.canisianum.at/zeitbegrenzt/100-jahre-ukrainer-vortrag.html.

Dańczura-Dynowska, Jolanta. "Alfred Biesiadecki, wybitny polski histopatolog i jego wkład do dermatologii." *Archiwum Historii i Filozofii Medycyny* 58, no. 4 (1995): 379–97.

Danylenko, Andrii. "Myxajlo Luckaj: A Dissident Forerunner of Literary Rusyn?" *Slavonic and East European Review* 87, no. 2 (2009): 201–26.

Datsenko, Ihor. "Das Wörterbuch der juridischpolitischen Terminologie (1851) als Index der galizischen Realität." In *Galizien: Fragmente eines diskursiven Raums*, edited by Doktoratskolleg Galizien, 131–41. Innsbruck: StudienVerlag, 2009.

Deák, István. *Beyond Nationalism: A Social and Political History of the Habsburg Officer Corps, 1848–1918*. Oxford: Oxford University Press, 1990.

Deak, John. *Forging a Multinational State: State Making in Imperial Austria from the Enlightenment to the First World War*. Stanford, CA: Stanford University Press, 2015.

Dejmek, Jindřich. "Učňovská a vandrovní léta Edvarda Beneše (1904–1913)." *Moderní Dějiny* 11 (2003): 5–64.

Dekleva, Tatjana. "Ustanovitev univerze v Ljubljani." In *Ustanovitev Univerze v Ljubljani v letu 1919*, edited by Jože Ciperle and Tatjana Dekleva, 29–42. Ljubljana: Univerza, 2009.

Denkschrift der kaiserlichen Akademie der Wissenschaften: Philosophisch-historische Klasse. Vol. 1. Vienna: Aus der Kaiserlich-Königlichen Hof- und Staatsdruckerei, 1850.

Denkschrift für die Vervollständigung der k.k. Karl-Franzens-Universität zu Graz. Graz: Josef A. Kienreich, 1861.

Die deutsche Karl-Ferdinands-Universität in Prag unter der Regierung Sr. Majestät des Kaisers Franz Josef I. Prague: J. G. Calve, 1899.

Dietl, Józef. *O reformie szkół krajowych*. Vol. 1, *Stanowisko szkoły, rada szkolna krajowa, język*. Cracow: Drukarnia Uniwersytetu Jagiellońskiego, 1865.

Un diplomate étranger qui a longtemps résidé dans ce pays [Liudwig Tengoborskij]. *De l'instruction publique en Autriche*. Paris: Cousin, 1841.

Dobija-Witczak, Olga. "František Tomáš Bratránek (1815–1884)." In *Złota Księga Wydziału Filologicznego*, edited by Jan Michalik and Wacław Walecki, 50–58. Cracow: Księgarnia Akademicka, 2000.

Drunen, Jeroen van. "'A Sanguine Bunch': Regional Identification in Habsburg Bukovina, 1774–1919." PhD diss., University of Amsterdam, 2013.

Dumreicher, Armand. *Die Verwaltung der Universitäten seit dem letzten politischen Systemwechsel in Oesterreich*. Vienna: Alfred Hölder, 1873.

Dumreicher, Johann-Heinrich. *Über die Nothwendigkeit von Reformen des Unterrichtes an den medicinischen Facultäten Österreichs*. Vienna: Alfred Hölder, 1878.

———. *Zeitfragen betreffend die Universität mit besonderer Beziehung auf Medicin, von Carl Rokitansky*. Vienna: Carl Gerold, 1864.

Durand, Charles. *La mise en place des monopoles du savoir*. Paris: Harmattan, 2001.

Ďurčanský, Marek. "Członkostwo zagraniczne polskich i czeskich uczonych w akademiach narodowych: PAU i ČAVU." *Prace Komisji Historii Nauki PAU* 6 (2004): 177–211.

———. "Władysław Mieczysław Kozłowski a česká filosofie." In *Slovanství a věda v 19. a 20. století*, edited by Hana Barvíková, 55–76. Prague: Archiv Akademie věd České republiky, 2005.

Dybiec, Julian. *Uniwersytet Jagielloński 1918–1939*. Cracow: Polska Akademia Umiejętności, 2000.

———. "Związki Akademii Umiejętności w Krakowie z nauką czeską i słowacką w latach 1873–1918." In *Z dziejów polsko-czeskich i polsko-słowackich kontaktów naukowych*, edited by Irena Jasiukowa-Stasiewicz and Jan Janko, 34–61. Warsaw: Polska Akademia Nauk, 1990.

Ebert, Andreas D. *Jüdische Hochschullehrer an preußischen Universitäten (1870–1924): Eine quantitative Untersuchung mit biografischen Skizzen.* Frankfurt am Main: Mabuse, 2008.

Editorial. *Časopis Společnosti wlastenského museum w Čechách* 1, no. 1 (1827): 3–8.

Egglmaier, Herbert H. *Ernst Freiherr von Feuchtersleben als Bildungspolitiker.* Klagenfurt: Abteilung für Historische und Vergleichende Pädagogik, Universität Klagenfurt, 2000.

———. "Graf Thun und das Rechtsstudium an den oberitalienischen Universitäten Padua und Pavia." In *Forschungen zur Geschichte des Alpen-Adria-Raumes: Festgabe für em.o.Univ.-Prof. Dr. Othmar Pickl zum 70. Geburtstag*, edited by Herwig Ebner, Paul Roth, and Ingeborg Wiesflecker-Friedhuber, 107–18. Graz: Selbstverlag des Instituts für Geschichte der Karl-Franzens-Universität, 1997.

Ehrhard, Albert. *Der Katholizismus und das zwanzigste Jahrhundert im Lichte der kirchlichen Entwicklung der Neuzeit.* 1901. Stuttgart: Jos. Roth'sche Verlagsbuchhandlung, 1902.

Eiselberg, Anton von. *Lebensweg eines Chirurgen: Eine Autobiographie aus der großen Zeit der Wiener Medizin 1860–1937.* Vienna: Christian Brandstätter, 1991.

Eisenbach, Arthur. *Emancypacja Żydów na ziemiach polskich 1785–1870 na tle europejskim.* Warsaw: Państowy Instytut Wydawniczy, 1988.

Elster, Ludwig. *Die Gehälter der Universitäts-Professoren und die Vorlesungshonorare unter Berücksichtigung der in Aussicht genommenen Reformen in Preussen und Oesterreich.* Jena: Fischer, 1897.

d'Elvert, Christian. *Geschichte der Studien- Schul- und Erziehungs-Anstalten in Mähren und Österr. Schlesien insbesondere der olmützer Universität, in den neueren Zeiten.* Brno: Rohrer, 1857.

"Das Ende der deutschen Universität in Prag." *Wiener Medizinische Wochenschrift* 32, no. 7 (1882): 197–98.

Engelbrecht, Helmut. *Geschichte des österreichischen Bildungswesens: Erziehung und Unterricht auf dem Boden Österreichs.* Vol. 5, *Von 1918 bis zur Gegenwart.* Vienna: Österreichischer Bundesverlag, 1988.

———. *Geschichte des österreichischen Bildungswesens.* Vol. 3, *Von der frühen Aufklärung bis zum Vormärz.* Vienna: Österreichischer Bundesverlag, 1984.

Engemann, Iris. *Die Slowakisierung Bratislavas: Universität, Theater und Kultusgemeinden 1918–1948.* Wiesbaden: Harrassowitz, 2012.

Entwurf des Österreichischen Reichstages welcher in der Zeit vom 22. Juli 1848 bis 4. März 1849 getagt hat, zuerst in Wien, ab dem 22. November 1848 in Kremsier ("Kremsier Entwurf"). http://www.verfassungen.de/at/kremsier49.htm.

Estreicher, Karol. *Dr. Jan Kanty Szlachtowski: Rzecz czytana na posiedzeniu c. k. Towarzystwa Naukowego Krakowskiego dnia 5 lutego 1872 roku.* Cracow: Kraj, 1872.

Eulenburg, Franz. *Der "Akademische Nachwuchs": Eine Untersuchung über die Lage und die Aufgaben der Extraordinarien und Privatdozenten*. Leipzig: B. G. Teubner, 1908.

Eulner, Hans-Heinz. *Die Entwicklung der medizinischen Spezialfächer an den Universitäten des deutschen Sprachgebietes*. Stuttgart: Enke, 1970.

Evans, R. J. W. "Széchenyi and Austria." In *History and Biography: Essays in Honour of Derek Beals*, edited by T. C. W. Blanning and David Cannadine, 113–41. Cambridge: Cambridge University Press, 2002.

Exner, Franz. *Die Stellung der Studierenden auf der Universität: Eine Rede gehalten an der k.k. Universität zu Prag*. Prague: Gottlieb Haase Söhne, 1837.

Farkas, Gábor. "A birodalmi felsőoktatás modernizációja: A kultuszminiszter oktatási tervezete (1849–1854)." In *Österreichisch-ungarische Beziehungen auf dem Gebiet des Hochschulwesens*, edited by Zsolt K. Lengyel, József Zsigmond Nagy, and Gábor Ujváry, 163–78. Székesfehérvár: Kodolányi János Főiskola, Eötvös Loránd Tudományegyetem Könyvtára, 2010.

Fasora, Lukáš, ed. *Moravské vyrovnání z roku 1905: Možnosti a limity národnostního smíru ve střední Evropě*. Brno: Matice moravská, 2006.

Feferman, Anita Burdman, and Solomon Feferman. *Alfred Tarski: Life and Logic*. Cambridge: Cambridge University Press, 2004.

Feichtinger, Johannes. "Der erinnerte Feind und nationale Integration: Zentraleuropa im langen 19. Jahrhundert." In *Der erinnerte Feind*, edited by Johannes Feichtinger and Johann Heiss, 300–322. Vienna: Mandelbaumverlag, 2013.

———. "Positivismus in der österreichischen Philosophie: Ein historischer Blick auf die frühe Positivismusrezeption." *Newsletter Moderne: Zeitschrift des Spezialforschungsbereichs Moderne—Wien und Zentraleuropa um 1900* 7, no. 2 (2004). http://www-gewi.kfunigraz.ac.at/moderne/heft13f.htm.

———. *Wissenschaft als reflexives Projekt: Von Bolzano über Freud zu Kelsen. Österreichische Wissenschaftsgeschichte 1848–1938*. Bielefeld: Transcript, 2010.

Feichtinger, Johannes, and Franz Leander Fillafer. "Leo Thun und die Nachwelt: Der Wissenschaftsreformer in der österreichischen Geschichts- und Kulturpolitik des 19. und 20. Jahrhunderts." In *Die Thun-Hohenstein'schen Universitätsreformen 1849–1860: Konzeption—Umsetzung—Nachwirkungen*, edited by Christof Aichner and Brigitte Mazohl, 347–78. Vienna: Böhlau, 2017.

Fellerer, Jan. *Mehrsprachigkeit im galizischen Verwaltungswesen (1772–1914): Eine historisch-soziolinguistische Studie zum Polnischen und Ruthenischen (Ukrainischen)*. Cologne: Böhlau, 2005.

Fellner, Fritz. *". . . ein wahrhaft patriotisches Werk": Die Kommission für Neuere Geschichte Österreichs 1897–2000*. Vienna: Böhlau, 2001.

Feyen, Benjamin, and Ewa Krzaklewska, eds. *The ERASMUS Phenomenon—Symbol of a New European Generation?* Bern: Peter Lang, 2013.

Fichtner, Paula Sutter. "History, Religion, and Politics in the Austrian *Vormärz*." *History and Theory* 10, no. 1 (1971): 33–48.

Fillafer, Franz Leander. "Hermann Bonitz: Philologe, Mitschöpfer der Universitätsreform." In *Universität—Politik—Gesellschaft: 650 Jahre Universität Wien—Aufbruch in das neue Jahrhundert*, edited by Mitchell G. Ash and Josef

Ehmer, 189–95. Vienna: Vienna University Press; Göttingen: V&R Unipress, 2015.

———. "The 'Imperial Idea' and Civilizing Missions." *Historyka: Studia metodologiczne* 42 (2012): 37–60. https://www.academia.edu/15087613/_full_issue_Jan_Surman_Klemens_Kaps_Postcolonial_Galicia_Prospects_and_Possibilities_Historyka_42_2012.

Finkel, Ludwik. "Historya Uniwersytetu Lwowskiego do r. 1869." In *Historya Uniwersytetu Lwowskiego*, edited by Ludwik Finkel and Stanisław Starzyński, 3–351. L'viv: C. K. Uniwersytet Lwowski, 1894.

Finkel, Ludwik, and Stanisław Starzyński. *Historya Uniwersytetu Lwowskiego*. L'viv: C. K. Uniwersytet Lwowski, 1894.

Finnegan, Diarmid. "The Spatial Turn: Geographical Approaches in the History of Science." *Journal of the History of Biology* 41, no. 2 (2008): 369–88.

Fisette, Denis. "Carl Stumpf." In *The Stanford Encyclopedia of Philosophy (Spring 2009 Edition)*, edited by Edward N. Zalta. Stanford, CA: Metaphysics Research Lab, 2009. http://plato.stanford.edu/archives/spr2009/entries/stumpf/.

Fleck, Christian. "Autochthone Provinzialisierung: Universität und Wissenschaftspolitik nach dem Ende der nationalsozialistischen Herrschaft in Österreich." *Österreichische Zeitschrift für Geschichtswissenschaften* 7, no. 1 (1997): 67–92.

Floßmann, Ursula. *Österreichische Privatrechtsgeschichte*. 5th ed. Vienna: Springer, 2005.

Folta, Jaroslav, and Pavel Šišma. "Jan Sobotka, Literatura." In *Významní matematici v českých zemích*. Brno, 2003. http://web.math.muni.cz/biografie/.

Fořtová, Hana, and Doubravka Olšáková. "Úvodní studie." In *Lev Thun—Alexis de Tocqueville (Korespondence 1835–1856)*, edited by Hana Fořtová and Doubravka Olšáková, 11–139. Prague: Oikúmene, 2011.

Frajdl, Jiří. *Zápas s germanizací v období austroslavismu ve Slovinsku a Chorvatsku: studijní texty*. Prague: Křesťanskosociální hnutí ve spolupráci s Obvodní radou Klubu českého pohraničí, 2005.

Frankfurter, Salomon. *Graf Leo Thun-Hohenstein, Franz Exner und Hermann Bonitz: Beiträge zur Geschichte der österreichischen Unterrichtsreform*. Vienna: Alfred Hölder, 1893.

Frankl, Michal. *"Emancipace od židů" Český antisemitismus na konci 19. Století*. Prague: Paseka, 2007.

Franzos, Karl Emil. *Aus Halb-Asien*. Vol. 2. 3rd rev. and exp. ed. Stuttgart: Adolf Bonz, 1889.

Friedman, Rebecca. *Masculinity, Autocracy and the Russian University, 1804–1863*. Basingstoke: Palgrave Macmillan, 2005.

Friedmann, Filip. *Die galizischen Juden im Kampfe um ihre Gleichberechtigung (1848–1868)*. Frankfurt am Main / Łódź: J. Kaufmann / F. Friedmann, 1929.

[Frind, Wenzel]. *Eine eventuelle Theilung (Verdoppelung) der Prager theologischen Fakultät, ist mit dem Prinzipe und der Aufgabe der Kirche vereinbar, Zugleich als Studie über das Verhältnis der Kirche zur Nationalität*. Prague: Ambr. Opitz in Warnsdorf, 1884.

Fritsch, Nina-Sophie. "At the Leading Edge—Does Gender Still Matter? A Qualitative

Study of Prevailing Obstacles and Successful Coping Strategies in Academia." *Current Sociology* 63, no. 4 (2015): 1–19.

———. "Warum Wissenschaftlerinnen die Universität verlassen: Eine biografische Fallanalyse zu Ausstiegsgründen aus dem österreichischen Universitätssystem." *SWS-Rundschau* 54, no. 2 (2014): 159–80.

"Fullerton, George Stuart (1825–1925)." In *Pennsylvania Biographical Dictionary*, edited by Jan Onofrio, vol 1:424–26. St. Clair Shores: Somerset, 1999.

Fumasoli, Tatiana, Gaële Goastellec, and Barbara M. Kehm, eds. *Academic Work and Careers in Europe: Trends, Challenges, Perspectives*. Dordrecht: Springer, 2014.

Gartner, Theodor, and Stephan Smal-Stocki [Stepan Smal'-Stoc'kyj]. *Minoritätsvotum in der vom k.k. Bukowiner Landesschulrathe behufs Regelung der ruthenischen Schulortographie eingesetzten Commission abgegeben im November 1887*. Chernivtsi: Selbstverlag, 1888.

Gawrecki, Dan. "Pokus o založení univerzity v Opavě v roce 1870." *Acta historica et museologica Universitatis Silesianae Opaviensis* 2 (1995): 127–32.

Gehler, Michael. "Studentischer Antisemitismus an der Universität Innsbruck." In *Die Geschichte der Juden in Tirol von den Anfängen bis in die neueste Zeit*, edited by Günther Pallaver, 73–87. Bozen: Sturzflüge, 1986.

Geiger, Roger L. *The History of American Higher Education: Learning and Culture from the Founding to World War II*. Princeton, NJ: Princeton University Press, 2014.

Gellner, Ernest. *Language and Solitude: Wittgenstein, Malinowski, and the Habsburg Dilemma*. Cambridge: Cambridge University Press, 1998.

Gesemann, Gerhard. "Bericht über das Studienjahr 1933/34." In *Anstelle der Feierlichen Inauguration des Rektors der Deutschen Universität in Prag für das Studienjahr 1934/35*, 5–25. Prague: Selbstverlag, 1935.

Głąbiński, Stanisław. *Zamach na Uniwersytet Polski we Lwowie*. L'viv: Nakładem Autora, 1902.

Gliboff, Sander. "Evolution, Revolution, and Reform in Vienna: Franz Unger's Ideas on Descent and Their Post-1848 Reception." *Journal of the History of Biology* 31, no. 2 (1998): 179–209.

Goebl, Hans. "Theodor Gartner und das typologische Denken seiner Zeit." In *Akten der Theodor Gartner-Tagung*, edited by Guntram A. Plangg and Maria Iliescu, 13–23. Innsbruck: Institut für Romanistik, 1987.

Goethe, Johann Wolfgang von. *Faust: A Tragedy*. Translated by Taylor Bayard. London: Ward, Lock, 1889.

Goll, Jaroslav. *Der Hass der Völker und die österreichischen Universitäten*. Prague: Bursík & Kohout, 1902.

———. *Rozdělení Pražské university Karlo-Ferdinandovy roku 1882 a počátek samostatné University české*. Prague: Nakl. Klubu historického, 1908.

Goller, Peter. "Georg Schenach (1843–1853): 'Ein Vermittlungsphilosoph.'" In *Die Lehrkanzeln für Philosophie an der Philosophischen Fakultät der Universität Innsbruck 1848 bis 1945*, edited by Peter Goller, 17–34. Innsbruck: Wagner, 1989.

———. *Naturrecht, Rechtsphilosophie oder Rechtstheorie? Zur Geschichte der Rechtsphilosophie an Österreichs Universitäten (1848–1945)*. Frankfurt am Main: Peter Lang, 1997.

———. "Ein starkes Stück: Die versuchte Habilitation eines kommunistischen Juden . . . Universitäten im Lichte politischer u. rechtlicher Willkür am Beispiel des Habilitationsverfahrens von Karl Horovitz <1892–1958> an der Wiener Universität 1923–1925." *Jahrbuch 1998: Dokumentationsarchiv des österreichischen Widerstandes* (1998): 111–34.

Gomperz, Theodor. *Essays und Erinnerungen.* Stuttgart: Deutsche Verlags-Anstalt, 1905.

Gordin, Michael. *Scientific Babel: How Science Was Done before and after Global English.* Chicago: University of Chicago Press, 2015.

Granovetter, Mark. "The Strength of Weak Ties: A Network Theory Revisited." *Sociological Theory* 1 (1983): 201–33.

Grassauer, Ferdinand. "Universitätsbibliothek." In *Geschichte der Wiener Universität von 1848 bis 1898: Als Huldigungsfestschrift zum 50-jähr. Regierungsjubiläum seiner k.k. apostolischen Majestät des Kaisers Franz Josef I*, edited by Akademischer Senat der Wiener Universität, 367–80. Vienna: Alfred Hölder, 1898.

Graversen, Ebbe K. "Human Capital Mobility—a Comparable Knowledge Indicator for the Nordic Countries." In *Science and Technology Indicators for the Nordic Countries 2000: A Collection of Articles*, edited by Kirsten Wille Maus, 69–81. Copenhagen: Nordic Council of Ministers, 2001. http://www.oecd.org/dataoecd/35/25/2100200.pdf.

Gregory, Derek, Peter Meusburger, and Laura Suarsana. "Power, Knowledge, and Space: A Geographical Introduction." In *Geographies of Knowledge and Power*, edited by Peter Meusburger, Derek Gregory, and Laura Suarsana, 1–18. Dordrecht: Springer Science + Business Media, 2015.

Grosser, Otto. "Bericht über das Studienjahr 1929/30 [sic!]." In *Die feierliche Inauguration des Rektors der Deutschen Universität in Prag für das Studienjahr 1929/30*, 1–11. Prague: Selbstverlag, 1930.

Grünberg, Carl. "Anton Menger: Sa vie—son œuvre." *Revue d'histoire des doctrines économiques et socials* 2 (1909): 1–40.

Grzybowski, Waldemar. "Sprachwissenschaftliche Biographien aus Ostdeutschland: Deutsche und polnische Linguisten im sprachwissenschaftlichen und sprachpolitischen Diskurs der Zwischenkriegszeit." In *Grenzüberschreitende Biographien zwischen Ost- und Mitteleuropa: Wirkung—Interaktion—Rezeption*, edited by Tobias Werner, 285–309. Vienna: Peter Lang, 2009.

Guereña, Jean-Louis. "L'université espagnole vers 1900." In *Sozialer Raum und akademische Kulturen: Studien zur europäischen Hochschul- und Wissenschaftsgeschichte im 19. und 20. Jahrhundert = A la recherche de l'espace universitaire européen*, edited by Jürgen Schriewer, Edwin Keiner, and Christophe Charle, 113–31. Frankfurt am Main: Peter Lang, 1993.

Gürtler, Adolf. *Deutsche Hochschulnot in Österreich: Referat erstattet in der Monatsversammlung der Vereinigung deutscher Hochschullehrer in Graz am 12. Februar 1913.* Graz: Im Selbstverlage der Vereinigung, 1913.

Gutachten und Anträge zur Reform der medicinischen Studien- und Rigorosen-Ordnung. Vienna: k.k. Universitäts-Buchdruckerei Karl Gorischek, 1894.

Haas, Georg-Emmanuel. *Ueber den Zustand der österr. Universitäten mit besonderer Beziehung auf die Wiener Hochschule.* Augsburg: Schmid, 1853.

Haas, Theodor. "Die sprachlichen Verhältnisse der Juden in Österreich." *Zeitschrift für Demographie und Statistik der Juden* 11, no. 1 (1915): 1–12.

Hacohen, Malachi Haim. "The Culture of Viennese Science and the Riddle of Austrian Liberalism." *Modern Intellectual History* 6, no. 2 (2009): 369–96.

———. *Karl Popper, the Formative Years, 1902–1945: Politics and Philosophy in Interwar Vienna*. Cambridge: Cambridge University Press, 2002.

———. "Kosmopoliten in einer ethnonationalen Zeit? Juden und Österreicher in der Ersten Republik." In *Das Werden der Ersten Republik: . . . der Rest ist Österreich*, edited by Helmut Konrad and Wolfgang Maderthaner, 1:281–316. Vienna: Gerold, 2008.

Hafner, Stanislaus. "Geschichte der österreichischen Slawistik." In *Beiträge zur Geschichte der Slawistik in nichtslawischen Ländern*, edited by Josef Hamm and Günther Wytrzens, 11–88. Vienna: Österreichische Akademie der Wissenschaften, 1985.

Halbwidl, Dieter. "Life and Times of Count Leo Thun." *Mensch—Wissenschaft—Magie* 30 (2013): 111–31.

Hálek, Jan. "Jan Peisker (1851–1933)." *Akademický bulletin*, no. 5 (2008): 5.

Haller, Rudolf. "Gibt es eine *österreichische Philosophie*?" In *Fragen zu Wittgenstein und Aufsätze zu Österreichischer Philosophie*, edited by Rudolf Haller, 31–43. Amsterdam: Rodopi, 1986.

Hamann, Brigitte. "Anton Gindely—ein altösterreichisches Schicksal." In *Nationale Vielfalt und gemeinsames Erbe in Mitteleuropa: Vorträge anläßlich der Verleihung des Anton-Gindely-Preises für Geschichte der Donaumonarchie*, edited by Erhard Busek and Gerald Stourzh, 27–37. Vienna: Verlag für Geschichte und Politik; Munich: Oldenbourg, 1990.

Hanák, Péter. *The Garden and the Workshop: Essays on the Cultural History of Vienna and Budapest*. With a preface by Carl E. Schorske. Princeton, NJ: Princeton University Press, 1998.

Hankiewicz, Klemens [Klymentij/Klym Hankevyč]. *Zarys filozofii słowiańskiej*. Rzeszów: Wydawnictwo Uniwersytetu Rzeszowskiego, 2011.

Hartel, Wilhelm. *Festrede zur Enthüllung des Thun-Exner-Bonitz Denkmals gehalten in der 1. Hauptsitzung der 42. Versammlung deutscher Philologen und Schulmänner am 24. Mai 1893*. Vienna: A. Holzhausen, 1893.

Hausleitner, Marianna. *Die Rumänisierung der Bukowina: Die Durchsetzung des nationalstaatlichen Anspruchs Grossrumäniens 1918–1944*. Munich: Oldenbourg, 2001.

Hausmann, Guido. "Das Territorium der Ukraine: Stepan Rudnyc'kys Beitrag zur Geschichte räumlich-territorialen Denkens über die Ukraine." In *Die Ukraine: Prozesse der Nationsbildung*, edited by Andreas Kappeler, 145–58. Cologne: Böhlau, 2011.

Havelka, Miloš. "A Hundred Years of the 'Czech Question' and the Czech Question a Hundred Years On." *Czech Sociological Review* 3, no. 1 (1995): 7–19.

Havránek, Jan. "Anton Gindely, ein Historiker, der zwischen zwei Nationen stand." *Acta Universitatis Carolinae—Philosophica et Historica* 3 (1993): 101–9.

———. "Česká univerzita v jednání rakouských uřadů do roku 1881." *AUC-HUCP* 22, no. 1 (1982): 33–89.

———. "Karolinum v revoluci 1848." *AUC-HUCP* 26, no. 2 (1986): 35–75.

———. "Nineteenth Century Universities in Central Europe: Their Dominant Position in the Science and Humanities." In *Bildungswesen und Sozialstruktur in Mitteleuropa im 19. und 20. Jahrhundert = Education and Social Structure in Central Europe in the 19th and 20th Centuries*, edited by Victor Karady and Wolfgang Mitter, 9–26. Vienna: Böhlau, 1990.

———. "Univerzita Karlova, rozmach a perzekuce." In *Dějiny Univerzity Karlovy IV (1918–1990)*, edited by Jan Havránek and Zdeněk Pousta, 19–59. Prague: Karolinum, 1997.

Heindl, Waltraud. *Josephinische Mandarine: Bürokratie und Beamte in Österreich*. Vol. 2, *1848 bis 1914*. Vienna: Böhlau, 2013.

———. "Der Liberalismus scheiterte. Scheiterte der Liberalismus?" In *"Dürfen's denn das?" Die fortdauernde Frage zum Jahr 1848*, edited by Sigurd Paul Scheichl and Emil Brix, 85–95. Vienna: Passagen, 1999.

———. "Universitäten und Eliten im österreichischen Vormärz." *Études danubiennes* 3, no. 2 (1987): 117–28.

———. "Universitätsreform—Gesellschaftsreform: Bemerkungen zum Plan eines 'Universitätsorganisationsgesetzes' in den Jahren 1854/55." *Mitteilungen des Österreichischen Staatsarchivs* 35 (1983): 134–49.

———. "Universitätsreform und politisches Programm: Die Sprachenfrage an der Universität Krakau im Neoabsolutismus." *Österreichische Osthefte* 20 (1978): 79–98.

Heiss, Hans, and Thomas Götz. *Am Rand der Revolution: Tirol 1848/49*. Bozen: Folio, 1998.

Helcel, Antoni Zygmunt. *Starodawne prawa polskiego pomniki poprzedzone wywodem historyczno krytycznym tak zwanego Prawodawstwa Wiślickiego Kazimiérza Wielkiego w texcie ze starych rękopism krytycznie dobranym*. Cracow and Warsaw: Drukarnia C. K. Uniwersytetu; nakładem Księgarni Gustawa Sennewalda, 1856.

———. "O teoretycznéj i praktycznéj oświacie." *Kwartalnik naukowy, wydawany w połączeniu prac miłośników umiejętności* 1, no. 1 (1835): 1–26.

———. *Uwagi nad kwestyą językową w szkołach i uniwersytetach Galicyi i Krakowa, osnowane na liście odręcznym Jego C. K. Apostolskiej Mości z dnia 20 października 1860 r.* Cracow: D. E. Friedlein, 1860.

Helfert, Joseph-Alexander Freiherr von. "Katastrofa: Vlastní zkušenosti a pamětí. (dokončení)" *Osvěta: Listy pro rozhled v umĕne, vĕde a politice* 27, no. 8 (1897): 678–85.

———. *Die sprachliche Gleichberechtigung in der Schule und ihre verfassungsmäßige Behandlung*. Prague: Tempsky, 1861.

———. *Über Nationalgeschichte und den gegenwärtigen Stand ihrer Pflege in Österreich*. Prague: Calve, 1853.

Helvig, Jean-Michel. "Antijudaisme ou antisémitisme le procés Sebastian Brunner—Ignaz Kuranda (10 mai 1860)." PhD diss., University of Paris IV, Sorbonne, 1996.

Herben, Jan. *Masarykova sekta a Gollova škola*. Prague: Pokrok, 1912.

Herman, Eufemiusz Józef. *Historia neurologii polskiej*. Wrocław: Zakład narodowy im. Ossolińskich, Polska Akademia Nauk, 1975.

Hermann, Tomáš. *Emanuel Rádl a české dějepisectví: Kritika českého dějepisectví ve sporu o smysl českých dějin*. Prague: Univerzita Karlova v Praze, Filozofická fakulta, 2002.

Hermann-Otavský, Karl. "Zpráva o studijním roku 1918–19." In *Inaugurace Rektora dne 3. Prosince 1919*, 3–29. Prague: Univerzita Karlova, 1919.

Heřmanová, Darina. "Vedec prof. MUDr. J. Buchtala." In *Oddelenie histórie medicíny a zdravotníctva Ústavu sociálneho lekárstva LF UK v Bratislave 1990–2000*, edited by Eulália Sedláčková, Ľudmila Pavlíková, Darina Heřmanová, and Helena Rapantová, 48–51. Bratislava: Juga, 2001.

Himka, John Paul. "The Construction of Nationality in Galician Rus': Icarian Flights in Almost All Directions." In *Intellectuals and the Articulation of the Nation*, edited Ronald Grigor Suny and Michael D. Kennedy, 109–64. Ann Arbor: University of Michigan Press, 1999.

Hirsch, Marianne, and Leo Spitzer. *Ghosts of Home: The Afterlife of Czernowitz in Jewish Memory*. Berkeley: University of California Press, 2010.

Hlaváčková, Ludmila. "Budování klinických pracovišť české lékařské fakulty v době rozdělen pražské univerzity: II. Snahy o vybudování českých klinických pracovišť před rozdělením univerzity." *Sbornik lékařský* 85, no. 4 (1983): 110–15.

———. "Čeština v medicíně a na pražské lékařské fakultě (1784–1918)." In *Místo Národnich Jazyku ve Výuce, Vědě a Vzdělání v Habsburské Monarchii 1867–1918 / Position of National Languages in Education, Educational System and Science of the Habsburg Monarchy 1867–1918*, edited by Harald Binder, Luboš Velek, and Barbora Křivohlavá, 327–44. Prague: Výzkumné centrum pro dějiny vědy, 2003.

———. "Čtyřnásobné Jubileum Bohumila Eiselta (1831–1908)." *Časopis lékařův českých* 150, no. 5 (2011): 619–23.

Hlaváčková, Ludmila, and Petr Svobodný. *Dějiny pražskich lékarských fakult 1348–1990*. Prague: Karolinum, 1993.

Hnilica, Jiři. "Kulturní a intelektuální výměna mezi Čechami a Francií 1870–1925." *AUC-HUCP* 45, nos. 1–2 (2005): 95–126.

———. "'Missionaries of Science and Spirit': The Topic of Visiting Professors Illustrated by the Example of Czechoslovak-French Cultural Relations in the Years 1918–1948." In *Conference Proceedings: Universities in Central Europe—Crossroads of Scholars from All over the World*, edited by Milada Sekyrková and Petr Cajthaml, 207–13. Prague: Institute of the History of Charles University and Archive of Charles University Prague, 2012.

Hofer, Hans-Georg. "Jenseits der männlichen Abwehrfront: Arztberuf und Medizinstudium im Spiegel der Neurastheniedebatten um 1900." In *Medizinerinnen*, edited by Sonia Horn and Ingrid Arias, 45–53. Vienna: Verlagshaus der Ärzte, 2003.

Hofer, Veronika, and Michael Stöltzner. "What Is the Legacy of Austrian Academic Liberalism?" *NTM Zeitschrift für Geschichte der Wissenschaften, Technik und Medizin* 20, no. 1 (2012): 31–42.

Hoffmann, Roland. *T. G. Masaryk und die tschechische Frage: Nationale Ideologie und politische Tätigkeit bis zum Scheitern des deutsch-tschechischen Ausgleichsversuchs vom Februar 1909*. Munich: Oldenburg, 1988.

Höflechner, Walter. "Die Auswirkungen politischer und kultureller Veränderungen auf Forschungsorientierung und Wissenschaftsorganisation." In *Geschichte der österreichischen Humanwissenschaften*, vol. 1, *Historischer Kontext, wissenssoziologische Befunde und methodologische Voraussetzungen*, edited by Karl Acham, 149–214. Vienna: Passagen, 1999.

———. *Das Historische Seminar der Karl-Franzens-Universität Graz*. Graz: Elektronische Veröffentlichungen aus dem Zentrum für Wissenschaftsgeschichte der Karl-Franzens-Universität Graz, 2007. http://gams.uni-graz.at:8080/fedora/get/o:wissg-hs/bdef:Navigator.fs/get/.

———. *Materialien zur Entwicklung der Physik und ihrer "Randfächer" Astronomie und Meteorologie an den österreichischen Universitäten 1752–1938*. Graz: Institut für Geschichte der Karl-Franzens-Universität Graz, 2002. http://www-gewi.uni-graz.at/wissg/gesch_der_physik.

———. *Die österreichische Rektorenkonferenz 1911–1938, 1945–1969*. Vienna: Verein zur Förderung der Rektorenkonferenz, 1993.

———. "Zum Einfluss des deutschen Hochschulwesens auf Österreich in den Jahren 1875–1914." In *Wissenschaftsgeschichte und Wissenschaftspolitik im Industriezeitalter: Das "System Althoff" in historischer Perspektive*, edited by Bernhard vom Brocke, 155–83. Hildesheim: Lax, 1991.

Höfler, Karl Adolf Constantin, and Paul Joseph Šafařík. *Glagolitische Fragmente*. Prague: Gottlieb Haase Söhne, 1857.

Hofmann, Andreas C. "Deutsche Universitätspolitik im Vormärz zwischen Zentralismus, 'Transstaatlichkeit' und 'Eigenstaatlichkeitsideologien' (1815/19 bis 1848)." PhD diss., Ludwig-Maximilians-Universität München, 2014.

———. "Studium, Universität und Staat in Bayern 1825 bis 1848: Eine Skizze der Universitätspolitik Ludwigs I." *Aventinus Bavaric* 2 (2006). http://www.aventinus-online.de/no_cache/persistent/artikel/7760/.

Hohes Abgeordnetenhaus: Petition der Privatdozenten-Vertreter der österreichischen Hochschulen in Angelegenheit von Standesfragen. Vienna: Fischer, 1910.

Höhne, Steffen, ed. *August Sauer (1855–1926): Ein Intellektueller in Prag im Spannungsfeld von Kultur- und Wissenschaftspolitik*. Vienna: Böhlau, 2008.

Holovac'kyj, Ivan. *Ivan Horbačevs'kij: 1854–1942. Žyttiepysno-bibliografičnyj narys*. L'viv: Naukove Tovarystvo im. Ševčenka, 1995.

[Holovac'kyj, Jakiv (Яків Головацький), Markiyan Šaškevyč (Маркіян Шашкевич), and Ivan Vahylevyč (Іван Вагилевич)]. *Rusalka Dněstrovaja. Ruthenische Volkslieder*. Buda: Pysmom Korol. Vseučylyšča Peštanskoho, 1837.

Hruševs'kyj, Mychajlo. "Do svitu: Z podorožnich vražen'." *Visnyk*, no. 44 (1909): 11–23.

———. "Sprava ukraïns'kych katedr i nasi naukovi potreby." In *Mychajlo Hruševs'kyj: Tvory v 50 tomach, serija "Suspil'no-polityčni tvory,"* vol. 1, *1894–1907*, edited by Jaroslav Daškevyč, Ihor Hyryč, and Pavlo Sochań, 458–84. L'viv: Cvit, 2002.

Hrytsak, Iaroslav. "Ruslan, Bohdan and Myron: Three Constructed Identities among Galician Ruthenians/Ukrainians, 1830–1914." In *Extending the Borders of Russian History: Essays in Honor of Alfred J. Rieber*, edited by Marsha Siefert, 97–112. Budapest: Central European University Press, 2003.

Huber, Alfons. *Geschichte der Gründung und der Wirksamkeit der Kaiserlichen*

Akademie der Wissenschaften während der ersten fünfzig Jahre ihres Bestandes. Vienna: C. Gerold's Sohn, 1897.

Huber, Kurt A. "Die Prager theologischen Fakultäten von 1883/1891 bis 1945." In *Die Teilung der Prager Universität 1882 und die intellektuelle Desintegration in den böhmischen Ländern*, edited by Ferdinand Seibt, 37–54. Munich: Oldenbourg, 1984.

Huber, Wolfgang. "Zur Geschichte der Wissenschaften." In Österreich 1918–1938, edited by Erika Weinziert and Kurt Skalnik, 2:559–88. Vienna: Styria, 1983.

Hubmann, Bernhard. "Carl Ferdinand Peters (1825–1881): Beitrag zu seiner Biographie." *Berichte der Geologischen Bundesanstalt* 53 (2001): 31–48.

Hueppe, Ferdinand. *Kulturbedürfnisse und Universitäten in Oesterreich (Sonderdruck aus Heft 221/21 der "Hochschul-Nachrichten")*. Munich: Akademischer Verlag, 1909.

Hulewicz, Jan. *Akademia Umiejętności w Krakowie 1873–1918: Zarys Dziejów*. Wrocław: Ossolineum, 1958.

"107 Hörer der Medizin an der k. k. Universität zu Prag . . ." In *Stenografická zpráva XXXV. sezení třetího ročního zasedání sněmu českého od roku 1861, dne 11. května 1864 / Stenographischer Bericht über die XXXV. Sitzung der dritten Jahres-Session des böhmischen Landtages vom Jahre 1861, am 11. Mai 1864* (=Stenographische Berichte des böhmischen Landtages. Dritte Jahressession 1864 / *Stenografické zprávy sněmu Českého. Třetí zasedání* 1864). Prague: Statthalterei-Buchdruckerei, 1864.

Immeles, Henryk. *Antysemityzm w Austryi*. L'viv: Jedność, 1910.

Iordan, Iorgu. "Besprechungen und Anzeigen: V. Rumänisch." *Zeitschrift für romanische Philologie* 54, no. 2 (1934): 363–79.

Isaievych, Iaroslav. *Voluntary Brotherhood: Confraternities of Laymen in Early Modern Ukraine*. Edmonton: Canadian Institute of Ukrainian Studies, 2006.

Iwanicki, Mieczysław. *Oświata i szkolnictwo ukraińskie w Polsce w latach 1918–1939*. Siedlce: Wyższa Szkoła Pedagogiczna, 1975.

Jagić, Vatroslav. "A Survey of Slavistic Studies." *Slavonic Review* 1, no. 1 (1922): 40–58.

Jagodzińska, Agnieszka. "Asymilacja, czyli bezradność historyka: O krytyce terminu i pojęcia." In *Wokół akulturacji i asymilacji Żydów na ziemiach polskich*, edited by Konrad Zieliński, 15–31. Lublin: Uniwersytet Marii Curie-Skłodowskiej, 2010.

Janko, Jan, and Soňa Štrbáňová. "Uplatnění nového českého přírodovědného názvosloví na českých vysokých školách v průběhu 19. století." In *Místo Národnich Jazyku ve Výuce, Vědě a Vzdělání v Habsburské Monarchii 1867–1918 / Position of National Languages in Education, Educational System and Science of the Habsburg Monarchy 1867–1918*, edited by Harald Binder, Luboš Velek, and Barbora Křivohlavá, 297–312. Prague: Výzkumné centrum pro dějiny vědy, 2003.

———. *Věda Purkyňovy doby*. Prague: Academia, 1988.

Janowski, Maciej. *Narodziny inteligencji (1750–1831)*. Vol. 1 of *Dzieje inteligencji polskiej do roku 1918*, edited by Jerzy Jedlicki. Warsaw: Neriton, 2012.

Januszewski, Bernhard. "Geneza Katedry języków i literatur słowiańskich we Wrocławiu." *Rocznik Wrocławski* 7/8 (1963/64): 12–76.

Jastrzębski, Jarosław. "Reforma Jędrzejewicza w państwowym szkolnictwie akademickim II Rzeczypospolitej: Wzmocnienie prerogatyw władz państwowych." *Prace historyczne* 138 (2011): 159–75.

Jedlička, Jaroslav. "Eduard Albert—Jaroslav Goll—50 listů korespondence." *AUC-HUCP* 13, nos. 1–2 (1973): 215–67.

Jedlicki, Jerzy. *Błędne koło 1832–1864*. Vol. 2 of *Dzieje inteligencji polskiej do roku 1918*, edited by Jerzy Jedlicki. Warsaw: Neriton, 2012.

Jessen, Ralph, and Jacob Vogel, eds. *Wissenschaft und Nation in der europäischen Geschichte*. Frankfurt am Main: Campus, 2002.

Jindra, Jiří. "České vědecké společnosti exaktních věd v 19. století." In *Místo Národnich Jazyku ve Výuce, Vědě a Vzdělání v Habsburské Monarchii 1867–1918 / Position of National Languages in Education, Educational System and Science of the Habsburg Monarchy 1867–1918*, edited by Harald Binder, Luboš Velek, and Barbora Křivohlavá, 401–20. Prague: Výzkumné centrum pro dějiny vědy, 2003.

Jirásek, Arnold. *Eduard Albert: Pokus o kroniku a rozbor života, práce i významu E. Alberta, učiněný ke stému výročí jeho narození (20. ledna 1941)*. Prague: Československá chirurgická společnost, 1946.

Jiroušek, Bohumil. "Historik W. W. Tomek." In *W. W. Tomek, historie a politika (1818–1905): Sborník příspěvků královéhradecké konference k 100. výročí úmrtí W. W. Tomka*, edited by Miloš Řezník, 15–29. Pardubice: Univerzita Pardubice, 2006.

———. "Jazyky v životě a díle Antonína Rezka." In *Místo Národnich Jazyku ve Výuce, Vědě a Vzdělání* v Habsburské Monarchii 1867–*1918 / Position of National Languages in Education, Educational System and Science of the Habsburg Monarchy 1867–1918*, edited by Harald Binder, Luboš Velek, and Barbora Křivohlavá, 531–35. Prague: Výzkumné centrum pro dějiny vědy, 2003.

———. "Mimořádná profesura Josefa Pekaře (ve světle vztahů Antonína Rezka a Jaroslava Golla)." In *Proměny elit v moderní době*, edited by Milena Lenderová, Zdeněk Bezecný, and Jiří Kubeš, 167–78. České Budějovice: Historický ústav Jihočeské univerzity, 2003.

Jochen, August. *Sonderaktion Krakau: Die Verhaftung der Krakauer Wissenschaftler am 6. November 1939*. Hamburg: Hamburger Edition, HIS Verlagsgesellschaft, 1997.

Johler, Reinhard. "Vielfalt." In *Habsburg Neu Denken: Vielfalt und Ambivalenz in Zentraleuropa. 30 Kulturwissenschaftliche Stichworte*, edited by Johannes Feichtinger and Heidemarie Uhl, 229–36. Vienna: Böhlau, 2016.

John, Michael, and Albert Lichtblau. "Jewries in Galicia and Bukovina, in Lemberg and Czernowitz: Two Divergent Examples of Jewish Communities in the Far East of the Austro-Hungarian Monarchy." In *Jewries at the Frontier: Accommodation, Identity, Conflict*, edited by Sander L. Gilman and Milton Shain, 29–66. Urbana: University of Illinois Press, 1999.

———. "Mythos 'deutsche Kultur': Jüdische Gemeinden in Galizien und der Bukowina. Zur unterschiedlichen Ausformung kultureller Identität." In *Studien zur Geschichte der Juden in Österreich*, edited by Martina Keil and Eleonore Lappin, 81–121. Bodenheim: Philo, 1997.

Johnson, Owen. *Slovakia, 1918–1938: Education and the Making of a Nation*. Boulder, CO: East European Monographs, 1985.

Jöns, Heike. "Academic Travel from Cambridge University and the Formation of Centres of Knowledge, 1885–1954." *Journal of Historical Geography* 34, no. 2 (2008): 338–62.

———. "Grenzüberschreitende Mobilität und Kooperation in den Wissenschaften: Deutschlandaufenthalte US-amerikanischer Humboldt-Forschungspreisträger aus einer erweiterten Akteursnetzwerkperspektive." PhD diss., Heidelberg University, 2003.

Judson, Pieter M. "L'Autriche-Hongrie était-elle un empire?" *Annales: Histoire, Sciences Sociales* 6, no. 3 (2008): 563–96.

———. *Exclusive Revolutionaries: Liberal Politics, Social Experience, and National Identity in the Austrian Empire, 1848–1914*. Ann Arbor: University of Michigan Press, 1996.

———. *Guardians of the Nation: Activists on the Language Frontiers of Imperial Austria*. Cambridge, MA: Harvard University Press, 2006.

———. *The Habsburg Empire: A New History*. Cambridge, MA: Belknap Press of Harvard University Press, 2016.

———. *Wien brennt! Die Revolution von 1848 und ihr liberales Erbe*. Vienna: Böhlau, 1998.

Judson, Pieter M., and Marsha L. Rozenblit, eds. *Constructing Nationalities in East Central Europe*. New York: Berghahn Books, 2005.

Jung, Julius. *Julius Ficker (1826–1902): Ein Beitrag zur deutschen Gelehrtengeschichte*. Innsbruck: Wagner'sche Universitäts-Buchhandlung, 1907.

Kačkan, Volodymyr. "Anatol' Vachnjanyn i joho rol' v rozvytku kul'tury Zachidnoï Ukraïny." *Narodna tvorčist' ta etnohrafija*, nos. 1–2 (2003): 15–24.

Kačmar, Volodymyr. "Problema ukraïnśkoho universytetu u Lv'ovi v kinci XIX-na počatku XX st.: Suspil'no-polityčnyj aspekt." PhD diss., Ivan Franko State University in L'viv, 1999.

———. "Problema zasnuvannja ukraïns'koho universytetu u L'vovi n rubeži XIX–XX st. u konteksti nacional'noho žyttja." *Visnyk L'vivskoho universytetu: Seria žurnalistyka* 26 (2004). http://www.lnu.edu.ua/faculty/jur/publications/visnyk26/Statti_Kachmar.htm.

———. "Sprava ukraïns'koho universytetu na tli pol's'ko-ukraïns'kych superečnostej u Halyčyni 1901–1908 rr." *Problemy slov'janoznavstva* 42 (2002): 47–58.

———. "Suspil'no-polityčnyje vidlunnja secesiï ukraïns'kich studentiv z L'vivskoho Universitetu v grudni 1901 roku." *Visnyk L'vivskoho universytetu: Seria istoryčna* 34 (1999): 289–99.

Kadletz-Schöffel, Hedwig. "Metternich und die Wissenschaften." PhD diss., University of Vienna, 1989.

Kalousek, Joseph. *Geschichte der Königlichen Böhmischen Gesellschaft der Wissenschaften: Sammt einer kritischen Übersicht ihrer Publicationen aus dem Bereiche der Philosophie, Geschichte und Philologie, aus Anlass des hundertjährigen Jubelfestes der Gesellschaft 1884*. Prague: Königlich Böhmische Gesellschaft der Wissenschaften, 1885.

———. *Karel IV, otec vlasti: Ku pětistoleté památce jeho umrtí*. Prague: Jos. R. Vilímek, 1878.

———. Über die Nationalität Karl*'s IV: Entgegnung auf einen von Prof. Dr. J. Loserth unter demselben Titel in den "Mittheilungen des Vereins für die Geschichte der Deutschen Böhmen" veröffentlichten Aufsatz*. Prague: W. Nagel, Selbstverlag, 1879.

Kamiš, Adolf. "Tschechisch-deutsche Beziehungen in der politischen Terminologie zu Beginn der konstitutionellen Zeit." In *Deutsch-tschechische Beziehungen im Bereich der Sprache und Kultur: Aufsätze und Studien*, edited by Bohuslav Havránek and Rudolf Fischer, 71–83. Berlin: Akademie-Verlag, 1968.

Kanjuh, Vladimir. "Đorđe Joannović—First Serbian Oncologist-Scientist (on the Occasion of the 75 Anniversary of His Tragic Death)." *Archive of Oncology* 16, no. 12 (2008): 18–21.

Kappeler, Andreas. *Kleine Geschichte der Ukraine*. 2nd ed. Munich: Beck, 2000.

———. *Russland und die Ukraine: Verflochtene Biographien und Geschichten*. Vienna: Böhlau, 2012.

Karolewicz, Grażyna. *Nauczyciele akademiccy Katolickiego Uniwersytetu Lubelskiego w okresie międzywojennym*. Vol. 1. Lublin: Katolicki Uniwersytet Lubelski, 1996.

Der katholische Universitäts-Verein zu Salzburg. *Dem hochwürdigsten Episkopate und den Katholiken Österreichs zum 25-jähr. Vereins-Jubiläum 1884 bis 1909 gewidmet vom Zentral-Ausschuß*. Salzburg: Verlag der katholischen Universitäts-Verein, 1909.

Kernbauer, Alois. *Das Fach Chemie an der Philosophischen Fakultät der Universität Graz*. Graz: Akademische Druck- und Verlagsanstalt, 1985.

———. "Konzeptionen der Österreich-Geschichtsschreibung 1848–1938." In *Forschungen zur Geschichte des Alpen-Adria-Raumes: Festgabe für em.o.Univ.-Prof. Dr. Othmar Pickl zum 70. Geburtstag*, edited by Herwig Ebner, Paul Roth, and Ingeborg Wiesflecker-Friedhuber, 255–73. Graz: Selbstverlag des Instituts für Geschichte der Karl-Franzens-Universität, 1997.

Kienzl, Lisa. *Nation, Identität und Antisemitismus: Der deutschsprachige Raum der Donaumonarchie 1866 bis 1914*. Göttingen: V&R Unipress, 2014.

Kilcher, Andreas B. "Sprachendiskurse im jüdischen Prag um 1900." In *Franz Kafka im sprachnationalen Kontext seiner Zeit: Sprache und nationale Identität in öffentlichen Institutionen der böhmischen Länder*, edited by Marek Nekula, Ingrid Fleischmann, and Albrecht Greule, 61–86. Vienna: Böhlau, 2007.

King, Jeremy. "The Nationalisation of East Central Europe: Ethnicism, Ethnicity and Beyond." In *Staging the Past: The Politics of Commemoration in Habsburg Central Europe, 1848 to the Present*, edited by Maria Bucur and Nancy Meriwether Wingfield, 112–52. West Lafayette, IN: Purdue University Press, 2001.

Kink, Rudolf. *Geschichte der kaiserlichen Universität zu Wien*. 2 vols. Vienna: Carl Gerold & Sohn, 1854.

Kisch, Guido. *Die Prager Universität und die Juden, 1348–1848: Mit Beiträgen zur Geschichte des Medizinstudiums*. Amsterdam: John Benjamins, 1969.

Klanska [Kłańska], Maria. "Moritz Rappaport als Brückenbauer zwischen der deutschen, jüdischen und polnischen Kultur." *Trans: Internet-Zeitschrift für*

Kulturwissenschaften 15 (2004). http://www.inst.at/trans/15Nr/03_5/klanska15.htm.

Klebesberg, Raimund von. *Innsbrucker Erinnerungen, 1902–1952*. Innsbruck: Wagner, 1953.

Kleinert, Andreas. "Lampa, Anton." In *Neue Deutsche Biographie*, edited by Historische Kommission bei der Bayrischen Akademie der Wissenschaften, 13:453–54. Berlin: Duncker & Humblot, 1982.

"Kleinwächter, Friedrich von." In *Das Österreichische Biographische Lexikon*, edited by Österreichische Akademie der Wissenschaften, 3:392–93. Vienna: Verlag der Österreichischen Akademie der Wissenschaften, 1965.

Klemun, Marianne. "Wissenschaft und Kolonialismus—Verschränkungen und Figurationen." *Wiener Zeitschrift zur Geschichte der Neuzeit* 9, no. 1 (2009): 1–12.

Klíma, Arnošt. *Češi a Němci v revoluci 1848–1849*. Prague: Univerzita Karlova, 1988.

Klin, Eugeniusz, and Jaromir Loužil, eds. *František Tomáš Bratranek—ein Polonophiler Mittler zwischen den Nationen*. Zielona Góra: Wydawictwo Wyższej Szkoły Pedagogicznej, 1987.

Kohlrausch, Martin, Katrin Steffen, and Stefan Wiederkehr. "Expert Cultures in Central Eastern Europe: The Internationlization of Knowledge and the Transformation of Nation States since World War I—Introduction." In *Expert Cultures in Central Eastern Europe: The Internationalization of Knowledge and the Transformation of Nation States since World War I*, edited by Martin Kohlrausch, Katrin Steffen, and Stefan Wiederkehr, 9–30. Osnabrück: fibre, 2010.

Koja, Stephan, ed. *Gustav Klimt: Der Beethoven-Fries und die Kontroverse um die Freiheit der Kunst*. Munich: Prestel, 2006.

Kokešová, Helena. *Eduard Albert: Příspěvek k životopisu a edice korespondence*. Prague: Scriptorium: Výzkumné centrum pro dějiny vědy, Masarykův ústav Akademie věd České republiky, 2004.

Kolář, Pavel. *Geschichtswissenschaft in Zentraleuropa: Die Universitäten Prag, Wien und Berlin um 1900*. Leipzig: Leipziger Universitätsverlag, 2008.

Kolberg, Oskar, ed. *Lud: Jego zwyczaje, sposób życia, mowa, podania, przysłowia, obrzędy, gusła, zabawy, pieśni, muzyka i tańce*. 86 vols. Various publishers, 1857–.

Kolowrat, Franz Graf von. "An die vaterländischen Freunde der Wissenschaften." *ISIS oder Encyclopädische Zeitung*, nos. 1–4 (1818): 1100–1103.

———. "An die vaterländischen Freunde der Wissenschaften." *Verhandlungen der Gesellschaft des vaterländischen Museums in Böhmen* 1 (1823): 18–24.

Konarski, Stanisław. "'Zimmermanniada' w Uniwersytecie Jagiellońskim (1910–1911)." In *Postępowe tradycje młodzieży akademickiej w Krakowie*, edited by Henryk Dobrowolski, Mirosław Frančić, and Stanisław Konarski, 135–204. Cracow: Wydawnictwo Literackie, 1962.

Konrád, Ota. *Dějepisectví, germanistika a slavistika na Německé univerzitě v Praze 1918–1945*. Prague: Karolinum, 2011.

———. "Die Deutsche Universität in Prag in der Ersten Tschechoslowakischen Republik—zwischen Kooperation und Konfrontation." In *Universitäten in Zeiten des Umbruchs: Fallstudien über das mittlere und östliche Europa im 20.*

Jahrhundert, edited by Elmar Schübl and Harald Heppner, 29–42. Berlin: Lit, 2011.

Konstantinović, Zoran. "'In der praktischen Philosophie ist ihm die Menschenwürde das Wesentlichste . . .': Zu Franz Miklošič' Bewerbung um die Innsbrucker Lehrkanzel für Philosophie [1838/39]." *Österreichische Osthefte* 33 (1991): 95–103.

Kořalka, Jiří. *František Palacký (1798–1876): Der Historiker der Tschechen im österreichischen Vielvölkerstaat*. Vienna: Verlag der österreichischen Akademie der Wissenschaften, 2007.

Korb, Axel-Johannes. *Kelsens Kritiker: Ein Beitrag zur Geschichte der Rechts- und Staatstheorie (1911–1934)*. Tübingen: Mohr Siebeck, 2010.

Kossov [Kossow], Marjan. *Úkoly kulturní spolupráce československo-polské*. Mladá Boleslav: Národní Zájmy, [1932].

Kostlán, Antonín, Jan Janko, and Ladislav Niklíček. "Prosazování myšlenky akademie věd v českých zemích raného nvověku (16.–18. století)." In *Bohemia docta: K historickým kořenům vědy v českých zemích*, edited by Alena Míšková, Martin Franc, and Antonín Kostlán, 145–71. Prague: Academia, 2010.

Kostlán, Antonín, and Alice Velková, eds. *Wissenschaft im Exil: Tschechoslowakei als Kreuzweg 1918–1989*. Prague: Výzkumné centrum pro dějiny vědy, 2004.

Kostner, Maria. "Die Geschichte der italienischen Universitätsfrage in der österreichisch-ungarischen Monarchie von 1864–1914." PhD diss., University of Innsbruck, 1970.

Kotasék, Alfréd. *Karel Pawlík (1849–1914), Osobnost a dílo*. Prague: Univerzita Karlova, 1994.

Kotschubinský [Kočubinskij], Al[eksandr] [Александр Кочубинский]. "Miklosich und Šafárik: Ein Beitrag zu ihren wechselseitigen Beziehungen." *Archiv für slavische Philologie* 25 (1903): 621–27.

Kozik, Jan. *The Ukrainian National Movement in Galicia, 1815–1849*. Edmonton: Canadian Institute of Ukrainian Studies, University of Alberta, 1986.

Kożuchowski, Adam. *The Afterlife of Austria-Hungary: The Image of the Habsburg Monarchy in Interwar Europe*. Pittsburgh, PA: University of Pittsburgh Press, 2013.

Kozuschek, Waldemar. *Jan Mikulicz-Radecki 1850–1905: Współtwórca Nowoczesnej Chirurgii / Johann von Mikulicz-Radecki: Mitbegründer der modernen Chirurgie*. Wrocław: Wydawnictwo Uniwersytetu Wrocławskiego, 2003.

Kraml, Jiří, and Jiří Duchoň. "110 let české lékařské chemie." In *110 let české lékařské chemie a biochemie*, edited by Marie Balíková, 7–16. Prague: Galén, 1994.

Kratochvíl, Milan. *Jan Evangelista Purkyně a jeho snahy o reformu české školy*. Prague: Státní pedagogické nakladatelství, 1987.

Kril', Mychajlo. "Nevidomi lysty Ivana Holovac'koho do Antona Petruševyča." *Ukraïna moderna* 2–3 (1999): 350–72.

———. "Pedahohična i naukova dijal'nist' Jozefa Maussa." In *Wielokulturowe środowisko historyczne Lwowa w XIX i XX w. / Bahatokul'turne istoryčne seredovyšče L'vova v XIX i XX stolittjach*, edited by Jerzy Maternicki and Leonid Zaszkilniak, 4:93–106. L'viv: L'vivs'kij nacional'nyj un-t im. Ivana Franka, Žešuvs'kyj un-t, 2006.

———. "Z čes'koï korespondenciï Jakova Holovac'koho." *Problemy slov'janoznavstva* 50 (1999): 217–22.

Křivský, Pavel. *Augustin Smetana*. Prague: Univerzita Karlova, 1990.

Krofta, Kamil. "Anton Gindely." *Zprávy zemského archivu Království českého* 3 (1910): 147–395.

Krones, Franz Xaver Ritter von Marchland. *Geschichte der Karl-Franzens Universität in Graz: Festgabe zur Feier ihres dreihundertjährigen Bestandes*. Graz: Leuschner und Lubensky, 1886.

Krueger, Rita. *Czech, German, and Noble: Status and National Identity in Habsburg Bohemia*. Oxford: Oxford University Press, 2009.

Kryp'jakevyč, Ivan. "Istoryčno-filosofična sekcija NTŠ pod kerivnyctvom Mychajla Hruševśkoho u 1894–1913 rokach." *Zapysky Naukovoho tovarystv imeni Ševčenka: Praci istoryko-filosofs'koï sekciï* 122 (1991): 392–411.

Krzoska, Markus. "Ein Wissenschaftler zwischen Elfenbeinturm und Öffentlichkeit: Der Lemberger Rechtshistoriker Oswald Balzer (1859–1933)." In *Beruf und Berufung: Geschichtswissenschaft und Nationsbildung in Ostmittel- und Südosteuropa*, edited by Markus Krzoska and Hans-Christian Maner, 217–38. Münster: Lit, 2005.

Kulczykowski, Mariusz. *Żydzi—studenci Uniwersytetu Jagiellońskiego w dobie autonomicznej Galicji (1867–1918)*. Cracow: Księgarnia Akademicka, 1995.

Kuprian, Hermann. "'Machen Sie diesem Skandal ein Ende. Ihre Rektoren sind eine nette Gesellschaft.' Modernismusdiskussion, Kulturkampf und Freiheit der Wissenschaft: Die Wahrmund-Affäre." In *Politische Affären und Skandale in Österreich: Von Mayerling bis Waldheim*, edited by Michael Gehler and Hubert Sickinger, 99–127. 2nd ed. Vienna: Thaur, 1996.

Kutschker, Johann. "Erklärung des Kanzlers der k.k. Universität zu Wien über die Bitte der protestantisch-theologischen Fakultät um Einverleibung in die genannte Hochschule, abgegeben in der Sitzung Venerabilis Consistorii am 12. Mai 1863." In *Der katholische Charakter der Wiener Universität: Eine Denkschrift der theologischen Facultät*, edited by Anton Horny, 151–63. Vienna: Verlag der Mechitharisten-Congregations-Buchhandlung, 1863.

Kyjan, Oleksandr. *Volodymyr Antonovyč: Istoryk j orhanizator Kyïvss'koï istoryčnoï školy*. Kiev: Instytut istoriï Ukraïny Nacional'noï Akademiï Nauk Ukraïny, 2005.

Lambert, David, and Alan Lester, eds. *Colonial Lives across the British Empire: Imperial Careering in the Long Nineteenth Century*. Cambridge: Cambridge University Press, 2006.

Landau, Richard. *Geschichte der jüdischen Ärzte*. Berlin: Karger, 1895.

Lane, Hugo Victor. "State Culture and National Identity in a Multi-ethnic Context: Lemberg, 1772–1914." PhD diss., University of Michigan, 1999.

Lange, Ludwig. *Die klassische Philologie in ihrer Stellung zum Gesammtgebiete der Wissenschaften und in ihrer inneren Gliederung: Eine Antrittsvorlesung, gehalten am 24. April 1855 in Prag*. Prague: Calve, 1855.

Lanser, Edith. "Friedrich Jodl: Von Feuerbach zur Gesellschaft für ethische Kultur."

Newsletter Moderne: Zeitschrift des Spezialforschungsbereichs Moderne—Wien und Zentraleuropa um 1900 6, no. 2 (2003): 16–20.

Lappin, Eleonore. "Die Czernowitzer Sprachkonferenz (1908) und der Streit um die jüdische Nationalsprache." In *Minikosmos Bukowina—Kulturleistungen eines Landstriches: Beiträge der internationalen wissenschaftlichen Konferenz anlässlich des 130. Jahrestages der Czernowitzer Universität am 3.–4.10.2005 / Mini-kosmos Bukovyny: Material Mižnarodnoï konferenciï, pryscjačenoï 130-riččjo* Černivec'*koho un-tu (3–4 žovtnja 2005 r.)*, edited by Österreich Kooperation, Naukovo-doslidnyj Centr Bukovynoznavstva pry Bukowina Zentrum an der Nationalen Jurij-Fedkowitsch Universität Czernowitz, and Avstrijška kooperacia Černivećkomu nacional'nomu universyteti im. Jorija Fed'kovyča, 260–70. Chernivtsi: Selena Bukowina/Zelena Bukovyna, 2006.

Latour, Bruno. *Science in Action: How to Follow Scientists and Engineers through Society.* Cambridge, MA: Harvard University Press, 1987.

Lazarenko, J. K. *300 rokiv L'vivs'koho uniwersytetu.* L'viv: Vydavnyctvo L'vivs'koho uniwersytetu, 1961.

Leeb, Hedda. "Geschichte der Universität Innsbruck von 1898 bis 1908." 2 vols. PhD diss., University of Innsbruck, 1967.

Leitner, Rainer. "Das Reformerwerk von Exner, Bonitz und Thun: Das österreichische Gymnasium in der zweiten Hälfte des 19. Jahrhunderts—Kaderschmiede der Wiener Moderne." In *Zwischen Orientierung und Krise: Zum Umgang mit Wissen in der Moderne*, edited by Sonja Rinhofner-Kreidl, 17–69. Vienna: Böhlau, 1998.

Lemberg, Hans. "Universität oder Universitäten in Prag—und der Wandel der Lehrsprache." In *Universitäten in nationaler Konkurrenz: Zur Geschichte der Prager Universitäten im 19. und 20. Jahrhundert*, edited by Hans Lemberg, 19–32. Munich: Oldenbourg, 2003.

Lemeškin, Ilja, and Jolanta Zabarskaitė, eds. *Lituanistinis Augusto Schleicherio palikimas./Das lituanistische Erbe August Schleichers.* Vol. 1. Vilnius: Lietuvių kalbos institutas, 2008.

Lentze, Hans. *Andreas Freiherr von Baumgartner und die Thunsche Studienreform.* Vienna: Rohrer, 1959.

———. "Graf Thun und die deutsche Rechtsgeschichte." *Mitteilungen des Instituts für Österreichische Geschichte* 63 (1955): 500–521.

———. *Die Universitätsreform des Ministers Graf Leo Thun-Hohenstein.* Vienna: Verlag der Österreichischen Akademie der Wissenschaften, 1962.

Leo XIII. "Quod Votis: Encyclical of Pope Leo XIII on the Proposed Catholic University." 30 April 1902. http://www.vatican.va/holy_father/leo_xiii/encyclicals/documents/hf_l-xiii_enc_30041902_quod-votis_en.html.

Lesky, Albin. "Gomperz, Theodor." In *Neue Deutsche Biographie*, edited by Historische Kommission bei der Bayrischen Akademie der Wissenschaften, 6:641–42. Berlin: Duncker & Humblot, 1964.

Lesky, Erna. *The Vienna Medical School of the 19th Century.* Translated by L. Williams and I. S. Levi. Baltimore: Johns Hopkins University Press, 1976.

Lewicki, Jakub. *Między tradycją a nowoczesnością: Architektura Lwowa lat 1893–1918*. Warsaw: Oficyna Wydawnicza Towarzystwa Opieki nad Zabytkami; Neriton, 2005.

Lewicki, Karol. "Katedra Literatury Polskiej na Uniwersytecie Jagiellońskim w latach 1803–1848." In *Dzieje Katedry Historii Literatury Polskiej*, edited by Tadeusz Ulewicz, 43–83. Cracow: Wydawnictwo Literackie, 1966.

Lhotsky, Alphons. "Das Ende des Josephinismus: Epilegomena zu Hans Lenzes Werk über die Reformen des Ministers Grafen Thun." *Mitteilungen des Österreichischen Staatsarchivs* 15 (1962): 526–49.

Lichtenberger-Fenz, Brigitte. *". . . deutscher Abstammung und Muttersprache": Österreichische Hochschulpolitik in der Ersten Republik*. Vienna: Geyer Edition, 1990.

———. "Österreichs Universitäten und Hochschulen—Opfer oder Wegbereiter der nationalsozialistischen Gewaltherrschaft? Am Beispiel der Universität Wien." In *Willfährige Wissenschaft: Die Universität Wien 1938–1945*, edited by Gernot Heiß, Siegfried Mattl, Sebastian Meissl, Edith Saurer, and Karl Stuhlpfarrer, 3–15. Vienna: Verlag für Gesellschaftskritik, 1989.

Liebeschütz, Hans. *Das Judentum im deutschen Geschichtsbild von Hegel bis Max Weber*. Tübingen: J. C. B. Mohr (Paul Siebeck), 1967.

Linsbauer, Karl, Ludwig Linsbauer, and Leopold Ritter von Portheim, eds. *Wiesner und seine Schule: Ein Beitrag zur Geschichte der Botanik. Festschrift anläszlich der dreiszigjährigen Bestandes des Pflanzenphysiologischen Institutes der Wiener Universität*. Vienna: A. Hölder, 1903.

Lipiński, Krzysztof. "Wilhelm Creizenach (1851–1919)." In *Uniwersytet Jagielloński: Złota Księga Wydziału Filologicznego*, edited by Jan Michalik and Wacław Walecki, 107–15. Cracow: Księgarnia Akademicka, 2000.

Liske, Xawer. *Der angebliche Niedergang der Universität Lemberg: Offenes Sendeschreiben an das Reichsrathsmitglied Herrn Eduard Dr. Suess, Prof. an der Universität Wien*. L'viv: Gubrynowicz & Schmidt, 1876.

Litsch, Karel. "F. L. Rieger o univerzitní otázce v českém sněmu roku 1866." *AUC-HUCP* 22, no. 1 (1982): 27–33.

———. "Zur Rechtsstellung der Prager Universitätsprofessoren in der ersten Hälfte des 19. Jahrhunderts." In *Bildungsgeschichte, Bevölkerungsgeschichte, Gesellschaftsgeschichte in den böhmischen Ländern und in Europa: Festschrift für Jan Havránek zum 60. Geburtstag*, edited by Hans Lemberg, Karel Litsch, Richard Georg Plaschka, and Györgi Ránki, 1–17. Vienna: Verlag für Geschichte und Politik, 1988.

Litwin, Dorota. "Udział Polaków w naczelnych organach państwa austriackiego w okresie dualistycznej monarchii." In *Polacy w austriackim parlamencie w 130. rocznicę Klubu Polskiego*, edited by Władysław Kucharski, 103–24. Lublin: Multico, 1997.

Livezeanu, Irina. *Cultural Politics in Greater Romania: Regionalism, Nation Building, and Ethnic Struggle, 1918–1930*. Ithaca, NY: Cornell University Press, 2000.

Livingstone, David N. *Putting Science in Its Place: Geographies of Scientific Knowledge*. Chicago: University of Chicago Press, 2003.

Lotman, Yuri M. *Universe of the Mind: A Semiotic Theory of Culture*. Bloomington: Indiana University Press, 2000.

Lotze, Hermann. *Briefe und Dokumente*. Edited by Reinhardt Pester and Ernst Wolfgang Orth. Würzburg: Königshausen & Neumann, 2003.

Loužil, Jaromír. "Franz Thomas Bratraneks Leben und Philosophie." *Bohemia* 13 (1972): 182–210.

———. *Ignác Jan Hanuš: Studie s ukázkami z díla*. Prague: Melantrich, 1971.

Löw, Martina. "The Constitution of Space: The Structuration of Spaces through the Simultaneity of Effect and Perception." *European Journal of Social Theory* 11, no. 1 (2008): 25–49.

Luft, Robert. "'Politische Professoren' in Böhmen 1861–1914." In *Bildungsgeschichte, Bevölkerungsgeschichte, Gesellschaftsgeschichte in den böhmischen Ländern und in Europa: Festschrift für Jan Havránek zum 60. Geburtstag*, edited by Hans Lemberg, Litsch Karel, Richard Georg Plaschka, and Györgi Ránki, 286–306. Vienna: Verlag für Geschichte und Politik, 1988.

Łypacewicz, Wacław. *Polnisch-Tschechische Beziehungen*. Warsaw: Polnisches Institut fuer Zusammenarbeit mit dem Auslande, 1936.

Maciak, Dariusz. *Próba porozumienia polsko-ukraińskiego w Galicji w latach 1888–1895*. Warsaw: Wydawnictwo Uniwersytetu Warszawskiego, 2006.

Mahler, Arthur. *Rede gehalten in der Universitäts-Debatte zum Dringlichkeitsantrag Prof. Masaryk's im oesterreichischen Abgeordnetenhaus, am 3. Dezember 1907*. Vienna: Unitas, 1908.

Maisel, Thomas. *Alma Mater auf den Barrikaden: Die Universität Wien im Revolutionsjahr 1848*. Vienna: Wiener Universitätsverlag, 1998.

Majewski, Przemysław. "Gwardie narodowe w Galicji w czasie Wiosny Ludów." In *Per aspera ad astra: Materiały z XVI Ogólnopolskiego Zjazdu Historyków Studentów w Krakowie 16–20 kwietnia 2008*, vol. 7, *Historia Polski pod zaborami*, edited by Adam Świątek, 31–48. Cracow: Koło Naukowe Historyków Uniwersytetu Jagiellońskiego, AT Group, 2008.

Malfèr, Stefan. "Die Sprachenfrage und der verstärkte Reichsrat von 1860." In *A Lajtán innen és túl / Jenseits und diesseits der Leitha: Elektronische Festschrift für Éva Somogyi zum 70. Geburtstag*, edited by Imre Ress and Dániel Szabó, 93–118. Budapest: Magyar Tudományos Akadémia Törtenettudományi Intezet, 2007.

Małecki, Antoni. *Gramatyka języka polskiego: Większa*. L'viv: Nakładem Autora, 1863.

Malíř, Jiří. "Otázka založení druhé české univerzity a její nacionální a politická instrumentalizace." In *Místo Národnich Jazyku ve Výuce, Vědě a Vzdělání v Habsburské Monarchii 1867–1918 / Position of National Languages in Education, Educational System and Science of the Habsburg Monarchy 1867–1918*, edited by Harald Binder, Luboš Velek, and Barbora Křivohlavá, 247–73. Prague: Výzkumné centrum pro dějiny vědy, 2003.

Mamić, Mile. "Das deutsch-slawische Wörterbuch der juridisch-politischen Terminologie (seine Konzeption und Realisierung)." In *Balten—Slaven—Deutsche: Aspekte und Perspektiven kultureller Kontakte. Festschrift für Friedrich Scholz zum 70. Geburtstag*, edited by Ulrich Obst and Gerhard Ressel, 131–38. Münster: Lit, 1999.

Mandlerová, Jana. "K boji za zřízeni 2. české university w Brně 1882–1918 (Příspěvek o impulzech a struktuře české vědecké politiky na sklonku Rakousko-uherské monarchie)." *AUC-HUCP* 10, no. 1 (1969): 95–116.

———. "K zahraničním cestám učitelů vysokých škol v českých zemích (1888–1918)." *Dějiny věd a techniky* 2, no. 4 (1969): 232–46.

Manteuffel, Tadeusz. *Uniwersytet Warszawski w latach 1915/16–1934/35: Kronika.* Warsaw: Uniwersytet Józefa Piłsudskiego, 1936.

Martinczak, Franciszek. "Geneza zjazdów Związku Lekarzy Słowiańskich." *Archiwum Historii Medycyny* 43, no. 1 (1980): 37–43.

———. "Profesor ortopedii Vitezslav Chlumsky (karta z działalności we Wrocławiu i Krakowie)." *Archiwum Historii Medycyny* 37, no. 1 (1974): 93–101.

Masaryk, Thomas Garrique. "Jak zvelebovati naši litaraturu náukovou? Clánek III." *Athenaeum: Listy pro literaturu a kritiku vědeckou* 2, no. 9 (1885): 270–75.

———. "Jak zvelebovati naši litaraturu náukovou? Článek IV." *Athenaeum: Listy pro literaturu a kritiku vědeckou* 2, no. 12 (1885): 76–78.

———. "O reformě právnických studií." *Athenaeum: Listy pro literaturu a kritiku vědeckou* 9, no. 4 (1892): 97–110.

Maternicki, Jerzy. "Polskie szkoly historyczne we Lwowie w XIX w." In *Wielokulturowe środowisko historyczne Lwowa w XIX i XX w.*, edited by Jerzy Maternicki and Leonid Zaszkilniak, 3:23–45. Rzeszów: Wydawnictwo Uniwersytetu Rzeszowskiego, 2005.

Mates, Pavel, Petr Prucha, and Jan Svatoň. *Vývoj organizace a řízení československých vysokých škol v letech 1918–1983.* Prague: Ústav školských informací při Ministerstvu Školství Česko-Slovenskej republiky, 1984.

Mathes, William Lloyd. "The Struggle for University Autonomy in the Russian Empire during the First Decade of the Reign of Alexander II (1855–1866)." PhD diss., Columbia University, 1966.

Maurer, Trude. "Der Weg zur Mündigkeit: Auslandsaufenthalte rußländischer Wissenschaftler im 19. und frühen 20. Jahrhundert." *Hyperboreus* 10, nos. 1–2 (2004): 60–77.

Mayer, Gottfried. Österreich als katholische Grossmacht*: Ein Traum zwischen Revolution und liberaler Ära.* Vienna: Verlag der Österreichischen Akademie der Wissenschaften, 1989.

Mazohl-Wallnig, Brigitte. "Der Einfluß Bolzanos und der Bolzanisten auf die österreichische Universitätsreform der Jahre 1848/49." In *Bernard Bolzano und die Politik: Staat, Nation und Religion als Herausforderung für die Philosophie im Kontext von Spätaufklärung, Frühnationalismus und Restauration*, edited by Helmut Rumpler, 221–46. Vienna: Böhlau, 2000.

McClellan, James E. *Science Reorganized: Scientific Societies in the Eighteenth Century.* New York: Columbia University Press, 1985.

McGonigle, Thomas C., and James F. Quigley. *A History of the Christian Tradition: From the Reformation to the Present.* Mahwah, NJ: Paulist, 1996.

Mecenseffy, Grete. *Evangelische Lehrer an der Universität Wien.* Vienna: Böhlau, 1967.

Mecherzyński, Karol. *Historya języka niemieckiego w Polsce.* Cracow, 1845.

Meister, Richard. *Entwicklung und Reformen des österreichischen Studienwesens.* Vol. 2, *Dokumente.* Vienna: Böhlhaus Nachf., Kommissionsverlag der Österreichischen Akademie der Wissenschaften, 1963.

———. *Geschichte der Akademie der Wissenschaften in Wien 1847–1947.* Vienna: Adolfs Holzhausens Nachfolger, 1947.

———. "Lehr- und Lernfreiheit in der Thunschen Universitätsreform und in der Gegenwart in Österreich." *Anzeiger der Österreichischen Akademie der Wissenschaften, philologisch-historische Klasse* 15 (1957): 207–32.

Meixner, Adolf. "Professor D. Karl Alfons Penecke zur 100. Wiederkehr seines Geburtstages." *Carinthia* 68, no. 2 (1958): 63–90.

Melik, Vasilij, and Peter Vodopivec. "Die slowenische Intelligenz und die österreichischen Hochschulen 1848–1918." In *Wegenetz europäischen Geistes II: Universitäten und Studenten. Die Bedeutung studentischer Migration in Mittel und Südosteuropa vom 18. bis zum 20. Jahrhundert,* edited by Richard Georg Plaschka and Karlheinz Mack, 134–54. Vienna: Verlag für Geschichte und Politik, 1987.

Mell, Wolfgang-Rüdiger. "Ein rechtsgeschichtlicher Beitrag zur 'Selbstbestimmung und Fremdbestimmung der österreichischen Universitäten.'" In *Selbstbestimmung und Fremdbestimmung der österreichischen Universitäten: Ein Beitrag zur Soziologie der Universität,* edited by Marina Fischer and Hermann Strasser, 1–78. Vienna: Abteilung Soziologie des Instituts für Höhere Studien und Wissenschaftliche Forschung, 1973.

Menzel, Adolf. "Bericht über das Studienjahr 1915/16." In *Die feierliche Inauguration des Rektors der Wiener Universität für das Jahr 1916/17 am 6. November 1916.* Vienna: Selbstverlag, 1916.

Mészáros, Andor. "Maďarština na (buda)peštské univerzitě a v uherském školství v 19. Století." In *Místo Národnich Jazyku ve Výuce, Vědě a Vzdělání v Habsburské Monarchii 1867–1918 / Position of National Languages in Education, Educational System and Science of the Habsburg Monarchy 1867–1918,* edited by Harald Binder, Luboš Velek, and Barbora Křivohlavá, 147–56. Prague: Výzkumné centrum pro dějiny vědy, 2003.

Meusburger, Peter. *Bildungsgeographie: Wissen und Ausbildung in der räumlichen Dimension.* Heidelberg: Spektrum Akademischer Verlag, 1998.

Meusburger, Peter, David N. Livingstone, and Heike Jöns, eds. *Geographies of Science: Academic Mobilities, Knowledge Spaces, and Public Encounters.* Dordrecht: Springer Science + Business Media, 2010.

Meusburger, Peter, Michael Welker, and Edgar Wunder, eds. *Clashes of Knowledge: Orthodoxies and Heterodoxies in Science and Religion.* Dordrecht: Springer Science + Business Media, 2008.

Michalewska, Krzysztofa. "Habilitacje w Uniwersytecie Jagiellońskim 1848–1918."

Zeszyty Naukowe Uniwersytetu Jagiellońskiego: Prace Historyczne 71, no. 12 (1963): 79–133.

———. "Próby utworzenia uniwersytetu ukraińskiego w Polsce 1919–26." PhD diss., Jagiellonian University, Cracow, 1974.

Michler, Werner. *Darwinismus und Literatur: Naturwissenschaftliche und literarische Intelligenz in Österreich, 1859–1914*. Vienna: Böhlau, 1998.

Mikoletzky, Juliane. "Vom Polytechnischen Institut zur Technischen Hochschule: Die Reform des technischen Studiums in Wien, 1850–1875." *Mitteilungen der österreichischen Gesellschaft für Wissenschaftsgeschichte* 15 (1995): 79–100.

Mildenberger, Florian. "'. . . als Conträrsexual und als Päderast verleumdet . . .'—der Prozess um den Naturforscher Theodor Beer (1866–1919) im Jahre 1905." *Zeitschrift für Sexualforschung* 18, no. 4 (2005): 332–51.

Miller, Aleksei I. *The Ukrainian Question: The Russian Empire and Nationalism in the Nineteenth Century*. Budapest: Central European University Press, 2003.

Miller, Aleksei I., and Alfred J. Rieber, eds. *Imperial Rule*. Budapest: Central European University Press, 2004.

Mills, Eric L. *Biological Oceanography: An Early History, 1870–1960*. Toronto: University of Toronto Press, 2012.

Miroslav, Michaela. "Plans for Slovak Autonomy in Hungarian Politics 1918–1920." *Historický časopis: Historického ústavu SAV* 58, suppl. (2010): 53–82.

Miskell, Louise. *Meeting Places: Scientific Congresses and Urban Identity in Victorian Britain*. Farnham: Ashgate, 2013.

Míšková, Alena. "'Heraus aus Prag!'—Pläne für den aufbau und Umzüge deutscher Hochschulen in Böhmen." In *Die Suche nach dem Zentrum: Wissenschaftliche Institute und Bildungseinrichtungen der Deutschen in Böhmen (1800–1945)*, edited by Kristina Kaiserová and Miroslav Kunštát, 119–44. Münster: Waxmann, 2014.

Míšková, Alena, and Michael Neumüller. *Společnost Pro Podporu Německé Vědy, Umění a Literatury v Čechách (Německá Akademie Věd v Praze): Materiály k Dějinám a Inventář Archivního Fondu = Die Gesellschaft zur Förderung deutscher Wissenschaft, Kunst und Literatur in Böhmen (Deutsche Akademie der Wissenschaften in Prag): Materialen zu ihrer Geschichte und Inventar des Archivbestandes: 1891–1945*. Prague: Archiv Akademie věd České republiky, 1994.

Moklak, Jarosław. *W walce o tożsamość Ukraińców: Zagadnienie języka wykładowego w szkołach ludowych i średnich w pracach galicyjskiego Sejmu Krajowego 1866–1892*. Cracow: Historia Iagellonica, 2004.

Molisch, Peter, ed. *Eduard Neussers studentische Erinnerungen aus dem Jahre 1848*. Vienna: Sonderabdruck aus den Mitteilungen des Vereines für Geschichte der Stadt Wien, Vol. XIII/XIV, 1933.

Molnár, László. "A pesti egyetemi orvoskar 1848/49-ben." *Semmelweis Egyetem* 9 (15 March 2008). http://www.orvostortenelem.hu/tankonyvek/tk-05/pdf/3.4.3/molnar_pesti_egyetemi_orvoskar.pdf.

Montgomery, Scott. "Impacts of a Global Language on Science: Are There Disadvantages?" In *Language as a Scientific Tool: Shaping Scientific Language across*

Time and National Traditions, edited by Miles MacLeod, Rocío G. Sumillera, Jan Surman, and Ekaterina Smirnova, 199–218. New York: Routledge, 2016.

Morée, Peter. "Jan Hus as a Threat to the German Future in Central Europe: The Bohemian Reformer in the Controversy between Constantin Höfler and František Palacký." In *The Bohemian Reformation and Religious Practice*, edited by Zdenek V. David and David R. Holeton, 4:295–307. Prague: Main Library, 2002.

Moser, Michael [Mozer, Michael']. "Movnyj svit 'Studium ruthenium.'" In *Ucrainica I. Současná ukrajinistika: Problémy jazyka, literatury a kultury. K 65. narozeninám prof. Josefa Anderše*, 316–25. Olomouc: Universita Palackého, 2004.

———. "Some Viennese Contributions to the Development of Ukrainian Terminologies." In *Ukraine's Re-integration into Europe: A Historical, Historiographical and Political Urgent Issue*, edited by Giovanna Brogi-Bercoff and Giulia Lami, 139–80. Alessandria: Edizioni dell'Orso, 2005.

———. "Die sprachliche Erneuerung der galizischen Ukrainer zwischen 1772 und 1848/1849 im mitteleuropäischen Kontext." In *Contemporary Cultural Studies in Central Europe*, edited by Ivo Pospíšil and Michael Moser, 81–118. Brno: Ústav slavistiky Filozofické fakulty Masarykovy univerzity v Brně, 2004.

Mozer, Michael' [Moser, Michael]. "Josyf Levyc'kyj jak borec' za kul'turu 'ruskoï' (ukraïns'koï) movy." In *Confraternitas: Jobilejnyj zbirnyk na počanu Jaroslava Isajevyča*, edited by Mykola Krykun and Ostap Sereda, 447–60. L'viv: Instytut ukraïnoznavstva im. I. Kryp'jakevyča Nacional'na Akademiya Nauk Ukraïny, 2007.

Mróz, Tomasz. *Wincenty Lutosławski (1863–1954): "Jestem Obywatelem Utopii."* Cracow: Polskia Akademia Umiejętności, 2008.

Mrozowska, Kamilla. "Okres ucisku i daremnych prób wyzwoleńczych, 1833–1850." In *Dzieje Uniwersytetu Jagiellońskiego w latach 1765–1850*, edited by Mirosława Chamcówna and Kamilla Mrozowska, vol. 2, pt. 1, 182–235. Cracow: Uniwersytet Jagielloński, 1965.

Mudryj, Vasyl'. *Zmahannja za ukraïns'kyj universytet v Halyčyni*. 1923. L'viv: Vydavnyctvo Naukovoho Tovarystva imeny Ševčenka, 1999.

Murko, Matthias. *Deutsche Einflüsse auf die Anfänge der böhmischen Romantik*. Graz: Styria, 1897.

———. *Deutsche Einflüsse auf die Anfänge der slavischen Romantik*. 2 vols. Graz: Styria, 1897.

Mühlberger, Kurt. "Das 'Antlitz' der Wiener Philosophischen Fakultät in der zweiten Hälfte des 19. Jahrhunderts: Struktur und personelle Erneuerung." In *Eduard Suess und die Entwicklung der Erdwissenschaften zwischen Biedermeier und Sezession*, edited by Johannes Seidl, 67–104. Göttingen: V&R Unipress, 2009.

———. "Pflanzstätten der Wissenschaft: Zur Einführung von Seminaren und Instituten in den Universitätsbetrieb in Österreich nach 1848." *Mitteilungen des Instituts für Österreichische Geschichtsforschung* (forthcoming).

Mühlfeld, Johann G. Megerle von, ed. *Handbuch für alle kaiserlich-königlichen, ständischen und städtischen Beamten, deren Wittwen und Waisen: Oder Darstellung aller ihnen durch die neuesten allerhöchsten Gesetze vom Jahre 1806 bis 1822 zustehenden Rechte und obliegenden Verbindlichkeiten*. Vol. 1. Vienna: Mösle, 1824.

Müller, Jiří. "'Insigniáda' a dobové pojetí nacionalismu na stránkách Přítomnosti." *Marginalia historica: Časopis pro dějiny vzdělanosti a kultury* 4, no. 1 (2013): 9–52.

Müller, Reinhard. "Das Leben Othmar Spanns: Ein Vortrag von Hans Riehl 1954." *Zyklos 1: Jahrbuch für Theorie und Geschichte der Soziologie* 1 (2015): 341–82.

———. "Maksymilian Ernest Gumplowicz (1861–1897)." In *Dwa życia Ludwika Gumplowicza: Wybór tekstów*, edited by Jan Surman and Gerald Mozetič, 95–101. Warsaw: Oficyna Naukowa, 2010.

Mycielska, Dorota. "Drogi życiowe profesorów przed objęciem katedr akademickich w niepodległej Polsce." In *Inteligencja polska pod zaborami: Studia*, edited by Ryszarda Czepulis-Rastenis, 243–90. Warsaw: Polskie Wydawnictwo Naukowe, 1979.

Mylik, Mirosław. *Stefan Pawlicki jeden z prekursorów nauki polskiej*. Warsaw: Wydawnictwo Uniwersytetu Kardynała Stefana Wyszyńskiego, 2005.

"Nachricht über die Fortsetzung dieser Zeitschrift im J. 1832." *Jahrbuch des Böhmischen Museums für Natur und Länderkunde, Geschichte, Kunst und Literatur* 2, no. 4 (1831).

Naegerle, August. "Bericht über das Studienjahr 1929/30." In *Die feierliche Inauguration des Rektors der Deutschen Universität in Prag für das Studienjahr 1930/31*, 5–43. Prague: Selbstverlag, 1931.

———. "Bericht über die Studienjahre 1918/19 u. 1919/20." In *An Stelle der feierlichen Inauguration des Rektors der Deutschen Universität in Prag für die Studienjahre 1919/20 und 1920/21*, 5–74. Prague: Selbstverlag, 1921.

Nedza, Maria Julita. *Polityka Stypendialna Akademii Umiejętności w Latach 1878–1920: Fundacje Gałęzowskiego, Pileckiego i Osławskiego*. Wrocław: Zakład Narodowy im. Ossolińskich. Wydawnictwo Polskiej Akademii Nauk, 1973.

Nemec, Birgit, and Klaus Taschwer. "Terror gegen Tandler: Kontext und Chronik der antisemitischen Attacken am I. Anatomischen Institut der Universität Wien 1910–1933." In *Der lange Schatten des Antisemitismus: Kritische Auseinandersetzungen mit der Geschichte der Universität Wien im 19. und 20. Jahrhundert*, edited by Oliver Rathkolb, 147–72. Göttingen: V&R Unipress, 2013.

Německo-český slovník vědeckého názvosloví pro gymnasia a reálné školy: Od komise k ustanovení vědeckého názvosloví pro gymnasia a reálné školy. Prague: Nákladem Kalve'ského knihkupectví; Bedřich Tempský, 1853.

Die Neugestaltung der österreichischen Universitäten über Allerhöchsten Befehl dargestellt von dem k.k. Ministerium für Kultus und Unterricht. Vienna: k.k. Hof- und Staatsdruckerei, 1853.

Niederstätter, Alois. "Feuerstein ist nach wie vor bei den Veilchenblauen: Die Briefe von Richard Wollek an den Vorarlberger Landeshauptmann Adolf Rhomberg (1897/98)." *Alemannia Studens: Mitteilungen des Vereins für Vorarlberger Bildungs- und Studenten-Geschichte* 2 (1992): 13–64.

Niedhammer, Martina. "'Slovozpyt' und 'filologie': Nationale Implikationen der sprachwissenschaftlichen Forschung in der Königlich böhmischen Gesellschaft der Wissenschaften im 19. Jahrhundert." In *Sprache, Gesellschaft und Nation in Ostmitteleuropa: Institutionalisierung und Alltagspraxis*, edited by Martina Niedhammer, Klaas-Hinrich Ehlers, Marek Nekula, and Hermann Scheuringer, 33–49. Göttingen: V&R Unipress, 2014.

Nies, Franz, ed. *Europa denkt mehrsprachig / L' Europe pense en plusieurs langues*. Tübingen: Narr, 2005.

Niklíček, Ladislav, Irena Manová, and Bohumil Hájek. "Profesor Vojtěch Šafařík a počátky výuky chemie na české univerzité v Praze." *AUC-HUCP* 22, no. 1 (1982): 71–93.

Nitsch, K[azimierz]. "Moje wspomnienia językowe VII." *Język Polski* 39, no. 5 (1959): 355–61.

Novak, Clemens, and Martin Haidinger. *Der Anteil der organisierten Studentenschaft an der Märzrevolution 1848 in Wien und die Bedeutung dieses Ereignisses für die Korporationen von Heute*. Vienna: Akademischer Corporations-Club, 1999.

Nowak, Ewa. *Polska młodzież w Austrii w XIX i XX wieku: Migracje-Edukacja-Stowarzyszenia*. Lublin: Wydawnictwo Uniwersytetu Marii Curie-Skłodowskiej, 2007.

Nurowski, Marcin. *Szymon Askenazy: Wielki Polak wyznania mojżeszowego*. Warsaw: Marcin Nurowski, 2005.

Nyomárkay, István. "Spracherneuerungen in Mitteleuropa im 19. Jahrhundert: Versuch der Herausbildung muttersprachlicher Terminologien in den mitteleuropäischen Sprachen." *Studia Slavica* 53, no. 2 (2008): 425–40.

Oberkofler, Gerhard. "Deutschnationalismus und Antisemitismus in der Innsbrucker Studentenschaft um 1920." *Tiroler Heimatblätter* 56 (1981): 65–71.

———. "Der Kampf der Universität um die Einheit des Landes Tirol (1918–1920)." *Tiroler Heimatblätter* 55 (1980): 78–89.

———. *Die Rechtslehre in italienischer Sprache an der Universität Innsbruck*. Innsbruck: Kommisionverlag der Österreichischen Kommisionsbuchhandlung, 1975.

———. *Samuel Steinherz (1857–1942): Biographische Skizze über einen altösterreichischen Juden in Prag*. Innsbruck: StudienVerlag, 2008.

Obermajer, Jarosław. "Zabroniony Zjazd Lekarzy i Przyrodników Polskich w roku 1898." *Archiwum Historii Medycyny* 28, nos. 1–2 (1965): 119–23.

Orel, Vítezslav. "Professor Alexander Zawadzki (1798–1868)—Mendel's Superior at the Technical Modern School in Brno." *Folia Mendeliana Musei Moraviae* 7 (1972): 13–20.

Orlevyč, Iryna Vasylivna. "Dijal'nist' L'vivs'koho Stavropihijs'koho Instytutu (kinec' XVIII—60-i rr. XIX st.)." PhD diss., Ivan Kryp'jakevyč Institute of Ukrainian Studies, Ukrainian National Academy of Science, 2000.

Osadczy, Włodzimierz. *Święta Ruś: Rozwój i oddziaływanie idei prawosławia w Galicji*. Lublin: Wydawnictwo Uniwersytetu Marii Curie-Skłodowskiej, 2007.

Österreichische Geschichte für das Volk. 17 vols. Vienna: Prandel, 1864–69.

Ottenthal, Emil von. "Theodor von Sickel." *Mitteilungen des Instituts für Österreichische Geschichtsforschung* 29 (1908): 545–59.

Ottner, Christine. "Zwischen Wiener Localanstalt und Centralpunct der Monarchie: Einzugsbereich und erste Geschichtsforschungsunternehmen der kaiserlichen Akademie der Wissenschaften." *Anzeiger der phil.-hist. Klasse der Österreichischen Akademie der Wissenschaften* 143, no. 1 (2008): 171–96.

Ottner-Diesenberger, Christine. "Joseph Chmel und Johann Friedrich Böhmer: Die Anfänge der Regesta Imperii im Spannungsfeld von Freundschaft und Wissenschaft." In *Wege zur Urkunde, Wege der Urkunde, Wege der Forschung*, edited by Karel Hruza and Paul Herold, 257–91. Vienna: Böhlau, 2005.

Pacholkiv, Svjatoslav. *Emanzipation durch Bildung: Entwicklung und gesellschaftliche Rolle der ukrainischen Intelligenz im habsburgischen Galizien (1890–1914)*. Munich: Oldenbourg, 2002.

———. "Die 'Ostjuden' als Begriff in der Geschichte." In *"Ostjuden"—Geschichte und Mythos: Juden in Mitteleuropa 2011*, edited by the Institut für jüdische Geschichte Österreichs, 2–11. St. Pölten: Institut für jüdische Geschichte Österreichs, 2011.

[Palacký, František]. "Adresse oder Petition des Slaven Congresses in Prag an Seine k.k. Majestät, 1848." In *Slovanský sjezd v Praze roku 1848: Sbírka dokumentů*, edited by Václav Žáček, 370–75. Prague: Nakl. Československé akademie věd, 1958.

Palacký, František. "Letter Sent by František Palacký to Frankfurt (1848)." Translated by William Beardmore. *Slavonic and East European Review* 26 (1947/48): 303–8. Originally published as "Hlas o připojení-se Rakauska k zemi německé," *Národní nowiny*, no. 10 (1848): 37; no. 14 (1848): 45–46.

———. *Urkundliche Beiträge zur Geschichte Böhmens und seiner Nachbarländer im Zeitalter Georg's von Podiebrad (1450–1471)*. Vienna: Kaiserlich-Königliche Hof- und Staatsdruckerei, 1860.

Paljoch, Ol'ha. "Ukraïns'ke knyhovydannja u L'vivi XIX sr.: Rol' drukaren' Stavropihijs'koho Instytutu ta Naukovoho tovarystva im Ševčenka." *Zapysky L'vivs'koï naukovoï biblioteky im. V. Stefanyka: Zbyrnyk naukovych prac'* 16, no. 1 (2008): 54–72.

Palló, Gábor. "Scientific Nationalism: A Historical Approach to Nature in Late Nineteenth-Century Hungary." In *The Nationalisation of Scientific Knowledge in the Habsburg Empire, 1848–1918*, edited by Mitchell G. Ash and Jan Surman, 102–12. Basingstoke: Palgrave Macmillan, 2012.

Palouš, Radim. Česká zkušenost: *Příspěvek k dějinám české filosofie. O Komenského škole stáří, o Bolzanově významu v našem duchovním vývoji a o Masarykově filosofickém mládí—se závěrečným odkazem k Patočkovi*. Prague: Academia, 1994.

Pamiętnik Towarzystwa Literacko-Słowiańskiego przy Uniwersytecie Wrocławskim wydany w roku złotego jubileuszu. Wrocław: Nakładem Drukarni Polskiej (Jan Szymański), 1886.

Papierzyńska-Turek, Mirosława. *Sprawa ukraińska w Drugiej Rzeczypospolitej 1922–1926*. Cracow: Wydawnictwo Literackie, 1979.

Papiór, Jan. "Zur Geschichte und Studienpläne der polnischen Germanistik." In *Daß eine Nation die andere verstehen möge: Festschrift für Marian Szyrocki zu seinem 60. Geburtstag*, edited by Norbert Honsza and Hans-Gert Roloff, 549–68. Amsterdam: Rodopi, 1988.

Paulinyiová, E. "Vznik a vývin filozofické fakulty v rokov 1921–1945." In *50 rokov Univerzity Komenského*, edited by Július Bartl, 284–321. Bratislava: Univerzita Komenského, 1969.

Pavlík, Jan. *Vzpomínky na zemřelé jezuity, narozené v Čechách, na Moravě a v Moravském Slezsku od roku 1814*. Olomouc: Refugium Velehrad-Roma s.r.o., 2011.

Pawelski, Jan. "Po anarchii uniwersyteckiej." *Przegląd Powszechny* 54, no. 2 (1911): 1–16.

Perkowska, Urszula. "Études scientifiques des universitaires de Cracovie à Vienne dans les années 1800–1918." *Zeszyty Naukowe Uniwersytetu Jagiellońskiego: Prace Historyczne* 88 (1989): 305–12.

———. "La genèse et la caractéristique de la loi sur les écoles supérieures du 13 juillet 1920." *Zeszyty Naukowe Uniwersytetu Jagiellońskiego: Prace Historyczne* 79 (1985): 95–107.

———. *Kształtowanie się zespołu naukowego w Uniwersytecie Jagiellońskim (1860–1920)*. Wrocław: Zakład Narodowy im. Ossolińskich, 1975.

———. *Uniwersytet Jagielloński w latach I wojny światowej*. Cracow: Universitas, 1990.

Pernes, Jiří. *Kapitoly z dějin Vysokého učení technického v Brně (cesta moravské techniky 20. stoletím)*. Brno: Vysoké Učeni Technické, Nakladatelstvi VUTIUM, 2009.

Pernter, Joseph Maria. *Voraussetzungslose Forschung, freie Wissenschaft und Katholizismus*. Vienna: Braumüller, 1902.

Perricelli, Magda. "'O Trieste o nulla!': I 'fatti di Innsbruck' nella stampa quotidiana del Regno d'Italia." In *Università e nazionalismi: Innsbruck 1904 e l'assalto alla Facoltà di giurisprudenza italiana*, edited by Günther Pallaver and Michael Gehler, 161–93. Trento: Fondazione Museo storico del Trentino, 2010.

Pešek, Jiři. "Jüdische Studenten an den Prager Universitäten 1882–1939." In *Franz Kafka im sprachnationalen Kontext seiner Zeit: Sprache und nationale Identität in öffentlichen Institutionen der böhmischen Länder*, edited by Marek Nekula, Ingrid Fleischmann, and Abrecht Greule, 213–27. Vienna: Böhlau, 2007.

Pešek, Jiři, Alena Míšková, Petr Svobodný, and Jan Janko. "Německá univerzita v Praze v letech 1918–1939." In *Dějiny Univerzity Karlovy*, vol. 4, *1918–1945*, edited by Jan Havránek and Zdeněk Pousta, 181–211. Prague: Karolinum, 1998.

Petr, Jan. "Přínos Univerzity Karlovy k rozvoji slovanské jazykovědy na univerzitě Klimenta Ochridského v Sofii od doby jejího založení do současnost." *Práce z dějin slavistiky* 13 (1989): 7–41.

Petráň, Josef. "Filozofická fakulta v letech 1918–1939 (1945)." In *Dějiny Univerzity Karlovy*, vol. 4, *1918–1945*, edited by Jan Havránek and Zdeněk Pousta, 121–61. Prague: Karolinum, 1998.

———. *Nástin dějin filozofické fakulty Univerzity Karlovy v Praze (do roku 1948)*. Prague: Univerzita Karlova, 1984.

Petritsch, Wolfgang. "Die slowenischen Studenten an der Universität Wien (1848–1890)." PhD diss., University of Vienna, 1972.

Pietsch, Tamson. *Empire of Scholars: Universities, Networks and the British Academic World, 1850–1939*. Manchester: Manchester University Press, 2013.

Pilch, Andrzej. "Ukraińcy na wyższych uczelniach Lwowa 1923–1926." *Annales Universitatis Mariae Curie-Skłodowska* 54/55 (1999/2000): 242–46.

Pisa, Karl. *Ernst Freiherr von Feuchtersleben: Pionier der Psychosomatik (Literatur und Leben)*. Vienna: Böhlau, 1998.

Plan einer zeitgemäßen Reform der österreichischen Universitäten: Entworfen und dem Ministerium des Unterrichts überreicht von dem Collegium der Professoren der medic-chirurg. Studien der k.k. Wiener Hochschule. Vienna, 1848.

Plattner, Irmgard. "La città di Innsbruck alla svolta del secolo." In *Università e nazionalismi: Innsbruck 1904 e l'assalto alla Facoltà di giurisprudenza italiana*, edited by Günther Pallaver and Michael Gehler, 47–80. Trento: Fondazione Museo storico del Trentino, 2010.

Podraza, Antoni. "Dobry, Lepszy, Bobińska: Wysłuchała i opracowała Rita Pagacz-Moczarska." *Alma Mater*, no. 63 (2004): 48–50.

Pohl, Walter. "National Origin Narratives in the Austro-Hungarian Monarchy." In *Manufacturing Middle Ages: Entangled History of Medievalism in Nineteenth-Century Europe*, edited by Patrick J. Geary and Gábor Klaniczay, 13–50. Leiden: Brill, 2013.

Pokorná, Magdaléna. "Královská česká společnost nauk." In *Bohemia docta: K historickým kořenům vědy v českých zemích*, edited by Alena Míšková, Martin Franc, and Antonín Kostlán, 58–144. Prague: Academia, 2010.

Polák, Stanislav. *T. G. Masaryk: Za ideálem a pravdou*. Vol. 4, *1900–1914*. Prague: Masarykův ústav Akademie věd České republiky, 2005.

———. *T. G. Masaryk: Za ideálem a pravdou*. Vol. 2, *1882–1893*. Prague: Masarykův ústav Akademie věd České republiky, 2001.

Polišenský, Josef. "Anton Gindely und die böhmische Geschichtswissenschaft." *Acta Universitatis Carolinae—Philosophica et Historica* 3 (1993): 13–21.

Polishchuk, Petro, and Bohdan Struminsky. "Gymnasium." In *Internet Ecyclopedia of Ukraine*. N.d. http://www.encyclopediaofukraine.com/display.asp?linkpath=pages\G\Y\Gymnasium.htm (accessed 1 December 2014).

Popper, Karl. "Prague Lecture." 25 May 1994. https://www.lf3.cuni.cz/3LFEN-255.html.

Porter, Brian A. *Faith and Fatherland: Catholicism, Modernity, and Poland*. Oxford: Oxford University Press, 2011.

———. *When Nationalism Began to Hate: Imagining Modern Politics in Nineteenth-Century Poland*. New York: Oxford University Press, 2000.

Posch, Herbert. "Studierende und die Universität Wien in der Dauerkrise 1918 bis 1938." In *"Anschluß" und Ausschluss 1938: Vertriebene und verbliebene Studierende der Universität Wien*, edited by Herbert Posch, Doris Ingrisch, and Gert Dressel, 61–99. Vienna: Lit, 2008.

"Prager Briefe." *Wiener Medizinische Wochenschrift* 32, no. 44 (1882): 1315–16.

Pražák, Albert. *Politika a revoluce: Paměti*. Prague: Academia, 2004.

———. "Zehn Jahre Universität in Bratislava: Die Entwicklung der Hochschule, ihr heutiger Stand und ihre Zukunftsaussichten (repr. from *Prager Presse* 2/7/1929)."

In *Ročenka University Komenského za studijní rok 1928–1929*, edited by Albert Pražák, 244–47. Bratislava: Universita Komenského, 1930.

Preglau-Hämmerle, Susanne. *Die politische und soziale Funktion der österreichischen Universität: Von den Anfängen bis zur Gegenwart*. Innsbruck: Inn-Verlag, 1986.

Presl, Jan Svatopluk. "Oznámenj." *Krok: Weřegný spis wšenaučný pro wzdělance národu Česko-Slowanského* 1, no. 1 (1821): 7–18.

Prinz, Friedrich. "Graf Leo Thun." In *Lebensbilder zur Geschichte der böhmischen Länder*, edited by Karl Bosl, 2:139–68. Munich: Oldenbourg, 1976.

Procopovici, Alexe. "[Eugen Herzog†]." *Revista Filologică* 2 (1928/29): 232–45.

Prokopovych, Markian. "Staging Empires and Nations: Politics in the Public Space of Habsburg Lemberg." In *Die Besetzung des öffentlichen Raumes: Politische Plätze, Denkmäler und Straßennamen im europäischen Vergleich*, edited by Peter Stachel and Rudolf Jaworski, 427–53. Berlin: Frank & Timme, 2007.

Prokopowitsch, Erich. *Gründung, Entwicklung und Ende der Franz-Josephs-Universität in Czernowitz*. Clausthal-Zellerfeld: Pieper, 1955.

Pumpla, Vaclav. "Snahy o zřízení české university na Moravě v 19. století." *Historická Olomouc* 11 (1998): 145–52.

Purkyně, Jan Evangelista. *Akademie*. Prague: Nakladatelství Československé akademie věd, 1962.

———. "Kritika: Carl Rokitansky, Die Conformität der Universitäten mit Rücksicht auf gegenwärtige österreichische Zustände." *Časopis lékařův českých* 3, no. 3 (1863): 22–24.

———. "Kritika: Carl Rokitansky, Zeitfragen betreffend die Universität mit bes. Beziehung auf Medizin." *Časopis lékařův českých* 2, no. 20 (1863): 256–58.

———. "Řeč na sjezdu německých přírodovědců a lékařů v Karlových Varech dne 25. záři 1862." In *Jan Evangelista Purkyně, Sebrané Spisy, Svazek IX: Věda, vychova, společnost; studie a úvahy*, edited by Vladislav Kruta and Zdeněk Hornof, 131–32. Prague: Nakladatelství Československé akademie věd, 1965.

Puşcariu, Sextil. "Eugen Herzog [Article nécrologique]." *Dacoromania* 5 (1929): 882–83.

Putz, Hannelore. "König Ludwig I. von Bayern und seine Universität." In *Domus Universitatis: Das Hauptgebäude der Ludwig-Maximilians-Universität 1835–1911–2011*, edited by Claudius Stein, 27–46. Munich: Herbert Utz, 2015.

Rabinbach, Anton G. "The Migration of Galician Jews to Vienna, 1857–1880." *Austrian History Yearbook* 11 (1975): 43–54.

Rabl, Hans. "Bericht des Prorektors." In *Die feierliche Inauguration des Rektors der Grazer Universität für das Studienjahr 1925/6*, 1–36. Graz: Leuschner, 1927.

———. "Die Verluste der Grazer Universität im Weltkrieg." In *Beiträge zur Geschichte der Karl-Franzens-Universität-Graz*, 51–60. Graz: Leuschner & Lubensky, 1927.

Raffler, Marlies. *Museum—Spiegel der Nation? Zugänge zur historischen Museologie am Beispiel der Genese von Landes- und Nationalmuseen in der Habsburgermonarchie*. Vienna: Böhlau, 2007.

Raj, Kapil. *Relocating Modern Science: Circulation and the Construction of Knowledge in South Asia and Europe, 1650–1900*. Basingstoke: Palgrave Macmillan, 2007.

Rak, Jiří. "Obraz Němce v české historiografii 19. století." In *Obraz Němců, Rakouska a Německa v české společnosti 19. a 20. století*, edited by Jan Křen and Eva Broklová, 49–75. Prague: Karolinum, 1998.

Rampley, Matthew. *The Vienna School of Art History: Empire and the Politics of Scholarship*. University Park: Pennsylvania State University Press, 2013.

Rappaport, Moritz. *Prolog zur Feier des 100-jährigen Geburtstages Friedrich Schillers*. L'viv: Poremba, 1859.

Raspa, Venanzio. "Bolzano e la filosofia austriaca." *Discipline filosofiche* 21, no. 2 (2011): 245–85.

Rathkolb, Oliver. "Gewalt und Antisemitismus an der Universität Wien." In *Der lange Schatten des Antisemitismus: Kritische Auseinandersetzungen mit der Geschichte der Universität Wien im 19. und 20. Jahrhundert*, edited by Oliver Rathkolb, 69–92. Göttingen: V&R Unipress, 2013.

Rebenich, Stefan. *Theodor Mommsen und Adolf Harnack: Wissenschaft und Politik im Berlin des ausgehenden 19. Jahrhunderts*. Berlin: de Gruyter, 1997.

Recchi, Ettore. *Mobile Europe: The Theory and Practice of Free Movement in the EU*. Basingstoke: Palgrave Macmillan, 2015.

Rederowa, Danuta. "Formy współpracy Polskiej Akademii Umiejętności z zagranicą." *Studia i materiały z dziejów nauki polskiej, Seria A* 10 (1966): 77–136.

Rehbinder, Manfred. "Die rechts- und staatswissenschaftliche Fakultät der Franz-Josephs-Universität Czernowitz: Ihr Beitrag zur Erforschung des Rechts in einer multikulturellen Gesellschaft." In *Festschrift Hans Stoll zum 75. Geburtstag*, edited by Gerhard Hohloch, Hans Stoll, Rainer Frank, and Peter Schlechtriem, 327–44. Tübingen: Mohr Siebeck, 2001.

Reinbothe, Roswitha. *Deutsch als internationale Wissenschaftssprache und der Boykott nach dem Ersten Weltkrieg*. Frankfurt am Main: Peter Lang, 2006.

Reisch, Emil. "Aufgaben unserer Universitäten nach dem Kriege: Inaugurationsrede, gehalten am 6. November 1916." In *Die feierliche Inauguration des Rektors der Wiener Universität für das Studienjahr 1916/17*, 57–96. Vienna: Holzhausen, 1916.

Reitterer, H. "Stalzer, Josef (1880–1914)." In *Das Österreichische Biographische Lexikon*, edited by Österreichische Akademie der Wissenschaften, 13:84–85. Vienna: Verlag der Österreichischen Akademie der Wissenschaften, 2007.

Rentetzi, Maria. *Trafficking Materials and Gendered Research Practices: Radium Research in Early 20th Century Vienna*. New York: Columbia University Press, 2008.

Richter, Franz Johannes. "Über öffentliche Erziehungs- und Unterrichtsanstalten in Europa, mit besonderen Rücksicht auf das Schul- und Studienwesen des österreichischen Kaiserstaates." *Österreichische Blätter für Litteratur und Kunst* 2, no. 128 (1845): 993–98; no. 141 (1845): 1097–101; no. 142 (1845): 1105–8; no. 143 (1845): 1115–18.

Rinnerthaler, Alfred. "Der Fall Wahrmund: Politische, rechtliche und diplomatische Turbulenzen im Umfeld von Modernismus und Antimodernismus in Österreich." In *Österreich und der Heilige Stuhl im 19. und 20. Jahrhundert*, edited by Hans Paarhammer and Alfred Rinnerthaler, 187–246. Frankfurt am Main: Peter Lang, 2001.

———. "Von der Benediktiner- zur Staatsuniversität: Vom Werden der Salzburger 'Alma Mater.'" In *Bürgerliche Freiheit und christliche Verantwortung: Festschrift für Christoph Link zum siebzigsten Geburtstag*, edited by Heinrich De Wall and Michael Germann, 787–826. Tübingen: Mohr, Siebeck, 2003.

Rokitansky, Carl. *Die Conformität der Universitäten mit Rücksicht auf gegenwärtige österreichische Zustände*. Vienna: Sallmayer, 1863.

Rollinger, Robin. "Anton Marty." In *The Stanford Encyclopedia of Philosophy*, Winter 2008 Edition, edited by Edward N. Zalta. Stanford, CA: Metaphysics Research Lab, 2008. http://plato.stanford.edu/archives/win2008/entries/marty/.

Romanczuk, Julian [Юліан Романчук]. *Die Ruthenen und ihre Gegner in Galizien*. Vienna: C. W. Stern, 1902.

Rose, Alison. *Jewish Women in Fin de Siècle Vienna*. Austin: University of Texas Press, 2008.

Röskau-Rydel, Isabel. *Kultur an der Peripherie des Habsburger Reiches: Die Geschichte des Bildungswesens und der kultureller Einrichtungen in Lemberg von 1772 bis 1848*. Vienna: Harrassowitz, 1993.

———. *Niemiecko-austriackie rodziny urzędnicze w Galicji 1772–1918: Kariery zawodowe—środowisko—akulturacja i asymilacja*. Cracow: Wydawnictwo Naukowe Uniwersytetu Pedagogicznego, 2011.

Rosner, Robert. *Chemie in Österreich 1740–1914: Lehre—Forschung—Industrie*. Vienna: Böhlau, 2004.

Rossbacher, Karlheinz. *Literatur und Bürgertum: Fünf Wiener jüdische Familien von der liberalen Ära zum Fin de Siècle*. Vienna: Böhlau, 2003.

Rozenblit, Marsha L. "The Assertion of Identity: Jewish Student Nationalism at the University of Vienna before the First World War." *Leo Baeck Institute Yearbook* 27, no. 1 (1982): 171–86.

———. "A Note on Galician Jewish Migration to Vienna." *Austrian History Yearbook* 19 (1983): 143–52.

Rüdegger, Erika. "Die philosophischen Studien an der Wiener Universität 1800 bis 1848." PhD diss., University of Vienna, 1964.

Rudnitsky, Stephen. *Ukraine, the Land and Its People: An Introduction to Its Geography*. 1910. New York: Rand McNally, 1918.

Rudnytsky, Ivan L. "Polish-Ukrainian Relations: The Burden of History." In *Essays in Modern Ukrainian History*, edited by Peter L. Rudnytsky, 49–76. Edmonton: Canadian Institute of Ukrainian Studies, University of Alberta, 1987.

Rus'ka čytanka dla nyžčoï gymnaziï. Vienna: Nakladom Pravytels'tva, 1852.

Saart, Tõnis. "Construction of Peripheries: Foucault vs. Lotman and Potential Peripherization of New Member States in the EU." In *European Peripheries*, edited by György Andrássy, Jyrki Käkönen, and Noémi Nagy, 21–38. Pécs: Publikon, 2012.

Sablik, Karl. *Julius Tandler, Mediziner und Sozialreformer: Eine Biographie*. Vienna: A. Schendl, 1983.

Šafárik, Pavel Jozef. "Myšlénky o prowedení stejného práwa českého i německého jazyka w školách českych." *Časopis českého musea* 22, no. 2 (1848): 171–97.

Salvini-Plawen, Luitfried, and Matthias Svojtka. *Fische, Petrefakten und Gedichte: Rudolf Kner (1810–1869)—ein Streifzug durch sein Leben und Werk.* Linz: Biologiezentrum der Oberösterreichischer Landesmuseen, 2008.

Šamberger, Zdeněk. "Časopis Vídeňský Denník a jeho poslání v letech 1850–1851 (Ke ztroskotanému pokusu Leo Thuna o založení české konzervativní strany)." *Slovanský přehled* 71, no. 1 (1985): 26–40.

Sartori, Franz. *Historisch-ethnographische Übersicht der wissenschaftlichen Cultur, Geistesthätigkeit und Literatur des österreichischen Kaiserthumes nach seinen mannigfaltigen Sprachen und deren Bildungsstufen in skizzierten Umrissen bearbeitet: Erster Theil.* Vienna: Carl Gerold, 1830.

Schaefer, Richard. "Infallibility and Intentionality: Franz Brentano's Diagnosis of German Catholicism." *Journal of the History of Ideas* 68, no. 3 (2007): 477–99.

Scheichl, Sigurd Paul. "The Context and Nuances of Anti-Jewish Language: Were All the 'Antisemites' Antisemites?" In *Jews, Antisemitism, and Culture in Vienna*, edited by Ivar Oxaal, Michael Pollak, and Gerhard Both, 89–110. London: Routledge & Kegan Paul, 1987.

Schelsky, Helmut. *Einsamkeit und Freiheit, Idee und Gestalt der deutschen Universität und ihrer Reformen.* Hamburg: Rowohlt, 1963.

Schenk, Frithjof Benjamin. *Russlands Fahrt in die Moderne: Mobilität und sozialer Raum im Eisenbahnzeitalter.* Stuttgart: Steiner, 2014.

Schleicher, August. *Die Formenlehre der kirchenslawischen Sprache, erklärend und vergleichend dargestellt.* Bonn: H. B. König, 1852.

Schmeiser, Martin. *Akademischer Hasard: Das Berufsschicksal des Professors und das Schicksal der deutschen Universität 1870–1920. Eine verstehend soziologische Untersuchung.* Stuttgart: Klett-Cotta, 1994.

Schmied, Erich. "Die altösterreichische Gesetzgebung zur Prager Universität: Ein Beitrag zur Geschichte der Prager Universität bis 1918." In *Die Teilung der Prager Universität 1882 und die intellektuelle Desintegration in den böhmischen Ländern*, edited by Ferdinand Seibt, 11–24. Munich: Oldenbourg, 1984.

Schmied-Kowarzik, Wolfdietrich. "Vergessene Impulse der Wiener Philosophie um die Jahrhundertwende: Eine philosophische Skizze wider den main stream verdrängenden Erinnerns." In *Die Wiener Jahrhundertwende: Einflüsse—Umwelt—Wirkungen*, edited by Jürgen Nautz and Richard Vahrenkamp, 181–201. Vienna: Böhlau, 1993.

Schmidt-Hartmann, Eva. "Die philosophische Fakultät der tschechischen Universität um 1882: Kontinuität und Wandel." In *Die Teilung der Prager Universität 1882 und die intellektuelle Desintegration in den böhmischen Ländern*, edited by Ferdinand Seibt, 95–110. Munich: Oldenbourg, 1984.

Schneider, Alfred. "Briefe österreichischer Gelehrter aus den Jahren 1849–1862: Beiträge zur Geschichte der österreichischen Unterrichtsreform." *Archiv für österreichischen Geschichte* 113 (1936): 171–304.

Schneller, Joseph. *Historische Darstellung der Entwickelung der medicinischen Facultät zu Wien, nebst einer kurzen Uebersicht der wissenschaftlichen Leistungen des medicinischen Doctoren-Collegiums.* Vienna: Zamarski, 1856.

Schober, Richard. *Theodor Freiherr von Kathrein (1842–1916), Landeshauptmann von Tirol: Briefe und Dokumente zur katholisch-konservativen Politik um die Jahrhundertwende*. Innsbruck: Wagner, 1992.

Schöpf, Joseph Anton. *Denkschrift des Comites für Wiederherstellung der Universität in Salzburg*. Salzburg: Zaunrath'sche Buchdruckerei, 1870.

Schorske, Karl. *Fin-de-Siècle Vienna: Politics and Culture*. New York: Knopf, 1980.

Schröder, Tilman M. *Naturwissenschaften und Protestantismus im deutschen Kaiserreich: Die Versammlungen der Gesellschaft Deutscher Naturforscher und Ärzte und ihre Bedeutung für die evangelische Theologie*. Stuttgart: Franz Steiner, 2008.

Schweizer, Claudia. "Migrating Objects: The Bohemian National Museum and Its Scientific Collaborations in the Early Nineteenth Century." *Journal of the History of Collections* 18, no. 2 (2006): 187–99.

Sdvižkov, Denis A. *Das Zeitalter der Intelligenz: Zur vergleichenden Geschichte der Gebildeten in Europa bis zum Ersten Weltkrieg*. Göttingen: Vandenhoeck & Ruprecht, 2006.

[Sealsfield, Charles]. *Austria as It Is, or Sketches of Continental Courts, by an Eye-Witness*. London: Hurst, Chance, 1828.

Seebacher, Felicitas. *"Freiheit der Naturforschung!" Carl Freiherr von Rokitansky und die Wiener Medizinische Schule: Wissenschaft und Politik im Konflikt*. Vienna: Verlag der Österreichischen Akademie der Wissenschaften, 2006.

———. *Das Fremde im "deutschen" Tempel der Wissenschaften*. Vienna: Verlag der Österreichischen Akademie der Wissenschaften, 2011.

———. "'Der operierte Chirurg': Theodor Billroths Deutschnationalismus und akademischer Antisemitismus." *Zeitschrift für Geschichtswissenschaft* 56, no. 4 (2006): 317–38.

Seibt, Ferdinand, ed. *Die Teilung der Prager Universität 1882 und die intellektuelle Desintegration in den böhmischen Ländern*. Munich: Oldenbourg, 1984.

Seidl, Johannes, Franz Pertlik, and Matthias Svojtka. "Franz Xaver Maximilian Zippe (1791–1863): Ein böhmischer Erdwissenschaftler als Inhaber des ersten Lehrstuhls für Mineralogie an der Philosophischen Fakultät der Universität Wien." In *Eduard Suess und die Entwicklung der Erdwissenschaften zwischen Biedermeier und Sezession*, edited by Johannes Seidl, 161–209. Göttingen: V&R Unipress, 2009.

Seiler, Martin. "Das 'Manifest der österreichischen Philosophie': Die Materialien Kurt Blaukopfs über die Berufung Robert Zimmermanns an die Universität Wien (1860–1861), im Spiegel von Philosophiegeschichte, Universitätsreform, Berufungspolitik, staatlicher Religions- und Konfessionsgesetzgebung und Verfassungsgeschichte." In Wissenschaftliche Weltauffassung und Kunst: Kunst, Kunsttheorie und Kunstforschung im wissenschaftlichen Diskurs. Ein Projekt des Instituts Wiener Kreis, n.d. http://www.univie.ac.at/ivc/WWUK/manifest.html (accessed 21 October 2014).

Sekyrková, Milada. "Rozdělení polytechniky—precedens pro univerzitu?" In *Místo Národnich Jazyku ve Výuce, Vědě a Vzdělání v Habsburské Monarchii 1867–1918 / Position of National Languages in Education, Educational System and Science of the Habsburg Monarchy 1867–1918*, edited by Harald Binder, Luboš Velek, and Barbora Křivohlavá, 231–40. Prague: Výzkumné centrum pro dějiny vědy, 2003.

Sexl, Lore. "Gottfried Wilhelm Leibniz und die geplante Kaiserlicher Akademie der Wissenschaften in Wien." In *Theoria cum praxi: Aus der Welt des Gottfried Wilhelm Leibniz*, edited by Hermann Hunger, 69–239. Vienna: Verlag der Österreichischen Akademie der Wissenschaften, 2012.

Shapin, Steven, and Adi Ophir. "The Place of Knowledge: A Methodological Survey." *Science in Context* 4, no. 1 (1991): 3–22.

Simonek, Stefan. "Möglichkeiten und Grenzen postkolonialistischer Literaturtheorie aus slawistischer Sicht." In *Habsburg postcolonial: Machtstrukturen und kollektives Gedächtnis*, edited by Ursula Prutsch, Johannes Feichtinger, and Moritz Csáky, 129–39. Innsbruck: Studien Verlag, 2003.

Simons, Thomas W., Jr. "The Prague Origins of the Guntherian Converts (1800–1850)." *Leo Baeck Institute Yearbook* 22, no. 1 (1977): 245–56.

Sinel, Alen. *The Classroom and the Chancellery: State Educational Reform in Russia under Count Dmitry Tolstoi*. Cambridge, MA: Harvard University Press, 1973.

Šišma, Pavel. *Matematika na německé technice v Brně*. Prague: Prometheus, 2002.

Slabošpyc'kyj, Mychajlo. *Ukraïns'ki mecenaty: Narysy z istoriï ukraïns'koï kul'tury*. Kiev: Vydavnyctvo M.P. Koc', Jaroslaviv Val, 2001.

Slapnicka, Helmut. "Die juridischen Fakultäten der Prager Universitäten 1900–1939." In *Universitäten in nationaler Konkurrenz: Zur Geschichte der Prager Universitäten im 19. und 20. Jahrhundert*, edited by Hans Lemberg, 63–84. Munich: Oldenbourg, 2003.

Slawitschek, Dozent Dr. "Eine allgemeine Deutsch Hochschule." *Deutsche Hochschul-Zeitung*, 30 March 1918, 1.

Šlechtová, Alena, and Josef Levora. *Členové České akademie věd a umění 1890–1952*. Prague: Academia, 2004.

Słoczyński, Henryk. "Z dziejów czarnej legendy krakowskiej historiografii konserwatywnej: Józef Szujski w opiniach współczesnych i potomnych." *Kwartalnik Historyczny* 102, nos. 3–4 (1995): 209–44.

Smal'-Stoc'kyj, Stepan. "Fedir Gartner." *Zapysky Naukovoho tovarystva imeni Ševčenka: Praci filolohičnoï sekciï* 86–87 (1925): 239–48.

Smetana, Augustin. *Geschichte eines Excommunicirten: Eine Selbstbiographie von Augustin Smetana*. Edited by Alfred Meißner. Leipzig: Grunow, 1863.

———. *Příběh jedné exkomunikace a doprovodné texty*. Edited and translated by Irena Šnebergová. Prague: Filosofia, 2008.

Smith, Barry. *Austrian Philosophy: The Legacy of Franz Brentano*. Chicago: Open Court, 1994.

Smith, Helmut Walser. *German Nationalism and Religious Conflict: Culture, Ideology, Politics, 1870–1914*. Princeton, NJ: Princeton University Press, 1995.

Soboń, Marcin. *Polacy wobec Żydów w Galicji doby autonomicznej w latach 1868–1914*. Cracow: Verso, 2011.

Sommaruga, Franz Freiherr von. *Rede des Ministers des öffentlichen Unterrichtes Dr. Franz Freiherrn von Sommaruga gehalten in der Aula der Wiener Universität am 30. März 1848*. N.p., 1848.

Sosnowska, Danuta. *Inna Galicja.* Warsaw: Dom Wydawniczy "Elipsa," 2008.

Spector, Scott. "Where Personal Fate Turns to Public Affair: Homosexual Scandal and Social Order in Vienna, 1900–1910." *Austrian History Yearbook* 38 (2007): 15–24.

Srbik, Heinrich Ritter von. *Heinrich Ritter von Srbik: Die wissenschaftliche Korrespondenz des Historikers, 1912–1945.* Edited by Jürgen Kammerer. Boppard am Rhein: Harald Boldt, 1988.

Stachel, Peter. "Das österreichische Bildungssystem zwischen 1749 und 1918." In *Geschichte der österreichischen Humanwissenschaften,* vol. 1, *Historischer Kontext, wissenschaftssoziologische Befunde und methodologische Voraussetzungen,* edited by Karl Acham, 115–46. Vienna: Passagen, 1999.

Stadler, Friedrich. *The Vienna Circle: Studies in the Origins, Development, and Influence of Logical Empiricism.* Vienna: Springer, 2001.

Štaif, Jiří. "Palackýs Partei der tschechischen Liberalen und die konservative Variante der böhmischen Politik." In *1848/49 Revolutionen in Ostmitteleuropa: Vorträge der Tagung des Collegiums Carolinum in Bad Wiessee vom 30. November bis 1. Dezember 1990,* edited by Rudolf Jaworski and Robert Luft, 57–73. Munich: Oldenburg, 1996.

Staliński, Zbigniew. "Leopold Adametz (1861–1941)." In *Złota Księga Akademii Rolniczej,* edited by Zbigniew Staliński, 62–67. Cracow: Wydawnictwo Akademii Rolniczej, 2000.

Staliūnas, Darius. *Making Russians: Meaning and Practice of Russification in Lithuania and Belarus after 1863.* Amsterdam: Rodopi, 2007.

Starnawski, Jerzy. "Sylwetki lwowskich historyków literatury: Richard Maria Werner (14 VIII 1854–13 I 1913)." *Przegląd Wschodni* 2, no. 2 (1993): 485–90.

———. "Towarzystwa naukowe z zakresie humanistyki na terenie Galicji." In *Galicja i jej dziedzictwo,* vol. 3, *Nauka i oświata,* edited by Andrzej Meissner and Jerzy Wyrozumski, 51–68. Rzeszów: Wydawnictwo Wyższej Szkoły Pedagogicznej, 1995.

Starzyński, Stanisław. "Historya Uniwersytetu Lwowskiego 1869–1894." In *Historya Uniwersytetu Lwowskiego,* edited by Ludwik Finkel and Stanisław Starzyński, 3–417. L'viv: C. K. Uniwersytet Lwowski, 1894.

Štátni úrad štatistický. *Štatistická príručka republiky československé IV.* Prague: Bursık & Kohout, 1932.

Staudacher, Anna L. *Jüdische Konvertiten in Wien 1782–1868.* Part 1. Frankfurt am Main: Peter Lang, 2002.

———. *". . . meldet den Austritt aus dem mosaischen Glauben": 18000 Austritte aus dem Judentum in Wien, 1868–1914. Namen—Quellen—Daten.* Frankfurt am Main: Peter Lang, 2009.

———. "Zwischen Emanzipation und Assimilation—jüdische Juristen in Wien des Fin-de-Siècle." In *Hans Kelsen: Leben—Werk—Wirksamkeit,* edited by Robert Walter, Werner Ogris, and Thomas Olechowski, 41–53. Vienna: Manz, 2009.

Steif, Yvonne. *Wenn Wissenschaftler feiern: Die Versammlungen deutscher Naturforscher und Ärzte 1822 bis 1913.* Stuttgart: Wissenschaftliche Verlags-Gesellschaft, 2003.

Stein, Lorenz von. *Lehrfreiheit, Wissenschaft und Collegiengeld*. Vienna: Alfred Hölder, 1875.

Steltner, Ulrich. "Grenzgänger zwischen der deutschen und der polnischen Literatur: Tadeusz Rittner und Stanisław Przybyszewski." In *Auf der Suche nach einer größeren Heimat . . . : Sprachwechsel/Kulturwechsel in der slawischen Welt*, edited by Ulrich Steltner, 105–15. Jena: Collegium Europaeum Jenense, 1999.

Stempin, Arkadiusz. "Die Wiedererrichtung einer polnischen Universität: Warschau unter deutscher Besatzung." In *Kollegen—Kommilitonen—Kämpfer: Europäische Universitäten im Ersten Weltkrieg*, edited by Trude Maurer, 127–46. Stuttgart: Franz Steiner, 2006.

Ein Stich in's Wespennest oder Der 6. allgemeine österreichische Katholikentag (1907) und die katholische Universitätsfrage. Salzburg: Katholischer Universitäts-Verein, [1907].

"Stosunki Naukowe Polski z zagranicą w latach 1919–1925." *Nauka Polska* 6 (1927): 391–425.

Stourzh, Gerald. *Der Umfang der österreichischen Geschichte: Ausgewählte Studien 1990–2010*. Vienna: Böhlau, 2011.

Štrbáňová, Soňa. "Congresses of the Czech Naturalists and Physicians in the Years 1880–1914 and the Czech-Polish Scientific Collaboration." In *Studies of Czechoslovak Historians for the 18th International Congress of the History of Science*, edited by Jan Janko, 79–122. Special issue of *Acta historiae rerum naturalium necnon technicarum*, vol. 21. Prague: Institute for Czechoslovak and General History, Institute for Czechoslovak and General History, 1989.

———. "Turning 'Province' to a 'Centre'? Ambitions to Establish an Institutionalized Network of Slavic Scientists at the Turn of the 19th Century." *Dějiny věd a techniky* 48, no. 4 (2015): 274–305.

Stremayr, Carl von. *Erinnerungen aus dem Leben*. Vienna: Eigenverlag, 1899.

Struve, Henryk. "Życie i twórczość Józefa Kremera." In *Wykład systematyczny filozofii obejmujący wszystkie jéj części w zarysie: Dla miłośników téj umiejętności pragnących dokładniéj się z nią obeznać*, vol. 1, *Fenomenologia, logika*, by Józef Kremer, edited by Henryk Struve, 1–175. Warsaw: S. Lewental, 1877.

Studnička, F[rantišek] J[osef]. "O rozvoji naši literatury fysikální za posledních padesáte let." *Časopis musea království českého* 50, no. 1 (1876): 35–46.

Stumpf, Karl. "Erinnerungen an Franz Brentano." In *Franz Brentano: Zur Kenntnis seines Lebens und seiner Lehre*, edited by Oskar Kraus, 87–149. Munich: Beck, 1919.

Suchodolski, Bogdan, ed. *Historia nauki polskiej*. Vol. 3, *1795–1862*. Wrocław: Zakład Narodowy im. Ossolińskich, Polska Akademia Nauk, Zakład Historii Nauki, Oświaty i Techniki, 1977.

Suk, Jiři. "Studie o počátích Gollovy školy." *Acta Universitatis Carolinae—Philosophica et Historica* 3 (1993): 147–69.

Šula, K. "Otakar Srdínko." *Almanach* České Akademie věd a umění 41 (1930): 957–75.

Suppan, Arnold. "Katholische Volksparteien in Ostmitteleuropa in der Zwischenkriegszeit am Beispiel der Tschechen und Slovaken." In *Christdemokratie in Europa im 20. Jahrhundert*, edited by Michael Gehler, Wolfram Kaiser, and Helmut Wohnout, 273–93. Vienna: Böhlau, 2001.

Suppan, Arnold, and Marija Wakounig. "Hans Uebersberger (1877–1962)." In *Osteuropäische Geschichte in Wien: 100 Jahre Forschung und Lehre an der Universität*, edited by Arnold Suppan, Marija Wakounig, and Georg Kastner, 91–165. Innsbruck: StudienVerlag, 2007.

Surman, Jan. "University Scandals and the Public Sphere of Imperial Austria: The Wahrmund and Zimmermann Affairs." *European Review of History—Revue Européene d'histoire*. Forthcoming.

———. "Cisleithanisch und transleithanisch oder habsburgisch? Ungarn und das Universitätssystem der Doppelmonarchie." In *Österreichisch-ungarische Beziehungen auf dem Gebiet des Hochschulwesens*, edited by Zsolt K. Lengyel, József Zsigmond Nagy, and Gábor Ujváry, 235–52. Székesfehérvár: Kodolányi János Főiskola, Eötvös Loránd Tudományegyetem Könyvtára, 2010.

———. "Du 'barbarisme' et 'civilisation': Le conflit entre les étudiantes polonais et ruthènes en 1907 et sa construction journalistique." In *La Galicie au temps des Habsbourg (1772–1918): Histoire, société, cultures en contact*, edited by Jacques Le Rider and Heinz Raschel, 175–88. Tours: Presses universitaires François-Rabelais, 2010.

———. "Imperial Knowledge? Die Wissenschaften in der späten Habsburgermonarchie zwischen Kolonialismus, Nationalismus und Imperialismus." *Wiener Zeitschrift zur Geschichte der Neuzeit* 9, no. 2 (2009): 119–33.

———. "Internationality in *Nauka Polska*: Infrastructure, Translation, Language." In *Science Studies in Interwar Poland*, edited by Friedrich Cain and Bernhard Kleeberg. Tubingen: Mohr Siebeck, 2018.

———. "Leon (sic!) Thun in polnischer Historiographie: Zur Tradition der Geschichte." In *Die Thun-Hohenstein'schen Universitätsreformen 1849–1860: Konzeption—Umsetzung—Nachwirkungen*, edited by Christof Aichner and Brigitte Mazohl, 317–46. Vienna: Böhlau, 2017.

———. "Objektivität, Bestandsaufnahme, Territorium: Galizische Quelleneditionen und ihre Verortung zwischen wissenschaftlichen und ideologischen Ansprüchen." In *Geschichtsforschung in Deutschland und Österreich im 19. Jahrhundert: Ideen—Akteure—Institutionen*, edited by Christine Ottner and Klaus Ries, 198–221. Stuttgart: Franz Steiner, 2014.

———. "Symbolism, Communication and Cultural Hierarchy: Galician Discourses of Language Hegemony at the Beginning of the Second Half of the Nineteenth Century." *Historyka: Studia Metodologiczne* 42 (2012): 151–74. https://www.academia.edu/15087613/_full_issue_Jan_Surman_Klemens_Kaps_Postcolonial_Galicia_Prospects_and_Possibilities_Historyka_42_2012.

Surman, Jan, and Klemens Kaps, eds. *Galicia Postcolonial: Prospects and Possibilities*. Special issue, *Historyka: Studia metodologiczne* 42 (2012). https://www.academia.edu/15087613/_full_issue_Jan_Surman_Klemens_Kaps_Postcolonial_Galicia_Prospects_and_Possibilities_Historyka_42_2012.

Surman, Jan, and Gerald Mozetič, eds. *Dwa życia Ludwika Gumplowicza: Wybór tekstów*. Warsaw: Oficyna Naukowa, 2010.

Šváb, Jan. "I. česká chirurgická kliniká a její vliv na rozvoj chirugie v Českích zemích." In *120 let 1. české chirurgické kliniky 1. lékařské fakulty Univerzity Karlovy v Praze*, edited by Pavel Kleiner, 9–46. Prague: Karolinum, 2002.

Svatoš, Michal. "Univerzitní Působení Filologa Josefa Krále." *AUC-HUCP* 22, no. 2 (1982): 65–108.

Svobodný, Petr. "Contacts between Bohemian and Hungarian Medical Faculties (14th–20th Centuries)." In *Universitas Budensis 1395–1995, International Conference for the History of Universities on the Occasion of the 600th Anniversary of the Foundation of the University of Buda*, edited by László Szögi and Júlia Varga, 251–60. Budapest: Bak-Fisch KTF, 1997.

———. "Lékařská fakulta 1918–1945." In *Dějiny Univerzity Karlovy*, vol. 4, *1918–1945*, edited by Jan Havránek and Zdeněk Pousta, 93–119. Prague: Karolinum, 1998.

Svojtka, Matthias, Johannes Seidl, and Barbara Steininger. "Von Neuroanatomie, Paläontologie und slawischem Patriotismus: Leben und Werk des Josef Victor Rohon (1845–1923)." *Mensch—Wissenschaft—Magie* 26 (2009): 123–59.

Syllaba, Theodor. *Jan Gebauer na pražské Univerzitě*. Prague: Karlova Univerzita, 1983.

Szögi, Laszlo. "Der Ausbau des modernen Universitätsnetzes in der bürgerlichen Gesellschaft Ungarns 1848–1918." *Mensch—Wissenschaft—Magie* 26 (2009): 93–106.

Szporluk, Roman. *The Political Thought of Thomas G. Masaryk*. Boulder, CO: East European Monographs, 1981.

Szujski, Józef. "Mowa na pogrzebie Antoniego Walewskiego prof. Uniw. Jagiell., Członka Akademii." *Rocznik Zarządu Akademii Umiejętności w Krakowie* (1877): 137–42.

Szybalski, Wacław. "Professor Alexander Zawadzki of Lvov University—Gregor Mendel's Mentor and Inspirer." *Biopolymers and Cell* 26, no. 2 (2010): 83–86.

Szymiczkowa, Maryla. *Tajemnica domu Helclów*. Cracow: Znak, 2015.

Taschenbuch der Wiener k.k. Universität für das Jahr 1847. Vienna: Besck'sche Universitätsbuchhandlung, 1847.

Taschwer, Klaus. "Geheimsache Bärenhöhle: Wie ein antisemitisches Professorenkartell der Universität Wien nach 1918 jüdische und linke Forscherinnen und Forscher vertrieb." In *Alma Mater Antisemitica: Akademisches Milieu, Juden und Antisemitismus an den Universitäten Europas zwischen 1918 und 1939*, edited by Regina Fritz, Grzegorz Rossoliński-Liebe, and Jana Starek, 221–44. Vienna: New Academic Press, 2016.

———. "Universitätskrise und wissenschaftliche Volksbildung: Über die Lage der Wissenschaften in Wien um 1900 und ihre Wechselwirkungen mit der Universitätsausdehnungsbewegung." *Spurensuche*, n.s., 12, no. 1 (2001): 4–23.

———. "Wissenschaft für viele: Zur Wissensvermittlung in der Wiener Volksbildungsbewegung rund um 1900." PhD diss., University of Vienna, 2002.

Tel'vak, Vitalij. "Mychajlo Hruševs'kij contra Volodymyr Mil'kovyč (do problemy formuvannja dyskusijnoho polja ukraïns'koï istoriohrafiï počatku XX stolittja)." *Archivoznavstvo. Archeohrafija. Džereloznavstvo* 9 (2008): 225–69.

Těšínská, Emilie. "K česko-polským vědeckým stykům v oblasti matematicko-fyzikálních věd." In *Semináře a studie Výzkumneho Centra pro Dêjiny Vêdy z Let 2002–2003*, edited by Antonín Kostlán, 341–76. Prague: Výzkumne Centrum pro dêjiny vêdy, 2003.

Teske, Armin. *Marian Smoluchowski: Leben und Werk.* 1955. Wrocław: Zakład Narodowy im. Ossolińskich, Polska Akademia Nauk, 1977.

Tham, Karel Ignac. *Obrana gazyka českého protí zlobiwym geho vtrhačům, též mnohym wlastencům, w cwičenj se w něm liknawym a nedbalym sepsaná.* Prague: J. F. ze Schönfeldu, 1783.

Ther, Philipp. "Das Europa der Nationalkulturen: Die Nationalisierung und Europäisierung der Oper im 'langen' 19. Jahrhundert." *Journal of Modern European History* 5, no. 1 (2007): 39–66.

Thienen-Adlerflycht, Christoph. *Graf Leo Thun im Vormärz: Grundlagen des böhmischen Konservativismus im Kaisertum Österreich.* Vienna: Böhlau, 1967.

Thon, Jakob. "Anteil der Juden am Hochschulstudium in Oesterreich seit dem Jahre 1851." *Zeitschrift für Demographie und Statistik der Juden*, no. 3 (1907): 33–38.

[Thun-Hohenstein, Leo]. "Bericht von Thun-Hohenstein an Doblhoff, 28. July 1848." In *Slovanský sjezd v Praze roku 1848: Sbírka dokumentů*, edited by Václav Žáček, 487–90. Prague: Nakl. Československé akademie věd, 1958.

Thun-Hohenstein, Leo. *Betrachtungen über die Zeitverhältnisse, insbesondere im Hinblicke auf Böhmen.* Prague: J. G. Calve, 1849.

———. Úwahy o nynějších poměrech hledíc zwláště k Čechám. Prague: J. B. Kalwe, 1849.

Tichy, Robert, and Johannes Wallner. "Johannes Frischauf—eine schillernde Persönlichkeit in Mathematik und Alpinismus." *Internationale Mathematische Nachrichten* 63, no. 210 (2009): 21–32.

Tomek, Václav Vladivoj. *Paměti z mého žiwota.* Vol. 1. Prague: František Řivnáč, 1905.

Tomusk, Voldemar. "European Higher Education Considering Gellner, Malinowski und Wittgenstein." In *The Bologna Process in Central and Eastern Europe*, edited by Tamás Kozma, Magdolna Rébay, Andrea Óhidy, and Éva Szolár, 33–63. Wiesbaden: Springer Fachmedien, 2014.

Trencsényi, Balázs, and Michal Kopeček, eds. *Discourses of Collective Identity in Central and Southeast Europe 1770–1945.* Vol. 1, *Late Enlightenment—Emergence of the Modern "National Idea."* Budapest: Central European University Press, 2006.

Tretera, Ivo. *J. F. Herbart a jeho stoupenci na pražské univerzitě.* Prague: Univerzita Karlova, 1989.

Trnka, Miloslav. "Profesor Bohumil Karel Vydra ze Lštění na Šumavě." *Rodopisná revue on-line* 9–10 (2010): 20–21. http://rodopisna-revue-online.tode.cz/9-10-10_soubory/20-21_trnka-vydra.pdf.

Turczynski, Emanuel. "Czernowitz, eine vom Bildungsbürgertum errungene Universität im Dienst staatlicher Bildungs- und Wissenschaftsförderung." In *Universitäten im östlichen Mitteleuropa: Zwischen Kirche, Staat und Nation. Sozialgeschichtliche und politische Entwicklungen*, edited by Peter Wörster, 209–25. Munich: Oldenbourg, 2008.

Tureček, Dalibor. "Murkovy 'Deutsche Einflüsse' a jejich české přijetí." In *Matija Murko v myšlenkovém kontextu evropské slavistiky: Sborník studií*, edited by Ivo Pospíšil and Miloš Zelenka, 87–99. Brno: Masarykova univerzita, 2005.

Tvrdík, Milan. "August Sauer und die Prager tschechische Germanistik." In *August Sauer (1855–1926): Ein Intellektueller in Prag im Spannungsfeld von Kultur- und Wissenschaftspolitik*, edited by Steffen Höhne, 133–46. Vienna: Böhlau, 2008.

Twardowski, Kazimierz. *Die Universität Lemberg: Materialien zur Beurteilung der Universitätsfrage*. Vienna: Selbstverlag, 1907.

"Twenty-Five Years of A-B-C: Miroslav Hroch's Impact on Nationalism Studies." Special issue, *Nationalities Papers: The Journal of Nationalism and Ethnicity* 38, no. 6 (2010).

Ukrajinská svobodná univerzita (1921–1996). Prague: Národní knihovna České republiky, Slovanská knihovna, 1998.

Unger, Joseph. *Zur Reform der Wiener Universität: Ein Votum erstattet in der Sitzung des Unterrichtsrathes am 29. Dezember 1865*. Vienna: Manz, 1869.

Universitati Leopoliensi, Trecentesimum Quinquagesimum Anniversarium Suae Fundationis Celebranti: In Memoriam. Cracow: Polska Akademia Umiejętności, 2011.

Unowsky, Daniel. "'Our Gratitude Has No Limit': Polish Nationalism, Dynastic Patriotism, and the 1880 Imperial Inspection Tour of Galicia." *Austrian History Yearbook* 34 (2003): 145–71.

Uray, Johannes. "Czernowitz—Salzburg: Die Idee zum Transfer einer Universität (1916–1920)." In *Universitäten in Zeiten des Umbruchs: Fallstudien über das mittlere und östliche Europa im 20. Jahrhundert*, edited by Elmar Schübl and Harald Heppner, 69–82. Berlin: Lit, 2011.

Ureña, Enrique M. *El Krausismo alemán: Los congresos de filósofos y el krausofrobelismo (1833–1881)*. Madrid: Universidad Pontificia de Comillas de Madrid, 2002.

Vacek, Jiří. "Institucionální základna ukrajinské emigrace v Československu v letech 1919–1945." In *Ruská a ukrajinská emigrace v ČSR v letech 1918–1945*, edited by Václav Veber, 1:35–45. Prague: Seminář pro dějiny východní Evropy při Ústavu světových dějin Filozofické fakultě Univerzity Karlovy, 1993.

Vakarčuk, Ivan, and Jaroslav Isaievyč, eds. *Listuvannja Ivana Franka ta Mychajla Drahomanova*. L'viv: Vydavnyčyj centr L'vivs'koho Nacional'noho Universytetu im. Ivana Franka, 2006.

Vaněček, Václav. "'V Praze budou . . . dvě university . . .' (Zák. z. 28.2.1882, č.24, Ř.z., § 1)." *AUC-HUCP* 22, no. 1 (1982): 7–14.

Vietor, Martin. "Die Gründung der Comenius-Universität in Bratislava (Pressburg) und die österreichischen Universitätsgesetze." In *Festschrift Hans Lentze: Zum 60. Geburtstage dargebracht von Fachgenossen und Freunden*, edited by Nikolaus Grass and Werner Ogris, 587–97. Innsbruck: Wagner, 1969.

———. "Pravnická fakulta Univerzity Komenskeho." In *50 rokov Univerzity Komenského*, edited by Július Bartl, 159–280. Bratislava: Univerzita Komenského v Bratislave, 1969.

Vodrážková-Pokorná, Lenka. "Die Anfänge der tschechischen Germanistik und ihre ersten Repräsentanten an der Prager Universität." In *Universitäten in nationaler Konkurrenz: Zur Geschichte der Prager Universitäten im 19. und 20. Jahrhundert*, edited by Hans Lemberg, 115–34. Munich: Oldenbourg, 2003.

———. *Die Prager Germanistik nach 1882: Mit besonderer Berücksichtigung des Lebenswerkes der bis 1900 an die Universität berufenen Persönlichkeiten.* Frankfurt am Main: Peter Lang, 2006.

Volkov, Shulamit. "Soziale Ursachen des Erfolgs in der Wissenschaft—Juden im Kaiserreich." *Historische Zeitschrift* 245, no. 2 (1987): 315–42.

Völlmecke, Irma. "Österreichische Hochschulstatistik 1829 bis 1979." In *Geschichte und Ergebnisse der zentralen amtlichen Statistik in Österreich 1829–1979*, edited by Österreichisches Statistisches Zentralamt, 479–512. Vienna: Österreichische Staatsdruckerei, 1979.

"Vorbericht." In *Urkunden zur Geschichte von Österreich, Steiermark, Kärnten, Krain, Görz, Triest, Istrien, Tirol: Aus den Jahren 1246–1300; aus den Originalen des Kais. Kön. Haus- Hof- und Staats-Archives*, edited by Joseph Chmel (= *Fontes rerum Austriacarum. Österreichische Geschichtsquellen, herausgegeben von der Historischen Commission der Kaiserlichen Akademie der Wissenschaften in Wien. Zweite Abtheilung Diplomataria et Acta I. Band. Diplomatatium Miscellum Seculi XIII*). Vienna: k.k. Hof- und Staatsdruckerei, 1849.

Voznjak, Mychajlo [Mychaĭlo Vozniak]. "Nedopuščennja Ivana Franka do docentury u L'vivs'komu universyteti." *Ivan Franko: Statti i materialy* 1 (1948): 43–63.

———. "Z-za redakcijnych kulis videns'koho Vistnyka ta Zori Halyc'koï." *Zapysky Naukovoho Tovarystva imeny Ševčenka* 107 (1912): 73–109.

Wachholz, Leon. "Bryk Antoni." In *Polski Słownik Biograficzny*, edited by Władysław Konopczyński, 3:27. Cracow: Polska Akademia Umiejętności, Gebethner i Wolff, 1937.

———. "Dwie obsady katedr lekarskich w Uniwersytecie Jagiellonskim w wieku XIX." *Archiwum Historii i Filozofii Medycyny* 10 (1930): 226–33.

Wahrmund, Ludwig. *Katholische Weltanschauung und freie Wissenschaft: Ein populärwissenschaftlicher Vortrag unter Berücksichtigung des Syllabus Pius X und der Enzyklika "Pascendi Dominici gregis."* Munich: J. F. Lehmann, 1908.

———. *Religion und Klerikalismus: Rede, gehalten bei der Gründung der Ortsgruppe Innsbruck des Vereines "Freie Schule."* Innsbruck: A. Edlinger / Verlag der Ortsgruppe Innsbruck des Vereines "Freie Schule," 1902.

———. *Universität und Kirche: Akten zum Fall Wahrmund.* Frankfurt am Main: Neuer Frankfurter Verlag, 1902.

Wakounig, Maria. "Wissenschaft und Kariere? Polnische Mediziner an der Wiener Uni zwischen 1870 und 1914." In *Polen im alten Österreich: Kultur und Politik*, edited by Walter Leitsch and Stanisław Trawkowski, 107–15. Vienna: Böhlau, 1993.

Wallaschek, Michael. *Ludwig Karl Schmarda (1819–1908): Leben und Werk.* Halle: Publisher unknown, 2014.

Wangermann, Ernst. *Aufklärung und staatsbürgerliche Erziehung: Gottfried van Swieten als Reformator des österreichischen Unterrichtswesens 1781–1791.* Vienna: Verlag für die Geschichte und Politik, 1978.

Wasserman, Janek. *Black Vienna: The Radical Right in the Red City, 1918–1938.* Ithaca, NY: Cornell University Press, 2014.

Wassilko, Theophila. "Der Unterrichtsrat: Ein vergessener Zwischenakt aus der Geschichte des österreichischen Unterrichtswesens." *Mitteilungen des Österreichischen Staatsarchivs* 6 (1953): 312–26.

Wedekind, Michael. "The Sword of Science: German Scholars and National Socialist Annexation Policy in Slovenia and Northern Italy." In *German Scholars and Ethnic Cleansing (1920–1945)*, edited by Ingo Haar and Michael Fahlbusch, 110–38. New York: Berghahn Books, 2005.

Weeks, Theodore R. "Assimilation, Nationalism, Modernization, Antisemitism: Notes on Polish-Jewish Relations, 1855–1905." In *Antisemitism and Its Opponents in Modern Poland*, edited by Robert Blobaum, 20–38. Ithaca, NY: Cornell University Press, 2005.

———. "Jan Baudouin de Courtenay: The Linguist as Anti-nationalist and Imperial Citizen." In *Eliten im Vielvölkerreich: Imperiale Biographien in Russland und Österreich-Ungarn (1850–1918) / Elites and Empire: Imperial Biographies in Russia and Austria-Hungary (1850–1918)*, edited by Malte Rolf and Tim Buchen, 338–54. Berlin: de Gruyter, 2015.

Weinzierl, Erika. "Aehrenthal and the Italian University Question." In *Intellectual and Social Developments in the Habsburg Empire from Maria Theresa to World War I: Essays Dedicated to Robert S. Kann*, edited by Stanley B. Winters and Robert S. Kann, 241–70. Boulder, CO: East European Quarterly, 1975.

———. "Helfert, Joseph Freiherr von." In *Neue Deutsche Biographie*, edited by Historische Kommission bei der Bayrischen Akademie der Wissenschaften, 8:469–70. Berlin: Duncker & Humblot, 1969.

———, ed. *Der Modernismus: Beiträge zu seiner Erforschung.* Graz: Styria, 1974.

Weiß, Otto. "Bolzanisten und Güntherianer in Wien 1848–1851." In *Bernard Bolzano und die Politik: Staat, Nation und Religion als Herausforderung für die Philosophie im Kontext von Spätaufklärung, Frühnationalismus und Restauration*, edited by Helmut Rumpler, 247–82. Vienna: Böhlau, 2000.

———. *Der Modernismus in Deutschland: Ein Beitrag zur Theologiegeschichte.* Regensburg: Pustet, 1995.

[Werburg, Viktor von Andrian]. *Österreich und dessen Zukunft.* Hamburg: Hoffmann und Campe, 1843.

White, Richard. *Railroaded: The Transcontinentals and the Making of Modern America.* New York: W. W. Norton, 2011.

Widmann, Eva S. "Vormärzliches Studium im Spiegel autobiographischer Quellen." In *Österreichische Bildungs- und Schulgeschichte von der Aufklärung bis zum Liberalismus*, edited by Gerda Mraz, 118–37. Eisenstadt: Institut für Österreichische Kulturgeschichte, 1974.

Wierzbicki, Andrzej. "Wokół 'czarnej legendy' historiografii krakowskich konserwatystów." *Kwartalnik Historyczny* 104, no. 2 (1997): 63–87.

Wildauer, Tobias. *Festrede zu Schillers hundertjährigen Geburtstag bei der von der k.k. Universität zu Innsbruck veranstalteten Feier in der Aula am 10 November 1859*. Innsbruck: Wagner, 1859.

Wilde, William. *Austria, Its Literary, Scientific and Medical Institutions with Notes and a Guide to the Hospitals and Sanatory Establishments of Vienna*. Dublin: Curry, 1843.

Wilfing, Alexander. "Kant and 'Austrian Philosophy': An Introduction." In *Detours: Approaches to Immanuel Kant in Vienna, in Austria, and in Eastern Europe*, edited by Violetta L. Waibel, 19–26. Göttingen: V&R Unipress, 2015.

Wilke, Carsten. *"Den Talmud und den Kant": Rabbinerausbildung an der Schwelle zur Moderne*. Hildesheim: Olms, 2003.

Winter, Eduard. *Der Bolzanoprozess: Dokumente zur Geschichte der Prager Karlsuniversität im Vormärz*. Brno: Rohrer, 1944.

Winter, Eduard, and Liane Zeil, eds. *Wissenschaft und Religion im Vormärz: Der Briefwechsel Bernard Bolzanos mit Michael Josef Fesl, 1822–1848*. Berlin: Akademie-Verlag, 1965.

Wistrich, Robert S. *Die Juden Wiens im Zeitalter Kaiser Franz Josephs*. Vienna: Böhlau, 1999.

[Wittelshöfer, Leopold]. "Das Attentat auf die Prager Universität." *Wiener Medizinische Wochenschrift* 16, no. 20 (1866): 323–26.

Wittmann, Helga, and Erich Ziegler. *Die Entwicklung chemischer Wissenschaften an der Universität Graz: Ein Leistungsbericht*. Graz: Akademische Druck- u. Verlagsanstalt, 1985.

Wojtaszek, Zdzisław, Halina Kuzyk, Alojzy Morzyniec, Jerzy Dubowy, and Krystyna Łopata. *Karol Olszewski*. Warsaw: Państwowe Wydawnictwo Naukowe, 1990.

Wojtkiewicz-Rok, Wanda. *Dzieje Wydziału Lekarskiego Uniwersytetu Lwowskiego w latach 1894–1918*. Wrocław: Wydawnictwo Uniwersytetu Wrocławskiego, 1992.

Woleński, Jan. *Filozoficzna szkoła lwowsko-warszawska*. Warsaw: Państwowe Wydawnictwo Naukowe, 1995.

———. "The Reception of the Lvov-Warsaw School." In *The Lvov-Warsaw School and Contemporary Philosophy*, edited by Katarzyna Kijania-Placek and Jan Woleński, 3–19. Dordrecht: Kluwer, 1998.

Wolff, Larry. "Dynastic Conservatism and Poetic Violence in Fin-de-Siècle Cracow: The Habsburg Matrix of Polish Modernism." *American Historical Review* 106, no. 3 (2001): 735–64.

Wöller, Burkhard. *"Europa" als historisches Argument: Nationsbildungsstrategien polnischer und ukrainischer Historiker im habsburgischen Galizien*. Bochum: Verlag Dr. Dieter Winkler, 2014.

———. "Joseph Mauss (1778–1858)." In *Złota księga historiografii lwowskiej XIX i XX w.*, edited by Jerzy Maternicki, Paweł Sierżęga, and Leonid Zaszkilniak, 2:21–30. Rzeszów: Wydawnictwo Uniwersytetu Rzeszowskiego, 2014.

———. "Ludwig Zehnmark (1753–1814)." In Złota księga historiografii lwowskiej XIX i XX w., edited by Jerzy Maternicki, Paweł Sierżęga, and Leonid Zaszkilniak, 2:11–20. Rzeszów: Wydawnictwo Uniwersytetu Rzeszowskiego, 2014.

Wozniak, Peter. "Count Leo Thun: A Conservative Savior of Educational Reform in the Decade of Neoabsolutism." *Austrian History Yearbook* 26 (1995): 61–81.

Wróbel, Piotr. "The Jews of Galicia under Austrian-Polish Rule, 1869–1918." *Austrian History Yearbook* 25 (1994): 97–138.

———. "Szymon Askenazy." In *Nation and History: Polish Historians from the Enlightenment to the Second World War*, edited by Peter Brock, Piotr Wróbel, and John Stanley, 221–45. Toronto: University of Toronto Press, 2006.

Wurzbach, Constantin von. "Ludwig Karl Schmarda." In *Biographisches Lexikon des Kaiserthums Oesterreich*, edited by Constantin von Wurzbach, 30:155–58. Vienna: Kaiserlich-königliche Hof- und Staatsdruckerei, 1875.

Wyka, Maria. "Stanisław Tarnowski jako historyk literatury polskiej." In *Stanisław Tarnowski (1837–1917): Materiały z Posiedzenia Naukowego PAU w dniu 14.XI.1997 r.*, edited by Rita Majkowska, 9–16. Cracow: Wydawnictwo Polskiej Akademii Umiejętności, 1999.

Wýtah z auctů českého Museum, týkajících se příjmů a wydání Matice české roku 1847. Prague: Tiskem c. k. dworní knihtiskárny synů Bohumila Háze, 1848.

[Zach, František Alexandr]. "Manifest des zu Prag versammelten Slaven-Congresses an alle europäischen Völker, 1848." In *Slovanský sjezd v Praze roku 1848: Sbírka dokumentů*, edited by Václav Žáček, 365–68. Prague: Nakl. Československé akademie věd, 1958.

Zahra, Tara. *Kidnapped Souls: National Indifference and the Battle for Children in the Bohemian Lands, 1900–1948.* Ithaca, NY: Cornell University Press, 2008.

Závodský, Artur. *František Ladislav Čelakovsky.* Prague: Melantrich, 1982.

[Zawadzki, Władysław]. *Literatura w Galicji (1772–1848): Ustęp z pamiętników Władysława Zawadzkiego.* L'viv: Nakładem Władysława Webera, 1878.

Zayarnyuk, Andriy. *Framing the Ukrainian Peasantry in Habsburg Galicia, 1846–1914.* Edmonton: Canadian Institute of Ukrainian Studies Press, 2013.

———. "Mapping Identities: The Popular Base of Galician *Russophilism* in the 1890s." *Austrian History Yearbook* 41 (2010): 117–42.

Zeil, Wilhelm. *Slawistik an der deutschen Universität in Prag (1882–1945).* Munich: Otto Sagner, 1995.

Zilynská, Blanka. "Karl Adolf Constantin Ritter von Höfler jako univerzitní učitel." In *Německá medievistika v českých zemích do roku 1945*, edited by Pavel Soukup and František Šmahel, 193–224. Prague: Výzkumné centrum pro dějiny vědy, 2004.

Zur Frage der Collegiengelder: Denkschrift verfasst von Mitgliedern des Medizinischen Professoren-Collegiums der Wiener Universität. Vienna: Bergmann, 1896.

Index

www.ingramcontent.com/pod-product-compliance
Ingram Content Group UK Ltd.
Pitfield, Milton Keynes, MK11 3LW, UK
UKHW020653040726
472824UK00001B/37